A GUIDED TOUR OF MATHEMATICAL METHODS FOR THE PHYSICAL SCIENCES
THIRD EDITION

Mathematical methods are essential tools for all physical scientists. This book provides a comprehensive tour of the mathematical knowledge and techniques that are needed by students across the physical sciences. In contrast to more traditional textbooks, all the material is presented in the form of exercises. Within these exercises, basic mathematical theory and its applications in the physical sciences are well integrated. In this way, the mathematical insights that readers acquire are driven by their physical-science insight. This third edition has been completely revised: new material has been added to most chapters, and two completely new chapters on probability and statistics and on inverse problems have been added. This guided tour of mathematical techniques is instructive, applied, and fun. This book is targeted for all students of the physical sciences. It can serve as a stand-alone text or as a source of exercises and examples to complement other textbooks.

We dedicate this book to our loving and beloved families:
Idske, Hylke, Hidde, and Julia,
and
Mila, Sasha, and Niels.
We also dedicate this book to our "scientific families":
our teachers and mentors, in particular Guust Nolet and John Scales,
our colleagues, and our students.

A GUIDED TOUR OF MATHEMATICAL METHODS FOR THE PHYSICAL SCIENCES
THIRD EDITION

ROEL SNIEDER
Colorado School of Mines

KASPER VAN WIJK
University of Auckland

CAMBRIDGE
UNIVERSITY PRESS

CAMBRIDGE
UNIVERSITY PRESS

32 Avenue of the Americas, New York NY 10013-2473, USA

Cambridge University Press is part of the University of Cambridge.

It furthers the University's mission by disseminating knowledge in the pursuit of
education, learning and research at the highest international levels of excellence.

www.cambridge.org
Information on this title: www.cambridge.org/9781107084964

© Roel Snieder 2001, 2004
© Roel Snieder and Kasper van Wijk 2015

First published 2001
Second edition published 2004
Third edition published 2015

A catalogue record for this publication is available from the British Library

Library of Congress Cataloguing in Publication data
Snieder, Roel, 1958–
A guided tour of mathematical methods for the physical sciences / Roel Snieder,
Colorado School of Mines, Kasper van Wijk, University of
Auckland. – Third edition.
pages   cm
Includes bibliographical references and index.
ISBN 978-1-107-08496-4 (alk. paper)
1. Mathematical analysis.   2. Physical sciences–Mathematics.
3. Mathematical physics.   I. Van Wijk, Kasper.   II. Title.
QA300.S794   2015
515–dc23
2015000908

ISBN 978-1-107-08496-4 Hardback
ISBN 978-1-107-64160-0 Paperback

Cover image: The image shows a soap bubble on a rotating table. The internal reflection of light
within the soap film causes variations in the thickness of the film to show up as different colors.
This laboratory experiment is used to study vortices in rotating systems; the experiment can be
seen as a small-scale version of a hurricane. This figure is taken with permission from the
following publication: Meuel, T., Y.L. Xiong, P. Fischer, C.H. Bruneau, M. Bessafi, and H. Kellay,
Intensity of vortices: from soap bubbles to hurricanes, Scientific Reports, 3, p3455, 2013.

Cover design by James F. Brisson

Contents

Figures

Tables

About the authors

Roel Snieder.

Roel Snieder holds the Keck Foundation Endowed Chair of Basic Exploration Science at the Colorado School of Mines. In 1984 he received a master's degree in geophysical fluid dynamics from Princeton University, and in 1987 a PhD in seismology from Utrecht University. For this work he received the Vening Meinesz award from the Netherlands Organization for Scientific Research. In 1988 he worked as a postdoctoral Fellow in the "Equipe de Tomographie Géophysique" at the Université Paris VI and was appointed in 1989 as associate professor at Utrecht University. In 1993 he was promoted to full professor of seismology at Utrecht University, where from 1997 to 2000 he served as dean of the Faculty of Earth Sciences and spearheaded the integration of the research of the department with the Netherlands Institute of Applied Geoscience TNO. Roel served on the editorial boards of *Geophysical Journal International*, *Inverse Problems*, *Reviews of Geophysics*, and the *European Journal of Physics*. In 2000 he was elected as Fellow of the American Geophysical Union for important contributions to geophysical inverse theory, seismic tomography, and the theory of surface waves. He is author of the textbook *The Art of Being a Scientist*, published by Cambridge University Press. He teaches his class "The Art of Science" to universities and industry. From 2003 to 2011 he was a member of the Earth Science Council of the U.S. Department of Energy. In 2008 Roel worked for the Global Climate and Energy Project at Stanford University on outreach and education on global energy. In 2008 he was a founding member of the humanitarian organization "Geoscientists Without Borders." Since 2010 he is director of the Center for Professional Education at the Colorado School of Mines. In 2011 he was elected as Honorary Member of the Society of Exploration Geophysicists. He received three teaching awards at the Colorado School of Mines. Roel was a volunteer firefighter from 2000 to 2014 and served as fire chief with Genesee Fire Rescue. For more information, visit his website `http://www.mines.edu/~rsnieder`.

Kasper van Wijk

Kasper van Wijk is a senior lecturer in the physics department at the University of Auckland in Auckland, New Zealand. Kasper is the director of the Physical Acoustics Laboratory at the University of Auckland (`http://physics.auckland.ac.nz/research/pal`). He studied geophysics in Utrecht (in the Netherlands) from 1991 to 1996, where he specialized in inverse theory. After teaching outdoor education in the mountains of Colorado, Kasper obtained his PhD in the Department of Geophysics at the Colorado School of Mines. After completing his PhD-degree, Kasper became a research faculty member at Colorado School of Mines. In 2006 Kasper van Wijk joined the geosciences department at Boise State University as an assistant professor. In 2011 he was awarded tenure and promoted to associate professor. In 2012, he moved to the Department of Physics at the University of Auckland, New Zealand, as a senior lecturer. His research interests center around elastic wave propagation in disordered media with applications ranging from medical imaging to global earth seismology. Kasper serves on the editorial board for the journal *Geophysics* and the *European Journal of Physics*.

He serves on the continuing education committee, as well as on the publications committee of the Society of Exploration Geophysicists. Kasper (co-)organized and taught geophysical field camps in Colorado, Oregon, and Thailand. His outreach efforts in Seismometers in Schools have exposed kids from 8–80 to the dynamic processes of our Earth.

1

Introduction

The topic of this book is the application of mathematics to physical problems. Mathematics and physics are often taught separately. Despite the fact that education in physics relies on mathematics, many students consider mathematics to be disjointed from physics. Although this point of view may strictly be correct, it becomes problematic when it concerns an education in the sciences. The reason for this is that mathematics is the *only* language at our disposal for quantifying physical processes. Furthermore, common mathematical foundations often form the link between different topics in physics.

A language cannot be learned by just studying a textbook. In order to truly learn how to use a language, one has to go abroad and start using that language. By the same token, one cannot learn how to use mathematics in the physical sciences by just studying textbooks or attending lectures; the only way to achieve this is to venture into the unknown and apply mathematics to physical problems. The goal of this book is to do exactly that; we present exercises to apply mathematical techniques and knowledge to physical concepts. These are not presented as well-developed theory, but as a number of problems that elucidate the issues at stake. In this sense, the book offers a guided tour: material for learning is presented, but true learning will only take place by active exploration. The interplay of mathematics and physics is essential in the process; mathematics is the natural language to describe physics, while physical insight allows for a better understanding of the mathematics that is presented.

How can you use this book most efficiently?

Since this book is written as a set of problems, you may frequently want to consult other material as well to refresh or deepen your understanding of material. In many places we refer to the book by Boas (2006). In addition, the books by Butkov (1968), Riley et al. (2006), and Arfken and Weber (2005) on mathematical physics are excellent.

In addition to books, colleagues in either the same field or other fields can be a great source of knowledge and understanding. Therefore, do not hesitate to work together with others on these problems if you are in the fortunate position to do so. This may not only make the work more enjoyable, it may also help you in getting "unstuck" at difficult moments and the different viewpoints of others may help to deepen yours.

For whom is this book written?

This book is set up with the goal of obtaining a good working knowledge of mathematical physics that is needed for students in physics or geophysics. A certain basic knowledge of calculus and linear algebra is required to digest the material presented here. For this reason, this book is meant for upper-level undergraduate students or lower-level graduate students, depending on the background and skill of the student. In addition, teachers can use this book as a source of examples and illustrations to enrich their courses.

This book is evolving

This book will be improved regularly by adding new material, correcting errors, and making the text clearer. The feedback of both teachers and students who use this material is vital in improving this text, please send your remarks to

Roel Snieder
Department of Geophysics, Colorado School of Mines, Golden, CO 80401, USA
email: rsnieder@mines.edu

Errata can be found at the following website: `http://www.mines.edu/~rsnieder/Errata.html`

Preface to the third edition

The third edition is a thorough revision and expansion of previous editions. We have added material throughout the text to help strengthen the learning process and provide essential material, and we have modified the problems so that the level of difficulty between problems is more balanced. We have added new chapters that cover Probability and Statistics (Chapter 21) and Inverse Problems (Chapter 22).

Acknowledgments

This book resulted from courses taught at Utrecht University, the Colorado School of Mines, Boise State University, and the University of Auckland. The remarks, corrections, and the encouragement of a large number of students have been very important in its development. It is impossible to thank all the students, but we especially want to thank the feedback from Jojanneke van den Berg, Jehudi Blom, Sterre Dortland, Thomas Geenen, Wiebe van Driel, Luuk van Gerven, Aaron Girard, Nathan Hancock, Noor Hogeweg, Wouter Kimman, and Frederiek Siegenbeek. In their role as teaching assistants, Dirk Kraaipoel and Jesper Spetzler have helped greatly in improving this book. Huub Douma has spent numerous hours at sea correcting earlier drafts. A number of colleagues have helped us very much with their comments; we especially want to mention Jami Johnson, Philip Cheng, Freeman Gilbert, Minhhuy Hô, Ted Jacobson, Alexander Kaufman, Antoine Khater, Rajmohan Kombiyil, Ken Larner, Sum Mak, Thijs Ruigrok, Hidetosi Sirai, Alex Summer, Jeannot Trampert, and John Wettlaufer.

We also want to acknowledge all the people who contributed to the open-source software used to help create this book. Typesetting was done in LaTeX, and python was used for the computations (including obspy for the seismology applications). Many of the figures were first drafted by Barbara McLenon in Xfig, who we thank for her support, but most of the figures in the 3rd edition were made in Inkscape.

The support and advice of Matt Lloyd, Adam Black, Eoin O'Sullivan, Jayne Aldhouse, Maureen Storey, Simon Capelin, and Philip Alexander of Cambridge University Press has been very helpful and stimulating during the preparation of this work. Lastly, we want to thank everybody who helped us in numerous ways to make writing this book such a joy.

2
Dimensional analysis

The material of this chapter is usually not covered in a book on mathematics. The field of mathematics deals with numbers and numerical relationships. It does not matter what these numbers are; they may account for physical properties of a system, but they may equally well be numbers that are not related to anything physical. Consider the expression $g = df/dt$. From a mathematical point of view these functions can be anything, as long as g is the derivative of f, with respect to t. The situation is different in physics. When $f(t)$ defines the position of a particle, and t denotes time, then $g(t)$ is a velocity. This relation fixes the physical dimension of $g(t)$. In mathematical physics, the physical dimension of variables imposes constraints on the relation between these variables. In this chapter we explore these constraints. In Section 2.2 we show that this provides a powerful technique for spotting errors in equations. In the remainder of this chapter we show how the physical dimensions of the variables that govern a problem can be used to find physical laws. Surprisingly, while most engineers learn about dimensional analysis, this topic is not covered explicitly in many science curricula.

2.1 Two rules for physical dimensions

In physics every physical parameter is associated with a physical dimension. The value of each parameter is measured with a certain physical unit. For example, when we measure how long a table is, the result of this measurement has dimension "length." This length is measured in a certain unit, which may be meters, inches, furlongs, or whatever length unit we prefer to use. The result of this measurement can be written as

$$l = 3 \text{ m}. \tag{2.1}$$

The variable l has the physical dimension of length. In this chapter we write this as

$$l \sim [L]. \tag{2.2}$$

4

The square brackets are used in this chapter to indicate a physical dimension. The capital letter L denotes length, T denotes time, and M denotes mass. Other physical dimensions include electric charge and temperature. When dealing with physical dimensions two rules are useful.

Rule 1 When two variables are added, subtracted, or set equal to each other, they must have the same physical dimension.

To see the logic of this rule, consider the following example. Suppose we have an object with a length of 1 meter and a time interval of 1 second. This means that

$$l = 1 \text{ m,}$$
$$t = 1 \text{ s.} \tag{2.3}$$

Since both variables have the same numerical value, we might be tempted to declare that

$$l = t. \tag{2.4}$$

It is, however, important to realize that the physical units that we use are arbitrary. Suppose, for example, that we had measured the length in feet rather than meters. In that case the measurements (2.3) would be given by

$$l = 3 \text{ ft,}$$
$$t = 1 \text{ s.} \tag{2.5}$$

Now the numerical value of the same length measurement is different! Since the choice of the physical units is arbitrary, we can scale the relation between variables of different physical dimensions in an arbitrary way. For this reason these variables cannot be equal to each other. This implies that they cannot be added or subtracted either.

The first rule implies the following rule.

Rule 2 Mathematical functions, other than $f(\xi) = \xi^N$, can act on dimensionless numbers only.

To see this, let us consider as an example the function $f(\xi) = e^\xi$. Using a Taylor series (a topic we will discuss in Chapter 3), this function can be written as an expansion in powers ξ^n:

$$f(\xi) = 1 + \xi + \frac{1}{2}\xi^2 + \cdots \tag{2.6}$$

According to Rule 1 the different terms in this expression must have the same physical dimension. The first term (the number 1) is dimensionless; hence, all the other terms in the series must be dimensionless. This means that ξ must be a dimensionless number as well. This argument can be used for any function $f(\xi)$ whose Taylor expansion contains different powers of ξ. Note that the argument does not hold for a function that contains only a single power of ξ, such as $f(\xi) = \xi^2$. The argument does, however, hold for the function $f(\xi) = \xi + \xi^2$ because it consists of a superposition of different powers of ξ.

These rules have several applications in mathematical physics. Suppose we want to find the physical dimension of a force, as expressed in the basic dimensions of mass, length, and time. The only thing we need to do is take one equation that contains a force. In this case Newton's law $F = ma$ comes to mind. The mass m has physical dimension $[M]$, while the acceleration has dimension $[L/T^2]$. Rule 1 implies that force has the physical dimension $[ML/T^2]$.

Problem a   The force F in a linear spring is related to the extension x of the spring by the relation $F = -kx$. Show that the spring constant k has dimension $[M/T^2]$.

Problem b   The angular momentum $\mathbf{L}$ of a particle with momentum $\mathbf{p}$ at position $\mathbf{r}$ is given by

$$\mathbf{L} = \mathbf{r} \times \mathbf{p}, \tag{2.7}$$

where $\times$ denotes the cross-product of two vectors. Show that angular momentum has the dimension $[ML^2/T]$.

Problem c   A plane wave is given by the expression

$$u(\mathbf{r}, t) = e^{i(\mathbf{k} \cdot \mathbf{r} - \omega t)}, \tag{2.8}$$

where $\mathbf{r}$ is the position vector and t denotes time. Show that $\mathbf{k}$ has dimensions $[L^{-1}]$ and ω has dimensions $[T^{-1}]$.

In quantum mechanics, the behavior of a particle is characterized by a wave equation, called the Schrödinger equation. When the wave propagates along the x-axis as a function of time, i.e., the wave propagation is one-dimensional, this equation is given by

$$i\hbar \frac{\partial \psi}{\partial t} = -\frac{\hbar^2}{2m} \frac{\partial^2 \psi}{\partial x^2} + V(x)\psi, \tag{2.9}$$

where x denotes the position, t denotes the time, m the mass of the particle, and $V(x)$ the potential energy of the particle. At this point it is not clear what the wave function $\psi(x, t)$ is, and how this equation should be interpreted. Also, the meaning of the symbol $\hbar$ is not yet defined. We can, however, determine the physical dimension of $\hbar$ without knowing the meaning of this variable.

Problem d Compare the physical dimensions of the left-hand side of equation (2.9) with the first term on the right-hand side and show that the variable $\hbar$ has the physical dimension of angular momentum. You can use Problem b to show this.

Chapter 8 has more on this variable, which turns out to be the reduced Planck's constant in the Schrödinger equation.

When using dimensional arguments, one needs to be aware of a curious property of the logarithm. Suppose L_1 and L_2 are both lengths, then the ratio L_1/L_2 is dimensionless. We can let the logarithm act on this ratio and use the following property of the logarithm

$$\ln (L_1/L_2) = \ln (L_1) - \ln (L_2). \tag{2.10}$$

This equation is mathematically correct, but note that in the left-hand side the logarithm acts on a dimensionless function, while in the right-hand side it acts on a length! This means that Rule 2 can be relaxed for the logarithm. Doing so, however, can be a source of confusion and mistakes, and it is best to let all mathematical functions act on dimensionless variables only.

2.2 A trick for finding mistakes

The requirement that all terms in an equation have the same physical dimension is an important tool for spotting mistakes. Cipra (2000) gives many useful tips for spotting errors in his delightful book *Misteaks* [sic] ... *and How to Find Them Before the Teacher Does.* As an example of using dimensional analysis for spotting mistakes, we consider the erroneous equation

$$E = mc^3, \tag{2.11}$$

where E denotes energy, m denotes mass, and c is the speed of light. Let us first find the physical dimension of energy. The work done by a force $\mathbf{F}$ over a displacement $d\mathbf{r}$ is given by $dE = \mathbf{F} \cdot d\mathbf{r}$. We showed in Section 2.1 that force has the dimension $[ML/T^2]$. This means that energy has the dimension $[ML^2/T^2]$. The speed of light has dimension $[L/T]$, which means that the right-hand side of equation (2.11) has physical dimension $[ML^3/T^3]$. This is not equal to the dimensions of energy on the left-hand side. Therefore, expression (2.11) is wrong.

8 *A Guided Tour of Mathematical Methods for the Physical Sciences*

Problem a   Now that we have determined that expression (2.11) is incorrect, we can use the requirement that the dimensions of the different terms must match to guess how to set it right. Show that the right-hand side must be divided by a velocity to match the dimensions.

It is not clear that the right-hand side must be divided by the speed of light to give the correct expression $E = mc^2$. Dimensional analysis tells us only that it must be divided by something with the dimension of velocity. For all we know, that could be the speed at which the average snail moves.

Problem b   Is the following equation dimensionally correct?

$$(\mathbf{v} \cdot \nabla \mathbf{v}) = -\nabla p. \tag{2.12}$$

In this expression $\mathbf{v}$ is the velocity of fluid flow, p is the pressure, and ∇ is the gradient vector (which essentially is a derivative with respect to the space coordinates, as explained in Chapter 5). To find the dimension of the right-hand side, you can use the fact that pressure is force per unit area.

Problem c   Answer the same question for the expression that relates the particle velocity v to the pressure p in an acoustic medium:

$$v = \frac{p}{\rho c}, \tag{2.13}$$

where ρ is the mass density and c is the velocity of propagation of acoustic waves.

Problem d   In quantum mechanics, the energy E of the harmonic oscillator is given by

$$E_n = \hbar \omega (n + 1/2), \tag{2.14}$$

where ω is a frequency, n is a dimensionless integer, and $\hbar$ is Planck's constant divided by 2π, as introduced in Problem d of the previous section. Verify if this expression is dimensionally correct.

In general, it is a good idea to carry out a dimensional analysis while working in mathematical physics. This may help in finding the mistakes that we all make while doing derivations. It takes a little while to become familiar with the dimensions of properties that are used most often, but this is an investment that pays off in the long run.

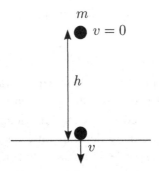

Figure 2.1 Definition of the variables for a falling ball.

2.3 Buckingham pi theorem

In this section we introduce the Buckingham pi theorem. This theorem can be used to find the relation between physical parameters, based on dimensional arguments. As an example, let us consider a ball shown in Figure 2.1 with mass m that is dropped from a height h. We want to find the velocity with which it strikes the ground. The potential energy of the ball before it is dropped is mgh, where g is the acceleration of gravity. This energy is converted into kinetic energy $\frac{1}{2}mv^2$ as it strikes the ground. Equating these quantities and solving for the velocity gives

$$v = \sqrt{2gh}. \tag{2.15}$$

Let us suppose we did not know about classical mechanics. In that case, dimensional analysis could be used to guess relation (2.15). We know that the velocity is some function of the acceleration of gravity, the initial height, and the mass of the particle: $v = f(g, h, m)$. The physical dimensions of these properties are given by

$$v \sim [L/T], \quad g \sim [L/T^2], \quad h \sim [L], \quad m \sim [M]. \tag{2.16}$$

Let us consider the dimension mass first. The dimension mass enters only the variable m. We cannot combine the variable m with the parameters g and h in any way to arrive at a quantity that is independent of mass. Therefore, the velocity does not depend on the mass of the particle. Next we consider the dimension time. The velocity depends on time as $[T^{-1}]$, the acceleration of gravity as $[T^{-2}]$, and h is independent of time. This means that we can match the dimension time only when

$$v \sim \sqrt{g}. \tag{2.17}$$

In this expression the left-hand side depends on the length as $[L]$, while the right-hand side varies with length as $[L^{1/2}]$. We have, however, not used the height h yet. The dimension length can be made to match if we multiply the right-hand side with $\sqrt{h}$. This means that the only combination of g and h that gives a velocity is given by

$$v \sim \sqrt{gh}. \tag{2.18}$$

This result agrees with expression (2.15), which was derived using classical mechanics. Note that in order to arrive at expression (2.18) we used only dimensional arguments, and did not need to have any knowledge from classical mechanics other than the fact that the velocity depends only on g and h. The dimensional analysis that led to expression (2.18), however, does not tell us what the proportionality constant in that expression is. Because the proportionality constant is dimensionless, it cannot be found by dimensional analysis.

The treatment given here may appear to be cumbersome. This analysis, however, can be carried out in a systematic fashion using the *Buckingham pi theorem* (Buckingham, 1914):

Buckingham pi theorem   If a problem contains N variables that depend on P physical dimensions, then there are $N - P$ dimensionless numbers that describe the physics of the problem.

The original paper of Buckingham is very clear, but as we will see at the end of this section, this theorem is not foolproof. Let us first apply the theorem to the problem of the falling ball. We have four variables: v, g, h, and m, so that $N = 4$. These variables depend on the physical dimensions $[M]$, $[L]$, and $[T]$, hence $P = 3$. According to the Buckingham pi theorem, $N - P = 1$ dimensionless number characterizes the problem. We want to express the velocity in the other parameters; hence, we seek a dimensionless number of the form

$$v g^\alpha h^\beta m^\gamma \sim [1], \tag{2.19}$$

where the notation in the right-hand side means that it is dimensionless. Let us seek the exponents α, β, and γ that make the left-hand side dimensionless. Inserting the dimensions of the different variables then gives the following dimensions

$$\left[\frac{L}{T}\right]\left[\frac{L^\alpha}{T^{2\alpha}}\right][L^\beta][M^\gamma] \sim [1]. \tag{2.20}$$

The left-hand side depends on length as $[L^{1+\alpha+\beta}]$. The left-hand side can only be independent of length when the exponent is equal to zero. Applying the same reasoning to each of the dimensions length, time, and mass, then gives

$$\begin{aligned}\text{dimension } [L]: \quad & 1 + \alpha + \beta = 0, \\ \text{dimension } [T]: \quad & -1 - 2\alpha = 0, \\ \text{dimension } [M]: \quad & \gamma = 0.\end{aligned} \tag{2.21}$$

This constitutes a system of three equations with three unknowns.

Problem a Show that the solution of this system is given by

$$\alpha = \beta = -\frac{1}{2}, \qquad \gamma = 0. \tag{2.22}$$

Inserting these values into expression (2.19) shows that the combination $vg^{-1/2}h^{-1/2}$ is dimensionless. This implies that

$$v = C\sqrt{gh}, \tag{2.23}$$

where C is the one dimensionless number in the problem as dictated by the Buckingham pi theorem. The approach taken here is systematic. In his original paper, Buckingham applied this treatment to a number of problems: the thrust provided by the screw of a ship, the energy density of the electromagnetic field, the relation between the mass and radius of the electron, the radiation of an accelerated electron, and heat conduction (Buckingham, 1914). There is, however, a catch that we introduce with an example. When air (or water) has a stably stratified mass–density structure, it can support oscillations where the restoring force is determined by the density gradient in the air. These oscillations occur with the *Brunt-Väisälä* frequency ω_B given by (Holton and Hakim, 2012; Pedlosky, 1982),

$$\omega_B = \sqrt{\frac{g}{\theta}\frac{d\theta}{dz}}, \tag{2.24}$$

where g is the acceleration of gravity, z is the height, and θ is the potential temperature (a measure of the thermal structure of the atmosphere).

Problem b Verify that this expression is dimensionally correct.

Problem c Check that this expression is also dimensionally correct when θ is replaced by the air pressure p, or the mass density ρ.

The result of Problem c indicates that the potential temperature θ can be replaced by any physical parameter, and expression (2.24) is still dimensionally correct. This means that a dimensional analysis alone can never be used to prove that θ should be the potential temperature. In order to show this, we need to know more of the physics of the problem.

Another limitation of the Buckingham pi theorem as formulated in its original form is that the theorem assumes that physical parameters need to be multiplied or divided to form dimensionless numbers (see equation 3 of Buckingham, 1914). The derivative of one variable with respect to another, however, has the same dimension as the ratio of these variables. For example, consider a problem where dimensional analysis shows that the variable of interest depends on the ratio of the acceleration of gravity and the height: g/h. The derivative of g with height dg/dz has the same physical dimension as g/h. Therefore, a dimensional analysis alone cannot completely describe the physics of the problem. Nevertheless, as we will see in the following section, it may provide valuable insights.

2.4 Lift of a wing

In this section we study the lift of a wing. Because in stationary flight the lift provided by a wing equals the weight of the aircraft or bird that is carried by the wing, we denote the lift of the wing with the symbol W. Since the lift is a force, this quantity has the dimension of force: $[W] = [F] = [ML/T^2]$. The lift depends on the mass density ρ of the air, the velocity v of the airflow, and the surface area S of the wing.

Problem a   Derive the following dimensional relations: $\rho \sim [M/L^3]$, $v \sim [L/T]$, and $S \sim [L^2]$.

Problem b   Count the number of variables and the number of physical dimensions to show that in the jargon of the Buckingham pi theorem $N = 4$ and $P = 3$.

This means that there is $N - P = 1$ dimensionless number that characterizes the lift of the wing. We want to express the lift W in the other parameters; therefore, we seek a dimensionless number of the form

$$W\rho^\alpha v^\beta S^\gamma \sim [1]. \qquad (2.25)$$

Problem c   Show that the requirement that the left-hand side does not depend on mass, length, and time, respectively, leads to the following linear equations:

$$1 + \alpha = 0,$$
$$1 - 3\alpha + \beta + 2\gamma = 0, \qquad (2.26)$$
$$2 + \beta = 0.$$

Problem d   Solve this system to derive that

$$\alpha = \gamma = -1, \qquad \beta = -2. \qquad (2.27)$$

Inserting this result in expression (2.25) implies that $W/(\rho v^2 S)$ is a dimensionless number. When this number is denoted by C_L, this means that the lift is given by

$$W = C_L \rho v^2 S. \qquad (2.28)$$

The coefficient C_L is called the *lift coefficient* (Kermode et al., 1996). This coefficient depends on the shape of the wing, and on the angle of *attack* (a measure of the orientation of the wing to the airflow). Let us think about the solution (2.28) for a moment. This expression states that the lift is proportional to the surface area. This makes sense: a larger wing produces more lift. The lift depends on the square of the velocity. It stands to reason that a larger flow velocity gives a larger lift, but the fact that the lift increases quadratically with the velocity is not easy to see. Lastly, the lift is proportional to the mass density of the air: for a given velocity, heavier air provides a larger lift, because the airflow deflected by the wing has a larger momentum. This has implications for the design of airports. For example, the airport of Denver is located at an elevation of about 1,600 meters. This high elevation, in combination with the warm temperatures in summertime, leads to a relatively small mass density of the air. As the surface area of the wings of aircraft is fixed by their design, the relatively small mass density must be compensated by a larger take-off velocity v. In order to achieve this large take-off velocity, aircraft need a longer runway to accelerate to the required take-off velocity. For this reason, the airport in Denver has extra long runways. All these conclusions follow from dimensional analysis only!

2.5 Scaling relations

We can take the dimensional analysis of the previous section even a step further. Suppose we consider different flying objects, and that each object is characterized by a linear dimension l.

Problem a   Use dimensional arguments to show that the volume V scales with the size as $V \sim l^3$, and that the surface area scales as $S \sim l^2$. The volume V should not be confused with the velocity v.

Problem b   Show that this implies that

$$S \sim V^{2/3}. \tag{2.29}$$

The mass of the flying object is proportional to its mass density ρ_f by the relation $m = \rho_f V$. The lift required to support this mass is given by

$$W = g\rho_f V. \tag{2.30}$$

Problem c   Insert the relations (2.29) and (2.30) into expression (2.28) to show that

$$g\rho_f V^{1/3} \approx C_L \rho v^2. \tag{2.31}$$

Problem d   Solve this expression for V, and insert this result into expression (2.30) to derive the following relation between the lift and the velocity

$$W \approx \frac{C_L^3 \rho^3}{g^2 \rho_f^2} v^6. \tag{2.32}$$

This expression predicts that the lift varies with the velocity to the sixth power. Figure 2.2 shows a compilation of the weight versus the cruising speed for various aircraft (top right), birds (middle), insects, and butterflies (bottom left). This figure is reproduced from the wonderful book by Tennekes (2009) about the science of flight. The points in this figure cluster around a straight line, when the weight and the cruising speed are plotted on a double logarithmic scale.

Problem e   Use the logarithmic power rule $\log(x^n) = n \log(x)$ to show that the clustering of data around a straight line in Figure 2.2 implies a power law relation of the form $W \sim v^n$.

Problem f   Measure the slope of the line in Figure 2.2 and show that this slope is close to the value $n = 6$ predicted by expression (2.32).

It is impressive that Figure 2.2 presents a quasi-linear relation between the logarithm of cruising speed and the logarithm of weight, for objects as disparate as a damsel fly to a jumbo jet: their weights span 11 orders of magnitude! Despite this extreme range in parameter values, the scaling law (2.32) holds remarkably well. Nevertheless, individual points show departures from the best-fitting line. This can have a number of reasons. For example, the density ρ_f and the lift coefficient C_L vary among different flying objects: the shape of the wing of a Boeing 747 is different from the shape of the wing of a butterfly. And one study suggests that in the time of flying dinosaurs such as the pteranodon, the air density ρ may have been significantly larger (Levenspiel, 2006). In any case, one can quantify the distance between data points and the straight line, to express the properties of the data points in a statistical sense. The topic of statistics is treated in Chapter 21.

This example shows that dimensional arguments can be useful in explaining the relationship between different physical parameters. Such relationships are also of importance in the design of *scale experiments*. An example of a scale experiment is a model of an aircraft in a wind tunnel. All physical parameters need to be scaled appropriately with the size of the model aircraft so that the physics is unaltered by the scaling. This is the case when the dimensionless numbers determined with the Buckingham pi theorem are the same for the scaled model as for the real aircraft. In this way the Buckingham pi theorem provides a systematic procedure for the design of scale experiments as well (Buckingham, 1914).

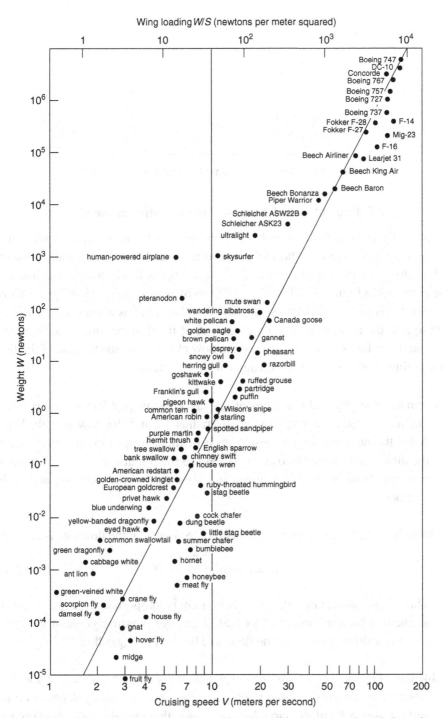

Figure 2.2   The weight of many flying objects (vertical axis) against their cruising speed (horizontal axis) on a log–log plot. This figure is reproduced from Tennekes (2009) with permission from MIT Press.

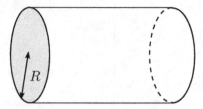

Figure 2.3   The geometry of a pipe through which a fluid flows.

2.6 Dependence of pipe flow on the radius of the pipe

The flow of a viscous fluid through a porous medium is important for understanding and managing aquifers and hydrocarbon reservoirs. Here we use dimensional analysis to study the dependence of flow of a viscous fluid through a cylindrical pipe as shown in Figure 2.3. The flow is driven by a pressure gradient $\partial p/\partial x$ along the center axis of the cylinder. We assume that the fluid has a viscosity μ, and we want to find the relation between the strength of the flow per unit time through a pipe and the radius R. As a measure of the flow rate, we use the volume of the flow per unit time, and designate this quantity with the symbol Φ.

Problem a   The physical quantities of relevance to this problem are the pressure gradient $\partial p/\partial x$, the viscosity μ, the radius R, and the flow rate Φ. Write down the physical dimensions of each of these properties. In order to find the dimension of the viscosity, you can use the relation $\tau = \mu \partial v/\partial z$, where τ is the shear stress (with dimension pressure), v the velocity, and z the distance.

Problem b   Use the Buckingham pi theorem to show that the flow rate is given by

$$\Phi = constant \; \frac{\partial p/\partial x}{\mu} R^4. \tag{2.33}$$

This expression states that the flow rate is proportional to the pressure gradient, which reflects the fact that a stronger pressure gradient generates a stronger driving force for the flow, and hence a stronger flow.

Problem c   Give a similar physical explanation for the dependence of the flow rate on the viscosity and the radius. The flow rate obviously depends on the surface area πR^2 of the pipe, but this effect alone would make the flow rate proportional to R^2 instead of the R^4-dependence predicted by scale analysis. There must thus be another physical factor, other than an increased surface area, that increases the flow rate. Can you think of a physical reason why the flow velocity would increase with the width of the pipe?

The result (2.33) can also be obtained by solving the Navier–Stokes equation (25.48) for the appropriate boundary condition, and by integrating the flow velocity over the pipe to give the flow rate Φ. This treatment is more cumbersome than the analysis of this section, but it does provide the proportionality constant in expression (2.33).

3

Power series

Some of the mathematical functions we often invoke in physics, $\sin(x)$, $\cos(x)$, $\exp(x)$, for example, can be written as series of powers of their argument x. This is discussed in Section 3.1, including the important issue of convergence of the terms in the series. We show in Section 3.2 a paradoxical example that when small numbers are concerned, an approximate result derived from a Taylor series can be more accurate than a numerical evaluation of an exact expression. Frequently the result of a calculation can be obtained by summing a series. In Section 3.3 this is used to study the behavior of a bouncing ball. The bounces are "natural" units for analyzing the problem at hand. In Section 3.4 the reverse is done when studying the reflection of waves by a stack of reflective layers. In this case, a series expansion actually gives physical insight into a complex expression.

3.1 Taylor series

In many applications in mathematical physics it is useful to write the quantity of interest as a sum of a number of terms. To fix our mind, let us consider the motion of a particle that moves along a line as time progresses. The motion is completely described by giving the position $x(t)$ of the particle as a function of time. Consider the four different types of motion that are shown in Figure 3.1. The simplest motion is a particle that does not move, as shown in panel (a). In this case, the position of the particle is constant:

$$x(t) = x_0. \qquad (3.1)$$

The value of the parameter x_0 follows by setting $t = 0$ in this expression; this immediately gives

$$x_0 = x(0). \qquad (3.2)$$

18

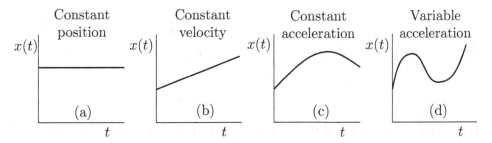

Figure 3.1   Four different kinds of motion of a particle along a line as a function of time.

In panel (b), the situation for a particle that moves with a constant velocity is shown, thus the position is a linear function of time:

$$x(t) = x_0 + v_0 t. \tag{3.3}$$

Again, setting $t = 0$ gives the parameter x_0, which is given again by (3.2). The value of the parameter v_0 follows by differentiating (3.3) with respect to time and by setting $t = 0$.

Problem a   Do this and show that

$$v_0 = \frac{dx}{dt}(t = 0). \tag{3.4}$$

This expression reflects that the velocity v_0 is given by the time-derivative of the position. Next, consider a particle moving with a constant acceleration a_0 as shown in panel (c). As you probably know from classical mechanics, the motion in that case is a quadratic function of time:

$$x(t) = x_0 + v_0 t + \frac{1}{2} a_0 t^2. \tag{3.5}$$

Problem b   Evaluate this expression at $t = 0$ to show that x_0 is given by (3.2). Differentiate (3.5) once with respect to time and evaluate the result at $t = 0$ to show that v_0 is again given by (3.4). Differentiate (3.5) twice with respect to time, set $t = 0$ to show that a_0 is given by

$$a_0 = \frac{d^2 x}{dt^2}(t = 0). \tag{3.6}$$

This result reflects the fact that the acceleration is the second derivative of the position with respect to time.

Let us now consider the motion shown in panel (d) where the acceleration changes with time. In that case the displacement as a function of time is not a linear function of time (as in (3.3) for the case of a constant velocity) nor is it a quadratic function of time (as in (3.5) for the case of a constant acceleration). Instead, the displacement is in general a function of all possible powers in t:

$$x(t) = c_0 + c_1 t + c_2 t^2 + \cdots = \sum_{n=0}^{\infty} c_n t^n. \tag{3.7}$$

This series, in which a function is expressed as a sum of terms with increasing powers of the independent variable, is called a *Taylor series*. At this point we do not know what the constants c_n are. These coefficients can be found in exactly the same way as in Problem b, in which you determined the coefficients a_0 and v_0 in the expansion (3.5).

Problem c Determine the coefficient c_m by differentiating expression (3.7) m times with respect to t and by evaluating the result at $t = 0$ to show that

$$c_m = \frac{1}{m!} \frac{d^m x}{dt^m} (t = 0). \tag{3.8}$$

Of course, there is no reason why the Taylor series can only be used to describe the displacement $x(t)$ as a function of time t. The Taylor series is frequently used to describe a function $f(x)$ that depends on x. Of course, it is immaterial what we call a function. By making the replacements $x \rightarrow f$ and $t \rightarrow x$, expressions (3.7) and (3.8) can also be written as:

$$f(x) = \sum_{n=0}^{\infty} c_n x^n, \tag{3.9}$$

with

$$c_n = \frac{1}{n!} \frac{d^n f}{dx^n} (x = 0). \tag{3.10}$$

You may also find this result in the literature written as

$$f(x) = \sum_{n=0}^{\infty} \frac{x^n}{n!} \frac{d^n f}{dx^n} (x = 0) = f(0) + x \frac{df}{dx} (x = 0) + \frac{x^2}{2} \frac{d^2 f}{dx^2} (x = 0) + \cdots . \tag{3.11}$$

Problem d By evaluating the derivatives of $f(x)$ at $x = 0$, show that the Taylor series of the following functions are given by:

$$\sin(x) = x - \frac{1}{3!} x^3 + \frac{1}{5!} x^5 - \cdots ; \tag{3.12}$$

$$\cos(x) = 1 - \frac{1}{2} x^2 + \frac{1}{4!} x^4 - \cdots ; \tag{3.13}$$

$$e^x = 1 + x + \frac{1}{2!}x^2 + \frac{1}{3!}x^3 + \cdots = \sum_{n=0}^{\infty} \frac{1}{n!}x^n; \qquad (3.14)$$

$$\frac{1}{1-x} = 1 + x + x^2 + \cdots = \sum_{n=0}^{\infty} x^n; \qquad (3.15)$$

$$(1-x)^\alpha = 1 - \alpha x + \frac{1}{2!}\alpha(\alpha-1)x^2 - \frac{1}{3!}\alpha(\alpha-1)(\alpha-2)x^3 + \cdots . \quad (3.16)$$

Up to this point we made the Taylor expansion around the point $x = 0$. However, one can make a Taylor expansion of $f(x + h)$ around any arbitrary point x. The associated Taylor series can be obtained by replacing the distance x that we move from the expansion point by a distance h and by replacing the expansion point 0 by x. Making the replacements $x \to h$ and $0 \to x$ expansion (3.11) is given by

$$f(x + h) = \sum_{n=0}^{\infty} \frac{h^n}{n!} \frac{d^n f}{dx^n}(x). \qquad (3.17)$$

Problem e Truncate this series after the second term and show that this leads to the following approximations:

$$f(x + h) - f(x) \approx h\frac{df}{dx}(x), \qquad (3.18)$$

$$\frac{df}{dx} \approx \frac{f(x + h) - f(x)}{h}. \qquad (3.19)$$

These expressions may appear to be equivalent in a trivial way. However, we will make extensive use of them in different ways. Equation (3.18) makes it possible to estimate the change in a function when the independent variable is changed slightly, whereas (3.19) is useful for estimating the derivative of a function given its values at neighboring points. The issue of estimating the derivative of a function is treated in much more detail in Section 11.2. Figure 11.2 makes it possible to also derive the estimate (3.19) geometrically, recognizing that the derivative of a function is just the slope of that function.

The Taylor series can also be used for functions of more than one variable. As an example, consider a function $f(x, y)$ that depends on the variables x and y. The generalization of the Taylor series (3.9) to functions of two variables is given by

$$f(x, y) = \sum_{n,m=0}^{\infty} c_{nm} x^n y^m. \qquad (3.20)$$

At this point the coefficients c_{nm} are not yet known. They follow in the same way as the coefficients of the Taylor series of a function that depends on a single variable:

by taking the partial derivatives of the Taylor series and evaluating the result at the point where the expansion is made.

Problem f   Take all the partial derivatives of (3.20) with respect to x and y up to second order, including the mixed derivative $\partial^2 f/\partial x \partial y$, and evaluate the result at the expansion point $x = y = 0$ to show that up to second order the Taylor expansion (3.20) is given by

$$
\begin{aligned}
f(x, y) = f(0, 0) &+ \frac{\partial f}{\partial x}(0, 0) \, x + \frac{\partial f}{\partial y}(0, 0) \, y \\
&+ \frac{1}{2}\frac{\partial^2 f}{\partial x^2}(0, 0) \, x^2 + \frac{\partial^2 f}{\partial x \partial y}(0, 0) \, xy \\
&+ \frac{1}{2}\frac{\partial^2 f}{\partial y^2}(0, 0) \, y^2 + \cdots
\end{aligned}
\tag{3.21}
$$

This is the Taylor expansion of $f(x, y)$ around the point $x = y = 0$.

Problem g   Make suitable substitutions in this result to show that the Taylor expansion around an arbitrary point (x, y) is given by

$$
\begin{aligned}
f(x + h_x, y + h_y) = f(x, y) &+ \frac{\partial f}{\partial x}(x, y) \, h_x + \frac{\partial f}{\partial y}(x, y) \, h_y \\
&+ \frac{1}{2}\frac{\partial^2 f}{\partial x^2}(x, y) \, h_x^2 + \frac{\partial^2 f}{\partial x \partial y}(x, y) \, h_x h_y \\
&+ \frac{1}{2}\frac{\partial^2 f}{\partial y^2}(x, y) \, h_y^2 + \cdots
\end{aligned}
\tag{3.22}
$$

Let us return to the Taylor series (3.9) with the coefficients c_m given by (3.10). This series hides an intriguing result. Equations (3.9) and (3.10) suggest that a function $f(x)$ is specified for all values of its argument x when all the derivatives are known at a single point $x = 0$. This means that the global behavior of a function is completely contained in the properties of the function at a single point. This is, in fact, not always true.

First, the series (3.9) is an infinite series, and the sum of infinitely many terms does not necessarily lead to a finite answer. As an example, look at the series (3.15). A series can only converge when the terms go to zero as $n \to \infty$, because otherwise every additional term changes the sum. The terms in the series (3.15) are given by x^n; these terms only go to zero as $n \to \infty$ when $|x| < 1$. In general, the Taylor series (3.9) only *converges* when x is smaller than a certain critical value called the *radius of convergence*. Details on the criteria for the convergence of series can be found in Boas (2006) or Butkov (1968), for example.

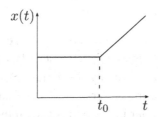

Figure 3.2   The motion of a particle that suddenly changes character at time t_0.

The second reason why the derivatives at one point do not necessarily constrain the function everywhere is that a function may change its character over the range of parameter values that is of interest. As an example, let us return to a moving particle and consider a particle at position $x(t)$ that is at rest until a certain time t_0 and that then starts moving with a uniform velocity $v \neq 0$:

$$x(t) = \begin{cases} x_0 & \text{for } t \leq t_0, \\ x_0 + v(t - t_0) & \text{for } t > t_0. \end{cases} \tag{3.23}$$

The motion of the particle is sketched in Figure 3.2. A straightforward application of (3.8) shows that all the coefficients c_n of this function vanish except c_0, which is given by x_0. The Taylor series (3.7) is therefore given by $x(t) = x_0$, which clearly differs from (3.23). The reason for this is that the function (3.23) changes its character at $t = t_0$ in such a way that nothing in the behavior for times $t < t_0$ predicts the sudden change in the motion at time $t = t_0$. Mathematically things go wrong because the first and higher derivatives of the function are not defined at time $t = t_0$.

Problem h   What is the first derivative of $x(t)$ just to the left and just to the right of $t = t_0$? What is the second derivative at that point?

The function (3.23) is said to be not analytic at the point $t = t_0$. The issue of analytic functions is treated in more detail in Sections 15.1 and 16.1.

Problem i   Try to compute the Taylor series of the function $x(t) = 1/t$ using (3.7) and (3.8). Draw this function and explain why the Taylor series cannot be used for this function.

Problem j   Do the same for the function $x(t) = \sqrt{t}$.

The examples in the last two problems show that when a function is not analytic at a certain point, the coefficients of the Taylor series are not defined. This signals that such a function cannot be represented by a Taylor series around that point.

3.2 Growth of the Earth by cosmic dust

In this section we use the growth of the Earth by the accretion of cosmic dust as an example to illustrate the usefulness of the (first order) Taylor series. The Earth is continuously bombarded from space by meteorites. Some of these meteorites can be large and lead to massive impact craters. As an example the gravity anomaly over the Chicxulub impact crater in Mexico is shown in Figure 20.1. The diameter of this impact crater is about 100 km. However, the bulk of the cosmic dust that falls from space onto the Earth is in the form of many small particles. The total mass of all the cosmic dust that falls on the Earth is estimated by Love and Brownlee (1993) to be approximately 5×10^7 kg/a. (The unit "a" stands for annum (or year); this means that the unit used here is kilograms per year.) This estimate is, however, not very accurate and we can probably only trust the first decimal of this number. This means that in subsequent calculations it is pointless to aim for an accuracy of more than one significant figure.

Since the cosmic dust increases the mass of the Earth, the size of the Earth will increase. In this section we determine the growth of the Earth's radius per year due to the bombardment of our planet by cosmic dust.

Problem a   Assuming that a density of meteorites is given by $\rho = 2.5 \times 10^3$ kg/m^3 (Love and Brownlee, 1993), show that the annual growth of the volume of the Earth is given by

$$\delta V = 2 \times 10^4 \text{ m}^3. \tag{3.24}$$

Also show that this corresponds to a block of $(27 \times 27 \times 27)$ m^3.

We assume that the Earth is a perfect sphere so that the volume and the radius r of the Earth are related by the relation

$$V = \frac{4\pi}{3} r^3. \tag{3.25}$$

From this relation we can deduce that the annual change δr of the radius of the Earth can be computed from the expression

$$\delta r = \left[\frac{3(V + \delta V)}{4\pi} \right]^{1/3} - \left(\frac{3V}{4\pi} \right)^{1/3}. \tag{3.26}$$

Problem b   Assume the Earth's radius is given by $r = 7000$ km. Insert this number and the value of δV from expression (3.24) into (3.26) and use a calculator to compute the increase, δr, of the radius of the Earth.

You have probably found that the annual increase in the Earth's radius is equal to zero. This cannot be true because we know that δV is not equal to zero. The reason

that your calculator has given you a wrong answer is that in (3.26) we are subtracting two very large numbers that differ by a small amount. The volume V of the
Earth is of the order 10^{21} m^3, while according to (3.24) the annual increase of the
volume is of the order 10^4 m^3. Most calculators carry out all the calculations with
a relatively small number of digits; most use between 6 and 10 decimals. When
you subtract two numbers that are very large and that have a very small difference, this difference will be truncated after say 6 or 10 decimals. In our problem,
the first 10 decimals of both terms in (3.26) are identical; hence, your calculator
tells you that the radius of the Earth is not growing because of the accretion of
cosmic dust.

In general, subtracting two large numbers that have a small difference leads
to numerical inaccuracies. Clearly a trick is needed to obtain the desired growth
of the Earth's radius. The cause of this problem is that the annual change in the
volume is very small, compared to the total volume. We can turn this problem
to our advantage by using that the Taylor series that we introduced in the previous section is extremely accurate when the independent variable is changed by
a very small amount. Here we compute the increase of the Earth's radius using
expression (3.18).

Problem c Show that this expression can also be written as

$$\delta f \approx \frac{\partial f}{\partial x}\delta x, \tag{3.27}$$

where δf is the change in the function $f(x)$ due to a change δx in the
independent variable x.

Problem d Apply this result to the function $r(V) = (3V/4\pi)^{1/3}$ that gives the
radius as a function of the volume to derive that

$$\delta r \approx \frac{1}{3}r\frac{\delta V}{V}. \tag{3.28}$$

Problem e Use this result to show that the annual increase of the radius of the
Earth due to the accretion of cosmic dust is on the order of 1 ångström per
year (1 ångström is 10^{-10} m).

Problem f Can you think of an object that is the size of 1 ångström?

Problem g How much has the Earth's radius increased over the age of the Earth?
In this calculation you may assume that the age of the Earth is 4.5 billion
years.

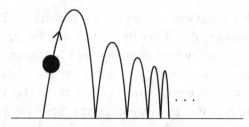

Figure 3.3   The motion of a bouncing ball that loses energy with every bounce. To visualize the motion of the ball better, the ball is given a constant horizontal velocity that is conserved during the bouncing.

The upshot of this calculation is that the growth of the Earth due to the present-day accretion of cosmic dust is negligible. However, the technique of using the first-order Taylor series to determine the small change in a quantity is extremely powerful. In fact, you have encountered in this section an example that demonstrates that an approximation can provide a more meaningful answer than a calculation carried out using a calculator or computer.

3.3 Bouncing ball

In this section we study a rubber ball that bounces on a flat surface and slowly comes to rest, as sketched in Figure 3.3. You may know from experience that the ball bounces more and more rapidly with time. The question we address is whether the ball can actually bounce infinitely many times in a finite amount of time. This problem is not an easy one. In general with large difficult problems it is a useful strategy to divide the large and difficult problem that you cannot solve into smaller and simpler problems that you can solve. By assembling these smaller subproblems, one can then often solve the large problem. This is exactly what we will do here. First, we will find how much time it takes for the ball to bounce once given its velocity. Given a prescription of the energy loss in one bounce, we will determine a relation between the velocity of subsequent bounces. From these ingredients we can determine the relation between the times needed for subsequent bounces. By summing this series over an infinite number of bounces we can determine the total time that the ball has bounced. *Keep this general strategy in mind when solving complex problems. Almost all of us are better at solving a number of small problems rather than a single large problem!*

Problem a   A ball moves upward from the level $z = 0$ with velocity v and is subject to a constant gravitational acceleration g. Determine the height the ball reaches and the time it takes for the ball to return to its starting point.

At this point we have determined the relevant properties for a single bounce. During each bounce the ball loses energy due to the fact that the ball is deformed inelastically during the bounce. We assume that during each bounce the ball loses a fraction γ of its energy.

Problem b Let the velocity at the beginning of the nth bounce be v_n. Show that with the assumed rule for energy loss, this velocity is related to the velocity v_{n-1} of the previous bounce by

$$v_n = \sqrt{1 - \gamma}\, v_{n-1}. \tag{3.29}$$

Hint: When the ball bounces upward from $z = 0$, all its energy is kinetic energy $\frac{1}{2}mv^2$.

In Problem a you determined the time it took the ball to bounce once, given the initial velocity, while expression (3.29) gives a recursive relation for the velocity between subsequent bounces. The relation between v_n and t_n can be derived from the expression $z(t) = v_n t - (1/2)gt_n^2$ by setting $z(t_n) = 0$. Here g denotes the acceleration of gravity. By assembling these results we can find a relation between the time t_n for the nth bounce and the time t_{n-1} for the previous bounce.

Problem c Determine this relation. In addition, let us assume that the ball is thrown up the first time from $z = 0$ to reach a height $z = H$. Compute the time t_0 needed for the ball to make the first bounce and combine these results to show that

$$t_n = \sqrt{\frac{8H}{g}}(1 - \gamma)^{n/2}, \tag{3.30}$$

where g is the acceleration of gravity.

We can use this expression to determine the total time T_N it takes to carry out N bounces. This time is given by $T_N = \sum_{n=0}^{N} t_n$. By setting N equal to infinity, we can compute the time T_∞ it takes to bounce infinitely often.

Problem d Determine this time by carrying out the summation and show that it is given by:

$$T_\infty = \sqrt{\frac{8H}{g}}\frac{1}{1 - \sqrt{1 - \gamma}}. \tag{3.31}$$

Hint: Write $(1 - \gamma)^{n/2}$ as $\left(\sqrt{1 - \gamma}\right)^n$ and treat $\sqrt{1 - \gamma}$ as the parameter x in the appropriate Taylor series of Section 3.1.

This result shows that the time it takes to bounce infinitely often is indeed finite. For the special case that the ball loses no energy, $\gamma = 0$ and T_∞ is infinite. This reflects that a ball that loses no energy will bounce forever.

Expression (3.31) looks messy. It often happens in mathematical physics that the final expression resulting from a calculation is so complex that it is difficult to understand it. However, often we know that certain terms in an expression can be assumed to be very small (or very large). This may allow us to obtain an approximate expression that is of a simpler form. In this way we trade accuracy for simplicity and understanding. In practice, this often turns out to be a good deal! In our example of the bouncing ball we assume that the energy loss γ at each bounce is small.

Problem e   Show that in this case $T_\infty \approx \sqrt{(8H/g)}\ 2/\gamma$ by using the leading terms of the appropriate Taylor series of Section 3.1.

In this example we have solved the problem in little steps. In general we take larger steps in the problems in this book, and you will have to discover how to divide a large step into smaller steps. The next problem is a "large" problem; solve it by dividing it into smaller problems. First, formulate the smaller problems as ingredients for the large problem before you actually start working on the smaller problems.

> *Make it a habit whenever you are solving problems to first formulate a strategy for how you are going to attack a problem before you actually start working on the subproblems. Make a list if this helps you and do not be deterred if you cannot solve a particular subproblem. Perhaps you can solve the other subproblems and somebody else can help you with the one you cannot solve.*

Keeping this in mind, solve the following "large" problem:

Problem f   Let the total distance traveled by the ball in the vertical direction during infinitely many bounces be denoted by S_∞. Show that $S_\infty \approx 2H/\gamma$.

The results of Problems e and f are actually quite useful. They tell us *how* the total bounce time and height approach infinity when the energy loss γ goes to zero.

Problem g   What are S_∞ and T_∞, in the case where γ approaches 1? Do the answers make physical sense to you?

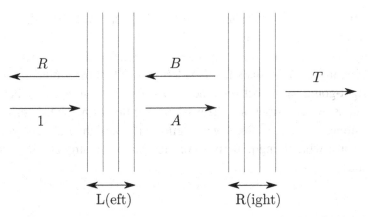

Figure 3.4 Geometry of the problem where two stacks of thin reflective layers are combined. The coefficients for the various left- and right-going waves are indicated.

3.4 Reflection and transmission by a stack of layers

In the early part of the twentieth century, Lord Rayleigh addressed the question of why some birds and insects have beautiful iridescent colors (Rayleigh, 1917). He explained this by studying the reflective properties of a stack of thin reflective layers. This problem is also of interest in geophysics; in exploration seismology one is also interested in the reflection and transmission properties of stacks of reflective layers in the Earth. Lord Rayleigh solved this problem in the following way.

Suppose we have one stack of layers on the left with reflection coefficient R_L and transmission coefficient T_L and another stack of layers on the right with reflection coefficient R_R and transmission coefficient T_R. If we add these two stacks together to obtain a larger stack of layers, what are the reflection coefficient R and transmission coefficient T of the total stack of layers? See Figure 3.4 for the scheme of this problem. The reflection coefficient is defined as the ratio of the strengths of the reflected and the incident waves, similarly the transmission coefficient is defined as the ratio of the strengths of the transmitted wave and the incident wave. To highlight the essential arguments, we simplify the analysis and ignore that the reflection coefficient for waves incident from the left and the right are in general not the same. However, this simplification does not change the essence of the coming arguments.

Before we start solving the problem, let us speculate what the transmission coefficient of the combined stack is. It may seem natural to assume that the transmission coefficient of the combined stack is the product of the transmission coefficient of the individual stacks:

$$T \overset{?}{=} T_L T_R. \tag{3.32}$$

However, this result is wrong and we will discover why. Consider Figure 3.4 again. The unknown quantities are R, T, and the coefficients A and B for the right-going and left-going waves between the stacks. An incident wave with strength 1 impinges on the stack from the left. Let us first determine the coefficient A of the right-going waves between the stacks. The right-going wave between the stacks contains two contributions: the wave transmitted from the left (this contribution has a strength $1 \times T_L$) and the wave reflected toward the right due the incident left-going wave with strength B (this contribution has a strength $B \times R_L$). This implies that:

$$A = T_L + B R_L. \tag{3.33}$$

Problem a   Using similar arguments show that:

$$B = A R_R, \tag{3.34}$$

$$T = A T_R, \tag{3.35}$$

$$R = R_L + B T_L. \tag{3.36}$$

This is all we need to solve our problem. The system of equations (3.33)–(3.36) consists of four linear equations with four unknowns A, B, R, and T. We could solve this system of equations by brute force, but some thought will make life easier for us. Note that the last two equations immediately give T and R once A and B are known. The first two equations give A and B.

Problem b   Show that

$$A = \frac{T_L}{(1 - R_L R_R)}, \tag{3.37}$$

$$B = \frac{T_L R_R}{(1 - R_L R_R)}. \tag{3.38}$$

This is a puzzling result. The right-going wave A between the layers not only contains the transmission coefficient of the left layer T_L but also an additional term $1/(1 - R_L R_R)$.

Problem c   Make a series expansion of $1/(1 - R_L R_R)$ in the quantity $R_L R_R$ and show that this term accounts for the waves that bounce back and forth between the two stacks. Hint: Use that R_L is the reflection coefficient for a wave that reflects from the left stack and R_R is the reflection coefficient for one that reflects from the right stack so that $R_L R_R$ is the total reflection coefficient for a wave that bounces once between the left and the right stacks.

This implies that the term $1/(1 - R_L R_R)$ accounts for the waves that bounce back and forth between the two stacks of layers. It is for this reason that we call this a *reverberation* term. It plays an important role in computing the response of layered media.

Problem d Show that the reflection and transmission coefficients of the combined stack of layers are given by:

$$R = R_L + \frac{T_L^2 R_R}{(1 - R_L R_R)}, \qquad (3.39)$$

$$T = \frac{T_L T_R}{(1 - R_L R_R)}. \qquad (3.40)$$

At the beginning of this section we conjectured that the transmission coefficient of the combined stacks is the product of the transmission coefficient of the separate stacks, see expression (3.32).

Problem e Is this conjecture correct? Under which conditions is it approximately correct?

Equations (3.39) and (3.40) are useful for computing the reflection and transmission coefficients of a large stack of layers. The reason for this is that it is extremely simple to determine the reflection and transmission coefficients of a very thin layer using the Born approximation. (The Born approximation is treated in Section 23.2.) Let the reflection and transmission coefficients of a *single* thin layer n be denoted by r_n and t_n, respectively, and let the reflection and transmission coefficients of a *stack* of n layers be denoted by R_n and T_n, respectively. Suppose that the left stack consists of n layers and that we want to add an $(n + 1)$th layer to the stack. In that case the right stack consists of a single $(n + 1)$th layer so that $R_R = r_{n+1}$ and $T_R = t_{n+1}$ and the reflection and transmission coefficients of the left stack are given by $R_L = R_n$, $T_L = T_n$. Using this in expressions (3.39) and (3.40) yields

$$R_{n+1} = R_n + \frac{T_n^2 r_{n+1}}{(1 - R_n r_{n+1})}, \qquad (3.41)$$

$$T_{n+1} = \frac{T_n t_{n+1}}{(1 - R_n r_{n+1})}. \qquad (3.42)$$

This means that given the known response of a stack of n layers, one can easily compute the effect of adding the $(n + 1)$th layer to this stack. In this way one can recursively build up the response of the complex reflector out of the known response of very thin reflectors. Computers are pretty stupid, but they are ideally suited for applying the rules (3.41) and (3.42) a large number of times. Of course this process has to start with a medium in which no layers are present.

Problem f   What are the reflection coefficient R_0 and the transmission coefficient T_0 when there are as yet no reflective layers present? Describe how one can compute the response of a thick stack of layers once we know the response of a very thin layer.

So far, we did not address that the reflection and transmission from a layer – or a stack of layers – depend on the wavelength. It is the thickness of a layer measured in wavelengths that determines how a wave interacts with that layer. For light, or electromagnetic waves, this means that particular layering on the feathers of birds and bodies of insects will reflect certain colors more than others. In explaining the colors of insects and birds, Lord Rayleigh prepared the foundations for a theory that later became known as *invariant embedding*, which turns out to be extremely useful for a number of scattering and diffusion problems (Bellman et al., 1960; Tromp and Snieder, 1989).

The main conclusion of the treatment of this section is that the transmission of a combination of two stacks of layers is not the product of the transmission coefficients of the two separate stacks, because the waves that repeatedly reflect

Figure 3.5   These swirls are grayscale representations of color changes in a soap bubble made by T. Meuel and H. Kellay, Université de Bordeaux and CNRS (Meuel et al., 2013), and is shown on the front cover of this book. The colors are an example of wave interference caused by reflections off the front and back of the soap layer.

between the two stacks leave an imprint on the transmission coefficient as well. Paradoxically, Berry and Klein (1997) showed in their analysis of "transparent mirrors" that for the special case of a large stack of layers with random transmission coefficients, the total transmission coefficient *is* the product of the transmission coefficients of the individual layers, despite the fact that multiple reflections play a crucial role in this process.

To illustrate the concept of wave interference, we end this chapter with a photo of a soap bubble (Figure 3.5; the color version features the cover of this book). The color patterns in the bubble are from the interference of light bouncing between the front and back surface of the soap film. Depending on the wavelength (i.e., color) of the light, and the distance traveled in the film, the wave interference can be destructive or constructive. The travel distance in the film depends on the (spatially varying) thickness of the film, as well as the angle at which the light enters the soap film. Now we understand the origin of the beautiful colors of the bubble, we will tackle the swirly pattern and the shape of the soap bubble in Sections 7.3 and 10.4, respectively.

4

Spherical and cylindrical coordinates

Many problems in mathematical physics exhibit a spherical or cylindrical symmetry. For example, the gravity field of the Earth is to first order spherically symmetric. Waves excited by a stone thrown into water are usually cylindrically symmetric. Although there is no reason why problems with such a symmetry cannot be analyzed using Cartesian coordinates (i.e., (x, y, z)-coordinates), it is usually not very convenient to use such a coordinate system. The reason for this is that the theory is usually much simpler when one selects a coordinate system with symmetry properties that are the same as the symmetry properties of the physical system that one wants to study. It is for this reason that spherical coordinates and cylindrical coordinates are introduced in this chapter. It takes a certain effort to become acquainted with these coordinate systems, but this effort is well spent because it makes solving a large class of problems much easier.

4.1 Introducing spherical coordinates

Figure 4.1 depicts a Cartesian coordinate system with its x-, y-, and z-axis, as well as the location of a point $\mathbf{r}$. This point can be described either by its x-, y-, and z-components, or by the radius r and two angles θ and φ. The latter description of $\mathbf{r}$ is in *spherical coordinates*, where the angle φ runs from 0 to 2π, while θ has values between 0 and π. Even though you are probably more comfortable with Cartesian coordinates, the angles θ and φ are closely related to the familiar geographical coordinates that define a point on the globe. In that case, φ can be compared with *longitude* and θ with *colatitude*, which is defined as (90 degrees minus *latitude*).

Problem a The city of Utrecht in the Netherlands is located at 52 degrees north and 5 degrees east. Show that the angles θ and φ (in radians) that correspond to this point on the sphere are given by $\theta = 0.663$ radians, and $\varphi = 0.087$ radians. (Don't forget that θ is the colatitude, not the latitude.)

In Cartesian coordinates the position vector can be written as

$$\mathbf{r} = x\hat{\mathbf{x}} + y\hat{\mathbf{y}} + z\hat{\mathbf{z}}, \tag{4.1}$$

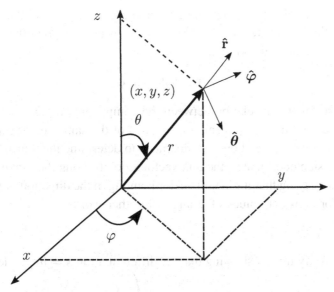

Figure 4.1   The definition of (x, y, z) in terms of the radius r and angles (θ, φ) of spherical coordinates.

where the caret (^) is used to denote a vector that is of unit length. An arbitrary vector can be expressed in a superposition of these basis vectors:

$$\mathbf{u} = u_x\hat{\mathbf{x}} + u_y\hat{\mathbf{y}} + u_z\hat{\mathbf{z}}. \tag{4.2}$$

We want also to express the same vector in basis vectors that are related to the spherical coordinate system. Before we can do so, we must first establish the connection between the Cartesian coordinates (x, y, z) and the spherical coordinates (r, θ, φ).

Problem b   Use Figure 4.1 to show that the Cartesian coordinates can be written as

$$\left. \begin{array}{l} x = r \sin\theta \cos\varphi, \\ y = r \sin\theta \sin\varphi, \\ z = r \cos\theta. \end{array} \right\} \tag{4.3}$$

Problem c   Use these expressions to derive the following expression for the spherical coordinates in terms of the Cartesian coordinates:

$$\left. \begin{array}{l} r = \sqrt{x^2 + y^2 + z^2}, \\ \theta = \arccos\left(z/\sqrt{x^2 + y^2 + z^2}\right), \\ \varphi = \arctan\left(y/x\right). \end{array} \right\} \tag{4.4}$$

We have now obtained the relation between the Cartesian coordinates (x, y, z) and the spherical coordinates (r, θ, φ). Suppose we want to express the vector **u** of equation (4.2) in spherical coordinates:

$$\mathbf{u} = u_r\hat{\mathbf{r}} + u_\theta\hat{\boldsymbol{\theta}} + u_\varphi\hat{\boldsymbol{\varphi}}, \tag{4.5}$$

and we want to know the relation between the components (u_x, u_y, u_z) in Cartesian coordinates and the components $(u_r, u_\theta, u_\varphi)$ of the same vector expressed in spherical coordinates. To find out, we first need to determine the unit vectors $\hat{\mathbf{r}}$, $\hat{\boldsymbol{\theta}}$, and $\hat{\boldsymbol{\varphi}}$. In Cartesian coordinates, the unit vector $\hat{\mathbf{x}}$ points along the x-axis. This is a different way of saying that it is a unit vector pointing in the direction of increasing values of x for constant values of y and z; in other words, $\hat{\mathbf{x}}$ can be written as: $\hat{\mathbf{x}} = \partial\mathbf{r}/\partial x$.

Problem d   Verify this by showing that the differentiation $\hat{\mathbf{x}} = \partial\mathbf{r}/\partial x$ leads to the

correct unit vector in the x-direction: $\hat{\mathbf{x}} = \begin{pmatrix} 1 \\ 0 \\ 0 \end{pmatrix}$.

Now consider the unit vector $\hat{\boldsymbol{\theta}}$. Using the same argument as for the unit vector $\hat{\mathbf{x}}$, we know that $\hat{\boldsymbol{\theta}}$ is directed toward increasing values of θ for constant values of r and φ. This means that $\hat{\boldsymbol{\theta}}$ can be written as $\hat{\boldsymbol{\theta}} = C\partial\mathbf{r}/\partial\theta$. The constant C follows from the requirement that $\hat{\boldsymbol{\theta}}$ is of unit length.

Problem e   Use this reasoning for all the unit vectors $\hat{\mathbf{r}}$, $\hat{\boldsymbol{\theta}}$, and $\hat{\boldsymbol{\varphi}}$ and expression (4.3) to show that:

$$\hat{\mathbf{r}} = \frac{\partial\mathbf{r}}{\partial r}, \quad \hat{\boldsymbol{\theta}} = \frac{1}{r}\frac{\partial\mathbf{r}}{\partial\theta}, \quad \hat{\boldsymbol{\varphi}} = \frac{1}{r\sin\theta}\frac{\partial\mathbf{r}}{\partial\varphi}, \tag{4.6}$$

and that this result can also be written as

$$\hat{\mathbf{r}} = \begin{pmatrix} \sin\theta\cos\varphi \\ \sin\theta\sin\varphi \\ \cos\theta \end{pmatrix}, \quad \hat{\boldsymbol{\theta}} = \begin{pmatrix} \cos\theta\cos\varphi \\ \cos\theta\sin\varphi \\ -\sin\theta \end{pmatrix}, \quad \hat{\boldsymbol{\varphi}} = \begin{pmatrix} -\sin\varphi \\ \cos\varphi \\ 0 \end{pmatrix}. \tag{4.7}$$

These equations give the x, y, and z coordinates of the unit vectors $\hat{\mathbf{r}}$, $\hat{\boldsymbol{\theta}}$, and $\hat{\boldsymbol{\varphi}}$. Note that on the right-hand side of (4.6) the derivatives of the position vector are divided by 1, r, and $r\sin\theta$, respectively. These factors are usually shown in the following notation:

$$h_r = 1, \quad h_\theta = r, \quad h_\varphi = r\sin\theta. \tag{4.8}$$

These scale factors play an important role in the general theory of curvilinear coordinate systems (see Butkov, 1968, for details). The material presented in the remainder of this chapter, as well as the derivation of vector calculus in spherical coordinates, can be based on the scale factors given in (4.8). However, this approach will not be taken here.

Problem f Verify explicitly that the vectors $\hat{\mathbf{r}}$, $\hat{\boldsymbol{\theta}}$, and $\hat{\boldsymbol{\varphi}}$ defined in this way form an orthonormal basis, that is, they are of unit length and perpendicular to each other:

$$\left(\hat{\mathbf{r}} \cdot \hat{\mathbf{r}}\right) = \left(\hat{\boldsymbol{\theta}} \cdot \hat{\boldsymbol{\theta}}\right) = \left(\hat{\boldsymbol{\varphi}} \cdot \hat{\boldsymbol{\varphi}}\right) = 1, \tag{4.9}$$

$$\left(\hat{\mathbf{r}} \cdot \hat{\boldsymbol{\theta}}\right) = \left(\hat{\mathbf{r}} \cdot \hat{\boldsymbol{\varphi}}\right) = \left(\hat{\boldsymbol{\theta}} \cdot \hat{\boldsymbol{\varphi}}\right) = 0. \tag{4.10}$$

The dot denotes the inner product of two vectors.

Problem g Using expressions (4.7) for the unit vectors $\hat{\mathbf{r}}$, $\hat{\boldsymbol{\theta}}$, and $\hat{\boldsymbol{\varphi}}$, show by calculating the cross-products explicitly that

$$\hat{\mathbf{r}} \times \hat{\boldsymbol{\theta}} = \hat{\boldsymbol{\varphi}}, \quad \hat{\boldsymbol{\theta}} \times \hat{\boldsymbol{\varphi}} = \hat{\mathbf{r}}, \quad \hat{\boldsymbol{\varphi}} \times \hat{\mathbf{r}} = \hat{\boldsymbol{\theta}}. \tag{4.11}$$

The Cartesian basis vectors $\hat{\mathbf{x}}$, $\hat{\mathbf{y}}$, and $\hat{\mathbf{z}}$ have similar cross-products, but these vectors point in the same direction at every point in space. This is not true for the spherical basis vectors $\hat{\mathbf{r}}$, $\hat{\boldsymbol{\theta}}$, and $\hat{\boldsymbol{\varphi}}$. Expression (4.7) reveals that for different values of the angles θ and φ, these unit vectors in spherical coordinates point in different directions. For several applications it is necessary to know how the basis vectors change with θ and φ. This change is described by the derivative of the unit vectors with respect to the angles θ and φ.

Problem h Show by direct differentiation of expressions (4.7) that the derivatives of the unit vectors with respect to the angles θ and φ are given by:

$$\left.\begin{array}{ll}
\partial\hat{\mathbf{r}}/\partial\theta = \hat{\boldsymbol{\theta}}, & \partial\hat{\mathbf{r}}/\partial\varphi = \sin\theta\,\hat{\boldsymbol{\varphi}}, \\
\partial\hat{\boldsymbol{\theta}}/\partial\theta = -\hat{\mathbf{r}}, & \partial\hat{\boldsymbol{\theta}}/\partial\varphi = \cos\theta\,\hat{\boldsymbol{\varphi}}, \\
\partial\hat{\boldsymbol{\varphi}}/\partial\theta = 0, & \partial\hat{\boldsymbol{\varphi}}/\partial\varphi = -\sin\theta\,\hat{\mathbf{r}} - \cos\theta\,\hat{\boldsymbol{\theta}}.
\end{array}\right\} \tag{4.12}$$

4.2 Changing coordinate systems

Now that we have derived the properties of the unit vectors $\hat{\mathbf{r}}$, $\hat{\boldsymbol{\theta}}$, and $\hat{\boldsymbol{\varphi}}$, we are in the position to derive how the components $(u_r, u_\theta, u_\varphi)$ of the vector $\mathbf{u}$ defined in equation (4.5) are related to the Cartesian coordinates (u_x, u_y, u_z). This can most easily be achieved by writing expressions (4.7) in the following form:

$$\left.\begin{aligned} \hat{\mathbf{r}} &= \sin\theta\cos\varphi\,\hat{\mathbf{x}} + \sin\theta\sin\varphi\,\hat{\mathbf{y}} + \cos\theta\,\hat{\mathbf{z}}, \\ \hat{\boldsymbol{\theta}} &= \cos\theta\cos\varphi\,\hat{\mathbf{x}} + \cos\theta\sin\varphi\,\hat{\mathbf{y}} - \sin\theta\,\hat{\mathbf{z}}, \\ \hat{\boldsymbol{\varphi}} &= -\sin\varphi\,\hat{\mathbf{x}} + \cos\varphi\,\hat{\mathbf{y}}. \end{aligned}\right\} \tag{4.13}$$

Problem a Convince yourself that this expression can also be written in a symbolic form as

$$\begin{pmatrix} \hat{\mathbf{r}} \\ \hat{\boldsymbol{\theta}} \\ \hat{\boldsymbol{\varphi}} \end{pmatrix} = \mathbf{M}\begin{pmatrix} \hat{\mathbf{x}} \\ \hat{\mathbf{y}} \\ \hat{\mathbf{z}} \end{pmatrix}, \tag{4.14}$$

with the matrix $\mathbf{M}$ given by

$$\mathbf{M} = \begin{pmatrix} \sin\theta\cos\varphi & \sin\theta\sin\varphi & \cos\theta \\ \cos\theta\cos\varphi & \cos\theta\sin\varphi & -\sin\theta \\ -\sin\varphi & \cos\varphi & 0 \end{pmatrix}. \tag{4.15}$$

Of course, expression (4.14) can only be considered to be a shorthand notation for equations (4.13) since the entries in (4.14) are vectors rather than single components.

The relation between the spherical components $(u_r, u_\theta, u_\varphi)$ and the Cartesian components (u_x, u_y, u_z) of the vector $\mathbf{u}$ can be obtained by inserting expressions (4.13) for the spherical coordinate unit vectors into the relation $\mathbf{u} = u_r\hat{\mathbf{r}} + u_\theta\hat{\boldsymbol{\theta}} + u_\varphi\hat{\boldsymbol{\varphi}}$.

Problem b Do this and collect all terms multiplying the unit vectors $\hat{\mathbf{x}}, \hat{\mathbf{y}}$, and $\hat{\mathbf{z}}$, respectively, to show that expression (4.5) for the vector $\mathbf{u}$ is equivalent to:

$$\begin{aligned} \mathbf{u} = {}&\left(u_r\sin\theta\cos\varphi + u_\theta\cos\theta\cos\varphi - u_\varphi\sin\varphi\right)\hat{\mathbf{x}} \\ &+ \left(u_r\sin\theta\sin\varphi + u_\theta\cos\theta\sin\varphi + u_\varphi\cos\varphi\right)\hat{\mathbf{y}} \\ &+ \left(u_r\cos\theta - u_\theta\sin\theta\right)\hat{\mathbf{z}}. \end{aligned} \tag{4.16}$$

Problem c Show that this relation can also be written as:

$$\begin{pmatrix} u_x \\ u_y \\ u_z \end{pmatrix} = \mathbf{M}^T\begin{pmatrix} u_r \\ u_\theta \\ u_\varphi \end{pmatrix}, \tag{4.17}$$

where the matrix $\mathbf{M}$ is given by (4.15). In this expression, $\mathbf{M}^T$ is the transpose of the matrix $\mathbf{M}$; that is, it is the matrix obtained by interchanging rows and columns of the matrix $M_{ij}^T = M_{ji}$.

With equation (4.17), we have not yet reached our goal of expressing the spherical coordinate components $(u_r, u_\theta, u_\varphi)$ of the vector $\mathbf{u}$ in the Cartesian components (u_x, u_y, u_z). This is most easily achieved by multiplying (4.17) with the inverse matrix $(\mathbf{M}^T)^{-1}$, which gives:

$$\begin{pmatrix} u_r \\ u_\theta \\ u_\varphi \end{pmatrix} = (\mathbf{M}^T)^{-1} \begin{pmatrix} u_x \\ u_y \\ u_z \end{pmatrix}. \tag{4.18}$$

However, now we have only shifted the problem because we do not know the inverse $(\mathbf{M}^T)^{-1}$. We could of course painstakingly compute this inverse, but this would be a laborious process that we can avoid. It follows by inspection of (4.15) that all the columns of $\mathbf{M}$ are of unit length and that the columns are orthogonal. This implies that $\mathbf{M}$ is an orthogonal matrix. Orthogonal matrices have the useful property that the transpose of the matrix is identical to the inverse of the matrix: $\mathbf{M}^{-1} = \mathbf{M}^T$.

Problem d The property $\mathbf{M}^{-1} = \mathbf{M}^T$ can be verified explicitly by showing that $\mathbf{M}\mathbf{M}^T$ and $\mathbf{M}^T\mathbf{M}$ are equal to the identity matrix; do this!

Problem e Use these results to show that the spherical coordinate components of $\mathbf{u}$ are related to the Cartesian coordinates by the following transformation rule:

$$\begin{pmatrix} u_r \\ u_\theta \\ u_\varphi \end{pmatrix} = \begin{pmatrix} \sin\theta\cos\varphi & \sin\theta\sin\varphi & \cos\theta \\ \cos\theta\cos\varphi & \cos\theta\sin\varphi & -\sin\theta \\ -\sin\varphi & \cos\varphi & 0 \end{pmatrix} \begin{pmatrix} u_x \\ u_y \\ u_z \end{pmatrix}. \tag{4.19}$$

4.3 Acceleration in spherical coordinates

You may wonder whether we really need all these transformation rules between a Cartesian coordinate system and a system of spherical coordinates. The answer is yes! An important example can be found in meteorology where air moves along a spherical surface. The velocity $\mathbf{v}$ of the air can be expressed in spherical coordinates as:

$$\mathbf{v} = v_r\hat{\mathbf{r}} + v_\theta\hat{\boldsymbol{\theta}} + v_\varphi\hat{\boldsymbol{\varphi}}. \tag{4.20}$$

The motion of the air is governed by Newton's law, as we will see in Chapter 5. But when the velocity $\mathbf{v}$ and the pressure force $\mathbf{F}$ are both expressed in spherical coordinates, it would be wrong to express the θ-component of Newton's law as: $\rho dv_\theta/dt = F_\theta$. The reason is that the basis vectors of the spherical coordinate system depend on the position, as we saw in the discussion of expression (4.7). When a particle moves, the directions of the basis vectors change, as well. This is a different way of saying that the spherical coordinate system is not like the Cartesian system, where the orientations of the coordinate axes are independent of the position. When computing the acceleration in such a system, additional terms appear that account for the fact that the coordinate system is curvilinear. The results of Section 4.1 contain all the ingredients we need.

Let us follow a particle moving over a sphere. The position vector **r** has an obvious expansion in spherical coordinates:

$$\mathbf{r} = r\hat{\mathbf{r}}. \tag{4.21}$$

Its velocity is obtained by taking the time derivative of this expression. However, the unit vector $\hat{\mathbf{r}}$ is, according to equation (4.7), a function of the angles θ and φ. This means that when we take the time derivative of (4.21) to obtain the particle velocity we also need to differentiate $\hat{\mathbf{r}}$ with time. Note that this is not the case with the Cartesian expression $\mathbf{r} = x\hat{\mathbf{x}} + y\hat{\mathbf{y}} + z\hat{\mathbf{z}}$, because the unit vectors $\hat{\mathbf{x}}$, $\hat{\mathbf{y}}$, and $\hat{\mathbf{z}}$ are constant. Hence, they do not change when the particle moves and they thus have a vanishing time derivative.

As an example, let us compute the time derivative of $\hat{\mathbf{r}}$. At first you may be surprised that we compute this time derivative. After all, the position vector and time seem to be independent variables. But what if the position vector describes the position of a particle, such as an orbiting satellite? In that case the position vector changes with time; as the satellite orbits the co-latitude θ and longitude φ change with time, and hence the position vector changes with time as well. The time derivative as we follow the satellite is called a *total time derivative*, and is discussed in more detail in Section 5.5. Using the chain rule it thus follows that:

$$\frac{d\hat{\mathbf{r}}}{dt} = \frac{d\hat{\mathbf{r}}(\theta, \varphi)}{dt} = \frac{d\theta}{dt}\frac{\partial\hat{\mathbf{r}}}{\partial\theta} + \frac{d\varphi}{dt}\frac{\partial\hat{\mathbf{r}}}{\partial\varphi}. \tag{4.22}$$

Problem a   Use expressions (4.12) to eliminate the derivatives $\partial\hat{\mathbf{r}}/\partial\theta$ and $\partial\hat{\mathbf{r}}/\partial\varphi$, and carry out a similar analysis for the time derivatives of the unit vectors $\hat{\boldsymbol{\theta}}$ and $\hat{\boldsymbol{\varphi}}$ to show that:

$$\left.\begin{array}{l} \dfrac{d\hat{\mathbf{r}}}{dt} = \dot{\theta}\hat{\boldsymbol{\theta}} + \sin\theta\,\dot{\varphi}\,\hat{\boldsymbol{\varphi}}, \\[2mm] \dfrac{d\hat{\boldsymbol{\theta}}}{dt} = -\dot{\theta}\hat{\mathbf{r}} + \cos\theta\,\dot{\varphi}\,\hat{\boldsymbol{\varphi}}, \\[2mm] \dfrac{d\hat{\boldsymbol{\varphi}}}{dt} = -\sin\theta\,\dot{\varphi}\,\hat{\mathbf{r}} - \cos\theta\,\dot{\varphi}\,\hat{\boldsymbol{\theta}}. \end{array}\right\} \tag{4.23}$$

In these and other expressions in this section a dot is used to denote the time derivative: $\dot{F} \equiv dF/dt$.

Problem b Use (4.21), the first line of (4.23), and the definition $\mathbf{v} = d\mathbf{r}/dt$ to show that in spherical coordinates:

$$\mathbf{v} = \dot{r}\hat{\mathbf{r}} + r\dot{\theta}\hat{\boldsymbol{\theta}} + r\sin\theta\,\dot{\varphi}\,\hat{\boldsymbol{\varphi}}. \qquad (4.24)$$

In spherical coordinates the components of the velocity are thus given by

$$\left.\begin{array}{l} v_r = \dot{r}, \\ v_\theta = r\dot{\theta}, \\ v_\varphi = r\sin\theta\,\dot{\varphi}. \end{array}\right\} \qquad (4.25)$$

This result can be interpreted geometrically. Let us consider the radial component of the velocity and Figure 4.2. To obtain the radial component of the velocity, we keep the angles θ and φ fixed and let the radius $r(t)$ change to $r(t + \Delta t)$ over a time Δt. The particle has moved a distance $r(t + \Delta t) - r(t) = (dr/dt)/\Delta t$ in a time Δt, so that the radial component of the velocity is given by $v_r = dr/dt = \dot{r}$. This is the result given by the first line of (4.25).

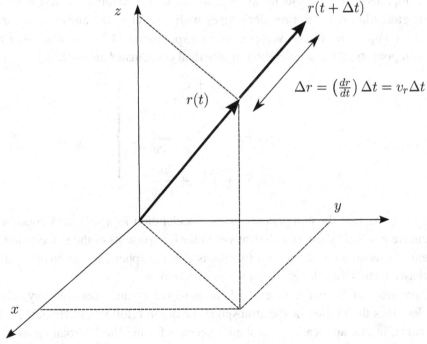

Figure 4.2 Definition of the geometric variables used to derive the radial component of the velocity.

Problem c   Use similar arguments to explain the form of the velocity components v_θ and v_φ given in (4.25).

Expression (4.21) may suggest to some readers that the position vector **r** depends on the radius r only, so it may come as a surprise that the velocity vector, the time derivative **r**, depends on the time derivative of the angles as well. Note, however, that the angles θ and φ are hidden in the unit vector $\hat{\mathbf{r}}$ in equation (4.21). The dependence of $\hat{\mathbf{r}}$ on the polar angles in given in expression (4.7).

Problem d   We are now in the position to compute the acceleration in spherical coordinates. To do this, differentiate (4.24) with respect to time and use expression (4.23) to eliminate the time derivatives of the basis vectors. Use this to show that the acceleration **a** is given by

$$\mathbf{a} = \left(\dot{v}_r - \dot{\theta} v_\theta - \sin\theta \, \dot{\varphi} v_\varphi\right)\hat{\mathbf{r}}$$
$$+ \left(\dot{v}_\theta + \dot{\theta} v_r - \cos\theta \, \dot{\varphi} v_\varphi\right)\hat{\boldsymbol{\theta}}$$
$$+ \left(\dot{v}_\varphi + \sin\theta \, \dot{\varphi} v_r + \cos\theta \, \dot{\varphi} v_\theta\right)\hat{\boldsymbol{\varphi}}. \tag{4.26}$$

Problem e   This expression is not quite satisfactory because it contains both the components of the velocity as well as the time derivatives $\dot{\theta}$ and $\dot{\varphi}$ of the angles. Eliminate the time derivatives with respect to the angles in favor of the components of the velocity using expressions (4.25) to show that the components of the acceleration in spherical coordinates are given by:

$$\left.\begin{aligned}
a_r &= \dot{v}_r - \frac{v_\theta^2 + v_\varphi^2}{r}, \\[2mm]
a_\theta &= \dot{v}_\theta + \frac{v_r v_\theta}{r} - \frac{v_\varphi^2}{r\tan\theta}, \\[2mm]
a_\varphi &= \dot{v}_\varphi + \frac{v_r v_\varphi}{r} + \frac{v_\theta v_\varphi}{r\tan\theta}.
\end{aligned}\right\} \tag{4.27}$$

It thus follows that the components of the acceleration in a spherical coordinate system are not simply the time derivatives of the components of the velocity in that system. As mentioned, the reason for this is that the spherical coordinate system uses basis vectors that change when the particle moves.

Expressions (4.27) play a crucial role in meteorology and oceanography, where one describes the motion of the atmosphere or ocean (Holton and Hakim, 2012). Of course, in that application one should account for the Earth's rotation as well, so that the terms accounting for the Coriolis effect and the centrifugal force need to be added (see Section 12.2). It should also be noted that the analysis of this section has been oversimplified when applied to the ocean or atmosphere. The reason for this is that the total time derivative of a moving parcel of air or water contains the explicit time derivative $\partial/\partial t$, as well as a term $\mathbf{v} \cdot \nabla$. The latter term accounts for the fact that the properties change, because the parcel moves to another location in space. The meaning of ∇ will be discussed in the next chapter. The difference between the partial derivative $\partial/\partial t$ and the total derivative d/dt is treated in Section 5.5; this distinction has not been taken into account in the analysis in this section. A complete treatment is given by Holton and Hakim (2012).

4.4 Volume integration in spherical coordinates

Carrying out a volume integration in Cartesian coordinates involves multiplying the function to be integrated by an infinitesimal volume element $dxdydz$ and integrating overall volume elements:

$$\iiint F dV = \iiint F(x, y, z) \, dx dy dz. \tag{4.28}$$

Although this seems to be a simple procedure, it can be quite complex when the function F depends in a complex way on the coordinates (x, y, z) or when the limits of integration are not simple functions of x, y, and z.

Problem a Compute the volume of a sphere of radius R by taking $F = 1$ in equation (4.28). Show first that in Cartesian coordinates the volume of the sphere can be written as

$$volume = \int_{-R}^{R} \int_{-\sqrt{R^2-x^2}}^{\sqrt{R^2-x^2}} \int_{-\sqrt{R^2-x^2-y^2}}^{\sqrt{R^2-x^2-y^2}} dz dy dx, \tag{4.29}$$

and then carry out the integrations.

After carrying out this exercise, you have probably become convinced that using Cartesian coordinates is not the most efficient way to derive that the volume of a sphere with radius R is given by $4\pi R^3/3$. Using spherical coordinates appears to be the way to go, but for this we need to be able to express an infinitesimal volume element dV in spherical coordinates. In doing this, we will use that the volume spanned by three vectors $\mathbf{a}$, $\mathbf{b}$, and $\mathbf{c}$ is given by (Boas, 2006):

$$volume = \det(\mathbf{a}, \mathbf{b}, \mathbf{c}) = \begin{vmatrix} a_x & b_x & c_x \\ a_y & b_y & c_y \\ a_z & b_z & c_z \end{vmatrix}. \tag{4.30}$$

If we change the spherical coordinate θ by an increment $d\theta$, the position vector changes from $\mathbf{r}(r, \theta, \varphi)$ to $\mathbf{r}(r, \theta + d\theta, \varphi)$, and this corresponds to a change $\mathbf{r}(r, \theta + d\theta, \varphi) - \mathbf{r}(r, \theta, \varphi) = \partial\mathbf{r}/\partial\theta \, d\theta$ in the position vector. Using the same reasoning for the variation of the position vector with r and φ, it follows that the infinitesimal volume dV corresponding to increments dr, $d\theta$, and $d\varphi$ is given by

$$dV = \det\left(\frac{\partial\mathbf{r}}{\partial r}dr, \ \frac{\partial\mathbf{r}}{\partial\theta}d\theta, \ \frac{\partial\mathbf{r}}{\partial\varphi}d\varphi\right). \tag{4.31}$$

Problem b   Show that this can be written as:

$$dV = \begin{vmatrix} \frac{\partial x}{\partial r} & \frac{\partial x}{\partial\theta} & \frac{\partial x}{\partial\varphi} \\ \frac{\partial y}{\partial r} & \frac{\partial y}{\partial\theta} & \frac{\partial y}{\partial\varphi} \\ \frac{\partial z}{\partial r} & \frac{\partial z}{\partial\theta} & \frac{\partial z}{\partial\varphi} \end{vmatrix} dr d\theta d\varphi = J dr d\theta d\varphi. \tag{4.32}$$

$$\underbrace{\qquad\qquad\qquad}_{J}$$

The determinant J is called the *Jacobian*, which is also sometimes written as

$$J = \frac{\partial(x, y, z)}{\partial(r, \theta, \varphi)}. \tag{4.33}$$

It should be kept in mind that this is nothing more than a new notation for the determinant in (4.32).

Problem c   Use expressions (4.3) and (4.32) to show that

$$J = r^2 \sin\theta. \tag{4.34}$$

Note that the Jacobian J in (4.34) is the product of the scale factors defined in equation (4.8): $J = h_r h_\theta h_\varphi$. This is not a coincidence; in general, the scale factors contain all the information needed to compute the Jacobian for an orthogonal curvilinear coordinate system (see Butkov, 1968, for details).

Problem d   A volume element dV is thus given in spherical coordinates by $dV = r^2 \sin\theta dr d\theta d\varphi$. Consider the volume element dV in Figure 4.3 that is defined by infinitesimal increments dr, $d\theta$, and $d\varphi$. Give an alternative derivation of this expression for dV by multiplying the sides of the infinitesimal volume shown in Figure 4.3.

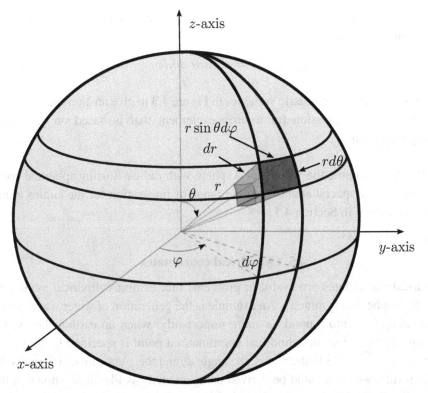

Figure 4.3 Definition of the geometric variables for an infinitesimal volume element dV in spherical coordinates.

In some applications, we want to integrate over the surface of a sphere rather than over a volume. For example, to compute the cooling of the Earth, we need to integrate the heat flow over the Earth's surface. The treatment used for deriving the volume integral in spherical coordinates can also be used to derive the surface integral. A key element in the analysis is that the surface spanned by two vectors **a** and **b** is given by $|\mathbf{a} \times \mathbf{b}|$. Again, an increment $d\theta$ of the angle θ corresponds to a change $(\partial \mathbf{r} \partial \theta) d\theta$ of the position vector. A similar result holds when the angle φ is changed.

Problem e Use these results to show that the surface element dS corresponding to infinitesimal changes $d\theta$ and $d\varphi$ is given by

$$dS = \left| \frac{\partial \mathbf{r}}{\partial \theta} \times \frac{\partial \mathbf{r}}{\partial \varphi} \right| d\theta d\varphi. \tag{4.35}$$

In deriving this you can use that the area spanned by two vectors **v** and **w** is given by $A = |\mathbf{v} \times \mathbf{w}| = |\mathbf{v}| \, |\mathbf{w}| \sin \psi$, where ψ is the angle between these vectors.

Problem f   Use expression (4.3) to compute the vectors in the cross-product and use this to derive that

$$dS = r^2 \sin\theta \; d\theta d\varphi. \qquad (4.36)$$

Problem g   Use the geometric variables in Figure 4.3 to give an alternative derivation of this expression for a surface element that is based on geometric arguments only.

Problem h   Compute the volume of a sphere with radius R using spherical coordinates. Pay special attention to the range of integration for the angles θ and φ, as defined in Section 4.1.

4.5 Cylindrical coordinates

Cylindrical coordinates are useful in problems that exhibit cylindrical symmetry rather than spherical symmetry. An example is the generation of water waves when a stone is thrown into a pond, or more importantly: when an earthquake excites a tsunami in the ocean. In cylindrical coordinates a point is specified by giving its distance $r = \sqrt{x^2 + y^2}$ to the z-axis, the angle φ, and the z-coordinate (Figure 4.4). All the results we need could be derived using an analysis like those shown in the previous sections. However, in such an approach we would do a large amount of unnecessary work. The key is to realize that at the equator of a spherical coordinate system (i.e., at the locations where $\theta = \pi/2$), the spherical coordinate system and the cylindrical coordinate system are identical (Figure 4.5).

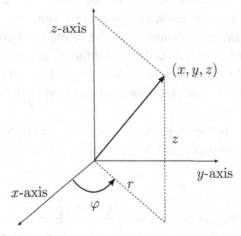

Figure 4.4   Definition of the geometric variables r, φ, and z used in cylindrical coordinates.

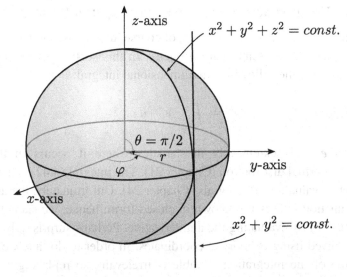

Figure 4.5 The cylindrical coordinate system has the same properties as the spherical coordinate system at the equator.

This figure shows that all results obtained for spherical coordinates can be used for cylindrical coordinates, by making the following substitutions:

$$\left.\begin{array}{l} r = \sqrt{x^2 + y^2 + z^2} \to \sqrt{x^2 + y^2}, \\ \theta \to \pi/2, \\ \hat{\boldsymbol{\theta}} \to -\hat{\mathbf{z}}, \\ r d\theta \to -dz. \end{array}\right\} \tag{4.37}$$

Problem a Convince yourself of this. To derive the third line, consider the unit vectors pointing in the direction of increasing values of θ and z at the equator.

Problem b Use the results of the previous sections and the substitutions (4.37) to show the following properties for a system of cylindrical coordinates:

$$\left.\begin{array}{l} x = r \cos \varphi, \\ y = r \sin \varphi, \\ z = z. \end{array}\right\} \tag{4.38}$$

$$\hat{\mathbf{r}} = \begin{pmatrix} \cos \varphi \\ \sin \varphi \\ 0 \end{pmatrix}, \qquad \hat{\boldsymbol{\varphi}} = \begin{pmatrix} -\sin \varphi \\ \cos \varphi \\ 0 \end{pmatrix}, \qquad \hat{\mathbf{z}} = \begin{pmatrix} 0 \\ 0 \\ 1 \end{pmatrix}, \tag{4.39}$$

$$dV = r dr d\varphi dz, \tag{4.40}$$

$$dS = r dz d\varphi. \tag{4.41}$$

4.6 Using cylindrical coordinates to compute a 1D integral

Spherical and cylindrical coordinates are, of course, mostly used to solve problems that have a spherical or cylindrical symmetry. In this section we use cylindrical coordinates to solve the following one-dimensional integral:

$$I = \int_{-\infty}^{+\infty} e^{-x^2} dx. \tag{4.42}$$

The function e^{-x^2} is extensively in statistics where it occurs in the Gauss distribution or normal distribution (Chapter 21). The integral (4.42) is also used in the asymptotic evaluation of integrals (Chapter 24). Unfortunately, the antiderivative of the function e^{-x^2} is not known in closed form; hence, we cannot compute the integral (4.42) by evaluating the antiderivative. Perhaps surprisingly, this integral can be solved using cylindrical coordinates. In order to do this, we first note that the name of the integration variable is irrelevant, so replacing $x \rightarrow y$ in expression (4.42) gives $I = \int_{-\infty}^{+\infty} e^{-y^2} dy$. Multiplying this with equation (4.42) gives

$$I^2 = \int_{-\infty}^{+\infty} e^{-(x^2+y^2)} dxdy. \tag{4.43}$$

We next interpret the variables x and y as coordinates in the x, y-plane. In that case $x^2 + y^2 = r^2$ and $dxdy$ equals a surface element dS, hence

$$I^2 = \int_{-\infty}^{+\infty} e^{-r^2} dS. \tag{4.44}$$

The surface element dS in the x, y-plane is not equal to dS in expression (4.41), because that surface element is for the curved outer wall of the cylinder. Instead we need the surface element for the top and bottom of the cylinder. An expression for this surface element follows by taking expression (4.40) and removing the height increment dz, so that $dS = rdrd\varphi$.

Problem a Use these results to show that

$$I^2 = \int_0^{2\pi} \int_0^{+\infty} e^{-r^2} rdrd\varphi. \tag{4.45}$$

Pay attention to the integration limits.

Problem b Carry out the integration over φ to show that

$$I^2 = 2\pi \int_0^{+\infty} e^{-r^2} rdr. \tag{4.46}$$

Problem c   In contrast to the original integral (4.42), this integral can be evaluated in closed form because $r\,dr = (1/2)dr^2$. Use this to evaluate the integral and show that $I^2 = \pi$.

It follows from the previous derivation that $I = \sqrt{\pi}$, hence

$$\int_{-\infty}^{+\infty} e^{-x^2}\,dx = \sqrt{\pi}. \tag{4.47}$$

And because of the symmetry of the integrand in x, we also have

$$\int_{0}^{+\infty} e^{-x^2}\,dx = \sqrt{\pi}/2. \tag{4.48}$$

    This example shows how a one-dimensional integral can be solved by extending it to two dimensions and then switching to cylindrical coordinates. Sometimes a tortuous route is the best one!

Problem d   In applications we often need instead of the integral (4.47) the related integral $\int_{-\infty}^{+\infty} e^{-bx^2}\,dx$, where b is a positive constant. Use a new integration variable $x' = \sqrt{b}\,x$ and expression (4.47) to derive that

$$\int_{-\infty}^{+\infty} e^{-bx^2}\,dx = \sqrt{\frac{\pi}{b}}. \tag{4.49}$$

5

Gradient

This chapter introduces the gradient of a function, which is important in the differentiation and integration of functions in more than one dimension (Section 5.3). Newton's law is derived in Section 5.4 from the concept of energy conservation. As a by-product of this derivation, it follows that the force is the (negative) gradient of the potential energy. The gradient plays a crucial role in the distinction between the partial time derivatives and the total time derivatives. This resulting distinction between an Eulerian and a Lagrangian formulation of problems involving fluid flow is shown in Section 5.5. In Section 5.6, we derive expressions for the gradient in spherical and cylindrical coordinates. In Section 7.7 we introduce examples of vector calculus of products of quantities.

5.1 Properties of the gradient vector

Let us consider a function f that depends on the variables x and y in a plane. We want to describe how this function changes when we move from point A in the plane to point C, as shown in Figure 5.1. The resulting change in the function f is denoted by $\delta f = f_C - f_A$, where f_A, f_B, and f_C denote the value of the function f at points A, B, and C, respectively. It follows from Figure 5.1 that

$$\left.\begin{array}{l} f_A = f(x, y), \\ f_B = f(x + \delta x, y), \\ f_C = f(x + \delta x, y + \delta y). \end{array}\right\} \tag{5.1}$$

It follows by adding and subtracting f_B that

$$\delta f = f_C - f_A = f_B - f_A + f_C - f_B. \tag{5.2}$$

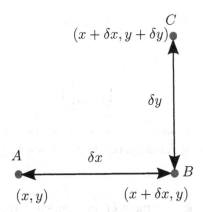

Figure 5.1   Definition of the points A, B, and C.

Problem a   Use (5.1) and expression (3.18) to derive that for small values of δx and δy:

$$\left. \begin{array}{r} f_B - f_A = \dfrac{\partial f}{\partial x}(x, y)\,\delta x, \\[2ex] f_C - f_B = \dfrac{\partial f}{\partial y}(x + \delta x, y)\,\delta y. \end{array} \right\} \tag{5.3}$$

Problem b   Insert this result in (5.2) and derive that to leading order in δx and δy, the result can be written as:

$$\delta f = \frac{\partial f}{\partial x}(x, y)\,\delta x + \frac{\partial f}{\partial y}(x + \delta x, y)\,\delta y. \tag{5.4}$$

Note that $\partial f / \partial y$ is evaluated at $(x + \delta x, y)$. However, using a Taylor expansion of $\partial f / \partial y$, we find that

$$\frac{\partial f}{\partial y}(x + \delta x, y)\,\delta y = \frac{\partial f}{\partial y}(x, y)\,\delta y + \frac{\partial^2 f}{\partial x \partial y}(x, y)\,\delta x \delta y. \tag{5.5}$$

The last term is of second order in δx and δy and can be ignored when these quantities are small, so that

$$\delta f = \frac{\partial f}{\partial x}(x, y)\,\delta x + \frac{\partial f}{\partial y}(x, y)\,\delta y. \tag{5.6}$$

This expression has the same form as the inner product between two vectors $\mathbf{a}$ and $\mathbf{b}$ in two dimensions: $(\mathbf{a} \cdot \mathbf{b}) = a_x b_x + a_y b_y$. Let us define $\delta \mathbf{r}$ as the difference in the location vector of the points C and A: $\delta \mathbf{r} \equiv \mathbf{r}_C - \mathbf{r}_A$. This vector has components δx and δy so that in two dimensions:

$$\delta \mathbf{r} = \begin{pmatrix} \delta x \\ \delta y \end{pmatrix}. \tag{5.7}$$

Similarly we define a vector whose x-component is the partial derivative of the function f with respect to x, and the y-component is the partial derivative with respect to y:

$$\nabla f \equiv \begin{pmatrix} \partial f / \partial x \\ \partial f / \partial y \end{pmatrix}. \qquad (5.8)$$

This vector is called the *gradient of f*. It is customary to denote the gradient of a function f by the symbol ∇f, but another notation you may find in the literature is grad f.

Problem c   Show that the increment δf of expression (5.6) can be expressed as the inner product of two vectors:

$$\delta f = (\nabla f \cdot \delta \mathbf{r}). \qquad (5.9)$$

This expression shows directly why the gradient is such a useful vector. Once we know the gradient ∇f, we can use (5.9) to compute the change in the function f when we change the point of evaluation with an arbitrary small step $\delta \mathbf{r}$. Note that this expression holds for *any* direction in which we can take this step. It should be noted that (5.9) is based on the first-order Taylor expansions in expression (5.3), which means that (5.9) only holds in the limit $\delta \mathbf{r} \to 0$. However, this expression is still extremely useful, because it forms the basis of the rules for differentiation and integration in more than one dimension. We will return to this issue in Section 5.3.

Problem d   The derivation up to this point has been for two dimensions in space only. Repeat this derivation for three dimensions and derive that (5.9) still holds when the gradient in three dimensions is defined as

$$\nabla f \equiv \begin{pmatrix} \partial f / \partial x \\ \partial f / \partial y \\ \partial f / \partial z \end{pmatrix} \quad \text{and} \quad \delta \mathbf{r} = \begin{pmatrix} \delta x \\ \delta y \\ \delta z \end{pmatrix}. \qquad (5.10)$$

Problem e   Compute ∇f when $f(x, y, z) = x\, e^{-y} \sin z$.

The gradient is a vector and it therefore has a direction and a magnitude. The direction of the gradient can be obtained from expression (5.9). If we change $\delta \mathbf{r}$ so that the value of the function f does not change in this direction, $\delta \mathbf{r}$ is a displacement along a surface where $f = const$. This situation is drawn in Figure 5.2. For such a change $\delta \mathbf{r}$, the corresponding change δf is by definition equal to zero, hence $(\nabla f \cdot \delta \mathbf{r}) = 0$.

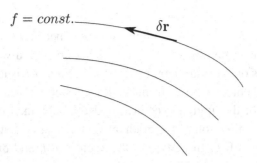

Figure 5.2   Contour lines (dashed) defined by the condition $f = const.$ and a perturbation $\delta \mathbf{r}$ in the position vector along a contour line.

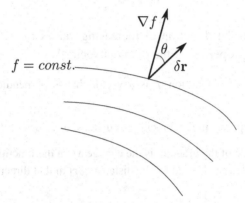

Figure 5.3   Contour lines (dashed) defined by the condition $f = const.$ and a perturbation $\delta \mathbf{r}$ in the position vector in an arbitrary direction.

Problem f   Use the properties of the inner product of two vectors to show that this identity implies that the vectors ∇f and $\delta \mathbf{r}$ are for this special perturbation perpendicular: $\nabla f \perp \delta \mathbf{r}$.

This last identity, of course, only holds when the step $\delta \mathbf{r}$ is taken along the surface where the function f is constant. This condition implies that the gradient vector is perpendicular to the surface $f = const.$ Now that we know this, the gradient can still point in two opposite directions perpendicular to a surface: toward increasing values of f or toward decreasing values of f. To determine whether ∇f points toward increasing or decreasing values of f, let us take a step $\delta \mathbf{r}$ in an arbitrary direction as shown in Figure 5.3.

Problem g   Use (5.9) to show that

$$\delta f = |\nabla f| \, |\delta \mathbf{r}| \cos \theta, \tag{5.11}$$

where θ is the angle between ∇f and $\delta \mathbf{r}$.

The change δf is largest when the vectors ∇f and $\delta \mathbf{r}$ point in the same direction. In that case $\cos \theta$ is equal to its maximum value of 1, because $\theta = 0$. We also know that δf increases most rapidly when $\delta \mathbf{r}$ is directed in such a way that one moves from smaller values of f toward larger values of f. Since δf is largest when $\theta = 0$, this means the gradient also points from small values of f toward high values of f.

The magnitude of the gradient vector can also be obtained from (5.11). Let us take a step $\delta \mathbf{r}$ in the direction of the gradient vector; that is, a step in the direction of increasing values of f. In that case, the vectors ∇f and $\delta \mathbf{r}$ are parallel, and $\cos \theta = 1$. This means that (5.11) implies that

$$|\nabla f| = \frac{\delta f}{|\delta \mathbf{r}|},\qquad(5.12)$$

where $\delta \mathbf{r}$ is a step in the direction of increasing values of f. Summarizing, we obtain the following properties of the gradient vector ∇f:

1.   The gradient of a function f is a vector that is perpendicular to the surface $f = const$.

2.   The gradient points in the direction of increasing values of f.

3.   The magnitude of the gradient is the change δf in the function in the direction of the largest increase divided by the distance $|\delta \mathbf{r}|$ in that direction.

5.2 Pressure force

An example of a two-dimensional function is shown in Figure 5.4, where the atmospheric pressure at sea level around New Zealand is shown in units of millibars.

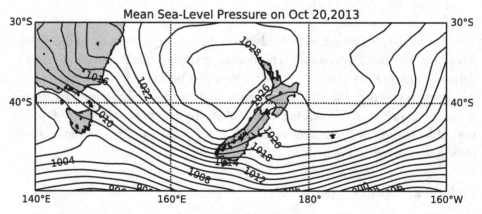

Figure 5.4   Contour map of the pressure at sea level in millibars. Data courtesy of NOAA.

Problem a   Draw the gradient vector of the pressure p in a number of locations on this map. When doing so, draw larger arrows where the gradient is larger.

Problem b   What is the approximate gradient $|\nabla p|$ over the northern tip of New Zealand's North Island? Is this gradient vector larger or smaller than over the southern part of the South Island?

There is a good reason why we use a map of the pressure to illustrate the concept of the gradient. When the air pressure is not constant, a parcel of air will experience a force that pushes it from a region of high pressure toward a region of lower air pressure. This force increases when the pressure varies more rapidly with distance. This can be described by the following relation between pressure force $\mathbf{F}_p$ and pressure p:

$$\mathbf{F}_p = -\nabla p. \tag{5.13}$$

As is customary in fluid mechanics, the pressure force is defined as the force per unit volume.

Problem c   The pressure in the weather map of Figure 5.4 is shown in units of millibars. If you have worked out Problem b properly, you will have found that the units of the pressure force are in millibars per kilometer. A pressure is a force per unit area, because the pressure times the area gives the total force that acts on this area. Deduce from this result that the pressure force has the dimensions *force/volume*.

Problem d   Use expression (5.13) to show that the pressure force points from regions of high pressure toward regions of lower pressure. Draw the direction of the pressure force at the high-pressure area in Figure 5.4 over New Zealand.

You may have noticed that we have not really derived expression (5.13) for the pressure force. Sometimes only physical arguments are used to state a physical law. However, one often needs to verify that the arguments employed have a proper mathematical basis. It is possible to derive the pressure force from the fact that the force that acts on a fluid on a hypothetical surface within that medium is perpendicular to that surface. We will see this next, and that its strength is given by the pressure times the surface area.

Problem e   Consider a small volume element with lengths δx, δy, and δz in the three coordinate directions as shown in Figure 5.5. Let us consider the net force in the x-direction. Use the concept of pressure as described above to show that the force acting on the left-hand plane of the volume is given by

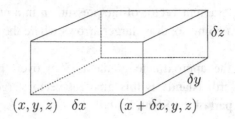

(x, y, z) δx $(x + \delta x, y, z)$

Figure 5.5   Definition of geometric variables for the derivation of the pressure force.

$p(x, y, z)\delta y\delta z \ \hat{\mathbf{x}}$ and that the force acting on the right-hand plane is given by $-p(x + \delta x, y, z)\delta y\delta z \ \hat{\mathbf{x}}$, where $\hat{\mathbf{x}}$ is the unit vector in the x-direction. Explain the minus sign in the second term.

The top, bottom, front, and back surfaces do not contribute to the x-component of the pressure force because the pressure force acting on these surfaces is directed in the y- or z-direction. This implies that the x-component of the total force is given by

$$f_x = -\big[p(x + \delta x, y, z) - p(x, y, z)\big]\delta y\delta z. \qquad (5.14)$$

Problem f   Use expression (3.18) to show that this can also be written as

$$f_x = -\frac{\partial p}{\partial x}\delta V, \qquad (5.15)$$

where the volume δV is equal to $\delta x\delta y\delta z$.

Problem g   Apply the same reasoning to obtain the y- and z-components of the force and show that the net force felt by the volume is given by

$$\mathbf{f} = -\nabla p \ \delta V. \qquad (5.16)$$

You may have been puzzled by the fact that in Problem c you deduced that the pressure force $\mathbf{F}_p$ has the dimensions $force/volume$. This means it is not really a force. In fact, we see in (5.16) that the net force that acts on the volume is given by $-\nabla p\delta V$. You have to keep in mind that the volume δV is not a physical entity, instead it is a mathematical volume used in our reasoning. The net force is proportional to δV, and since this volume is physically meaningless, the net force also has no physical meaning. However, when ρ is the mass-density of the gas or fluid, then $\delta m = \rho \ \delta V$ is the mass of the volume. This means that both the mass of the volume and the net force that acts on the volume are proportional to δV. When we apply Newton's law $\delta m\mathbf{a} = \mathbf{f}$ to this volume, we can divide both the left-hand side and the right-hand side by the arbitrary volume δV and Newton's law then takes the form $\rho\mathbf{a} = \mathbf{F}_p$, with the pressure force given by (5.13).

The previous derivation means that the pressure force $\mathbf{F}_p$ should be seen as a force per unit volume, just as the density ρ is the mass per unit volume. In fluid mechanics one always works with physical quantities per unit volume, for the simple reason that a fluid is not composed of physical small volumes. We return to a more rigorous treatment of Newton's law in fluid dynamics in Section 25.3 where the conservation of momentum in a continuous medium is treated.

The pressure force is one of the most important forces in fluid mechanics, meteorology, and oceanography, because it is the variation in the pressure that underlies the motion of fluids and gases. Physically it simply states that a fluid is pushed away from regions of higher pressure to regions of lower pressure.

5.3 Differentiation and integration

The results of Section 5.1 hold for infinitesimal changes $\delta\mathbf{r}$ only. However, a change over a finite distance can be thought of as being built up from many infinitesimal steps. Let us consider two points A and B that may be far apart as shown in Figure 5.6. Between the points A and B we can insert a large number of points $P_1, P_2, \ldots, P_N$. The difference in the function values at the points A and B can then be written as

$$f_B - f_A = (f_B - f_N) + (f_N - f_{N-1}) + \cdots + (f_2 - f_1) + (f_1 - f_A), \quad (5.17)$$

where f_j denotes the function evaluated at point P_j. What we are really doing is dividing the large distance from A to B into infinitesimally smaller intervals. If we take enough of these subintervals, we can apply (5.9) to each of these subintervals.

Problem a   Show that

$$f_B - f_A = \sum (\nabla f \cdot \delta\mathbf{r}), \quad (5.18)$$

where the sum is over the subintervals used in (5.17) and where $\delta\mathbf{r}$ is the increment in the position vector in each interval.

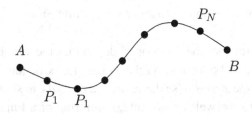

Figure 5.6   The definition of the points A and B at the end of an interval and the intermediary points $P_1, P_2, \ldots, P_N$.

The analysis in this section holds only in the limit $N \to \infty$, so that the interval is divided into infinitesimal intervals. In that case, the summation in (5.18) is replaced by an integration, and the notation $d\mathbf{r}$ is used rather than $\delta\mathbf{r}$:

$$f_B - f_A = \int_A^B (\nabla f \cdot d\mathbf{r}). \tag{5.19}$$

This expression is extremely useful, because it makes it possible to compute the change in a function between two points, once one knows the gradient of that function. It should be noted that (5.19) holds for *any* path that joins the points A and B, because the points $P_1, P_2, \ldots, P_N$ can be chosen arbitrarily, as long as they form a continuous path that joins the points A and B. This property can sometimes be exploited to choose the path for which the calculation of the integral is easiest.

Problem b   Use these results to show that the line integral of ∇f along any closed contour is equal to zero:

$$\oint (\nabla f \cdot d\mathbf{r}) = 0. \tag{5.20}$$

The gradient can also be used to determine the derivative of a function in a given direction. The *directional derivative* df/ds of the function in the direction of a unit vector $\hat{\mathbf{n}}$ is defined as the change of the function in the direction of $\hat{\mathbf{n}}$ normalized per unit distance:

$$\frac{df}{ds}(\mathbf{r}) = \lim_{\delta s \to 0} \frac{f(\mathbf{r} + \hat{\mathbf{n}}\delta s) - f(\mathbf{r})}{\delta s}. \tag{5.21}$$

Problem c   Use (5.9) in the numerator of this expression to derive that

$$\frac{df}{ds}(\mathbf{r}) = \left(\hat{\mathbf{n}} \cdot \nabla f\right). \tag{5.22}$$

This expression allows us to compute the derivative of a function in *any* direction, once the gradient is known.

Equations (5.19) and (5.22) generalize the rules for the integration and differentiation of functions of one variable to more space dimensions.

Problem d   To see this, let the function f depend on the variable x only and let the points A and B be located on the x-axis. Let $\hat{\mathbf{x}}$ denote the unit vector in the direction of the x-axis. Use the relation $d\mathbf{r} = \hat{\mathbf{x}}dx$ to show that in that case (5.19) reduces to the well-known integration rule of a function that depends only on one variable:

$$f_B - f_A = \int_A^B \frac{\partial f}{\partial x} dx. \tag{5.23}$$

Problem e   Show that the directional derivative along the x-axis in (5.22) is given by $\partial f / \partial x$, when the vector $\hat{\mathbf{n}}$ is directed along the x-axis ($\hat{\mathbf{n}} = \hat{\mathbf{x}}$).

5.4 Newton's law from energy conservation

In classical mechanics one can start from Newton's law and derive that the total energy of a mechanical system without friction is conserved. You may have discovered that in physics there is often no proof of the basic physical laws. For example, there is no "proof" of Newton's law. Its use is justified by the observation that it describes the motion of the Sun, Moon, and planets accurately. To a certain extent, it is arbitrary which physical law one uses as a starting point. In this section we use the concept of energy conservation as a starting point and then derive Newton's law.

Let us consider a mechanical system without friction. The total energy E is the sum of the kinetic energy $\frac{1}{2}mv^2$ and the potential energy $V(\mathbf{r})$:

$$E = \frac{1}{2}mv^2 + V(\mathbf{r}). \tag{5.24}$$

We assume that in such a system the total energy E is conserved. This means that the time derivative of this quantity equals zero:

$$\frac{dE}{dt} = 0. \tag{5.25}$$

To derive Newton's law from this expression, we need to compute the time derivative of both the kinetic and the potential energy.

Problem a   Show that the time derivative of the kinetic energy is given by

$$\frac{d}{dt}\left(\frac{1}{2}mv^2\right) = m\left(\mathbf{v} \cdot \frac{d\mathbf{v}}{dt}\right). \tag{5.26}$$

Hint: Write $v^2 = v_x^2 + v_y^2 + v_z^2$ and differentiate each term.

We also need to compute the time derivative of the potential energy. At first you may be tempted to conclude that this time derivative is equal to zero because the potential depends on the position $\mathbf{r}$ but not on time. However, as the particle moves through space, it is at different positions at different times. Therefore, the position of the particle is a function of time: $\mathbf{r} = \mathbf{r}(t)$. For this reason, the potential should be written as $V(\mathbf{r}(t))$. The time derivative of the potential then follows from the usual rule for taking a derivative:

$$\frac{dV(\mathbf{r})}{dt} = \lim_{\delta t \to 0} \frac{V(\mathbf{r}(t + \delta t)) - V(\mathbf{r}(t))}{\delta t}. \tag{5.27}$$

Problem b   Use (5.9) to show that this time derivative can be written as

$$\frac{dV(\mathbf{r})}{dt} = \lim_{\delta t \to 0} \frac{(\nabla V \cdot \delta \mathbf{r})}{\delta t}, \tag{5.28}$$

with $\delta \mathbf{r} = \mathbf{r}(t + \delta t) - \mathbf{r}(t)$.

Problem c   Use the definition of the velocity $\mathbf{v} = \lim_{\delta t \to 0} \delta \mathbf{r}/\delta t$ to show that

$$\frac{dV(\mathbf{r})}{dt} = (\mathbf{v} \cdot \nabla V). \tag{5.29}$$

Problem d   Use these results to show that the law of energy conservation (5.25) can be written as

$$\mathbf{v} \cdot \left(m\frac{d\mathbf{v}}{dt} + \nabla V \right) = 0. \tag{5.30}$$

Since $\mathbf{v}$ is arbitrary, (5.30) must hold for any velocity vector $\mathbf{v}$. This can only be the case when the term in brackets is equal to zero.

Problem e   Show that this implies that

$$m\frac{d\mathbf{v}}{dt} = \mathbf{F}, \tag{5.31}$$

with

$$\mathbf{F} = -\nabla V. \tag{5.32}$$

Equation (5.31) is Newton's law; we have derived it here from the requirement of energy conservation. As a by-product we have shown that the force in Newton's law is equal to $-\nabla V$. Note that the force (5.32) and the pressure force (5.13) share the common property that they follow from the gradient of a scalar function. You may find this analogy useful. We have argued in Section 5.2 that the pressure force pushes air from a region of higher pressure to a region of lower pressure. By the same reasoning the force $\mathbf{F} = -\nabla V$ pushes a particle from regions of higher potential energy to regions of lower potential energy. A simple example of this is gravity. The gravitational potential energy increases with the distance to the attracting body. This means that according to the reasoning given here, gravitation forces a particle closer to the attracting body. This is an observation that may be obvious to you when you see an apple fall from a tree. However, you may want to keep the analogy of the pressure force and the force associated with a general potential energy in mind because it helps to understand the concept of potential energy.

With all the elements we have assembled here, we can now derive the concept of power. This quantity is defined as the energy delivered to the particle per unit time. This means that the power is defined as the time derivative of the kinetic energy.

Problem f Use (5.25), (5.29), and (5.32) to derive that the power is given by

$$\frac{d}{dt}\left(\frac{1}{2}mv^2\right) = (\mathbf{F} \cdot \mathbf{v}).\tag{5.33}$$

How can we understand this result? Let the particle be displaced over a distance $\delta\mathbf{r}$ in a time increment δt. The work done by the force is given by $(\mathbf{F} \cdot \delta\mathbf{r})$. This means that the work per unit time is given by $(\mathbf{F} \cdot \delta\mathbf{r})/\delta t$. In the limit $\delta t \to 0$, the quantity $\delta\mathbf{r}/\delta t$ is the velocity $\mathbf{v}$, so that the power is given by $(\mathbf{F} \cdot \mathbf{v})$ as stated in (5.33). In Section 25.3 we will derive the time derivative of the kinetic energy of a fluid, and it will be shown in (25.22) that for such a system the energy delivered by the force per unit time is also given by $(\mathbf{F} \cdot \mathbf{v})$, with $\mathbf{F}$ the force per unit volume.

5.5 Total and partial time derivatives

We have seen in the previous section that the potential that acts on a particle changes with time when the particle moves to a different location. This principle of a temporal change that is caused by a motion in the system is very general. As an example, consider the situation in Figure 5.7, where an observer measures the temperature in an observation tower. The motion of the wind is toward the left, and a region of warm air is transported leftward with the wind. The observer detects an increase of

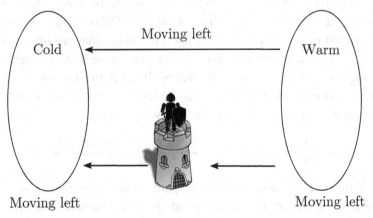

Figure 5.7   An observer standing on a fixed tower experiences an increase in temperature, because warm air moves toward the observer from right to left.

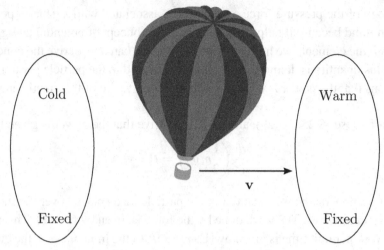

Figure 5.8   An observer in a balloon experiences an increase in temperature, because the balloon flies to a warmer region on the right.

the temperature with time, so that $\partial T/\partial t > 0$. In this expression the partial derivative symbol is used deliberately. In general, the temperature field is a function of both time and the space coordinates: $T = T(\mathbf{r}, t)$. For the observer in the observation tower the space coordinates are fixed, which means that any change detected by the observer is due to the temporal change of the local temperature only, but that the location is fixed. This is expressed by the definition of the partial derivative:

$$\frac{\partial T(\mathbf{r}, t)}{\partial t} \equiv \lim_{\delta t \to 0} \frac{T(\mathbf{r}, t + \delta t) - T(\mathbf{r}, t)}{\delta t}. \tag{5.34}$$

The time is varied in this derivative, but all other variables are kept fixed.

In contrast to the previous example let us now consider an observer that is carried along in a balloon by the wind, as in Figure 5.8. We assume in this situation that the temperature field is fixed in time, so that $\partial T/\partial t = 0$. The observer is carried by the wind from a cold region to a warmer region. This means that just like the observer in Figure 5.7, this second observer experiences an increase in the temperature. The rate of change of the temperature is denoted by the total time derivative dT/dt. We can see from this example that dT/dt can be larger than zero while $\partial T/\partial t = 0$. The total time derivative is defined in the following way:

$$\frac{dT(\mathbf{r}, t)}{dt} = \lim_{\delta t \to 0} \frac{T(\mathbf{r}(t + \delta t), t + \delta t) - T(\mathbf{r}(t), t)}{\delta t}. \tag{5.35}$$

The only difference between this and the partial time derivative (5.34) is that in (5.35) the position of the observation point changes, whereas in the partial time derivative (5.34) the position is fixed.

Problem a   Assuming that the temperature field in Figure 5.8 depends on position and not on time, show that dT/dt has the same form as expression (5.27), and use the results of Section 5.4 to show that

$$\frac{dT(\mathbf{r})}{dt} = (\mathbf{v} \cdot \nabla T) \quad \text{(in the special case of Figure 5.8).} \tag{5.36}$$

Problem b   Draw the gradient vector ∇T in Figure 5.8 at the location of the observer and show from the geometry of this vector and the velocity vector that $dT/dt > 0$.

This means that both observers in Figures 5.7 and 5.8 detect a rise in temperature, but their description of this temperature change is completely different. The first observer feels an increase in temperature, because the local temperature changes at her location; this is described by the partial time derivative $\partial T/\partial t$. The second observer detects an increase in temperature, because he is transported to a warmer region in a fixed temperature field; this is described by the total time derivative that is for this special case given by $dT/dt = (\mathbf{v} \cdot \nabla T)$.

In general the temperature field may change because of a combination of a change in the temperature at a fixed location and a movement of the observer. We will assume now that the temperature field depends on all three space coordinates and on time: $T = T(x(t), y(t), z(t), t)$. In this notation it is explicit that the space coordinates of the observation point depend on time as well. The total time derivative is given by (5.35) and can be written as:

$$\frac{dT}{dt} = \lim_{\delta t \to 0} \frac{T(x(t + \delta t), y(t + \delta t), z(t + \delta t), t + \delta t) - T(x(t), y(t), z(t), t)}{\delta t}. \tag{5.37}$$

Let the x-coordinate over the time interval change from x to $x + \delta x$, and the y- and z-coordinates change in a similar way.

Problem c   Use the first-order Taylor series to show that

$$\delta x = x(t + \delta t) - x(t) \approx \frac{\partial x}{\partial t}\delta t = v_x \delta t. \tag{5.38}$$

In the limit $\delta t \to 0$, the approximation becomes an identity. Explain that v_x is the x-component of the velocity vector.

Problem d   Generalize (3.22) for the case of a function that depends both on x, y, z, and on t to the following first-order Taylor series:

$$T(x + \delta x, y + \delta y, z + \delta z, t + \delta t) \approx$$
$$T(x, y, z, t) + \frac{\partial T}{\partial x}\delta x + \frac{\partial T}{\partial y}\delta y + \frac{\partial T}{\partial z}\delta z + \frac{\partial T}{\partial t}\delta t. \qquad (5.39)$$

Problem e   Insert (5.38) and (5.39) in (5.37) and take the limit $\delta t \to 0$ to derive that

$$\frac{dT}{dt} = \frac{\partial T}{\partial x}v_x + \frac{\partial T}{\partial y}v_y + \frac{\partial T}{\partial z}v_z + \frac{\partial T}{\partial t}. \qquad (5.40)$$

Problem f   Use the definition of the gradient vector to rewrite this as

$$\frac{dT}{dt} = \frac{\partial T}{\partial t} + (\mathbf{v} \cdot \nabla T). \qquad (5.41)$$

Problem g   This is the general expression of the total time derivative. Show that the time derivatives (5.34) and (5.36) seen by the observers in Figures 5.7 and 5.8 can both be obtained from this general expression for the total time derivative.

The temperature field was used in this section only for illustrative purposes. The analysis is of course applicable to any function that depends on both time and the space coordinates. Also, in the analysis we used "observers" to fix our minds. However, in general there are no observers and it is not essential that there is anybody present to "observe" the change in temperature. The total time derivative is always related to the movement of a certain quantity. In Section 5.4 this was the movement of a particle that is subjected to a potential $V(\mathbf{r})$. In the description of a gas it may concern the motion of a parcel of material in the gas that is carried around by the flow.

Whenever one describes a continuous system, such as the motion in the atmosphere, one has a choice in how to set up this description. One option is to consider every quantity at a fixed location in space, and specify how these quantities change with time. In that case, the change of the properties with time is described by the partial time derivative $\partial/\partial t$. This is called an *Eulerian* description. Alternatively, one may follow a certain property while it is being carried around by the flow. In that case the change of this property with time is given by the total time derivative d/dt. This is called a *Lagrangian* description. Which description is most convenient depends on the problem. In numerical applications the Eulerian description is usually most convenient because one can work with a fixed system of space coordinates and one only needs to specify how a property changes with time in that coordinate system. On the other hand, if one wants to study, for example, the spreading of pollution by wind, one aims at tracking particles, and for this the Lagrangian formulation is most convenient. For this reason it is important that one can transform the physical laws back and forth between an Eulerian and a Lagrangian formulation. We return to this issue in Chapter 25.

5.6 Gradient in spherical coordinates

The gradient vector is defined in expression (5.10) using a system of Cartesian coordinates. In many applications it is much more convenient to use a system of curvilinear coordinates, especially spherical and cylindrical coordinates. It is therefore useful to obtain the form of the gradient in these coordinate systems, as well. In this section we consider the expression of the gradient in spherical coordinates, but first we rewrite the expression of the gradient in Cartesian coordinates.

Problem a   Show that the gradient in Cartesian coordinates can also be written as:

$$\nabla f = \hat{\mathbf{x}} \frac{\partial f}{\partial x} + \hat{\mathbf{y}} \frac{\partial f}{\partial y} + \hat{\mathbf{z}} \frac{\partial f}{\partial z}. \tag{5.42}$$

Problem b   Take the inner product of this expression with the unit vector in the x-direction to show that

$$\frac{\partial f}{\partial x} = \left(\hat{\mathbf{x}} \cdot \nabla f \right). \tag{5.43}$$

Expression (4.4) gives the spherical coordinates r, θ, and φ in terms of the Cartesian coordinates x, y, and z. Using the chain rule of differentiation, one can then use for example that

$$\frac{\partial}{\partial x} = \frac{\partial r}{\partial x} \frac{\partial}{\partial r} + \frac{\partial \theta}{\partial x} \frac{\partial}{\partial \theta} + \frac{\partial \varphi}{\partial x} \frac{\partial}{\partial \varphi}.$$

Together with expression (4.7) for the unit vectors $\hat{\mathbf{r}}$, $\hat{\boldsymbol{\theta}}$, and $\hat{\boldsymbol{\varphi}}$ this could be used to derive the expression for the gradient in spherical coordinates. However, this approach is algebraically very complex and does not give much insight. Here we derive the gradient vector in spherical coordinates using expression (5.9). The idea is that the component of the gradient along a certain coordinate axis is simply given by the rate of change of the function in that direction: $\delta f = (\nabla f \cdot \delta \mathbf{s})$. Suppose that the change in position $\delta \mathbf{s}$ is in the direction of a unit vector $\hat{\mathbf{e}}$, then the change in position is given by $\delta \mathbf{s} = \hat{\mathbf{e}} \, \delta s$, where δs is given by: $\delta s \equiv |\delta \mathbf{s}|$. This means that $\delta f = (\nabla f \cdot \hat{\mathbf{e}}) \, \delta s$. However, as $(\nabla f \cdot \hat{\mathbf{e}})$ is the component of the gradient vector in the direction of the unit vector $\hat{\mathbf{e}}$, this means that the component $\nabla_e f$ of the gradient vector in the direction of $\hat{\mathbf{e}}$ is given by:

$$\nabla_e f \equiv \left(\nabla f \cdot \hat{\mathbf{e}} \right) = \frac{\delta f}{\delta s} = \lim_{\delta s \to 0} \frac{f(\mathbf{r} + \hat{\mathbf{e}} \delta s) - f(\mathbf{r})}{\delta s}. \tag{5.44}$$

Consider the spherical coordinate system shown in Figure 5.9.

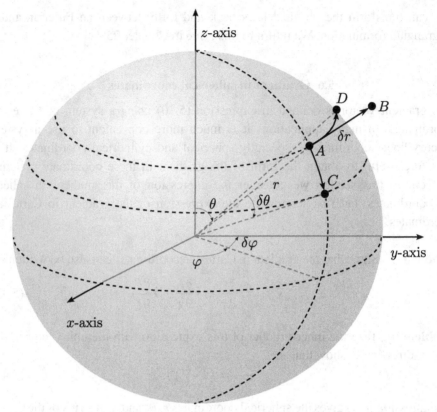

Figure 5.9   Definition of the geometric variables and the points A, B, C, and D in spherical coordinates.

Problem c   A change from point A to point B involves moving in the $\hat{\mathbf{r}}$-direction by an amount $\delta s = \delta r$. Use this and expression (5.44) to show that the radial r-component of the gradient vector is

$$\nabla_r f = \frac{\partial f}{\partial r}. \qquad (5.45)$$

Problem d   Use Figure 5.9 to show that the distance δs between the points A and C is in the $\hat{\boldsymbol{\theta}}$-direction with magnitude $\delta s = r\delta\theta$. Hint: How long would the arc s be if we extended $\delta\theta$ to a full circle?

Problem e   Use this and expression (5.44) to show that the θ-component of the gradient vector is

$$\nabla_\theta f = \frac{1}{r}\frac{\partial f}{\partial \theta}. \qquad (5.46)$$

Problem f Use Figure 5.9 to show that a change from point A to point D involves a change in the $\hat{\boldsymbol{\varphi}}$-direction with magnitude $\delta s = r \sin \theta \delta \varphi$. Hint: You could solve this by computing the projection of $\delta s = AD$ on the $x - y$ plane.

Problem g Use this and expression (5.44) to show that the azimuthal dependence of the gradient vector is

$$\nabla_\varphi f = \frac{1}{r \sin \theta} \frac{\partial f}{\partial \varphi}. \tag{5.47}$$

Combining the results of the previous three problems, we obtain the expression for the gradient in spherical coordinates:

$$\nabla f = \hat{\mathbf{r}} \frac{\partial f}{\partial r} + \hat{\boldsymbol{\theta}} \frac{1}{r} \frac{\partial f}{\partial \theta} + \hat{\boldsymbol{\varphi}} \frac{1}{r \sin \theta} \frac{\partial f}{\partial \varphi}. \tag{5.48}$$

Note that the gradient in spherical coordinates is *not* given by $\nabla f = \hat{\mathbf{r}} \partial f / \partial r + \hat{\boldsymbol{\theta}} \partial f / \partial \theta + \hat{\boldsymbol{\varphi}} \partial f / \partial \varphi$. There is a simple reason why this expression must be wrong. The first term on the right-hand side has the dimension $f/length$ because the radius is a length, while the second term has the dimension f because the angle θ is expressed in radians, which is a dimensionless quantity. This means that this expression must be wrong. The factors $1/r$ and $1/r \sin \theta$ in (5.48) account for the scaling of the distance δs with the change in the angles $\delta \theta$ and $\delta \varphi$, respectively. These terms therefore account for the fact that the system is curvilinear.

Problem h Show that each of the terms in (5.48) has the dimension $f/length$.

Problem i Do the same analysis for directional derivatives in cylindrical coordinates, as defined in Figure 4.4. Show that the gradient in cylindrical coordinates is given by

$$\nabla f = \hat{\mathbf{r}} \frac{\partial f}{\partial r} + \hat{\boldsymbol{\varphi}} \frac{1}{r} \frac{\partial f}{\partial \varphi} + \hat{\mathbf{z}} \frac{\partial f}{\partial z}. \tag{5.49}$$

6

Divergence of a vector field

The physical meaning of the divergence cannot be understood without understanding the sources, sinks, and flux of a vector field. These concepts are introduced in this chapter, followed by the example of the stability of the Earth's orbit around the Sun.

6.1 The flux

In Section 2.6 we used dimensional analysis to estimate the rate of flow of a fluid through a pipe. The flow rate Φ in that application measures the flow through a surface perpendicular to the pipe's axis. If we are interested in the amount of fluid carried by the pipe, we only need to be concerned with the component in the direction of the pipe. This total flow is an example of a quantity called *flux*, and is next to be analyzed in more detail.

Consider a vector field $\mathbf{v}(\mathbf{r})$ that represents the flow of a fluid with a constant density (i.e., the fluid is *incompressible*). We define a surface S in this fluid. Of course, the surface has an orientation in space, and the unit vector perpendicular to S is denoted by $\hat{\mathbf{n}}$. Infinitesimal elements of this surface are denoted with $d\mathbf{S} \equiv \hat{\mathbf{n}} dS$. Now suppose we are interested in the volume of fluid that flows per unit time through the surface S; this quantity is called Φ. If we want to know the flow through the surface, we only need to consider the component of $\mathbf{v}$ perpendicular to the surface; the flow along the surface is not relevant.

Problem a Show that the component of the flow *across* the surface is given by $(\mathbf{v} \cdot \hat{\mathbf{n}})\hat{\mathbf{n}}$ and that the flow *along* the surface is given by $\mathbf{v} - (\mathbf{v} \cdot \hat{\mathbf{n}})\hat{\mathbf{n}}$. If you find this problem difficult, you may want to look ahead to Section 12.1.

Using this result, the volume of the flow through the surface per unit time is given by

$$\Phi_{\mathbf{v}} = \iint (\mathbf{v} \cdot \hat{\mathbf{n}}) \, dS = \iint \mathbf{v} \cdot d\mathbf{S}. \tag{6.1}$$

This expression defines the flux Φ_v of the vector field **v** through the surface S. The definition of a flux is not restricted to the flow of fluids: a flux can be computed for any vector field. However, the analogy of fluid flow often is useful to understand the meaning of the flux and divergence.

Another example of such a vector field is the electric field (in Vm^{-1}) generated by a point charge q at the origin:

$$\mathbf{E}(\mathbf{r}) = \frac{q\hat{\mathbf{r}}}{4\pi \epsilon_0 r^2}, \tag{6.2}$$

where $\hat{\mathbf{r}}$ is the unit vector in the radial direction and ϵ_0 is the permittivity. A surface element on a sphere of radius R is given by $dS = R^2 d\Omega$, with $d\Omega = \sin\theta d\theta d\varphi$ an increment in the solid angle on the unit sphere. The normal to the spherical surface is given by $\hat{\mathbf{r}}$, so that the oriented surface element is given by $d\mathbf{S} = \hat{\mathbf{r}} R^2 d\Omega$.

Problem b   Show that the flux of the electric field through a spherical surface of radius R with the point charge at its center is given by

$$\Phi_E = \iint \mathbf{E} \cdot d\mathbf{S} = \frac{q}{4\pi \epsilon_0 R^2} \iint \left(\hat{\mathbf{r}} \cdot \hat{\mathbf{r}} \right) R^2 d\Omega$$

$$= \frac{q}{4\pi \epsilon_0} \int_0^{2\pi} \int_0^{\pi} \sin\theta d\theta d\varphi. \tag{6.3}$$

Problem c   Carry out the integration over the angles to show that the flux is given by

$$\Phi_E = \frac{q}{\epsilon_0}. \tag{6.4}$$

This means that the flux depends on the charge q and the permittivity ϵ_0, but not on the radius of the surface.

As a next example, we compute the flux of the magnetic field of the Earth through the Earth's surface. To first order, the magnetic field of the Earth is a dipole field. This is the field generated by a magnetic north pole and magnetic south pole that are very close together. The dipole vector **m** points from the south pole to the north pole and its magnitude defines the strength of the dipole. The magnetic field **B(r)** for a dipole is given by (p. 182 in Jackson, 1998):

$$\mathbf{B}(\mathbf{r}) = \frac{3\hat{\mathbf{r}}(\hat{\mathbf{r}} \cdot \mathbf{m}) - \mathbf{m}}{r^3}. \tag{6.5}$$

The SI unit for magnetic field is the Tesla (T). Note that the magnetic dipole field has the same form as the electric dipole field given in equation (20.47).

Let us align the z-axis of our coordinate system with the dipole. In that case $\mathbf{m} = m\hat{\mathbf{z}}$, and

$$\mathbf{B}(\mathbf{r}) = m\frac{3\hat{\mathbf{r}}(\hat{\mathbf{r}} \cdot \hat{\mathbf{z}}) - \hat{\mathbf{z}}}{r^3}. \tag{6.6}$$

Problem d   According to Figure 4.1, $(\hat{\mathbf{r}} \cdot \hat{\mathbf{z}}) = \cos\theta$. Use this result and the relation $d\mathbf{S} = \hat{\mathbf{r}}R^2 \sin\theta d\theta d\varphi$ to show that the magnetic flux is given by

$$\Phi_B = m \int_0^{2\pi} \int_0^{\pi} \frac{3\cos\theta - \cos\theta}{R^3} R^2 \sin\theta d\theta d\varphi. \tag{6.7}$$

In these expressions, R is the radius of the (spherical) Earth.

Problem e   Carry out the integration over the angle θ to show that this flux vanishes.

This means that the total magnetic flux through the surface of the Earth from a dipole field is zero: as many magnetic field lines point out of the Earth as into the Earth. In contrast, the electric field of a positive charge points away from the charge. As shown in expression (6.4), this gives a nonzero flux of the electric field through a spherical surface around that charge. This situation is different for the magnetic field. Field lines point away from one part of the dipole, and toward the other end of the dipole, which causes the magnetic flux to vanish.

6.2 The divergence

To introduce the concept of divergence, consider an infinitesimal rectangular volume with sides dx, dy, and dz, defined in Figure 6.1. The outward flux through the right surface perpendicular to the x-axis is given by $v_x(x + dx, y, z)dydz$,

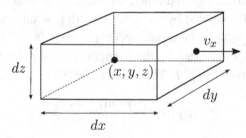

Figure 6.1   Definition of the geometric variables in the calculation of the flux of a vector field through an infinitesimal rectangular volume.

because $v_x(x + dx, y, z)$ is the component of the flow perpendicular to that surface and $dydz$ is the area of the surface. We have ignored the variation of v_x over the surfaces. This variation gives a correction proportional to dy or dz, which vanishes as dy and dz go to zero. By the same token, the flux through the left surface perpendicular through the x-axis is given by $-v_x(x, y, z)dydz$; the minus sign is due to the fact that the component of **v** in the direction *outward* from the cube is given by $-v_x$. (Alternatively one can say that for this surface the unit vector perpendicular to the surface and pointing outwards is given by $\hat{n} = -\hat{x}$.) This means that the total outward flux through the two surfaces is given by $v_x(x + dx, y, z)dydz - v_x(x, y, z)dydz = (\partial v_x/\partial x)\, dxdydz$. The same reasoning applies to the surfaces perpendicular to the y- and z-axes, so that the total outward flux through the sides of the cubes is

$$d\Phi_\mathbf{v} = \left(\frac{\partial v_x}{\partial x} + \frac{\partial v_y}{\partial y} + \frac{\partial v_z}{\partial z} \right) dV = (\nabla \cdot \mathbf{v})\, dV, \tag{6.8}$$

where dV is the volume $dxdydz$ of the cube. The quantity $(\nabla \cdot \mathbf{v})$ is called the *divergence* and is defined by

$$(\nabla \cdot \mathbf{v}) = \frac{\partial v_x}{\partial x} + \frac{\partial v_y}{\partial y} + \frac{\partial v_z}{\partial z}. \tag{6.9}$$

The above definition does not really tell us yet what the divergence really is. Dividing (6.8) by dV results in $(\nabla \cdot \mathbf{v}) = d\Phi_\mathbf{v}/dV$. This allows us to state in words that:

The divergence of a vector field is the outward flux of the vector field per unit volume.

6.3 The flow of a thin sheet of water

Let us now revisit the example of fluid flow, but in two dimensions. A thin sheet of fluid is pumped from the origin $\mathbf{r} = 0$. For simplicity, we assume again that the fluid is incompressible, which means that the mass-density is constant. We do not know yet what the resulting flow field is, but we know two things. First, away from the source at $\mathbf{r} = 0$, there are no sources or sinks of fluid flow. In that case, the flux of the flow through any closed surface S must be zero: "what goes in must come out." This in turn means that, the divergence of the flow is zero, except possibly near the source at $\mathbf{r} = 0$:

$$\nabla \cdot \mathbf{v} = 0 \quad \text{for} \quad \mathbf{r} \neq 0. \tag{6.10}$$

Second, we know that due to the symmetry of the problem the flow is directed in the radial direction and depends on the radius r only:

$$\mathbf{v}(\mathbf{r}) = f(r)\mathbf{r}. \tag{6.11}$$

This is enough information to determine the flow field. Unfortunately, we cannot immediately insert (6.11) in (6.10), because we only have an expression of the divergence in rectangular coordinates. However, there is a way to determine the flow from the expression above.

Problem a   Use the identity $r = \sqrt{x^2 + y^2}$ to show that

$$\frac{\partial r}{\partial x} = \frac{x}{r}, \tag{6.12}$$

and derive the corresponding equation for y. Using expressions (6.11), (6.12), and the chain rule for differentiation, show that

$$\nabla \cdot \mathbf{v} = 2f(r) + r\frac{df}{dr}. \tag{6.13}$$

Problem b   Insert this result in (6.10) and show that the flow field is given by $\mathbf{v}(\mathbf{r}) = A\mathbf{r}/r^2$.

The constant A is yet to be determined. Let $\dot{V}$ be the volume of liquid injected per unit time at the source $\mathbf{r} = 0$, this volume is carried away from the source by the sheet of fluid. Following an often used convention, the dot denotes the derivative with respect to time.

Problem c   Show that $\dot{V} = \int \mathbf{v} \cdot d\mathbf{S}$, where the integration is over an arbitrary surface around the source at $\mathbf{r} = 0$. Choosing a suitable surface, derive that

$$\mathbf{v}(\mathbf{r}) = \frac{\dot{V}}{2\pi}\frac{\hat{\mathbf{r}}}{r}. \tag{6.14}$$

Note that the unit vector $\hat{\mathbf{r}}$ is now used rather than the position vector $\mathbf{r}$. These vectors are related by the expression $\mathbf{r} = r\hat{\mathbf{r}}$. This flow field is shown in Figure 6.2. Note that the length of the flow vectors decreases when moving away from the origin, as predicted by the $1/r$ term in expression (6.14).

More complicated examples can be obtained from the results of the previous example of a single source at $\mathbf{r} = 0$. Suppose we have a source at $\mathbf{r}_+ = (L, 0)$, where a volume $\dot{V}$ is injected per unit time. In addition, a sink is located at $\mathbf{r}_- = (-L, 0)$, where a volume $-\dot{V}$ is removed per unit time. The total flow field can be obtained by superposition of flow fields of the form (6.14) for the source

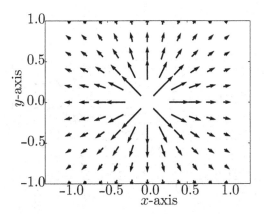

Figure 6.2 The vector field for fluid flow in a thin sheet with a source at the origin (expression (6.14)).

and the sink. The flow field from the source at $\mathbf{r}_+$ can be found by replacing the location vector $\mathbf{r}$ in (6.14) by the vector $\mathbf{r} - \mathbf{r}_+$. This amounts to the following replacements:

$$r = \sqrt{x^2 + y^2} \to \sqrt{(x - L)^2 + y^2}, \tag{6.15}$$

and

$$\hat{\mathbf{r}} = \mathbf{r}/r = \begin{pmatrix} x \\ y \end{pmatrix} / \sqrt{x^2 + y^2} \to \begin{pmatrix} x - L \\ y \end{pmatrix} / \sqrt{(x - L)^2 + y^2}. \tag{6.16}$$

Problem d Make similar replacements for the sink at $\mathbf{r}_-$ and show that the x- and y-components of the flow field in this case are given by:

$$v_x(x, y) = \frac{\dot{V}}{2\pi} \left[\frac{x - L}{(x - L)^2 + y^2} - \frac{x + L}{(x + L)^2 + y^2} \right], \tag{6.17}$$

$$v_y(x, y) = \frac{\dot{V}}{2\pi} \left[\frac{y}{(x - L)^2 + y^2} - \frac{y}{(x + L)^2 + y^2} \right]. \tag{6.18}$$

The flow field for a combination of a source and a sink is shown in Figure 6.3. Note that one can visually connect the arrows in this flow field to see the *streamlines* of the flow. These are the lines along which material particles flow. The streamlines can be found by using that the time derivative of the position of a material particle is the velocity:

$$d\mathbf{r}/dt = \mathbf{v}(\mathbf{r}). \tag{6.19}$$

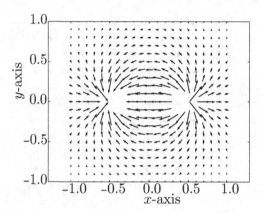

Figure 6.3   The vector field for fluid flow in a thin sheet as computed in Problem d for a source at $(0.5, 0)$ and a sink at $(-0.5, 0)$.

Inserting expressions (6.17) and (6.18) into equation (6.19) leads to two coupled differential equations for $x(t)$ and $y(t)$, which are difficult to solve. Fortunately, there are more intelligent ways of retrieving the streamlines. We return to this issue in Section 15.3.

6.4 Sources and sinks

In the previous example, a fluid flows outward from the source and converges on the sink. The terms "source" and "sink" have a clear physical meaning: they are directly related to the "source" of water as from a tap, and a "sink," as in the sink of a bathtub. The flow lines of the fluid diverge from the source, while they converge at the sink. This explains the term "divergence," because this quantity simply indicates to what extent flow lines originate (in the case of a source) or end (in the case of a sink).

This definition of sources and sinks is not restricted to fluid flow. For example, for the electric field the term "flow lines" should be replaced by the term "field lines." Electrical field lines originate at positive charges and end at negative charges. We first verify this by computing the divergence of the electrical field (6.2) for a point charge in three dimensions.

Problem a   Show that expression (6.2) can be written as

$$\mathbf{E}(\mathbf{r}) = \frac{q}{4\pi \epsilon_0 r^3} \begin{pmatrix} x \\ y \\ z \end{pmatrix}. \tag{6.20}$$

Problem b The x-derivative of the x-component of the electric field gives a contribution

$$\frac{\partial}{\partial x}\frac{x}{r^3} = \frac{1}{r^3} - \frac{3x}{r^4}\frac{\partial r}{\partial x}. \qquad (6.21)$$

Use expression (6.12) to show that this quantity is given by

$$\frac{\partial}{\partial x}\frac{x}{r^3} = \frac{1}{r^5}\left(r^2 - 3x^2\right). \qquad (6.22)$$

Problem c Now compute the y- and z-derivatives in the divergence of the electric field, and show that

$$\nabla \cdot \mathbf{E} = \frac{q}{4\pi\epsilon_0 r^5}\left(3r^2 - 3(x^2 + y^2 + z^2)\right) = 0. \qquad (6.23)$$

This means that away from the point charge ($r \neq 0$) the divergence of the electric charge vanishes. At the charge, $r = 0$ and it is not clear what the right-hand side of expression (6.23) is. It follows from equation (6.4) that the net flux through a spherical surface that includes a charge q is given by q/ϵ_0. For a positive charge the flux is positive, while for a negative charge the flux is negative.

Whilst the divergence vanishes for $r \neq 0$, as a consequence of equation (6.8) the flux can be nonzero only when the divergence is nonzero at the location of the point charge. Physically this means that the electric charge is the source of the electric field. This is reflected in the Maxwell equation for the divergence of the electric field:

$$\nabla \cdot \mathbf{E} = \rho(\mathbf{r})/\epsilon_0. \qquad (6.24)$$

The charge density $\rho(\mathbf{r})$ is the electric charge per unit volume, just as the mass-density denotes the mass per unit volume. In addition, expression (6.24) contains the electrical permittivity ϵ_0. This term serves as a *coupling constant* since it describes how "much" electrical field is generated by a given electrical charge density. It is obvious that a constant is needed here because the charge density and the electrical field have different physical dimensions; hence, a proportionality factor must be present. However, the physical meaning of a coupling constant goes much deeper, because it prescribes how strongly cause (the source) and effect (the field) are coupled.

In the previous example, the charge (density) gives rise to a monopole source. Next, we compute the divergence of the magnetic field (6.5) generated by a dipole $\mathbf{m}$. We are free to choose the orientation of the coordinate system, and opt to align the z-axis with the dipole vector $\mathbf{m}$; this means that $\mathbf{m} = m\hat{\mathbf{z}}$.

Problem d   Use this expression and the relation $\hat{\mathbf{r}} = \mathbf{r}/r$ to show that the magnetic field (6.5) can be written as

$$\mathbf{B} = \frac{3m\mathbf{r}\left(\mathbf{r}\cdot\hat{\mathbf{z}}\right)}{r^5} - \frac{m\hat{\mathbf{z}}}{r^3}. \tag{6.25}$$

Problem e   Write the vectors $\mathbf{r}$ and $\hat{\mathbf{z}}$ in component form in Cartesian coordinates to show that the components of the magnetic field are given by

$$\mathbf{B} = \begin{pmatrix} 3mxz/r^5 \\ 3myz/r^5 \\ 3mz^2/r^5 - m/r^3 \end{pmatrix}. \tag{6.26}$$

Problem f   Differentiate the different components to show that

$$\frac{\partial B_x}{\partial x} = 3m\left(\frac{z}{r^5} - \frac{5x^2z}{r^7}\right),$$
$$\frac{\partial B_y}{\partial y} = 3m\left(\frac{z}{r^5} - \frac{5y^2z}{r^7}\right), \tag{6.27}$$
$$\frac{\partial B_z}{\partial z} = 3m\left(\frac{3z}{r^5} - \frac{5z^3}{r^7}\right).$$

Problem g   Add these terms to show that the divergence of the magnetic field vanishes:

$$\nabla \cdot \mathbf{B} = 0. \tag{6.28}$$

By analogy with (6.24), one might have expected that the divergence of the magnetic field is nonzero at the source of the field and that it is related to a magnetic charge density:

$$\nabla \cdot \mathbf{B} = coupling\ const. \times \rho_B(\mathbf{r}),$$

where ρ_B would be the "density of magnetic charge." However, particles with a magnetic charge (usually called "magnetic monopoles") have not been found in nature, despite extensive searches. Therefore, the Maxwell equation for the divergence of the magnetic field is

$$(\nabla \cdot \mathbf{B}) = 0. \tag{6.29}$$

However, we should remember that this divergence is zero because of the absence of magnetic monopoles, rather than a vanishing coupling constant.

6.5 Divergence in cylindrical coordinates

In the previous analysis we expressed the divergence in Cartesian coordinates: $\nabla \cdot \mathbf{v} = \partial_x v_x + \partial_y v_y + \partial_z v_z$. As you may have discovered, the use of other coordinate systems such as cylindrical coordinates or spherical coordinates can often make life much simpler. Here we derive an expression for the divergence in cylindrical coordinates. In this system, the distance $r = \sqrt{x^2 + y^2}$ of a point to the z-axis, the azimuth $\varphi = \arctan(y/x)$ and z are used as coordinates, see Section 4.5. A vector $\mathbf{v}$ can be decomposed into components in this coordinate system:

$$\mathbf{v} = v_r \hat{\mathbf{r}} + v_\varphi \hat{\boldsymbol{\varphi}} + v_z \hat{\mathbf{z}}, \tag{6.30}$$

where $\hat{\mathbf{r}}$, $\hat{\boldsymbol{\varphi}}$, and $\hat{\mathbf{z}}$ are unit vectors in the direction of increasing values of r, φ, and z, respectively. As shown in Section 6.2, the divergence is the flux per unit volume. Let us consider the infinitesimal volume corresponding to increments dr, $d\varphi$, and dz shown in Figure 6.4. First, we need to find the flux of $\mathbf{v}$ through the surface elements perpendicular to $\hat{\mathbf{r}}$. The size of this surface is $r d\varphi dz$ and $(r + dr)d\varphi dz$ at r and $r + dr$, respectively. The normal components of $\mathbf{v}$ through these surfaces are $v_r(r, \varphi, z)$ and $v_r(r + dr, \varphi, z)$, respectively. Hence, the total flux through these two surface is given by $v_r(r + dr, \varphi, z)(r + dr)d\varphi dz - v_r(r, \varphi, z)(r)d\varphi dz$.

Problem a   Show that to first order in dr this quantity is equal to

$$\frac{\partial}{\partial r}(r v_r)\, dr d\varphi dz.$$

Hint: Use a first-order Taylor expansion for $v_r(r + dr, \varphi, z)$ in the quantity dr.

Problem b   Show that the flux through the surfaces perpendicular to $\hat{\boldsymbol{\varphi}}$ is to first order in $d\varphi$ given by $(\partial v_\varphi / \partial \varphi) dr d\varphi dz$.

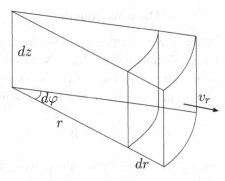

Figure 6.4   The geometric variables for the computation of the divergence in cylindrical coordinates.

Problem c   Show that the flux through the surfaces perpendicular to $\hat{\mathbf{z}}$ is to first order in dz given by $(\partial v_z/\partial z)rdrd\varphi dz$.

Problem d   Use Figure 6.4 to show that the volume dV is given by

$$dV = rd\varphi drdz. \tag{6.31}$$

Note that you had also obtained this result in expression (4.40) by assessing the spherical coordinate system in the equatorial plane.

Problem e   Use the fact that the divergence is the flux per unit volume, and the results of the previous problems to show that in cylindrical coordinates:

$$\nabla \cdot \mathbf{v} = \frac{1}{r}\frac{\partial}{\partial r}(rv_r) + \frac{1}{r}\frac{\partial v_\varphi}{\partial \varphi} + \frac{\partial v_z}{\partial z}. \tag{6.32}$$

Problem f   Use the expression above to re-derive (6.13) without using Cartesian coordinates as an intermediary. Make sure not to confuse $\mathbf{r}$ with $\hat{\mathbf{r}}$.

In the spherical coordinates defined in Chapter 4, a vector $\mathbf{v}$ can be expanded into the components v_r, v_θ, and v_φ in the directions of increasing values of r, θ, and φ, respectively. In spherical coordinates r has a different meaning to that in cylindrical coordinates because in spherical coordinates $r = \sqrt{x^2 + y^2 + z^2}$.

Problem g   Use the same reasoning as for the cylindrical coordinates to show that in spherical coordinates:

$$\nabla \cdot \mathbf{v} = \frac{1}{r^2}\frac{\partial}{\partial r}\left(r^2 v_r\right) + \frac{1}{r \sin\theta}\frac{\partial}{\partial \theta}(\sin\theta \, v_\theta) + \frac{1}{r \sin\theta}\frac{\partial v_\varphi}{\partial \varphi}. \tag{6.33}$$

6.6 The wave equation

In this section we derive the wave equation for acoustic waves, as compressive waves in a fluid. The first ingredient that is needed is Newton's law (5.31). Because a fluid is a continuous medium, we formulate Newton's law per unit volume. Since the mass per unit volume is the mass density ρ, Newton's law in this case is given by

$$\rho \frac{\partial \mathbf{v}}{\partial t} = -\nabla p + \mathbf{h}, \tag{6.34}$$

where $\mathbf{v}$ is the particle velocity. The term $-\nabla p$ is the pressure force that we derived in expression (5.13). In addition to the pressure force there may be other forces acting. These forces per unit volume are denoted by $\mathbf{h}$. For the moment, we assume that the material properties of the gas or fluid need not be constant throughout space.

Expression (6.34) gives the acceleration $\partial \mathbf{v}/\partial t$ for a given pressure p. To close the loop, we need to express the pressure in terms of the velocity. This is where the divergence comes in. We argued in Section 6.2 that the divergence is the net flux per unit volume. This means that $(\nabla \cdot \mathbf{v})$ is the net flux of the fluid flow per unit volume. When $(\nabla \cdot \mathbf{v}) > 0$, there is a net flow of fluid out of the volume, and as a result, the pressure drops. Conversely, when $(\nabla \cdot \mathbf{v}) < 0$, there is a net flow of fluid into the volume and the pressure rises. This means that the pressure is related to the divergence of the velocity by

$$\frac{\partial p}{\partial t} = -\kappa(\nabla \cdot \mathbf{v}). \tag{6.35}$$

The left-hand side contains a time derivative because the divergence of the velocity gives the *rate* of change with which fluid accumulates; hence, the left-hand side gives the rate of change of the pressure. The constant κ is the *bulk modulus*, this material constant determines how much the pressure increases for a given compression of the fluid. For water, κ is much larger than it is for air because it is much harder to compress water than it is to compress air. The reasoning used here was somewhat heuristic, but in expression (25.9) we more rigorously derive the connection between the change in volume of a fluid and $(\nabla \cdot \mathbf{v})$. Alternatively, the reader can also consult Middleton and Wilcock (1994), for example.

Problem a The acoustic wave equation follows by eliminating the velocity from equations (6.34) and (6.35). Do this by taking the time derivative of equation (6.35) and using expression (6.34) to eliminate $\partial \mathbf{v}/\partial t$. Show that this leads to

$$\frac{\partial^2 p}{\partial t^2} = \kappa \nabla \cdot \left(\frac{1}{\rho}\nabla p\right) - \kappa \nabla \cdot \left(\frac{\mathbf{h}}{\rho}\right). \tag{6.36}$$

Problem b Multiply this expression with ρ/κ and show that the result can be written as

$$\rho \nabla \cdot \left(\frac{1}{\rho}\nabla p\right) - \frac{\rho}{\kappa}\frac{\partial^2 p}{\partial t^2} = f, \tag{6.37}$$

with

$$f = \rho \nabla \cdot \left(\frac{\mathbf{h}}{\rho}\right). \tag{6.38}$$

The forcing term f depends in a complicated way on the body force $\mathbf{h}$, but in the end only f enters the acoustic wave equation.

Problem c   We next apply dimensional analysis as treated in Chapter 2 to the term ρ/κ. Use expression (6.35) to show that κ has the dimension of pressure, and that

$$\frac{\rho}{\kappa} \sim \frac{[T^2]}{[L^2]}. \tag{6.39}$$

Remember that pressure is force per unit surface area.

This means that ρ/κ has the dimension $1/velocity^2$. We next define a velocity c as

$$c = \sqrt{\frac{\kappa}{\rho}}. \tag{6.40}$$

With this definition, expression (6.37) can be written as

$$\rho \nabla \cdot \left(\frac{1}{\rho} \nabla p \right) - \frac{1}{c^2} \frac{\partial^2 p}{\partial t^2} = f. \tag{6.41}$$

This expression is called the *acoustic wave equation*.

Problem d   Show that for the special case where the density is constant, and the forcing term vanishes ($f = 0$), expression (6.41) reduces to

$$\nabla^2 p - \frac{1}{c^2} \frac{\partial^2 p}{\partial t^2} = 0. \tag{6.42}$$

In this expression ∇^2 is the Laplacian that is treated in more detail in Chapter 10. This differential operator in Cartesian coordinates is given by $\nabla^2 = \frac{\partial^2}{\partial x^2} + \frac{\partial^2}{\partial y^2} + \frac{\partial^2}{\partial z^2}$. Equation (6.42) is the *wave equation* for a medium with a constant density. This expression does not only hold for waves in a medium with constant density, it is ubiquitous in physics and also holds for electromagnetic waves (see, for example, expressions (6.37) and (6.38) of Jackson, 1998).

In the previous derivation the body force **h** acted as a source. In addition to this body force, one can also have a pressure source by modifying expression (6.35) into

$$\frac{\partial p}{\partial t} = -\kappa (\nabla \cdot \mathbf{v}) + q. \tag{6.43}$$

The term q describes a source that increases the pressure; an explosion is an example of such a source.

Problem e   Repeat the derivation of this section when the pressure source q is included and show that the acoustic waves still satisfy the acoustic wave equation (6.41), but that the pressure force in this case is given by

$$f = \rho \nabla \cdot \left(\frac{\mathbf{h}}{\rho}\right) - \frac{\rho}{\kappa}\frac{\partial q}{\partial t}. \tag{6.44}$$

This means that an explosive source and a body force combine into the single source term f.

6.7 Is life possible in a five-dimensional world?

Let us apply what we learned in this chapter to investigate whether the motion of the Earth around the Sun is stable or not. For that, we ask ourselves: When the position of the Earth is perturbed (for example, by the gravitational attraction of the other planets or by a passing asteroid), does the gravitational force cause the Earth to return to its original position (stability) or to spiral away from (or toward) the Sun? It turns out that these stability properties depend on the spatial dimension! We know that we live in a world of three spatial dimensions, but it is interesting to investigate whether the orbit of the Earth would also be stable in a world with a different number of spatial dimensions.

In the Newtonian theory the gravitational field $\mathbf{g}(\mathbf{r})$ satisfies (see Ohanian and Ruffini, 1976):

$$\nabla \cdot \mathbf{g} = -4\pi G\rho(\mathbf{r}), \tag{6.45}$$

where $\rho(\mathbf{r})$ is the mass-density and G is the gravitational constant that has a value of 6.67×10^{-8} cm^3 g^{-1} s^{-2}. The term $4\pi G$ plays the role of a coupling constant, just as the inverse permittivity ϵ_0 does in (6.24). Note that the right-hand side of the gravitational field equation (6.45) has the opposite sign to the right-hand side of the electric field equation (6.24). This is due to the fact that two electric charges of equal sign repel each other, while two masses of equal sign (mass being positive) attract each other. Zee (2005) gives an insightful explanation based on quantum field theory of this fundamental difference between the electrical field and the gravitational field. If the sign of the right-hand side of (6.45) were positive, masses would repel each other and structures such as planets, the solar system, and stellar systems would not exist.

Problem a We argued in Section 6.4 that electric field lines start at positive charges and end at negative charges. By analogy we expect that gravitational field lines end at the (positive) masses that generate the field. However, where do the gravitational field lines start?

Let us first determine the gravitational field of the Sun in N dimensions. Away from the Sun, the mass-density vanishes; this means that $\nabla \cdot \mathbf{g} = 0$. We assume that the mass-density of the Sun is spherically symmetric, then the gravitational field must also be spherically symmetric and is thus of the form:

$$\mathbf{g}(\mathbf{r}) = f(r)\mathbf{r}. \tag{6.46}$$

To make further progress, we must derive the divergence of a spherically symmetric vector field in N dimensions. Generalizing expression (6.33) to an arbitrary number of dimensions is not trivial, but fortunately this is not needed. We will make use of the property that in N dimensions: $r = \sqrt{\sum_{i=1}^{N} x_i^2}$.

Problem b Derive from this expression that

$$\partial r / \partial x_j = x_j / r. \tag{6.47}$$

Use this result to derive that for a vector field of the form (6.46):

$$\nabla \cdot \mathbf{g} = Nf(r) + r\frac{\partial f}{\partial r}. \tag{6.48}$$

Confirm this is consistent with the 2D problem of a thin sheet of fluid flow in Problem a of Section 6.2.

Outside the Sun, where the mass-density vanishes and $\nabla \cdot \mathbf{g} = 0$, we can use this result to determine the gravitational field.

Problem c Derive that

$$\mathbf{g}(\mathbf{r}) = -\frac{A}{r^{N-1}}\hat{\mathbf{r}}, \tag{6.49}$$

and check this result for three spatial dimensions.

The constant A is proportional to G, the mass of the Sun, and a constant that depends on the number of dimensions. Its precise value is, however, not important for the coming arguments. The minus sign has been added for convenience: the gravitational field points toward the Sun and hence $A > 0$.

Associated with the gravitational field is a gravitational force that attracts the Earth toward the Sun. If the mass of the Earth is denoted by m, this force is given by

$$\mathbf{F}_{grav} = -\frac{Am}{r^{N-1}}\hat{\mathbf{r}}, \tag{6.50}$$

and is directed toward the Sun. For simplicity we assume that the Earth is in a circular orbit. This means that the attractive gravitational force is balanced by the repulsive centrifugal force, which is given by

$$\mathbf{F}_{cent} = \frac{mv^2}{r}\hat{\mathbf{r}}. \tag{6.51}$$

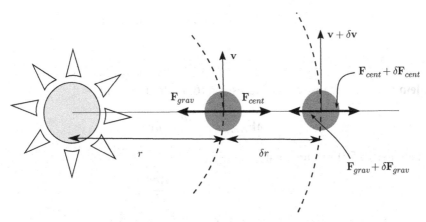

Figure 6.5   Definition of variables for the perturbed orbit of the Earth.

In equilibrium these forces balance: $\mathbf{F}_{grav} + \mathbf{F}_{cent} = 0$.

Problem d   Derive the velocity v from this requirement.

We now assume that the distance from the Earth to the Sun is changed from r to $r + \delta r$. The perturbation in the position $\delta\mathbf{r} = \delta r\,\hat{\mathbf{r}}$ perturbs the gravitational force and the centrifugal force. These perturbations are denoted by $\delta\mathbf{F}_{grav}$ and $\delta\mathbf{F}_{cent}$, respectively (Figure 6.5). The change in the total force is given by $\delta\mathbf{F}_{grav} + \delta\mathbf{F}_{cent}$. Let us first suppose that the Earth moves away from the Sun; in that case $\delta\mathbf{r}$ points from the Sun to the Earth. The orbit is stable when the net perturbation of the force draws the Earth back toward the Sun. In that case the vectors $\delta\mathbf{r}$ and $\left(\delta\mathbf{F}_{grav} + \delta\mathbf{F}_{cent}\right)$ point in opposite directions and

$$\left(\delta\mathbf{F}_{grav} + \delta\mathbf{F}_{cent}\right) \cdot \delta\mathbf{r} < 0 \quad \text{(stability)}. \tag{6.52}$$

Suppose on the other hand that the Earth moves toward the Sun so that $\delta\mathbf{r}$ points toward the Sun. The orbit is stable when the perturbation of the force pushes the Earth away from Sun. In that case the vectors $\delta\mathbf{r}$ and $\left(\delta\mathbf{F}_{grav} + \delta\mathbf{F}_{cent}\right)$ point in opposite directions as well so that (6.52) is satisfied. This means that the orbital motion is stable for perturbations when the gravitational field satisfies the criterion (6.52).

To compute the change in the centrifugal force, we use that angular momentum is conserved:

$$mrv = m(r + \delta r)(v + \delta v). \tag{6.53}$$

In what follows we consider small perturbations and retain only terms of first order in the perturbation. This means that we ignore higher-order terms such as the product $\delta r\,\delta v$.

Problem e   Use equation (6.53) to derive that to first order in δr

$$\delta v = -\frac{v}{r}\delta r. \tag{6.54}$$

Problem f   Use this result and expression (6.51) to derive that

$$\delta \mathbf{F}_{cent} = -\frac{3mv^2}{r^2}\delta \mathbf{r}, \tag{6.55}$$

then use (6.50) to show that

$$\delta \mathbf{F}_{grav} = (N-1)\frac{Am}{r^N}\delta \mathbf{r}. \tag{6.56}$$

Note that the perturbation of the centrifugal force does not depend on the number of spatial dimensions N, but that the perturbation of the gravitational force does.

Problem g   Use the value of the velocity derived in Problem d and expressions (6.55)–(6.56) to show that according to the criterion (6.52) the orbital motion is stable in less than four spatial dimensions. Show also that the requirement for stability is independent of the original distance r.

This intriguing result implies that orbital motion is unstable in more than four spatial dimensions. Because life seems to be tied to planetary systems with a central star that supplies the energy to sustain life on the orbiting planet(s), life would be impossible in a five-dimensional world! This stability requirement is independent of r: the stability properties of orbital motion do not depend on the size of the orbit. In other words, the gravitational field does not have "stable regions" and "unstable regions": the stability property depends only on the number of spatial dimensions.

7

Curl of a vector field

The concept of the curl operation is related to which degree a vector field has a sense of rotation. An example of a rotation is shown by the map of windspeeds in figure 7.1 modeled by the National Oceanic and Atmospheric Administration (NOAA) when Hurricane Katrina tormented the Gulf of Mexico in August 2005. Another example of swirling motion is shown on the cover of this book and in Figure 3.5, of the flow in a soap bubble that is placed on a rotating table (Meuel et al., 2013). The swirling, perhaps we should say curling, motion is mathematically captured by the *curl* of the flow field. In general, the curl of vector field has two sources: a rigid rotation and shear. If the vector field is fluid flow, both types of motion would cause a paddle wheel in the flow to rotate. This rotation is measured by the curl.

7.1 Introduction of the curl

We introduce the *curl* of a vector field $\mathbf{v}$ by its formal definition in terms of Cartesian coordinates (x, y, z) and its respective unit vectors $\hat{\mathbf{x}}, \hat{\mathbf{y}}$:

$$\text{curl } \mathbf{v} \equiv \begin{vmatrix} \hat{\mathbf{x}} & \hat{\mathbf{y}} & \hat{\mathbf{z}} \\ \partial_x & \partial_y & \partial_z \\ v_x & v_y & v_z \end{vmatrix} = \begin{pmatrix} \partial_y v_z - \partial_z v_y \\ \partial_z v_x - \partial_x v_z \\ \partial_x v_y - \partial_y v_x \end{pmatrix}. \tag{7.1}$$

It can be seen that the curl of a vector field is a *vector*. This is in contrast with the divergence of a vector field, which is a scalar. The notation with the determinant is informal, because the entries in a determinant should be numbers rather than vectors such as $\hat{\mathbf{x}}$ or differentiation operators such as $\partial_y = \partial/\partial y$. However, the notation in terms of a determinant is a simple way to remember the definition of the curl in Cartesian coordinates. We will write the curl of a vector field also as: curl $\mathbf{v} = \nabla \times \mathbf{v}$.

Problem a Verify that this notation with the curl expressed as the outer product of the operator ∇ and the vector $\mathbf{v}$ is consistent with definition (7.1).

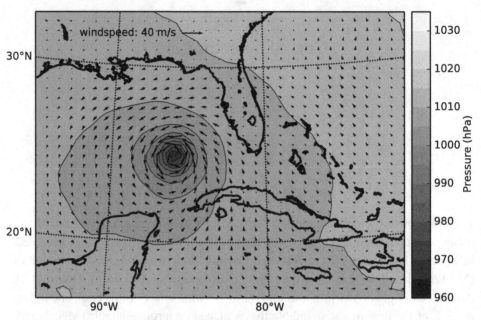

Figure 7.1   Modeled horizontal windspeed vectors around the eye of tropical cyclone Katrina in the Gulf of Mexico, as an example of a cylindrically symmetric source-free flow in the (x, y)-plane. Data courtesy of NOAA.

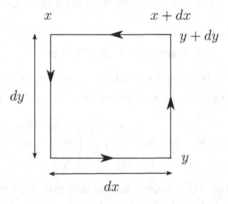

Figure 7.2   Definition of the geometric variables for the interpretation of the curl.

In general, the curl of a vector field is a three-dimensional vector. To see the physical interpretation of the curl, we will make life easy for ourselves by choosing a Cartesian coordinate system in which the z-axis is aligned with curl $\mathbf{v}$. In that coordinate system the curl is given by: curl $\mathbf{v} = (\partial_x v_y - \partial_y v_x)\hat{\mathbf{z}}$. Consider a small rectangular surface element oriented perpendicular to the z-axis with sides dx and dy, respectively; see Figure 7.2. We consider the line integral $\oint_{dxdy} \mathbf{v} \cdot d\mathbf{r}$ along a closed loop defined by the sides of this surface element, integrating in the counter-clockwise direction. This line integral can be written as the sum of the integral over the four sides of the surface element.

Problem b Show that the line integral is given by $\oint_{dxdy} \mathbf{v} \cdot d\mathbf{r} = v_x(x, y)dx + v_y(x+dx, y)dy - v_x(x, y+dy)dx - v_y(x, y)dy$, and use a first-order Taylor expansion to write this as

$$\oint_{dxdy} \mathbf{v} \cdot d\mathbf{r} = (\partial_x v_y - \partial_y v_x)\, dxdy, \tag{7.2}$$

with v_x and v_y evaluated at the point (x, y).

This expression can be rewritten as:

$$(\text{curl } \mathbf{v})_z = (\partial_x v_y - \partial_y v_x) = \frac{\oint_{dxdy} \mathbf{v} \cdot d\mathbf{r}}{dxdy}. \tag{7.3}$$

In this form we can express the meaning of the *curl* in words:

The component of curl **v** *in a certain direction is the closed line integral of* **v** *along a closed path perpendicular to this direction, per unit surface area.*

Note that this interpretation is similar to the interpretation of the divergence given in Section 6.2. There is, however, one major difference. As mentioned before, the *curl* of a vector field is a vector while the divergence is a scalar. This is reflected in our interpretation of the curl. Because a surface has an orientation defined by its normal vector, the curl is also a vector.

7.2 What is the curl of the vector field?

We consider again an incompressible fluid to study the curl of the velocity vector $\mathbf{v}$, and discover the physical meaning of the curl. It is not only for a didactic purpose that we consider the curl of fluid flow. In fluid mechanics this quantity plays such a crucial role that it is given a special name, the *vorticity* $\boldsymbol{\omega}$:

$$\boldsymbol{\omega} \equiv \nabla \times \mathbf{v}. \tag{7.4}$$

To simplify, we assume that the fluid moves in the (x, y)-plane only (i.e., $v_z = 0$) and that the flow depends only on x and y: $\mathbf{v} = \mathbf{v}(x, y)$.

Problem a Show that for such a flow

$$\boldsymbol{\omega} = \nabla \times \mathbf{v} = (\partial_x v_y - \partial_y v_x)\hat{\mathbf{z}}. \tag{7.5}$$

We first consider a cylindrically symmetric flow field. Such a flow field has rotational symmetry around one axis, and we will take this to be the z-axis. Because of the cylindrical symmetry and the earlier assumptions that the flow does not depend on z, the components v_r and v_φ depend neither on the azimuth φ ($= \arctan y/x$) used in the cylinder coordinates, nor on z. Instead, they only depend on the distance $r = \sqrt{x^2 + y^2}$ to the z-axis.

Problem b   Show that for a cylindrically symmetric flow field of an incompressible fluid (where $\nabla \cdot \mathbf{v} = 0$ everywhere, including $r = 0$), the radial component of the velocity must vanish: $v_r = 0$. Hint: Use expression (6.32) for the divergence in cylindrical coordinates, where for such a flow field the φ-derivative and the z-derivative both vanish.

This means that an incompressible flow with cylindrical symmetry has no net flow toward (or away from) the symmetry axis. The only nonzero component of the flow is therefore in the direction of $\hat{\varphi}$, which implies that the velocity field must be of the form:

$$\mathbf{v} = \hat{\varphi} v(r). \tag{7.6}$$

Figure 7.3 is a sketch of such a flow field. Note the similarity with the winds in Figure 7.1! The velocity depends only on r, because the symmetry of this problem dictates that $\partial v/\partial \varphi = \partial v/\partial z = 0$.

The problem we now face is that definition (7.1) is expressed in Cartesian coordinates, while the velocity in equation (7.6) is expressed in cylindrical coordinates.

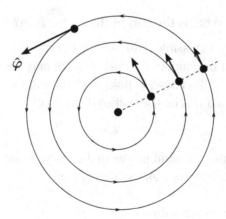

Figure 7.3   An axisymmetric source-free flow in the (x, y)-plane and the radial direction ($\hat{\varphi}$).

In Section 7.6 an expression for the curl in cylindrical coordinates will be derived. Alternatively, one can express the unit vector $\hat{\varphi}$ in Cartesian coordinates.

Problem c Verify that:

$$\hat{\varphi} = \begin{pmatrix} -y/r \\ x/r \\ 0 \end{pmatrix}. \tag{7.7}$$

Hints: Draw this vector in the (x, y)-plane; verify that this vector is perpendicular to the position vector **r** and that it is of unit length. Alternatively you can use expression (4.39).

Problem d Use (7.5), (7.7), and the chain rule for differentiation to show that for the flow field (7.6):

$$\nabla \times \mathbf{v} = (\nabla \times \mathbf{v})_z = \frac{\partial v}{\partial r} + \frac{v}{r}. \tag{7.8}$$

Hint: You have to use the derivatives $\partial r/\partial x$ and $\partial r/\partial y$ again. You determined these in Section 6.2.

We will use this result in Section 7.5 in a completely different problem that involves a vector field given by expression (7.6).

7.3 First source of vorticity: rigid rotation

In general, a nonzero curl of a vector field can have two origins: rigid rotation and shear. In this section we treat the effect of rigid rotation. Because we will use fluid flow as an example, we will speak about the vorticity. However, keep in mind that the results of this section (and the next) apply to any vector field. We consider a velocity field that describes a rigid rotation with the z-axis as the axis of rotation and an angular frequency Ω. This velocity field is of the form (7.6) with $v(r) = \Omega r$, because for a rigid rotation the velocity increases linearly with the distance r to the rotation axis.

Problem a Verify explicitly that every particle in the flow makes one revolution in a time $T = 2\pi/\Omega$ and that this time does not depend on the position of the particle. Hint: How long does it take to travel along the perimeter of a circle with radius r?

Problem b Use expression (7.8) to show that for this velocity field $\nabla \times \mathbf{v} = 2\Omega\hat{z}$.

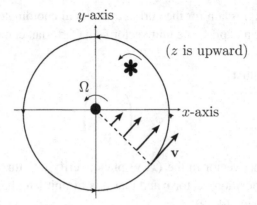

Figure 7.4   The vorticity of a rigid rotation. A paddle wheel moves with the flow, but also rotates around its own axis.

This means that the vorticity is twice the rotational vector $\Omega\hat{z}$. This result is derived here for the special case that the z-axis is the axis of rotation. (This can always be achieved, because any orientation of the coordinate system can be chosen.) The same result is obtained in Section 6.11 of Boas (2006), but with a different derivation. (Beware, the notation used by Boas, 2006, is different from that used in this book in a deceptive way!)

We saw that rigid rotation leads to a vorticity that is twice the rotation rate. Imagine we place a paddle wheel in the flow field that is associated with the rigid rotation, as in Figure 7.4. This paddle wheel moves with the flow and makes one revolution about its axis in a time $2\pi / \Omega$. Note that for the sense of rotation shown in Figure 7.4 the paddle wheel moves in the counterclockwise direction and that the curl points along the positive z-axis. This implies that the rotation of the paddle wheel not only denotes that the curl is nonzero, but also that the rotation vector of the paddle wheel is directed along the curl! This actually explains the origin of the word *vorticity*. In a vortex, the flow rotates around a rotational axis. The curl increases with the rotation rate; hence, it increases with the strength of the vortex. This strength of the vortex has been dubbed *vorticity*, and this term therefore reflects the fact that the curl of velocity denotes the (local) intensity of rotation in the flow.

Problem c Let us assume for the moment that the motion in the soap bubble on the cover of this book and in Figure 3.5 is a rigid rotation. What is the direction of the vorticity vector at the center of the vortex in the soap bubble?

7.4 Second source of vorticity: shear

In addition to rigid rotation, shear is another cause of vorticity. To see this we consider a fluid in which the flow is only in the x-direction and depends on the y-coordinate only: $v_y = v_z = 0$, $v_x = f(y)$.

Problem a Show that this flow does not describe a rigid rotation. Hint: How long does it take before a fluid particle returns to its original position?

Problem b Show that for this flow

$$\nabla \times \mathbf{v} = -\frac{\partial f}{\partial y}\hat{\mathbf{z}}. \qquad (7.9)$$

For example, consider the flow given by:

$$v_x = f(y) = v_0 e^{-y^2/L^2}. \qquad (7.10)$$

This flow field is sketched in Figure 7.5. The top paddle wheel rotates in the counterclockwise direction, because the flow at the top of the wheel is slower than the flow at the bottom of the paddle wheel. Likewise, the bottom paddle wheel rotates in the clockwise direction.

Problem c Compute $\nabla \times \mathbf{v}$ for this flow field and verify that both the curl and the rotational vector of the paddle wheels are aligned with the z-axis. Show that the vorticity is positive where the paddle wheels rotate in the counterclockwise direction and that the vorticity is negative where the paddle wheels rotate in the clockwise direction.

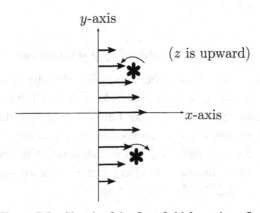

Figure 7.5   Sketch of the flow field for a shear flow.

It follows from the example in this section, and the example in Section 7.3, that both rotation and shear cause a nonzero vorticity. Both phenomena lead to the rotation of imaginary paddle wheels embedded in the vector field. Therefore, the curl of a vector field measures the local rotation of the vector field (in a literal sense). This explains why in some languages (e.g., Dutch) the notation *rot* **v** is used rather than *curl* **v**. This interpretation of the *curl* as a measure of (local) rotation is consistent with (7.3) in which the curl is related to the value of the line integral along the small contour. If the flow (locally) rotates and if we integrate along the fluid flow, the line integral $\oint \mathbf{v} \cdot d\mathbf{r}$ will be relatively large, so that this line integral indeed measures the local rotation.

Rotation and shear contribute both to the *curl* of a vector field. Let us consider once again a vector field of the form (7.6), which is axially symmetric around the z-axis. In the following we do not require the rotation around the z-axis to be rigid, so that $v(r)$ in (7.6) is arbitrary. We know that both the rotation around the z-axis and the shear are a source of vorticity.

Problem d   Show that for the flow

$$v(r) = \frac{A}{r} \qquad (7.11)$$

the vorticity vanishes for $r > 0$. The constant A is yet to be determined, but you can already make a sketch of this flow field.

The vorticity of this flow vanishes, even though the flow rotates around the z-axis (but not in a rigid rotation) and the flow has a nonzero shear. The reason that the vorticity vanishes is that the contribution of the rotation around the z-axis to the vorticity is equal but of opposite sign to the contribution of the shear, so that the total vorticity vanishes. Note that this implies that a paddle wheel does not rotate around its axis, as it moves with this flow!

7.5 Magnetic field induced by a straight current

At this point you may have gotten the impression that the flow field (7.11) has been contrived in an artificial way. However, keep in mind that all the arguments of the previous section apply to any vector field and that fluid flow was used only as an example. Let us now consider another example, where a magnetic field **B** is generated by an electrical current **J** that is independent of time. The Maxwell equation for the curl of the magnetic field in vacuum is for time-independent fields given by (see equation (5.22) in Jackson, 1998):

$$\nabla \times \mathbf{B} = \mu_0 \mathbf{J}, \qquad (7.12)$$

where μ_0 is the magnetic permeability of vacuum. This coupling constant governs the strength of the magnetic field, generated by a given current. It plays the same role as the inverse of permittivity ϵ_0 in (6.24), or the gravitational constant G in (6.45). The vector $\mathbf{J}$ denotes the electric current per unit volume (i.e., the electric current density).

For simplicity, we consider an electric current running through an infinite straight wire along the z-axis. Because of rotational symmetry around the z-axis, and because of translational invariance along the z-axis, the magnetic field strength does not depend on φ or z. Therefore, it must be of the form (7.6). In addition, the electrical current $\mathbf{J}$ vanishes away from the wire.

Problem a Use the results of Problem d of Section 7.2 to show that

$$\mathbf{B} = \frac{A}{r}\hat{\varphi}. \tag{7.13}$$

A comparison with equation (7.11) reveals that for this magnetic field the contribution of the "rotation" around the z-axis to $\nabla \times \mathbf{B}$ is exactly balanced by the contribution of the "magnetic shear" to $\nabla \times \mathbf{B}$. It should be noted that the magnetic field derived in this section is of great importance, because this field has been used to define the unit of electrical current, the Ampère. However, this can only be done when the constant A in expression (7.13) is known.

Problem b Why does the treatment of this section not tell us what the relation is between the constant A and the current $\mathbf{J}$ in the wire?

We return to this issue in Section 9.3, but for now we conclude that the magnetic field induced by a straight line current can be determined from the requirement that its curl vanishes away from the current line. In fact, Figure 7.3 is not only a model for the wind speed around the eye of a hurricane, it also describes the magnetic field $\mathbf{B}(x, y)$ around a straight line current directed along the z-axis.

7.6 The curl in spherical and cylindrical coordinates

In Section 6.5, expressions for the divergence in spherical coordinates and cylindrical coordinates were derived. Here we do the same for the curl, because these expressions are frequently very useful. It is possible to derive the curl in curvilinear coordinates by systematically transforming all the elements of the involved vectors and all the differentiations from Cartesian to curvilinear coordinates. As an alternative, we use the physical interpretation of the curl given by expression (7.3) to

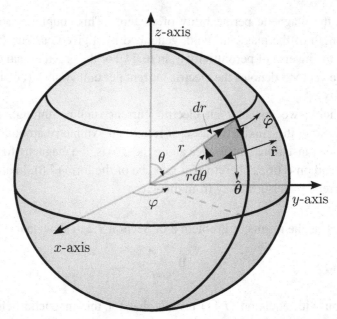

Figure 7.6   Definition of the geometric variables for the computation of the φ-component of the *curl* in spherical coordinates.

derive the curl in spherical coordinates. This expression states that a certain component of the curl of a vector field $\mathbf{v}$ is the line integral $\oint \mathbf{v} \cdot d\mathbf{r}$ along a contour perpendicular to the component of the curl in question, normalized by the surface area bound by that contour. As an example, we derive for a system of spherical coordinates the φ-component of the curl; see Figure 7.6 for the definition of the geometric variables.

Consider the dark shaded infinitesimal surface $r\,d\theta\,dr$ in Figure 7.6. We carry out the line integration along the surface in the direction shown in the figure. The reason for this is that the azimuth φ increases when we move into the figure; hence, $\hat{\boldsymbol{\varphi}}$ points into the figure. Following the rules of a right-handed screw, this corresponds to the indicated direction of integration.

Problem a   Sum the contributions of the four sides of the contour and use expression (7.3), to show that the φ-component of $\nabla \times \mathbf{v}$ is given by

$$(\nabla \times \mathbf{v})_\varphi = \frac{1}{r\,d\theta\,dr}[v_\theta(r + dr, \theta)(r + dr)\,d\theta - v_r(r, \theta + d\theta)\,dr$$
$$- v_\theta(r, \theta)r\,d\theta + v_r(r, \theta)\,dr], \tag{7.14}$$

where v_r and v_θ denote the components of $\mathbf{v}$ in the radial direction and in the direction of $\hat{\boldsymbol{\theta}}$, respectively.

Problem b Simplify the previous result with a Taylor expansion in the components of **v** in dr and $d\theta$ and linearizing the resulting expression in the infinitesimal increments dr and $d\theta$. You will find that

$$(\nabla \times \mathbf{v})_\varphi = \frac{1}{r}\frac{\partial}{\partial r}(rv_\theta) - \frac{1}{r}\frac{\partial v_r}{\partial \theta}. \tag{7.15}$$

Note that the final result does not depend on dr and $d\theta$. The same treatment can be applied to the other components of the curl. This leads to the following expression for the curl in spherical coordinates:

$$\nabla \times \mathbf{v} = \hat{\mathbf{r}}\frac{1}{r\sin\theta}\left[\frac{\partial}{\partial\theta}(\sin\theta\; v_\varphi) - \frac{\partial v_\theta}{\partial\varphi}\right] + \hat{\boldsymbol{\theta}}\frac{1}{r}\left[\frac{1}{\sin\theta}\frac{\partial v_r}{\partial\varphi} - \frac{\partial}{\partial r}(rv_\varphi)\right]$$
$$+ \hat{\boldsymbol{\varphi}}\frac{1}{r}\left[\frac{\partial}{\partial r}(rv_\theta) - \frac{\partial v_r}{\partial\theta}\right]. \tag{7.16}$$

Problem c Use a similar analysis to show that in cylindrical coordinates (r, φ, z) the curl is given by:

$$\nabla \times \mathbf{v} = \hat{\mathbf{r}}\left[\frac{1}{r}\frac{\partial v_z}{\partial\varphi} - \frac{\partial v_\varphi}{\partial z}\right] + \hat{\boldsymbol{\varphi}}\left[\frac{\partial v_r}{\partial z} - \frac{\partial v_z}{\partial r}\right]$$
$$+ \hat{\mathbf{z}}\frac{1}{r}\left[\frac{\partial}{\partial r}\left(rv_\varphi\right) - \frac{\partial v_r}{\partial\varphi}\right], \tag{7.17}$$

with $r = \sqrt{x^2 + y^2}$.

Problem d Use this last result to re-derive (7.8) for vector fields of the form $\mathbf{v} = v(r)\hat{\boldsymbol{\varphi}}$. Hint: In this case $v_r = v_z = 0$ and $v_\varphi = v(r)$.

7.7 Vector calculus of products

In many applications one needs to compute the gradient, divergence, or curl of products of functions or vectors. One can create such products in many different ways, especially if one takes into account that one can form inner products and outer products of vectors, and we show in this section a few examples how to apply vector calculus to such products.

We first compute the gradient of the product of two functions f and g. The gradient is given by

$$\nabla(fg) = \begin{pmatrix} \partial_x(fg) \\ \partial_y(fg) \\ \partial_z(fg) \end{pmatrix}, \tag{7.18}$$

where for brevity we use the notation ∂_x for the partial derivative ∂/∂_x, and a similar notation for the y and z derivatives.

Problem a   Each of the components in equation (7.18) consists of the derivative of the product of two functions. Apply the product rule to each of the components to show that

$$\nabla(fg) = \begin{pmatrix} \partial_x f \\ \partial_y f \\ \partial_z f \end{pmatrix} g + f \begin{pmatrix} \partial_x g \\ \partial_y g \\ \partial_z g \end{pmatrix}. \qquad (7.19)$$

The vectors in the right-hand side are the gradient of f and g, respectively, so this expression can also be written as

$$\nabla(fg) = (\nabla f)g + f\nabla g. \qquad (7.20)$$

This expression looks just like the product rule for derivatives $(\partial_x(fg) = (\partial_x f)g + f\partial_x g)$, which is no surprise because expression (7.20) was derived from the product rule.

Product rules that involve vectors are a little bit more complicated. As an example we first consider $\nabla \cdot (f\mathbf{v})$, where f is a function and $\mathbf{v}$ a vector. In component form this divergence can be written as $\nabla \cdot (f\mathbf{v}) = \nabla_x(fv_x) + \nabla_y(fv_y) + \nabla_z(fv_z)$.

Problem b   Apply the product rule to each of these derivatives, and rearrange the result to give

$$\nabla \cdot (f\mathbf{v}) = (\partial_x f)v_x + (\partial_y f)v_y + (\partial_z f)v_z + f\left(\partial_x v_x + \partial_y v_y + \partial_z v_z\right). \quad (7.21)$$

Problem c   Use the definition of the gradient and the divergence to write this as

$$\nabla \cdot (f\mathbf{v}) = (\nabla f) \cdot \mathbf{v} + f(\nabla \cdot \mathbf{v}). \qquad (7.22)$$

Note that this expression has the "look" of the product rule; it is the derivative of f times $\mathbf{v}$ plus f times the derivative of $\mathbf{v}$. However, because f is a function and $\mathbf{v}$ a vector, the derivative of f is a gradient and the derivative of $\mathbf{v}$ a divergence.

As a more complicated example we take the curl of a product $f\mathbf{v}$.

Problem d   Use the definition of the curl to show that

$$\nabla \times (f\mathbf{v}) = \begin{pmatrix} \partial_y(fv_z) - \partial_z(fv_y) \\ \partial_z(fv_x) - \partial_x(fv_z) \\ \partial_x(fv_y) - \partial_y(fv_x) \end{pmatrix}. \qquad (7.23)$$

Problem e Apply the product rule to each of the derivatives, and show that the result can be written as

$$\nabla \times (f\mathbf{v}) = \begin{pmatrix} (\partial_y f)v_z - (\partial_z f)v_y \\ (\partial_z f)v_x - (\partial_x f)v_z \\ (\partial_x f)v_y - (\partial_y f)v_x \end{pmatrix} + f \begin{pmatrix} \partial_y v_z - \partial_z v_y \\ \partial_z v_x - \partial_x v_z \\ \partial_x v_y - \partial_y v_x \end{pmatrix}. \tag{7.24}$$

Problem f A comparison with equation (7.1) shows that the last term is given by $f\nabla \times \mathbf{v}$. Use the definition of the gradient and the cross project to show that the first term in the right-hand side can be written as

$$\begin{pmatrix} (\partial_y f)v_z - (\partial_z f)v_y \\ (\partial_z f)v_x - (\partial_x f)v_z \\ (\partial_x f)v_y - (\partial_y f)v_x \end{pmatrix} = (\nabla f) \times \mathbf{v}. \tag{7.25}$$

Using these results in equation (7.24) gives the identity

$$\nabla \times (f\mathbf{v}) = (\nabla f) \times \mathbf{v} + f\nabla \times \mathbf{v}. \tag{7.26}$$

Note that equation (7.26) looks again like the product rule.

We show in the next example that a vector derivative applied to a product does not always give a result that "looks" like the product rule. Consider the divergence of the cross-product of two vectors: $\nabla \cdot (\mathbf{v} \times \mathbf{w})$. The cross-product of two vectors is another vector. When we take the divergence, we get a scalar; hence, the end product must be a scalar. The dot product $\mathbf{v} \cdot (\nabla \times \mathbf{w})$ is a scalar as well, and based on the experience with product rules gained in this section we might guess that $\nabla \cdot (\mathbf{v} \times \mathbf{w}) = \mathbf{v} \cdot (\nabla \times \mathbf{w}) + (\nabla \times \mathbf{v}) \cdot \mathbf{w}$. This expression is, however, not correct. Guesses can be great, but one can make mistakes!

Problem g Use the definitions of the cross-product and the divergence to show that

$$\begin{aligned} \nabla \cdot (\mathbf{v} \times \mathbf{w}) = & \ \partial_x(v_y w_z - v_z w_y) \\ + & \ \partial_y(v_z w_x - v_x w_z) \\ + & \ \partial_z(v_x w_y - v_y w_x). \end{aligned} \tag{7.27}$$

Problem h Apply the product rule to each of the terms in this expression and show that the result can be written as

$$\nabla \cdot (\mathbf{v} \times \mathbf{w}) = -v_x(\partial_y w_z - \partial_z w_y) - v_y(\partial_z w_x - \partial_x w_z) - v_z(\partial_x w_y - \partial_y w_x)$$
$$+ (\partial_y v_z - \partial_z v_y)w_x + (\partial_z v_x - \partial_x v_z)w_y + (\partial_x v_y - \partial_y v_z)w_z. \tag{7.28}$$

Problem i According to equation (7.1), the combination $(\partial_y v_z - \partial_z v_y)$ is the x-component of $\nabla \times \mathbf{v}$. Apply this reasoning to all terms in the right-hand side of equation (7.28) to show that

$$\nabla \cdot (\mathbf{v} \times \mathbf{w}) = -v_x (\nabla \times \mathbf{w})_x - v_y (\nabla \times \mathbf{w})_y - v_z (\nabla \times \mathbf{w})_z$$

$$(7.29)$$

$$+ (\nabla \times \mathbf{v})_x w_x + (\nabla \times \mathbf{v})_y w_y + (\nabla \times \mathbf{v})_z w_z,$$

where $(\nabla \times \mathbf{v})_y$ is, for example, the y-component of $\nabla \times \mathbf{v}$.

Each line in the right-hand side consists of the dot product of two vectors, and the previous result can be written as

$$\nabla \cdot (\mathbf{v} \times \mathbf{w}) = -\mathbf{v} \cdot (\nabla \times \mathbf{w}) + (\nabla \times \mathbf{v}) \cdot \mathbf{w}. \qquad (7.30)$$

Note the minus sign in the first term of the right-hand side. This minus sign did not appear in our initial guess just above Problem g. In hindsight it is logical that we should get a minus sign in one of the two terms. The cross-product of two vectors in antisymmetric in the sense that $\mathbf{v} \times \mathbf{w} = -\mathbf{w} \times \mathbf{v}$. Interchanging $\mathbf{v}$ and $\mathbf{w}$ gives a minus sign. Our final result should satisfy this property as well, and interchanging $\mathbf{v}$ and $\mathbf{w}$ in the right-hand side of expression (7.30) indeed gives a minus sign.

One can go on deriving more elaborate expressions for the gradient, divergence, and curl of different products of functions and vectors. This activity is like telling jokes about two Dutch authors walking down the street, or like eating peanuts ... it is hard to stop. But instead we refer the reader to online resources or textbooks (Jackson, 1998; Marsden and Tromba, 2003; Arfken and Weber, 2005) for tables of such expressions.

8

Theorem of Gauss

In Section 6.7 we determined that the divergence of the gravitational field in free space vanishes: $(\nabla \cdot \mathbf{g}) = 0$. This was sufficient to determine the gravitational field in N dimensions as expression (6.49). However, that expression is not quite satisfactory, because it contains an unknown constant A. In fact, at this point we have no idea how this constant is related to the mass M that causes the gravitational field! The reason for this is simple: to derive the gravitational field in (6.49) we have only used the field equation (6.45) for free space (where $\rho = 0$). However, if we want to find the relation between the mass and the resulting gravitational field, we must also use the field equation $(\nabla \cdot \mathbf{g}) = -4\pi G\rho$ at places where mass is present. More specifically, we have to *integrate* the field equation in order to find the total effect of the mass. The theorem of Gauss gives us an expression for the volume integral of the divergence of a vector field.

8.1 Statement of Gauss' theorem

Section 6.2 introduced the divergence as the flux per unit volume. In fact, expression (6.8) gives us the outward flux $d\Phi_v$ through an infinitesimal volume dV: $d\Phi_v = (\nabla \cdot \mathbf{v})dV$. We can integrate this expression to find the total flux through the surface S, which encloses the total volume V:

$$\oint_S \mathbf{v} \cdot d\mathbf{S} = \int_V (\nabla \cdot \mathbf{v})\, dV \quad \text{(in three dimensions)} \tag{8.1}$$

In deriving this, equation (6.8) has been used to express the total flux on the left-hand side. This expression is called the theorem of Gauss, or the divergence theorem.

We did not use the dimensionality of the space to derive (8.1); the relation holds in any number of dimensions. You may recognize the one-dimensional version of (8.1). In one dimension the vector $\mathbf{v}$ has only one component v_x; hence, $(\nabla \cdot \mathbf{v}) = \partial_x v_x$. A "volume" in one dimension is simply a line. Let this line run from $x = a$ to $x = b$. The "surface" of a one-dimensional volume consists of the endpoints of this line, so that the left-hand side of (8.1) is the difference of the function v_x at its endpoints. This implies that the theorem of Gauss in one dimension is

$$v_x(b) - v_x(a) = \int_a^b \frac{\partial v_x}{\partial x} dx \quad \text{(one dimension).} \tag{8.2}$$

This expression is hopefully familiar to you. We will use the two-dimensional version of the theorem of Gauss in Section 9.2 to derive the theorem of Stokes. The theorem of Gauss in two dimensions follows by replacing in expression (8.1) the surface integral in the left-hand side by $d\mathbf{l}$, which is a length increment along a bounding curve C that points outward perpendicular to the curve, and by replacing the volume integral in the right-hand side of expression (8.1) by a surface integral:

$$\oint_C \mathbf{v} \cdot d\mathbf{l} = \int_V (\nabla \cdot \mathbf{v}) \, dS \quad \text{(two dimensions).} \tag{8.3}$$

Problem a   Compute the flux of the vector field $\mathbf{v}(x, y, z) = (x+y+z)\hat{\mathbf{z}}$ through a sphere of radius R centered on the origin by explicitly computing the integral that defines the flux.

Problem b   Solve Problem a using Gauss' theorem (8.1).

8.2 Gravitational field of a spherically symmetric mass

In this section we use Gauss' theorem (8.1) to show that the gravitational field of a body with a spherically symmetric mass-density ρ depends only on the total mass and not on the distribution of the mass over that body. For a spherically symmetric body the mass-density depends only on radius: $\rho(\mathbf{r}) = \rho(r)$. Because of the spherical symmetry of the mass, the gravitational field is spherically symmetric and points in the radial direction:

$$\mathbf{g}(\mathbf{r}) = g(r)\hat{\mathbf{r}}. \tag{8.4}$$

Problem a   Use equation (6.45) for the gravitational field and Gauss' theorem for a surface completely enclosing the mass, to show that

$$\oint_S \mathbf{g} \cdot d\mathbf{S} = -4\pi GM, \tag{8.5}$$

where $M = \int_V \rho dV$ is the total mass of the body.

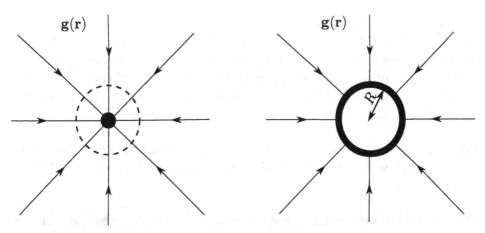

Figure 8.1 Two bodies with different mass distributions generate the same gravitational field $\mathbf{g}(\mathbf{r})$ for distances larger than radius R.

Problem b Use the surface of a sphere of radius r in (8.5) to show that the gravitational field is given in three dimensions by

$$\mathbf{g}(r) = -\frac{GM}{r^2}\hat{\mathbf{r}}. \tag{8.6}$$

This intriguing result implies that the gravitational field depends only on the total mass of the spherically symmetric body, but not on the distribution of the mass within that body. As an example, consider two bodies with the same mass. One body has all its mass located in a small ball near the origin and the other body has all its mass distributed on a thin spherical shell of radius R (Figure 8.1). According to (8.6) these bodies generate exactly the same gravitational field for distances $r > R$. This implies that gravitational measurements taken outside the two bodies cannot be used to distinguish between them. The nonunique relation between the gravity field and the underlying mass distribution is of importance for the interpretation of gravitational measurements taken in geophysical surveys.

Problem c Consider a mass located within a sphere of radius R, and a constant mass-density within that sphere. Integrate (6.45) over a sphere of radius $r < R$ to show that the gravitational field within the sphere is given by:

$$\mathbf{g}(\mathbf{r}) = -\frac{MGr}{R^3}\hat{\mathbf{r}}. \tag{8.7}$$

Plot the gravitational field as a function of r when the distance increases from zero to a distance larger than the radius R, where the gravitational field is given by equation (8.6). Verify that the gravitational field is continuous at the radius R of the sphere.

Note that all these conclusions hold identically for the electrical field when we replace the mass-density by the charge-density, because (6.24) for the divergence of the electric field has the same form as (6.45) for the gravitational field. As an example we consider a hollow spherical shell of radius R. On the spherical shell electrical charge is distributed with a constant charge-density: $\rho(r = R) = const.$

Problem d Use expression (6.24) for the electric field and Gauss' theorem to show that within the hollow sphere the electric field vanishes: $\mathbf{E}(\mathbf{r}) = 0$ for $r < R$.

This result implies that when a charge is placed within such a spherical shell, the electrical field generated by the charge distribution on the shell exerts no net force on this charge; the charge will not move. Since the electrical potential satisfies $\mathbf{E} = -\nabla V$, the result derived in Problem d implies that the potential is constant within the sphere. This property has actually been used to determine experimentally whether the electric field indeed satisfies (6.24). Measurements of the potential differences within a hollow spherical shell as described in Problem d can be carried out with great accuracy. Experiments based on this principle (usually in a more elaborate form) have been performed to ascertain the decay of the electric field of a point charge with distance. Writing the field strength as $1/r^{2+\epsilon}$ it has been shown that $\epsilon = (2.7 \pm 3.1) \times 10^{-16}$ (see Section I.2 of Jackson, 1998, for a discussion). The small value of ϵ is a remarkable experimental confirmation of (6.24) for the electric field.

8.3 Representation theorem for acoustic waves

Acoustic waves are waves that propagate through a fluid. You can hear the voice of others, because acoustic waves travel from their vocal tract to your ear. Acoustic waves are also commonly used to describe the propagation of waves through the Earth. Because the Earth is in large parts a solid body, this is strictly speaking not correct. However, under certain conditions (small scattering angles) the errors can be acceptable. In this problem we assume that the pressure oscillates in time as $p(\mathbf{r}, t) = p(\mathbf{r})e^{-i\omega t}$. (This amounts to analyzing the pressure field in the frequency domain. This is treated in Chapter 14.)

Problem a Use expression (6.37) to show that the pressure field $p(\mathbf{r})$ satisfies the following partial differential equation

$$\nabla \cdot \left(\frac{1}{\rho}\nabla p\right) + \frac{\omega^2}{\kappa}p = \frac{f}{\rho}. \tag{8.8}$$

In this expression $\rho(\mathbf{r})$ is the mass-density of the medium, ω is the angular frequency, and $\kappa(\mathbf{r})$ the bulk modulus (a factor that describes how strongly the medium resists changes in its volume). The right-hand side f/ρ describes the source of the acoustic waves. This term accounts, for example, for the action of your voice.

We now consider two pressure fields $p_1(\mathbf{r})$ and $p_2(\mathbf{r})$ that both satisfy (8.8) with sources $f_1(\mathbf{r})$ and $f_2(\mathbf{r})$ on the right-hand side of the equation, respectively.

Problem b Multiply equation (8.8) for p_1 with p_2, multiply equation (8.8) for p_2 with p_1 and subtract the resulting expressions. Integrate the result over a volume V to show that:

$$\int_V \left[p_2 \nabla \cdot \left(\frac{1}{\rho} \nabla p_1 \right) - p_1 \nabla \cdot \left(\frac{1}{\rho} \nabla p_2 \right) \right] dV = \int_V \frac{1}{\rho} (p_2 f_1 - p_1 f_2) \, dV.$$

(8.9)

Ultimately we want to relate the wave field at the surface S that encloses the volume V to the wave field within the volume. Obviously, Gauss' theorem is just the tool for doing this.

Problem c Use expression (7.22) with $f = p_2$ and $\mathbf{v} = \rho^{-1} \nabla p_1$ to show that

$$p_2 \nabla \cdot \left(\frac{1}{\rho} \nabla p_1 \right) = \nabla \cdot \left(\frac{1}{\rho} p_2 \nabla p_1 \right) - \frac{1}{\rho} (\nabla p_1 \cdot \nabla p_2).$$

(8.10)

What we are doing here is similar to the standard derivation of integration by parts. The easiest way to show that $\int_a^b f(\partial g / \partial x) dx = [f(x)g(x)]_a^b - \int_a^b (\partial f / \partial x) g \, dx$ is to integrate the identity $f(\partial g / \partial x) = \partial(fg)/dx - (\partial f / \partial x)g$ from $x = a$ to $x = b$. This last equation has exactly the same structure as (8.10).

Problem d Use (8.9), (8.10), and Gauss' theorem to derive that

$$\oint_S \frac{1}{\rho} (p_2 \nabla p_1 - p_1 \nabla p_2) \cdot d\mathbf{S} = \int_V \frac{1}{\rho} (p_2 f_1 - p_1 f_2) \, dV.$$

(8.11)

To see the power of this expression, consider the special case that the source f_2/ρ of p_2 is of unit strength and that this source is localized in a very small volume around a point $\mathbf{r}_0$ within the volume. This means that f_2 on the right-hand side of (8.11) is only nonzero at $\mathbf{r}_0$. The corresponding volume integral $\int_V p_1 f_2/\rho \, dV$ is in that case given by $p_1(\mathbf{r}_0)$. The wave field $p_2(\mathbf{r})$ generated by this point source is called the *Green's function*, and this special solution is denoted by $G(\mathbf{r}, \mathbf{r}_0)$. The concept of the Green's function is introduced in great detail in Chapter 17. The argument $\mathbf{r}_0$ is added to indicate that this is the wave field at location $\mathbf{r}$ due to a unit source at location $\mathbf{r}_0$. We now consider a solution p_1 that has no sources within the volume V (i.e., $f_1 = 0$). Let us simplify the notation further by dropping the subscript "1" in p_1.

Problem e Show by making all these changes that (8.11) can be written as

$$p(\mathbf{r}_0) = \oint_S \frac{1}{\rho} \left[p(\mathbf{r}) \nabla G(\mathbf{r}, \mathbf{r}_0) - G(\mathbf{r}, \mathbf{r}_0) \nabla p(\mathbf{r}) \right] \cdot d\mathbf{S}. \tag{8.12}$$

This result is called the "representation theorem," because it gives the wave field inside the volume when the wave field (and its gradient) are specified on the surface that bounds this volume. Expression (8.12) can be used to formally derive Huygens' principle, which states that every point on a wavefront acts as a source for other waves and that interference between these waves determines the propagation of the wavefront. Equation (8.12) also forms the basis for imaging techniques for seismic data (Schneider, 1978). In seismic exploration one records the wave field at the Earth's surface. This can be used as the surface S over which the integration is carried out. If the Green's function $G(\mathbf{r}, \mathbf{r}_0)$ is known, one can use expression (8.12) to compute the wave field in the interior of the Earth. Once the wave field in the interior of the Earth is known, one can deduce some of the properties of the material in the Earth. In this way, equation (8.12) (or its generalization to elastic waves in Aki and Richards, 2002) forms the basis of seismic imaging techniques.

This discussion on the Green's function seems to suggest that the problem of seismic imaging has been solved by the treatment of this section. There is, however, a catch. In order to apply (8.12) one must know the Green's function G. To know this function, one must know the medium. We thus have the interesting situation that once we know the Green's function, we can deduce the properties of the medium from (8.12), but that in order to use the Green's function we must know the medium. This suggests that seismic imaging cannot be carried out. Fortunately, it turns out that in practice it suffices to use a reasonable *estimate* of the Green's function in (8.12). Such an estimated Green's function can be computed once the velocity in the medium is reasonably well known. It is for this reason that *velocity estimation* is such a crucial step in seismic data processing (Claerbout, 1985; Yilmaz, 2001).

8.4 Flowing probability

In classical mechanics, the motion of a particle with mass m is governed by Newton's law: $m\ddot{\mathbf{r}} = \mathbf{F}$. When the force $\mathbf{F}$ is associated with a potential $V(\mathbf{r})$ the motion of the particle satisfies:

$$m \frac{d^2 \mathbf{r}}{dt^2} = -\nabla V(\mathbf{r}). \tag{8.13}$$

Newton's law, however, does not hold for very small particles. Atomic particles such as electrons are not described accurately by (8.13). One of the outstanding features of quantum mechanics is that an atomic particle is treated as a wave that describes the properties of that particle. This rather vague statement reflects the wave-particle duality that forms the basis of quantum mechanics. The wave function $\psi(\mathbf{r}, t)$ that describes a particle moving under the influence of a potential $V(\mathbf{r})$ satisfies the Schrödinger equation:[1]

$$i\hbar \frac{\partial \psi(\mathbf{r}, t)}{\partial t} = -\frac{\hbar^2}{2m} \nabla^2 \psi(\mathbf{r}, t) + V(\mathbf{r})\psi(\mathbf{r}, t). \tag{8.14}$$

In this expression, $\hbar$ is Planck's constant h divided by 2π; where Planck's constant has the numerical value $h = 6.626 \times 10^{-34} \text{kg m}^2/\text{s}$.

Suppose we are willing to accept that the motion of an electron is described by the Schrödinger equation, then the following question arises: What is the position of the electron as a function of time? According to the Copenhagen interpretation of quantum mechanics, this is a meaningless question because the electron behaves like a wave and does not have a definite location. Instead, the wave function $\psi(\mathbf{r}, t)$ dictates how likely it is that the particle is at location $\mathbf{r}$ at time t. Specifically, the quantity $|\psi(\mathbf{r}, t)|^2$ is the probability-density of finding the particle at location $\mathbf{r}$ at time t. This implies that the probability P_V that the particle is located within the volume V is given by $P_V = \int_V |\psi|^2 dV$. (Take care not to confuse the volume with the potential, because they are both indicated with the same symbol V.) This implies that the wave function is related to a probability. Instead of the motion of the electron, Schrödinger's equation dictates how the probability-density of the particle moves through space as time progresses. One expects that a "probability current" is associated with this movement. In this section we determine this current using the theorem of Gauss.

Problem a In the following we need the time derivative of $\psi^*(\mathbf{r}, t)$, where the asterisk denotes the complex conjugate. Derive the differential equation for $\psi^*(\mathbf{r}, t)$ by taking the complex conjugate of Schrödinger's equation (8.14).

Problem b Use this result to derive that for a volume V that is fixed in time:

$$\frac{\partial}{\partial t} \int_V |\psi|^2 dV = \frac{i\hbar}{2m} \int_V (\psi^* \nabla^2 \psi - \psi \nabla^2 \psi^*) dV. \tag{8.15}$$

---

[1] In this expression ∇^2 stands for the Laplacian, which is treated in Chapter 10. At this point you only need to know that the Laplacian of a function is the divergence of the gradient of that function: $\nabla^2 \psi = \text{div grad } \psi = \nabla \cdot \nabla \psi$.

Hint: Use that $\dfrac{\partial}{\partial t}|\psi|^2 = \dfrac{\partial}{\partial t}(\psi\psi^*) = \psi\dfrac{\partial\psi^*}{\partial t} + \psi^*\dfrac{\partial\psi}{\partial t}$, and use (8.14) and the result of Problem a.

Problem c Use Gauss' theorem to rewrite this expression as:

$$\frac{\partial}{\partial t}\int_V |\psi|^2 dV = \frac{i\hbar}{2m}\oint(\psi^*\nabla\psi - \psi\nabla\psi^*)\cdot d\mathbf{S}. \qquad (8.16)$$

Hint: Use a treatment similar to expression (8.10).

The left-hand side of this expression gives the time derivative of the probability that the particle is within the volume V. The only way the particle can enter or leave the volume is through the enclosing surface S. The right-hand side therefore describes the "flow" of probability through the surface S. More accurately, one can formulate this as the flux of the probability-density current. This means that expression (8.16) can be written as

$$\frac{\partial P_V}{\partial t} = -\int \mathbf{J}\cdot d\mathbf{S}, \qquad (8.17)$$

where the probability-density current $\mathbf{J}$ is given by:

$$\mathbf{J} = \frac{i\hbar}{2m}(\psi\nabla\psi^* - \psi^*\nabla\psi). \qquad (8.18)$$

As an example let us consider a plane wave:

$$\psi(\mathbf{r}, t) = A\,e^{i(\mathbf{k}\cdot\mathbf{r}-\omega t)}, \qquad (8.19)$$

where $\mathbf{k}$ is the wave vector and A an unspecified constant.

Problem d Show that the wavelength λ is related to the wave vector by the relation $\lambda = 2\pi/|\mathbf{k}|$. In which direction does the wave in (8.19) propagate?

Problem e Show that the probability-density current $\mathbf{J}$ for this wavefunction satisfies:

$$\mathbf{J} = \frac{\hbar\mathbf{k}}{m}|\psi|^2. \qquad (8.20)$$

This is a very interesting expression. The term $|\psi|^2$ gives the probability-density of the particle, while the probability-density current $\mathbf{J}$ physically describes the current of this probability-density. Since the probability-density current moves with the velocity of the particle, the remaining terms on the right-hand side of (8.20) must denote the velocity of the particle:

$$\mathbf{v} = \frac{\hbar\mathbf{k}}{m}. \qquad (8.21)$$

Since the momentum **p** is the mass times the velocity, (8.21) can also be written as $\mathbf{p} = \hbar \mathbf{k}$. This relation was proposed by de Broglie (1952), based on completely different arguments than those we have used here. Its discovery in 1924 was a major step in the development of quantum mechanics.

Problem f   Use this expression and the result of Problem e to compute your own wavelength while you are riding your bicycle. Are quantum-mechanical phenomena important when you ride your bicycle? Use your wavelength as an argument. Did you know you possessed a wavelength?

The connection between the wave function and momentum is discussed in detail by Berry (2013) who offers five different views of this connection.

9

Theorem of Stokes

In Chapter 8 we learned that in order to find the gravitational field of a mass we have to integrate the field equation (6.45) over the mass. Gauss' theorem can then be used to compute the integral of the divergence of the gravitational field. For the curl the situation is similar. In Section 7.5 we computed the magnetic field generated by a current in a straight infinite wire. The field equation

$$\nabla \times \mathbf{B} = \mu_0 \mathbf{J}, \tag{7.12}$$

was used to compute the field away from the wire. However, the solution (7.13) contained an unknown constant A. The reason for this is that the field equation (7.12) was only used outside the wire, where $\mathbf{J} = 0$. The treatment of Section 7.5 therefore did not provide us with the relation between the field $\mathbf{B}$ and its source $\mathbf{J}$. The only way to obtain this relation is to integrate the field equation. This implies we have to compute the integral of the curl of a vector field. The theorem of Stokes tells us how to do this.

9.1 Statement of Stokes' theorem

The theorem of Stokes is based on the principle that the curl of a vector field is the closed line integral of the vector field per unit surface area, as described in Section 7.1. Mathematically this statement is expressed by (7.2), which we write in a slightly different form as:

$$\oint_{dS} \mathbf{v} \cdot d\mathbf{r} = (\nabla \times \mathbf{v}) \cdot \hat{\mathbf{n}} \, dS = (\nabla \times \mathbf{v}) \cdot d\mathbf{S}. \tag{9.1}$$

The only difference with (7.2) is that in the above expression we have not aligned the z-axis with the vector $\nabla \times \mathbf{v}$. The infinitesimal surface is therefore not necessarily confined to the (x, y)-plane and the z-component of the curl is replaced by the component of the *curl* normal to the surface; hence, the occurrence of the terms $\hat{\mathbf{n}} dS$ in (9.1). Expression (9.1) holds for an infinitesimal surface area. However,

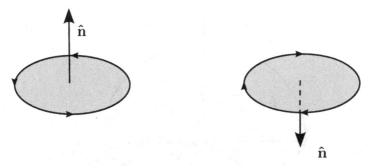

Figure 9.1 The relation between the sense of integration and the orientation of the surface.

this expression can immediately be integrated to give the surface integral of the curl over a finite surface S that is bounded by the curve C:

$$\oint_C \mathbf{v} \cdot d\mathbf{r} = \int_S (\nabla \times \mathbf{v}) \cdot d\mathbf{S}. \tag{9.2}$$

This result is known as the theorem of Stokes. The line integral on the left-hand side is over the curve that bounds the surface S. A proper derivation of Stokes' theorem can be found in Marsden and Tromba (2003).

A line integration along a closed surface can be carried out in two directions. What is the direction of the line integral on the left-hand side of Stokes' theorem (9.2)? To see this, we have to realize that Stokes' theorem was ultimately based on (7.2). The orientation of the line integration used in that expression is defined in Figure 7.2. Line integration is in the counterclockwise direction and the z-axis points out of the paper, which implies that the vector $d\mathbf{S}$ also points out of the paper. *This means that in Stokes' theorem the sense of the line integration and the direction of the surface vector $d\mathbf{S}$ are related through the rule for a right-handed screw. This orientation is indicated in Figure 9.1.*

There is something strange about Stokes' theorem. If we define a curve C over which we carry out the line integration, we can define many different surfaces S that are bounded by the same curve C. Apparently, the surface integral on the right-hand side of Stokes' theorem does not depend on the specific choice of the surface S as long as it is bounded by the curve C.

Let us verify this property with an example, where the vector field $\mathbf{v} = r\hat{\boldsymbol{\varphi}}$, and the curve C used for the line integral is a circle in the (x, y)-plane with radius R (Figure 9.2).

Problem a Compute the line integral on the left-hand side of (9.2) by direct integration.

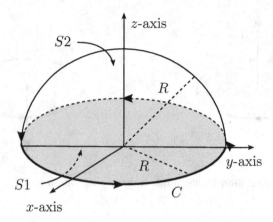

Figure 9.2   Definition of the geometric variables for Problem a.

Problem b   Compute the surface integral on the right-hand side of (9.2) by integrating over a disc of radius R in the (x, y)-plane (the surface S_1 in Figure 9.2).

Problem c   Compute the surface integral also by integrating over the upper half of a sphere with radius R (the surface S_2 in Figure 9.2).

Problem d   Verify that the three integrals are identical.

It is actually not difficult to prove that the surface integral in Stokes' theorem is independent of the specific choice of the surface S as long as it is bounded by the same contour C. Consider Figure 9.3 where the two surfaces S_1 and S_2 are bounded by the same contour C. We want to show that the surface integral of $\nabla \times \mathbf{v}$ is the same for the two surfaces, that is that:

$$\int_{S_1} (\nabla \times \mathbf{v}) \cdot d\mathbf{S} = \int_{S_2} (\nabla \times \mathbf{v}) \cdot d\mathbf{S}. \tag{9.3}$$

We can form a closed surface S by combining the surfaces S_1 and S_2.

Problem e   Show that (9.3) is equivalent to the condition

$$\oint_S (\nabla \times \mathbf{v}) \cdot d\mathbf{S} = 0, \tag{9.4}$$

where the integration is over the closed surfaces defined by the combination of S_1 and S_2. Pay particular attention to the signs of the different terms.

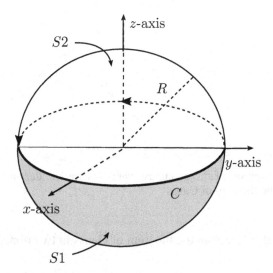

Figure 9.3   Two surfaces that are bounded by the same contour C.

Problem f   Use Gauss' theorem to convert (9.4) to a volume integral and show that the integral is indeed identical to zero. In doing so you need to use the identity $\nabla \cdot (\nabla \times \mathbf{v}) = 0$ (or in another notation div curl $\mathbf{v} = 0$). Make sure you can derive this identity.

The result you obtained in Problem c implies that condition (9.3) is indeed satisfied and that in the application of Stokes' theorem you can choose *any* surface as long as it is bounded by the contour over which the line integration is carried out. This is a very useful result, because the surface integration can often be simplified by choosing the surface carefully.

9.2 Stokes' theorem from the theorem of Gauss

Stokes' theorem is concerned with surface integrations. Since the curl is intrinsically a three-dimensional vector, Stokes' theorem is inherently related to three space dimensions. However, if we consider a vector field that depends only on the coordinates x and y ($\mathbf{v} = \mathbf{v}(x, y)$) with a vanishing component in the z-direction ($v_z = 0$), then $\nabla \times \mathbf{v}$ points along the z-axis.

Problem a   For a contour C confined to the (x, y)-plane, verify that for $\mathbf{v}$ Stokes' theorem (9.2) takes the form:

$$\oint_C (v_x dx + v_y dy) = \int_S \left(\partial_x v_y - \partial_y v_x \right) dx dy. \qquad (9.5)$$

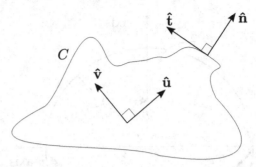

Figure 9.4   Definition of the geometric variables for the derivation of Stokes' theorem from the theorem of Gauss.

This result can be derived from the theorem of Gauss in two dimensions.

Problem b   Show that Gauss' theorem (8.1) for a vector field **u** in two dimensions can be written as

$$\oint_C (\mathbf{u} \cdot \hat{\mathbf{n}})ds = \int_S (\partial_x u_x + \partial_y u_y)dxdy, \tag{9.6}$$

where the unit vector $\hat{\mathbf{n}}$ is perpendicular to the curve C (see Figure 9.4) and where ds denotes the integration over the arclength of the curve C.

To derive the special form of Stokes' theorem (9.5) from Gauss' theorem (9.6), we have to define the relation between the vectors **u** and **v**. Let the vector **u** follow from **v** by a clockwise rotation over 90 degrees, as in Figure 9.4.

Problem c   Show that

$$v_x = -u_y \quad \text{and} \quad v_y = u_x. \tag{9.7}$$

We now define the unit vector $\hat{\mathbf{t}}$ to be directed along the curve C, as drawn in Figure 9.4. Since a rotation is an orthonormal transformation, the inner product of two vectors is invariant for a rotation over 90 degrees: $(\mathbf{u} \cdot \hat{\mathbf{n}}) = (\mathbf{v} \cdot \hat{\mathbf{t}})$.

Problem d   Verify this by expressing the components of $\hat{\mathbf{t}}$ in the components of $\hat{\mathbf{n}}$ and by using (9.7). The change in the position vector along the curve C is given by

$$d\mathbf{r} = \hat{\mathbf{t}}ds = \begin{pmatrix} dx \\ dy \end{pmatrix}. \tag{9.8}$$

Problem e   Use these results to show that (9.5) follows from (9.6).

What you have shown here is that Stokes' theorem is identical to the theorem of Gauss for two spatial dimensions, for the vector fields of this section.

9.3 Magnetic field of a current in a straight wire

We now return to the problem of the generation of the magnetic field induced by a current in an infinite straight wire, introduced in Section 7.5. Because of the cylindrical symmetry of the problem, we know that the magnetic field is oriented in the direction of the unit vector $\hat{\varphi}$ and that the field only depends on the distance $r = \sqrt{x^2 + y^2}$ to the wire:

$$\mathbf{B} = B(r)\hat{\varphi}. \tag{9.9}$$

In Section 7.5, we found the field up to an unknown constant A, but with Stokes' theorem we are ready to find this constant. We start by integrating the field equation $\nabla \times \mathbf{B} = \mu_0 \mathbf{J}$ over a disc of radius r perpendicular to the wire shown in Figure 9.5. When the disc is larger than the thickness of the wire, the surface integral of $\mathbf{J}$ gives the electric current I through the wire: $I = \int \mathbf{J} \cdot d\mathbf{S}$.

Problem a   Use these results and Stokes' theorem to show that

$$\mathbf{B} = \frac{\mu_0 I}{2\pi r}\hat{\varphi}. \tag{9.10}$$

We now have a relation between the magnetic field and the current that generates the field; hence, the constant A in expression (7.13) is now determined. Note that the magnetic field depends only on the total current through the wire: it does

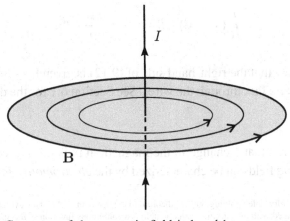

Figure 9.5   Geometry of the magnetic field induced by a current in a straight infinite wire.

not depend on the distribution of the electric current density **J** within the wire as long as the electric current density exhibits cylindrical symmetry. This situation is analogous to the analysis in Section 8.2 where we showed that a spherically symmetric mass distribution gives the same gravitational field outside the mass, regardless of the mass distribution within the body. Both for the magnetic field and the gravitational field, information about the detailed arrangement of their sources, either current or mass, does not influence the field outside the source region.

9.4 Magnetic induction and Lenz' law

The theory in the previous section deals with the generation of a magnetic field by a current. A magnet placed in this field will experience a force exerted by the magnetic field. This force is essentially the driving force in electric motors; using an electrical current that changes with time a time-dependent magnetic field is generated that exerts a force on magnets attached to a rotation axis.

In this section we study the reverse effect: What is the electrical field generated by a magnetic field that changes with time? In a dynamo, a moving part (e.g., a bicycle wheel) moves a magnet. This creates a time-dependent electric field. This process is called magnetic induction and is described by the following Maxwell equation (see Jackson, 1998):

$$\nabla \times \mathbf{E} = -\frac{\partial \mathbf{B}}{\partial t}. \tag{9.11}$$

To fix our mind, let us consider a wire with endpoints A and B, see Figure 9.6. The direction of the magnetic field is indicated in this figure. In order to find the electric field induced in the wire, integrate (9.11) over the surface enclosed by the wire[1]

$$\int_S (\nabla \times \mathbf{E}) \cdot d\mathbf{S} = -\int_S \frac{\partial \mathbf{B}}{\partial t} \cdot d\mathbf{S}. \tag{9.12}$$

Problem a   Show that the right-hand side of (9.12) is given by $-\partial\Phi/\partial t$, where Φ is the magnetic flux through the wire. (See Section 6.1 for the definition of the flux.)

We have discovered that a change in the magnetic flux is the source of an electric field. The resulting field can be characterized by the *electromotive force* $\mathcal{E}_{AB}$, which

---

[1]   You may feel uncomfortable applying Stokes' theorem to the open curve AB. However, remember that the electric field associated with the varying magnetic field is a continuous function of the space variables. The electric field is therefore the same at the points A and B where the wire is open. The line integral over the open curve C is therefore identical to the line integral along the closed contour.

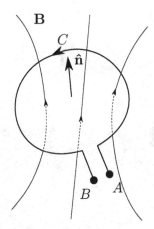

Figure 9.6   A wire-loop in a magnetic field **B**.

is a measure of the work done by the electric field on a unit charge when it moves
from point A to point B, as in Figure 9.6:

$$\mathcal{E}_{AB} \equiv \int_A^B \mathbf{E} \cdot d\mathbf{r}. \tag{9.13}$$

In fact, the word "electromotive force" is a misnomer. The electrical field is
related to an electric potential by the relation $\mathbf{E} = -\nabla V$. The line integral in the
right-hand side of equation (9.13) results in an electric potential. Therefore it would
be more appropriate to call $\mathcal{E}_{AB}$ the electromotive potential.

Problem b   Apply Stokes' theorem to the left-hand side of expression (9.12) to
show that the electromotive force satisfies

$$\mathcal{E}_{AB} = -\frac{\partial \Phi_{\mathbf{B}}}{\partial t}. \tag{9.14}$$

Problem c   Because of the electromotive force, an electric current will flow
through the wire. Determine the direction of this electric current. Show that
this electric current generates a magnetic field that opposes the change in
the magnetic field that generates the current. You learned in Section 9.3 the
direction of the magnetic field that is generated by an electric current in a
wire.

What we have discovered in Problem c is called *Lenz' law*, which states that induction currents lead to a secondary magnetic field that opposes the change in the
primary magnetic field that generates the electric current. This implies that coils in
electrical systems exhibit a certain inertia in the sense that they resist changes in

Figure 9.7   The TC1 seismograph consists of a magnet suspended from a slinky toy that hangs inside a coil. The system is damped with another magnet – below the first one – inside a short copper pipe.

the magnetic field that passes through the coil. The amount of inertia is described by a quantity called the inductance L. This quantity plays a role similar to mass in classical mechanics because the mass of a body also describes how strongly a body resists changing its velocity when an external force is applied.

The seismometer in Figure 9.7 functions on the principles discussed above: ground motion from an earthquake moves a magnet on the end of a slinky that is suspended in a coil. According to equation (9.11), this induces a current in the coil proportional to the ground motion. As discussed, the inductance of the coil provides resistance to the magnet movement, but this is not enough to stop the seismometer from ringing long after the ground motion has stopped. In the past, seismometers would further damp the motion by dragging it through a viscous fluid like oil. The TC1 seismometer in Figure 9.7 has a second magnet inside a copper tube. When set in motion, the electromotive force results in currents (so-called Eddy currents) in the copper tube that resist the motion, damping the whole system (van Wijk et al., 2013). Data from the TC1 seismometer is displayed in Figure 19.12.

9.5 Aharonov–Bohm effect

It was shown in Section 6.4 that because of the absence of magnetic monopoles the magnetic field is source-free: $(\nabla \cdot \mathbf{B}) = 0$. In electromagnetism one often expresses

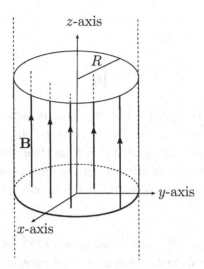

Figure 9.8   Geometry of the magnetic field.

the magnetic field as the curl of a vector field **A**:

$$\mathbf{B} = \nabla \times \mathbf{A}. \tag{9.15}$$

The advantage of writing the magnetic field this way is that for *any* field **A** the magnetic field satisfies $(\nabla \cdot \mathbf{B}) = 0$, because $\nabla \cdot (\nabla \times \mathbf{A}) = 0$, as you saw in Problem c of Section 9.1.

The vector field **A** is called the *vector potential*. The reason for this name is that it plays a role similar to the electric potential V. The electric and the magnetic fields follow from V and **A** respectively by differentiation: $\mathbf{E} = -\nabla V$ and $\mathbf{B} = \nabla \times \mathbf{A}$. The electric field vanishes when the potential V is constant. Since this constant may be nonzero, this implies that the electric field may vanish but that the potential is nonzero. Similarly, the vector potential can be nonzero (and variable) in parts of space where the magnetic field vanishes. For example, consider a magnetic field with cylindrical symmetry along the z-axis which is constant for $r < R$ and vanishes for $r > R$ (Figure 9.8):

$$\mathbf{B} = \begin{cases} B_0 \hat{\mathbf{z}} & \text{for} \quad r < R \\ 0 & \text{for} \quad r > R. \end{cases} \tag{9.16}$$

Because of the cylindrical symmetry, the vector potential **A** is a function of the distance r to the z-axis only and does not depend on z or φ.

Problem a   Use expression (7.17) to show that a vector potential of the form

$$\mathbf{A} = f(r)\hat{\boldsymbol{\varphi}} \tag{9.17}$$

gives a magnetic field in the required direction. Derive that $f(r)$ satisfies the following differential equation:

$$\frac{1}{r}\frac{\partial}{\partial r}(rf(r)) = \begin{cases} B_0 & \text{for} \quad r < R \\ 0 & \text{for} \quad r > R. \end{cases} \tag{9.18}$$

This differential equation for $f(r)$ can be immediately integrated. After integration, two integration constants are present. These constants follow from the requirements that the vector potential is continuous at $r = R$ and that $f(r = 0) = 0$. This latter requirement is needed because the direction of the unit vector $\hat{\boldsymbol{\varphi}}$ is undefined on the z-axis, where $r = 0$. The vector potential therefore only has a unique value at the z-axis when $f(r = 0) = 0$.

Problem b   Integrate the differential equation (9.18) subject to the boundary condition for $f(r = 0)$ to derive that the vector potential is given by

$$\mathbf{A} = \begin{cases} \dfrac{1}{2}B_0 r\hat{\boldsymbol{\varphi}} & \text{for} \quad r < R \\[2mm] \dfrac{1}{2}B_0\dfrac{R^2}{r}\hat{\boldsymbol{\varphi}} & \text{for} \quad r > R. \end{cases} \tag{9.19}$$

The importance of this expression is that although the magnetic field is only nonzero for $r < R$, the vector potential – and its gradient – is nonzero everywhere in space! The vector potential is thus much more nonlocal than the magnetic field. This leads to a very interesting effect in quantum mechanics, called the Aharonov–Bohm effect. Before introducing this effect we need to know more about quantum mechanics. As you have seen in Section 8.4, the behavior of atomic "particles" such as electrons is paradoxically described by a wave. The properties of this wave are described by Schrödinger's equation (8.14). When different waves propagate in the same region of space, interference can occur. In some parts of space the waves may enhance each other (constructive interference), while in other parts the waves cancel each other (destructive interference). This is observed for "particle waves" when electrons are sent through two slits and then detected on a screen behind these slits, as depicted in the left-hand panel of Figure 9.9. You might expect the electrons to propagate like bullets along straight lines. In that case, electrons would be detected only at the intersection of the screen with these straight lines. Instead, experiments display a pattern of fringes on the screen that is caused by the constructive and destructive interference of the electron waves. This interference pattern is sketched in Figure 9.9 on the right-hand side of the screens. This remarkable confirmation of the wave property of particles is described clearly by Feynman (1967). The situation is even more remarkable when one sends the electrons through the slits "one-by-one" so that only one electron passes through the slits at a time. In

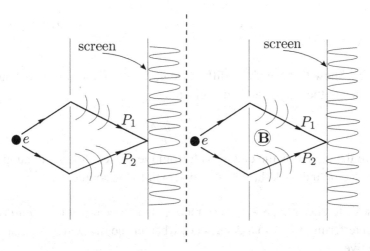

Figure 9.9 Experiment in which electrons travel through two slits and are detected on a screen behind the slits. The resulting interference pattern is sketched. The experiment without a magnetic field is shown on the left; the experiment with a magnetic field is shown on the right. Note the shift in the maxima and minima of the interference pattern between the two experiments.

that case one sees a dot at the detector for each electron. However, after many particles have arrived at the detector this pattern of dots forms the interference pattern of the waves (Silverman, 1993).

Let us consider the same experiment, but with a magnetic field given by (9.16) placed between the two slits. When the electrons propagate along the paths P_1 or P_2, they do not pass through this field; hence, one would expect that the electrons would not be influenced by this field and that the magnetic field would not change the observed interference pattern at the detector. However, observations reveal the magnetic field *does* change the interference pattern at the detector (see Silverman, 1993, for example). This surprising effect is called the Aharonov–Bohm effect (Aharonov and Bohm, 1959).

To understand this effect, we should note that a magnetic field in quantum mechanics leads to a phase shift of the wavefunction. If the wavefunction in the absence of a magnetic field is given by $\psi(\mathbf{r})$, the wavefunction in the presence of the magnetic field is given by $\psi(\mathbf{r}) \times \exp[(ie/\hbar c) \int_P \mathbf{A} \cdot d\mathbf{r}]$ (Sakurai, 1978). In this expression $\hbar$ is Planck's constant (divided by 2π), c is the speed of light, and $\mathbf{A}$ is the vector potential associated with the magnetic field. The integration is over the path P from the source of the particles to the detector. Consider now the waves that interfere in the two-slit experiment in the right-hand panel of Figure 9.9. The wave that travels through the upper slit experiences a phase shift $(e/\hbar c) \int_{P_1} \mathbf{A} \cdot d\mathbf{r}$, where the integration is over the path P_1 through the upper slit. The wave that travels through the lower slit obtains a phase shift $(e/\hbar c) \int_{P_2} \mathbf{A} \cdot d\mathbf{r}$ where the path P_2 runs through the lower slit.

Problem c   Show that the phase difference $\delta\varphi$ between the two waves due to the presence of the magnetic field is given by

$$\delta\varphi = \frac{e}{\hbar c} \oint_P \mathbf{A} \cdot d\mathbf{r}, \qquad (9.20)$$

where the path P is the closed path from the source through the upper slit to the detector and back through the lower slit to the source.

This phase difference affects the interference pattern, because it is the *relative* phase between interfering waves that determines whether the interference is constructive or destructive.

Problem d   Use Stokes' theorem and expression (9.15) to show that the phase difference can be written as

$$\delta\varphi = \frac{e\Phi_{\mathbf{B}}}{\hbar c}, \qquad (9.21)$$

where $\Phi_{\mathbf{B}}$ is the magnetic flux through the area enclosed by the path P.

This expression shows that the phase shift between the interfering waves is proportional to the magnetic field enclosed by the paths of the interfering waves, *despite the fact that the electrons never move through the magnetic field* **B**. Mathematically the reason for this surprising effect is that the vector potential is nonzero throughout space even when the magnetic field is confined to a small region of space, as we saw in expression (9.19). However, this explanation is purely mathematical and does not appear to be intuitive. These findings have led to speculation that the vector potential is actually a more "fundamental" quantity than the magnetic field (Silverman, 1993).

9.6 Wingtips vortices

If you have watched aircraft closely, you may have noticed that sometimes a stream of condensation is left behind by the wingtips (Figure 9.10). This is a different condensation trail than the thick contrails created by the engines. The condensation trails that start at the wingtips are due to a vortex (a spinning motion of the air) that is generated at the wingtips. This vortex is called the wingtip vortex. In this section we use Stokes' theorem to see that this wingtip vortex is closely related to the lift that is generated by the airflow along a wing.

Figure 9.10   Vortices trailing from the wingtips of a Boeing 727. Figure courtesy of NASA.

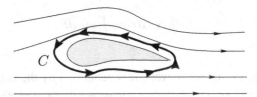

Figure 9.11   Sketch of the flow along an airfoil. The wing is shown in gray the contour C is shown by the thick solid line.

Let us first consider the airflow along a wing, sketched in Figure 9.11. The air traverses a longer path along the upper part of the wing than along the lower part. Because of the curved upper side of the wing, the velocity of the airstream along the upper part of the wing is larger than the velocity along the lower part.

Problem a   The *circulation* is defined as the line integral $\oint_C \mathbf{v} \cdot d\mathbf{r}$ of the air velocity along a curve. Is the circulation positive or negative for the curve C in Figure 9.11 for the indicated sense of integration? Use that the air moves faster over the upper side of the wing than under the lower side.

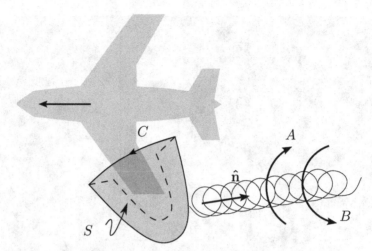

Figure 9.12   Geometry of the surface S and the wingtip vortex for an aircraft seen from above. The surface S encloses the wingtip of the aircraft. The edge of this surface is the same contour C as drawn in the previous figure.

Problem b   Consider now the surface S shown in Figure 9.12. Use Stokes' theorem and expression (7.4) for the vorticity $\boldsymbol{\omega}$ to show that the circulation satisfies

$$\oint_C \mathbf{v} \cdot d\mathbf{r} = \int_S \boldsymbol{\omega} \cdot d\mathbf{S}. \qquad (9.22)$$

This expression implies that whenever lift is generated by the net circulation of the air along the contour C around the wing, the integral of the vorticity over a surface that envelopes the wingtip is nonzero. The vorticity depends on the (spatial) derivative of the velocity. Since the flow is relatively smooth along the wing, the derivative of the velocity field is largest near the wingtips. Therefore, expression (9.22) states that vorticity is generated at the wingtips. As shown in Section 7.3, the vorticity is a measure of the local vortex strength. A wing can only produce lift when the circulation along the curve C is nonzero. The above reasoning implies that wingtip vortices are unavoidably associated with the lift produced by an airfoil.

Problem c   Consider the wingtip vortex shown in Figure 9.12. You obtained the sign of the circulation $\oint_C \mathbf{v} \cdot d\mathbf{r}$ in Problem a. Does this imply that the wingtip vortex rotates in the clockwise direction A of Figure 9.12 or in the counterclockwise direction B? Use equation (9.22) in your argumentation. You may assume that the vorticity is mostly concentrated at the trailing edge of the wingtips, as in Figure 9.12.

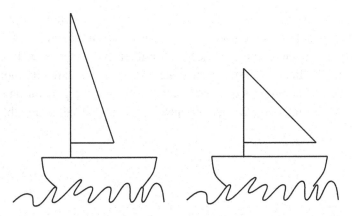

Figure 9.13 Two boats carrying sails with different aspect ratios.

The wingtip vortex carries kinetic energy. Since this energy is drawn from the moving aircraft, it is associated with the drag on the aircraft; this is called *induced drag* (Kermode et al., 1996). Modern aircraft often have wingtips that are turned upwards. These *winglets* modify the vorticity at the wingtip in such a way that the induced drag on the aircraft is reduced. In nature, a goose in a flock with the typical "V" formation reduces the wingtip vortex of the bird directly in front, and it has been proposed that this same goose "rides" the upward swing of the remaining wingtip vortex (Bajec and Heppner, 2009).

Just like aircraft, sailing boats suffer from an energy loss due to a vortex that is generated at the upper part of the sail (see the discussion of Marchaj, 2000); a sail can be considered to be a "vertical wing." Two boats shown in Figure 9.13 have sails with the same surface area but with different aspect ratios. The boat on the left will, in general, sail faster. The reason for this is that for the two boats the difference in the wind speed along the two sides of the sail will be roughly identical. This means that for the boat on the right the circulation $\oint_C \mathbf{v} \cdot d\mathbf{r}$ will be larger than that for the boat on the left, simply because the integration contour C is longer. The vorticity generated at the top of the sail of the boat on the right is therefore larger than for the boat on the left. Since this resulting "sailtip" vortex leads to a dissipation of energy, the sail of the boat on the left has a higher efficiency. For the same reason, planes that have to fly with a minimal energy loss – gliders, for instance – have thin and long wings. In contrast to this, planes that may waste energy in order to fly at a very high speed (such as the Concorde) have wings of a very different shape. Birds follow the same rule: birds that fly relatively slowly but that can glide efficiently over long distances – for example, the albatross – have long and thin wings, whereas birds that do not need to fly efficiently (such as a crow) have shorter and thicker wings.

The explanation that lift is generated by the increased airflow over the curved part of a wing is only partly correct. Planes can fly upside down and maintain lift, and some planes, such as the Lockheed Starfighter, fly perfectly well with wings that are essentially flat. The issue of the generation of lift is in reality more complicated than the theory presented here (Craig, 1997). Yet the generation of wingtip vortices and their impact on the performance of wings and sail is tangible.

10

The Laplacian

The Laplacian of a function consists of a special combination of the second partial derivatives of that function. Before we introduce this quantity, the relation between the second derivative and the curvature of a function is established in Section 10.1. The Laplacian is introduced in Section 10.3 using the physical example of a soap film that minimizes its surface area. The analysis used for this example is introduced in Section 10.2 where a proof is given that the shortest distance between two points is a straight line. The concept of the Laplacian is used in Section 10.5 to study the stability of matter, while in Section 10.6 the implications for the initiation of lightning is considered. Finally, the Laplacian in cylindrical and spherical coordinates is derived, and this is used in Section 10.8 to derive an averaging integral for harmonic functions.

10.1 Curvature of a function

Let us consider a function $f(x)$ that has an extremum. We are free to choose the origin of the x-axis, and the origin is chosen here at the location of the extremum. This means that the function $f(x)$ is stationary at the location $x = 0$ (Figure 10.1). The behavior of the function near its maximum can be studied using the Taylor series:

$$f(x) = \sum_{n=0}^{\infty} \frac{x^n}{n!} \frac{d^n f}{dx^n}(x = 0) = f(0) + x \frac{df}{dx}(x = 0) + \frac{x^2}{2} \frac{d^2 f}{dx^2}(x = 0) + \cdots .$$

$$(3.11)$$

The point $x = 0$ is a maximum or a minimum that implies by definition that the first derivative vanishes at $x = 0$, so that the function behaves near the extremum as

$$f(x) = f(0) + \frac{1}{2} \frac{d^2 f}{dx^2}(x = 0) \, x^2 + \cdots .$$

$$(10.1)$$

When the first derivative vanishes, the function can have either a maximum or a minimum. In more dimensions, the function can have a minimum in one direction

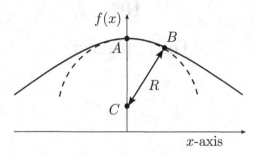

Figure 10.1   Definition of the geometric variables in the computation of the radius of curvature of a function.

and a maximum in another direction. For this reason the terminology "extremum" is not very accurate. Instead, one refers to a point at which the first derivative vanishes as a *stationary* point. This point is stationary, because according to (10.1) the function does not vary to first order with the independent variable. In more than one dimension a stationary point is defined by the requirement that the first partial derivatives with respect to *all* variables vanish.

The property of stationary points that the first derivative vanishes corresponds with the fact that at an extremum the slope of the function vanishes. The dominant behavior of the function near its stationary point is given by its *curvature*, described by the x^2-term in (10.1). In order to characterize this curvature one can define a circle as shown in Figure 10.1 that touches the function $f(x)$ near its extremum. This *tangent-circle* is shown by a dashed line in Figure 10.1. The radius R of this circle measures the curvature of the function at its extremum. For this reason, R is called the *radius of curvature*.

Problem a   Convince yourself that a small radius of curvature R implies a large curvature of $f(x)$ and a large radius R corresponds to a small curvature of $f(x)$. What is the shape of $f(x)$ when the radius of curvature tends to infinity $(R \to \infty)$? Hint: Draw this situation.

The radius of curvature can be determined using the points A, B, and C in Figure 10.1.

Problem b   Use Figure 10.1 to show that these points have the following coordinates in the (x, y)-plane

$$\mathbf{r}_A = \begin{pmatrix} 0 \\ f(0) \end{pmatrix}, \quad \mathbf{r}_B = \begin{pmatrix} x \\ f(x) \end{pmatrix}, \quad \mathbf{r}_C = \begin{pmatrix} 0 \\ f(0) - R \end{pmatrix}. \qquad (10.2)$$

Problem c   Since A and B are located on the same circle with C as center, the distance AC equals the distance BC. Use this condition, the coordinates (10.2), and the Taylor expansion (10.1) to derive the following expression

$$x^2 + \left(\frac{1}{2}f''x^2 + R\right)^2 = R^2, \qquad (10.3)$$

where f'' denotes the second derivative of the extremum: $f'' = d^2 f/dx^2$.

By expanding the square one can obtain an expression for the radius of curvature R. However, it is only necessary to account for terms up to order x^2. The reason for this is that in the Taylor expansion (10.1) the third-order terms – and higher – have been neglected. This is because the tangent-circle in Figure 10.1 approximates the function $f(x)$ well, only near its maximum. It is therefore not consistent to retain terms of third and higher order in the remainder of the calculation.

Problem d   Use this to derive from (10.3) the following relation between the radius of curvature and the second derivative

$$R = \frac{-1}{f''}. \qquad (10.4)$$

This expression relates the radius of the tangent-circle to the second derivative. The approximation of a function by its tangent-circle plays an important role in reflection seismology where it is used to derive that the *15-degrees approximation* accounts accurately for near-vertical wave propagation in the Earth (Claerbout, 1985; Yilmaz, 2001).

Of course, a stationary point can be either a minimum or a maximum: these two situations are shown in Figure 10.2.

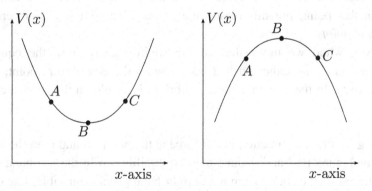

Figure 10.2   The behavior of a function near a minimum and a maximum.

Problem e   Use (10.4) to show that for the curve in the left-hand panel (a minimum) the radius of curvature is negative and that for the curve in the right-hand panel (a maximum) the radius of curvature is positive.

The result in the last problem is summarized by the following property of a stationary point:

$$
\left.
\begin{aligned}
\text{when } f(x) \text{ is a minimum: } & \frac{\partial^2 f}{\partial x^2} > 0; \\[2ex]
\text{when } f(x) \text{ is a maximum: } & \frac{\partial^2 f}{\partial x^2} < 0.
\end{aligned}
\right\}
\tag{10.5}
$$

The curve in the left-hand panel of Figure 10.2 is called *concave*, while the curve in the right-hand panel is called *convex*. This implies that a concave curve has a negative radius of curvature and a convex curve has a positive radius of curvature.

   The distinction between these different kinds of extrema is crucial to the stability properties of physical systems. Suppose that the function denotes a potential $V(x)$, then according to Section 5.4 the force associated with this potential is

$$
\mathbf{F} = -\nabla V.
\tag{5.32}
$$

In one dimension this expression is given by

$$
F(x) = -\frac{dV}{dx}.
\tag{10.6}
$$

Problem f   Show that at the points B in the panels of Figure 10.2 the force vanishes.

Suppose a particle that is influenced by the potential is at rest at one of the points B in Figure 10.2. Since the force vanishes at these points the particle will remain forever at that point. For this reason the points where $\nabla V = 0$ are called the *equilibrium points*.

   However, when we move the particle slightly away from the equilibrium points, the force may either push it back toward the equilibrium point, or push it further away. In the first case the equilibrium is *stable*; in the second case it is *unstable*.

Problem g   Draw the direction of the force at the points A and C in the two panels of Figure 10.2 and deduce that the equilibrium in the left-hand panel is stable, while the equilibrium in the right-hand panel is unstable. Use (10.5) to show that:

$$\left. \begin{array}{l} \text{the equilibrium is stable when } \dfrac{\partial^2 V}{\partial x^2} > 0; \\[12pt] \text{the equilibrium is unstable when } \dfrac{\partial^2 V}{\partial x^2} < 0. \end{array} \right\} \qquad (10.7)$$

If you find these arguments difficult, you can think of a ball that can move along the curves in the panels of Figure 10.2 in a gravitational field. At point B the ball is in an area that is flat. If it does not move, it will remain at that point forever. When the ball in the left-hand panel moves away from the equilibrium point, it rolls uphill and the gravitational force will send it back toward the equilibrium point B. However, when the ball in the right-hand panel moves away from point B, it will roll downhill further from the equilibrium point. The equilibrium at point B in the right-hand panel is unstable. These properties are all related to the second derivative of a function.

10.2 Shortest distance between two points

In this section we give a proof that the shortest distance between two points is a straight line. This mathematical problem corresponds to finding the shape of a rubber band that is spanned between two fixed points, because the tension in the rubber band minimizes the length of the rubber band. This may appear to be a trivial problem, but it is shown here because it sets the stage for the Laplacian. It also gives a brief introduction to *variational calculus*. This topic is treated in more detail in Chapter 27. The geometry of the problem is shown in Figure 10.3. Points A and B in the (x, y)-plane are the endpoints to the curve $y = h(x)$. We are looking for

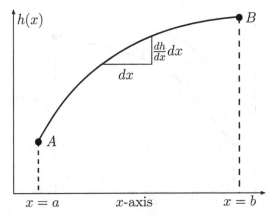

Figure 10.3 The relation between the derivative of a function and the arc-length of the corresponding curve.

the function $h(x)$ that describes the curve with the smallest length. We first need to determine the length of the curve given a certain shape $h(x)$. Consider an increment dx of the x-variable.

Problem a   Use the first-order Taylor expansion (3.18) to show that this increment corresponds to an increment $dy = (dh/dx)\ dx$ of the y-variable. Use this result to derive that the length L of the curve is given by:

$$L[h] = \int_a^b \sqrt{1 + h_x^2}\ dx, \tag{10.8}$$

where $h_x = dh/dx$.

Note that the length of the curve depends on the shape $h(x)$ of the curve; for this reason the notation $L[h]$ is used.

We want to find the function $h(x)$ that minimizes the length L. Unfortunately we cannot simply differentiate L with respect to h, because $h(x)$ is a function rather than a variable. We can, however, use the concept of stationarity as introduced in Section 10.1. When $L[h]$ is minimized, the length of the curve does not change to first order when $h(x)$ is perturbed. Consider Figure 10.4 in which a perturbation $\epsilon(x)$ is added to the original function $h(x)$. Since the endpoints of the curve are fixed, the perturbation is required to vanish at the endpoints:

$$\epsilon(a) = \epsilon(b) = 0. \tag{10.9}$$

To solve the problem, we need to find the change δL in the length of the curve that is caused by the perturbation $\epsilon(x)$.

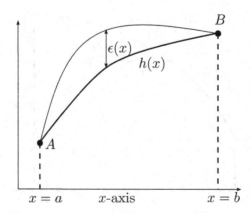

Figure 10.4   The unperturbed function $h(x)$ and the perturbation $\epsilon(x)$ that vanishes at the endpoint of the interval.

Problem b Replace h in (10.8) by $h + \epsilon$, carry out a first-order Taylor expansion of the integrand with respect to ϵ and use this to show that the perturbation of the length of the curve is given by

$$\delta L[h] = \int_a^b \frac{h_x \epsilon_x}{\sqrt{1 + h_x^2}} \, dx. \tag{10.10}$$

For the analysis to be as transparent as possible, we make an approximation to (10.10) by restricting ourselves to the special case for which the slope of the curve is small. Since $h_x = dh/dx$ denotes the slope of the curve, this corresponds to the condition $h_x \ll 1$. In this case the term h_x in the denominator can be ignored, and the perturbation is given by

$$\delta L[h] = \int_a^b h_x \epsilon_x dx. \tag{10.11}$$

We show in Section 27.4 that the results of this section also hold when the approximation $h_x \ll 1$ is not used; the only reason for making the approximation is that we do not want the principle of variational calculus to be hidden by analytical complexity.

Expression (10.11) gives the first-order perturbation of the length with respect to $\epsilon(x)$. The condition of stationarity tells us that for the shortest curve, (10.11) must equal zero for *any* small perturbation $\epsilon(x)$. We have not yet used the constraint (10.9), however, which states that the endpoints of the curve are fixed.

Problem c Carry out an integration by parts of (10.11) and use the constraint (10.9) to show that the first-order perturbation is given by

$$\delta L[h] = -\int_a^b \frac{d^2 h}{dx^2} \epsilon \, dx. \tag{10.12}$$

For the function $h(x)$ that minimizes the length of the curve this integral must vanish for *all* perturbations $\epsilon(x)$. This is the case when

$$\frac{d^2 h}{dx^2} = 0. \tag{10.13}$$

Problem d Let the y-coordinates of the points A and B be y_A and y_B, respectively. Integrate (10.13) subject to the condition that the curve goes through the points A and B and convince yourself that the solution is given by a straight line.

In the following section this minimization problem is generalized to two dimensions.

10.3 Shape of a soap film

In the previous section you derived that the shortest "curve" between two points has no curvature; hence, it is a straight line. Let us now consider a two-dimensional surface in three dimensions. We will seek the surface that minimizes the surface area, where the locations of the edges of the surface are fixed. Physically this problem describes the shape of a soap film, suspended in a wire frame. The surface tension of the soap film minimizes its surface area, subject to the constraint that the edges of the soap film are fixed. The shape of the soap film is described by giving the z-coordinate as a function of the x- and y-coordinates: $z = h(x, y)$.

At first sight one might guess that the soap film will be a plane, because any deviations of the soap film from this plane will increase its surface area. Since the curvature of a plane vanishes in the x-direction as well as in the y-direction, the analogy with expression (10.13) suggests that the soap film satisfies the following equations: $\partial^2 h/\partial x^2 = \partial^2 h/\partial y^2 = 0$. That this is not generally true can be seen in the soap film shown in Figure 10.5. In this example the wire frame that defines the edge of the soap film is not confined to a plane. This results in a soap film that

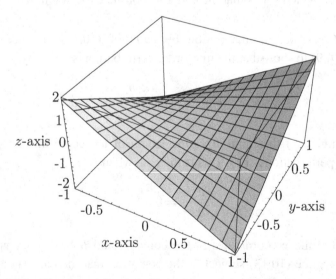

Figure 10.5   The shape of a soap film whose edges are fixed at the outer edges of the box.

is curved and hence it is not possible that both $\partial^2 h/\partial x^2$ and $\partial^2 h/\partial y^2$ are equal to zero. In this section we derive an equation for the shape of the soap film.

The position of a point on the soap film is given by the following position vector:

$$\mathbf{r} = \begin{pmatrix} x \\ y \\ h(x, y) \end{pmatrix}. \tag{10.14}$$

In Problem e of Section 4.4 you showed that the surface area dS of a sphere that corresponds to increments $d\theta$ and $d\varphi$ of the angles on the sphere is given by

$$dS = \left| \frac{\partial \mathbf{r}}{\partial \theta} \times \frac{\partial \mathbf{r}}{\partial \varphi} \right| d\theta d\varphi. \tag{4.35}$$

Problem a Use the reasoning in Problem e of Section 4.4 to show that increments of the surface area of the soap film satisfy the following expression

$$dS = \left| \frac{\partial \mathbf{r}}{\partial x} \times \frac{\partial \mathbf{r}}{\partial y} \right| dxdy. \tag{10.15}$$

Problem b Use (10.14) and (10.15) to derive that the total surface area is given by

$$S = \iint \sqrt{1 + |\nabla h|^2} \, dxdy. \tag{10.16}$$

Note the analogy between this expression and (10.8).

Since the surface tension that governs the shape of the soap film tends to minimize the surface area of the soap film, the shape $h(x, y)$ follows from the requirement that $h(x, y)$ is the function that minimizes the surface area (10.16). As in Section 10.2 the solution follows from the requirement that for the function $h(x, y)$ that minimizes the surface area, the surface area is stationary for perturbations of $h(x, y)$. This means that when $h(x, y)$ is replaced by $h(x, y) + \epsilon(x, y)$, the first-order change of the surface area S vanishes.

Problem c Make the substitution $h(x, y) \to h(x, y) + \epsilon(x, y)$, use the identity $|\nabla h|^2 = (\nabla h \cdot \nabla h)$ in (10.16) and linearize the result in $\epsilon(x, y)$ to derive that the perturbation of the surface area is to first order in $\epsilon(x, y)$ given by

$$\delta S[h] = \iint \frac{(\nabla h \cdot \nabla \epsilon)}{\sqrt{1 + |\nabla h|^2}} dxdy. \tag{10.17}$$

To concentrate on the essentials we assume that the deflection of the soap surface is small, which means that $|\nabla h| \ll 1$. Under this assumption the $|\nabla h|$-term in the denominator can be ignored and the perturbation is given by

$$\delta S[h] = \iint (\nabla h \cdot \nabla \epsilon) \, dx dy. \tag{10.18}$$

In this problem, the edge of the soap film is kept at a fixed location. This means that the perturbation $\epsilon(x, y)$ must vanish at the edge of the soap film:

$$\epsilon(x, y) = 0 \qquad \text{at the edge of the soap film.} \tag{10.19}$$

This constraint is incorporated in the following two problems.

Problem d   Use equation (7.22) to derive the identity $(\nabla h \cdot \nabla \epsilon) = \nabla \cdot (\epsilon \nabla h) - \epsilon \nabla \cdot \nabla h$.

Problem e   Insert this result in (10.18), apply Gauss' theorem (8.3) to the first term, and use (10.19) to show that the resulting integral over the edge of the surface vanishes, so that the perturbation of the surface area is given by

$$\delta S[h] = - \iint \epsilon(x, y) \, (\nabla \cdot \nabla h) \, dx dy. \tag{10.20}$$

The second derivative $(\nabla \cdot \nabla h)$ is called the *Laplacian* of h. This operator is often denoted by the notation Δ. However, since the term $\nabla \cdot \nabla$ is reminiscent of the square of the vector ∇, the notation ∇^2 is also used, and in this book this latter notation is the one we will mostly use for the Laplacian. The requirement that the first-order perturbation of the soap film vanishes for all functions $\epsilon(x, y)$ implies that $(\nabla \cdot \nabla h)$ must be equal to zero. The soap film therefore satisfies the following differential equation:

$$\nabla^2 h = 0. \tag{10.21}$$

In mathematical physics this equation is called the *Laplace equation*. Before analyzing this equation in more detail, we first define the Laplacian more precisely.

Problem f   Use the definition $\nabla^2 = \nabla \cdot \nabla$ to show that

$$\nabla^2 = \operatorname{div} \operatorname{grad} = \frac{\partial^2}{\partial x^2} + \frac{\partial^2}{\partial y^2} \qquad \text{(in two dimensions).} \tag{10.22}$$

Analogously, the Laplacian in three dimensions is defined as

$$\nabla^2 = \operatorname{div} \operatorname{grad} = \frac{\partial^2}{\partial x^2} + \frac{\partial^2}{\partial y^2} + \frac{\partial^2}{\partial z^2} \qquad \text{(in three dimensions).} \tag{10.23}$$

Let us now consider the shape of the soap film that is shown in Figure 10.5. At the beginning of this section it was argued that the curvatures $\partial^2 h / \partial x^2$ and $\partial^2 h / \partial y^2$ of the soap film in the x- and y-directions could not both be equal to zero.

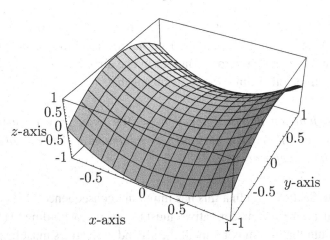

Figure 10.6   The shape of the same soap film as in the previous figure after a rotation through 45 degrees. (New edges define the area of this soap film.) The shape of the soap film is given by the function $h(x, y) = x^2 - y^2$; the shape of the soap film in the previous figure is given by $h(x, y) = 2xy$.

Instead, (10.21) implies that the *sum* of the curvatures in the x- and y-directions vanishes. This can be seen by rotating the soap film through 45 degrees as shown in Figure 10.6.

Problem g   Sketch cross sections of the surface in Figure 10.6 along the x- and y-axes, and show that for that surface the curvature in the x-direction is positive and the curvature in the y-direction is negative.

Problem h   The surface in Figure 10.6 is given by

$$h(x, y) = x^2 - y^2. \tag{10.24}$$

Show that for this surface $\nabla^2 h = 0$ and that the curvatures in the x- and y-direction cancel.

These results imply that the soap film behaves in a fundamentally different way than the rubber band between two fixed points that we treated in the previous section. The rubber band follows a straight line and the curvature is equal to zero, while for the soap film the *sum* of the curvatures in orthogonal directions vanishes. However, as shown in Figures 10.5 and 10.6, this does not imply that the soap film is a planar surface. The fact that the curvature terms $\partial^2 h/\partial x^2$ and $\partial^2 h/\partial y^2$ have opposite signs implies that when the function is concave in one direction, it must be convex in the other direction. This means that when the function is a maximum in one direction, it must be a minimum in the other direction: therefore, the surface has the shape of a saddle.

Let us now consider a stationary point of the function: that is, a point where $\partial h/\partial x = \partial h/\partial y = 0$. This point can be neither a minimum nor a maximum because when it is a minimum for variations in one direction, it *must* be a maximum for variations in the other direction. This means that:

Theorem   *A function that satisfies $\nabla^2 h = 0$ cannot have an extremum; the function can only have a maximum or minimum at the edge of the domain on which it is defined.*

We will see in Section 10.5 that this has important consequences. When a function is a minimum, all the second derivatives must be positive according to (10.5), while for an extremum that is a maximum all the second derivatives must be negative.

Problem i   Show that

$$\left.\begin{array}{l} \text{when } h(\mathbf{r}) \text{ is a minimum: } \nabla^2 h > 0; \\[2mm] \text{when } h(\mathbf{r}) \text{ is a maximum: } \nabla^2 h < 0. \end{array}\right\} \tag{10.25}$$

Problem j   The above condition is a necessary but not a sufficient condition. (This means that every minimum must satisfy the requirement $\nabla^2 h > 0$, but conversely when $\nabla^2 h > 0$ is satisfied, the function is not necessarily a minimum.) Show this by analyzing the stationary point of the function $h(x, y, z) = x^2 + y^2 - z^2$.

10.4 Sources of curvature

The main point of the previous section is that a function that satisfies the Laplace equation (10.21) can have nonzero curvature. In general there are two reasons why a soap film can be curved. The first reason is that the edges of the soap film are not confined to a common plane, as shown in Figures 10.5 and 10.6. Since the edges are part of the soap film, this necessarily implies that the soap film cannot be confined to a plane.

There is another reason why a soap film can be curved. Let us consider a soap film whose edges are located in the (x, y)-plane:

$$h(x, y) = 0 \qquad \text{at the edge.} \tag{10.26}$$

However, let us suppose that gravity pulls the soap film down. Such a soap film will bend down from the edges and will therefore be curved. This means that an external force can also lead to curvature of the soap film.

In this section we derive the differential equation for a soap film that is subject to a gravitational force. This is achieved by minimizing the total energy of the soap film. The gravitational energy is given by $\rho g h$, where ρ is the mass-density per unit area and g is the acceleration of gravity. The surface energy of the soap film is some positive constant k times the surface area of the soap film.

Problem a   Use these results and (10.16) to show that the total energy is given by:

$$E = \iint \left(k\sqrt{1 + |\nabla h|^2} + \rho g h \right) dx dy. \qquad (10.27)$$

The shape of the soap film is determined by the condition that the energy is a minimum.

Problem b   In the previous section we minimized the first term in the integrand. Generalize the derivation of the previous section to take the second term in (10.27) into account. Derive that the soap film under gravity satisfies

$$\nabla^2 h = \frac{\rho g}{k}. \qquad (10.28)$$

This equation is called *Poisson's equation*. The gravitational force has the effect that the total curvature of the soap film can be nonzero. This corresponds to the reasoning at the beginning of this section that gravity will make the soap film sag, so that the total curvature can be nonzero despite the fact that the edges of the soap film are confined to a plane. Note that gravity acts as a source term in the differential equation (10.28).

When you blow soap bubbles in your backyard, the force applied by blowing is the source of "bulging" the shape. With enough force, the soap will detach from the frame and form a soap bubble, as the one depicted in Figure 3.5. Now free from the frame that once restricted its shape, the bubble still prefers the shape with the smallest surface area for a given volume! In this case, that is a sphere. Only recently, a mathematical proof was given that two merged soap bubbles take on the characteristic double-bubble shape you surely have blown as a kid (or adult), with a flat surface joining the otherwise spherical shapes. This is called the *double-bubble conjecture* (Hutchings et al., 2002).

There is an interesting parallel between the soap film in a gravitational field and the electric potential that is generated by electric charges. According to equation (6.24) the electric field generated by a charge density ρ is given by

$$\nabla \cdot \mathbf{E} = \rho(\mathbf{r})/\epsilon_0. \tag{6.24}$$

According to (5.32) the potential V associated with this electric field is given by

$$\mathbf{E} = -\nabla V. \tag{10.29}$$

Problem c   Show that the potential satisfies

$$\nabla^2 V = -\rho(\mathbf{r})/\epsilon_0. \tag{10.30}$$

This means that the electric potential satisfies Poisson's equation (10.28), as well. The electric charge acts as the source of the electric potential just as the gravitational force acts as the source of the deflection of the soap film.

The analogy between the deflection of the soap film and the electric potential is interesting. Let us consider once more a soap film that is not subject to a gravitational force. Equation (10.28) of the soap film followed from the requirement that the squared-gradient $|\nabla h|^2$ integrated over the surface was minimized, because then the term $\sqrt{1 + |\nabla h|^2}$ in (10.16) is minimized as well. The analogy with the electric potential means that in free space ($\rho = 0$) the electric potential behaves in such a way that the squared-gradient $|\nabla V|^2$ of the potential integrated over the volume is minimized. However, according to (10.29) the gradient of the electric potential is the electric field. This means that the electric field behaves in such a way that the volume integral $\int |\mathbf{E}|^2 \, dV$ is minimized. However, this quantity is nothing but the energy of a static electric field (Jackson, 1998). This implies that the electric field is distributed in such a way that the energy of the electric field is minimized.

10.5   Instability of matter

The title of this section may surprise you, but the results that will be obtained here imply that according to the laws of classical physics the structure of matter and the mass distribution in the universe cannot be in a stable equilibrium. This result was derived by Earnshaw in 1842 in his work "On the nature of the molecular forces which regulate the constitution of the luminiferous ether" (Earnshaw, 1842). Let us consider a particle in three dimensions that is subject to a potential $V(\mathbf{r})$ that accounts for the gravitational attraction by other masses and the electrostatic force due to other charges. Let us suppose that at some point in space the particle is in equilibrium. This means that the force acting on the particle vanishes at that point, or equivalently that the gradient of the potential vanishes at that point: $\nabla V = 0$.

Problem a   Generalize expression (10.7) to three dimensions to show that this equilibrium point is only stable when the curvature of the potential in the three coordinate directions is positive:

$$\frac{\partial^2 V}{\partial x^2} > 0 \quad \text{and} \quad \frac{\partial^2 V}{\partial y^2} > 0 \quad \text{and} \quad \frac{\partial^2 V}{\partial z^2} > 0. \qquad (10.31)$$

According to (10.30), in free space ($\rho = 0$) the potential that corresponds to the gravitational force exerted by other masses, and the electrostatic force generated by other charges, satisfies Laplace's equation $\nabla^2 V = 0$.

Problem b Show that Laplace's equation implies that when the curvature of the potential is positive in one direction, the curvature must be negative in at least one other direction.

Problem c Use this to deduce that when the equilibrium point is stable to perturbations in one direction, it must be unstable for perturbations in at least one other direction.

An equilibrium point is in general unstable when it is unstable for perturbations in at least one of the directions. Problem c shows that an equilibrium point for the potential $V(\mathbf{r})$ that satisfies Poisson's equation cannot be stable.

This has far-reaching consequences. Let us consider a crystal. Within the framework of classical physics, each ion in the crystal moves in an electric potential that is generated by all the other ions in the crystal. This potential satisfies Laplace's equation at the location of the ion that we are considering because the net charge density ρ of the *other* ions is zero at that point. This means that the motion of the ion at its equilibrium point is not stable. When this is the case, the crystal is not stable because small perturbations of each ion from its equilibrium point lead to unstable motions of the ions. This implies that according to the laws of classical physics, matter is not stable! This result, known as *Earnshaw's theorem* (Earnshaw, 1842), states that the equilibrium points of any configuration of static electrical, magnetic, and gravitational fields are unstable.

Hopefully you are convinced that matter is stable, whatever the unsettling prediction of Earnshaw's theorem may be. The only way that we can resolve this paradox is if matter does not satisfy the equations of classical mechanics and electrostatics. This implies that quantum effects must play a crucial role in the stability of matter.

Earnshaw's theorem also applies to the gravitational field and it states that equilibrium points in the gravitational field are not stable. This means that a universe in equilibrium would be unstable. Let us first consider our solar system. It is essential for the solar system that the planets and their moons are in motion. This means that the solar system is not in a static equilibrium. In fact, the motion of the planets and moons is governed by the combination of the gravitational force plus the inertia force that is associated with the motion of the planets. The miracle of planetary motion is that the gravitational *motion* is stable. (You showed in Section 6.7 that the motion of planetary orbits is only stable in less than four spatial dimensions.) The situation is comparable to the movement of a bicycle, which derives its stability from the motion of the wheels that rotate. In the same way, it is the combination of the gravitational field and the inertial forces due to the motion of the planets that leads to stable planetary orbits. On scales much larger than the solar system, relativistic effects are important and Poisson's equation is not sufficient to describe the gravitational field on a cosmological scale (Ohanian and Ruffini, 1976).

In 1842, Earnshaw (1842) carried out his work not to study the structure of matter, but that of the ether: at the time "ether" was seen as the carrier of electromagnetic waves. He viewed the ether as a system of interacting particles and showed that the interaction between these particles could not be governed by Newton's law of gravitation using the analysis shown in this section. From the requirement of stability he derived conditions for the potential that governs the interaction between etheral particles. In 1842 Earnshaw could not invoke quantum mechanics in his description of interacting microscopic particles, because that theory had not yet been formulated.

10.6 Where does lightning start?

During a thunderstorm, the vertical motion of ice particles causes the separation of positive and negative electric charges in the atmosphere. This charge separation leads to an electric field within the atmosphere. When the field strength at a certain location is sufficiently strong, atoms are ionized and a current flows. This current induces further ionization, which extends the path of the current. This physical process describes the events that initiate a lightning bolt. In principle there seems to be no reason why a lightning bolt cannot start in mid-air, far away from the Earth's surface and far from the electrical charges that induce the electric field. In this section we show that lightning can only start at specific locations. To see this we study the Laplacian of the square of the length of a vector field $\mathbf{E}$. We first compute the Laplacian of the electric field.

Problem a   In vacuum, where $\rho = 0$, expression (10.30) for the electric potential reduces to $\nabla^2 V = 0$. Take the gradient of this expression and use equation (10.29) to show that

$$\nabla^2 \mathbf{E} = 0. \qquad (10.32)$$

Hint: You can use that the order of partial derivatives can be interchanged; hence, $\nabla \nabla^2 V = \nabla^2 \nabla V$.

Up to this point the Laplacian has always acted on a scalar; when it acts on a vector as in expression (10.32), it means that the Laplacian of every component is taken so that $\nabla^2 \mathbf{E}$ stands for a vector with components $\nabla^2 E_x$, $\nabla^2 E_y$, and $\nabla^2 E_z$, respectively.

Problem b Next we compute the Laplacian of $E^2 = E_x^2 + E_y^2 + E_z^2$. Show that that this quantity is given by

$$\nabla^2 E^2 = 2 \left(|\nabla E_x|^2 + |\nabla E_y|^2 + |\nabla E_z|^2 \right) + 2 \left(\mathbf{E} \cdot \nabla^2 \mathbf{E} \right). \qquad (10.33)$$

You may want to read Section 7.7 if you are not familiar with the vector calculus of products of functions.

Problem c Show that a vector field that satisfies the Laplace equation (10.32) also satisfies the following inequality:

$$\nabla^2 E^2 \geq 0. \qquad (10.34)$$

Problem d Use this to show that the strength of the vector field (E^2) cannot have a maximum in the region where (10.32) is satisfied.

Note that (10.34) does not exclude the possibility that E^2 has a minimum.

The result you derived in Problem d is all we need to solve our problem. Away from the electric charges in the atmosphere, a static electric field satisfies (10.32). According to Problem d, the electric field strength E^2 cannot have a maximum in this region. Because lightning will initiate where the field strength is largest, there are only two options for the origination of lightning: either lightning initiates where Laplace equation (10.32) does not hold (at the heights where electric charges that generate the electric field are present), or at the boundary of the area (the Earth's surface).

10.7 Laplacian in spherical and cylindrical coordinates

In many applications it is useful to consider the Laplacian in spherical or cylindrical coordinates. In principle this result can be derived by applying the transformation rules to the second partial derivatives in the Laplacian when changing from Cartesian to spherical or cylindrical coordinates. However, this route is unnecessarily complex, especially since we have done most of the work already. The key element in the derivation was derived in Problem f of Section 10.3 where you showed that the Laplacian is the divergence of the gradient: $\nabla^2 = \text{div grad}$. In Sections 5.6 and 6.5 you have already derived expressions for the gradient and the divergence in spherical and cylindrical coordinates, and all you need to do is to insert the expression for the gradient in curvilinear coordinates into the expression of the divergence in curvilinear coordinates.

Problem a   Use (5.49) and (6.32) to show that the Laplacian of a function f in cylindrical coordinates is given by

$$\nabla^2 f = \frac{1}{r}\frac{\partial}{\partial r}\left(r\frac{\partial f}{\partial r}\right) + \frac{1}{r^2}\frac{\partial^2 f}{\partial \varphi^2} + \frac{\partial^2 f}{\partial z^2}. \tag{10.35}$$

Problem b   Find the expressions in Sections 5.6 and 6.5 that allow you to derive that the Laplacian of a function f in spherical coordinates is given by

$$\nabla^2 f = \frac{1}{r^2}\frac{\partial}{\partial r}\left(r^2\frac{\partial f}{\partial r}\right) + \frac{1}{r^2\sin\theta}\frac{\partial}{\partial \theta}\left(\sin\theta\frac{\partial f}{\partial \theta}\right) + \frac{1}{r^2\sin^2\theta}\frac{\partial^2 f}{\partial \varphi^2}. \tag{10.36}$$

Later parts of this book make extensive use of these expressions of the Laplacian.

10.8 Averaging integrals for harmonic functions

In this section we focus on functions f that satisfy Laplace's equation (10.21), that is, $\nabla^2 f = 0$. Functions whose Laplacian is zero are called *harmonic functions*. Such functions play an important role in mathematical physics. We will show in Section 15.1 that the real and imaginary parts of analytic functions in the complex plane are harmonic functions. Let us first focus on a harmonic function $f(x, y)$ in two dimensions. In this section we derive that the function value at a certain point is equal to the average of that function over a circle centered around that point with an *arbitrary* radius. To see this we use a system of cylindrical coordinates and choose the origin of the cylindrical coordinates at the point that we consider. (Remember that we are free to choose the origin of the coordinate system.)

Problem a   Use expression (10.35) to show that f satisfies the following differential equation:

$$\frac{1}{r}\frac{\partial}{\partial r}\left(r\frac{\partial f}{\partial r}\right) + \frac{1}{r^2}\frac{\partial^2 f}{\partial \varphi^2} = 0. \tag{10.37}$$

Problem b   Integrate this expression over a disk of radius R centered at the origin and derive that

$$\int_0^R\int_0^{2\pi}\left[\frac{\partial}{\partial r}\left(r\frac{\partial f}{\partial r}\right) + \frac{1}{r}\frac{\partial^2 f}{\partial \varphi^2}\right]d\varphi\,dr = 0. \tag{10.38}$$

(Note the powers of r in this expression.)

Problem c   Carry out the φ-integration in the last term of the integrand to show that this term gives a vanishing contribution.

It is convenient to introduce at this point the average $\bar{f}(r)$ of the function f over a circle with radius r:

$$\bar{f}(r) \equiv \frac{1}{2\pi} \int_0^{2\pi} f(r, \varphi) \, d\varphi. \tag{10.39}$$

Problem d   Use (10.38) and the result of Problem c to derive that $\bar{f}$ satisfies the following equation:

$$\left[r \frac{\partial \bar{f}}{\partial r} \right]_{r=0}^{r=R} = 0. \tag{10.40}$$

Problem e   This expression holds for any radius R; hence, $r \partial \bar{f}/\partial r$ is independent of r, so that it is a constant: $r \partial \bar{f}/\partial r = C$. Integrate this expression to derive that

$$\bar{f}(r) = C \, \ln r + A, \tag{10.41}$$

where A is an unknown integration constant.

Problem f   The integration constant C must be equal to zero because $\bar{f}(r)$ is finite as $r \to 0$. Evaluate (10.41) at the origin, determine the constant A and show that $f(r)$ satisfies

$$f(r = 0) = \frac{1}{2\pi} \int_0^{2\pi} f(r, \varphi) \, d\varphi. \tag{10.42}$$

This expression states that the value of a harmonic function at the origin is the average of the function over a circle with arbitrary radius. The amazing property is that this holds for any value of the radius, provided f is harmonic everywhere within the circle with radius r. Note that we are free to choose the origin of the coordinate system so that the averaging integral (10.42) holds for any point.

The averaging integral can also be used to prove that a harmonic function cannot have a minimum or a maximum in the region where it is defined. Suppose the function had a maximum at a certain location. Then, there exists a circle around this point where the function has smaller values than at the maximum (otherwise it would not be a maximum). Equation (10.42) cannot hold for this circle because the right-hand side would be smaller than the left-hand side. Therefore, it is impossible for f to have a maximum.

Problem g   Generalize the derivation in this section to spherical coordinates and derive the following expression for harmonic functions in three dimensions

$$f(r = 0) = \frac{1}{4\pi} \int_0^\pi \int_0^{2\pi} f(r, \theta, \varphi) \, \sin\theta \, d\varphi d\theta. \qquad (10.43)$$

Problem h   Show that the right-hand side of this expression is the average of that function over a sphere with arbitrary radius r centered around that point $r = 0$.

This means that in three dimensions, the value of a harmonic function at a certain point is equal to the average of the function over a sphere with arbitrary radius r that is centered on that point.

11

Scale analysis

Many equations we encounter in mathematical physics are too complicated to solve analytically. One of the reasons can be that an equation contains too many terms to handle. In practice, however, these terms may vary in size. Ignoring the smaller terms may simplify the problem to the extent that it can be solved in closed form. Moreover, by deleting terms that are relatively small, we can focus on the terms that contain the significant physics. In this sense, ignoring the smaller terms can actually give a better physical insight into the processes that really do matter.

Scale analysis is a technique in which one estimates the different terms in an equation by considering the scale over which the relevant parameters vary. This is an extremely powerful tool for simplifying problems. A comprehensive overview of this technique with many applications is given by Kline (1986) and in Chapter 6 of Lin and Segel (1974). Interesting examples of the application of scaling arguments to biology are given by Vogel (1998).

With the application of scale analysis one caveat must be made. One of the major surprises of classical physics of the twentieth century was the discovery of chaos in dynamical systems (Tabor, 1989). In a chaotic system small changes in the initial conditions lead to a change in the time evolution of the system that grows exponentially with time. Deleting (initially) small terms from the equation of motion of such a system can have a similar effect; this can lead to changes in the system that may grow exponentially with time. This means that for chaotic systems one must be careful in omitting small terms from the equations.

The principle of scale analysis is first applied to the problem of determining the sense of rotation of a vortex in an emptying bathtub. Many of the equations that are used in physics are differential equations. For this reason it is crucial in scale analysis to be able to estimate the order of magnitude of derivatives. The estimation of derivatives is therefore treated in Section 11.2. In subsequent sections this is then applied to a variety of different problems.

11.1 Vortex in a bathtub

When emptying your bathtub, you can observe a beautiful vortex above the drain. When you ask people about the sense of rotation of this vortex, many will reply that the sense of rotation is always in the counterclockwise direction on the northern hemisphere and in the clockwise direction in the southern hemisphere. To verify that this is not consistently the case, take a bath, and empty the bathtub. Observe that it is as easy to obtain a vortex that rotates in the clockwise direction as it is to create a vortex that rotates in the counterclockwise direction. In fact, if you have ever had the opportunity to take a bath in Nairobi, you would observe that at the equator a bathtub drains in exactly the same way as at higher latitudes.

You may find it difficult to convince your friends of these facts. In that case, maybe a calculation will convince them. One approach would be to take the equations of fluid flow on a rotating Earth and compute numerically in all details the flow in your bathtub. However, this approach is not only terribly impractical, but it is also extremely difficult to carry out; you know neither the detailed shape of your bathtub, nor the initial conditions of the water in the bathtub before you drain it. An even more serious drawback of this approach is that this numerical simulation does not give much physical insight. At best it gives a perfect simulation of the real bathtub, but in that case it would be better to do the experiment in the real bathtub.

As an alternative we can estimate the relative strength of the forces that are acting on the fluid in the vortex. In Figure 11.1 the forces are shown that act on a fluid parcel in a vortex on the Earth that rotates with angular velocity $\mathbf{\Omega}$. The dominant forces are the pressure force $\mathbf{F}_{pres}$ that is associated with the pressure gradient in the fluid, the Coriolis force $\mathbf{F}_{Cor}$ due to the rotation of the Earth, and the centrifugal force $\mathbf{F}_{cent}$ that is due to the circular motion of the fluid in the vortex. We

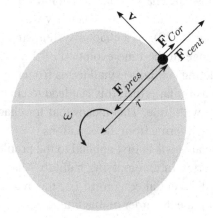

Figure 11.1   The forces that act in a fluid vortex on the northern hemisphere of a rotating Earth.

derive the Coriolis in Section 12.2. The rotational velocity of the vortex is denoted by ω. The only force that depends on the position on the Earth is the Coriolis force $\mathbf{F}_{Cor}$. In Figure 11.1 this force points to the right-hand side of the velocity vector; this is the case in the northern hemisphere, whereas in the southern hemisphere this force would point to the left-hand side of the velocity vector. The centrifugal force always points outward from the vortex, while the pressure force always points into the vortex. This means that any potential asymmetry between the behavior of a vortex in the northern hemisphere and one in the southern hemisphere must be due to the Coriolis force.

In order to test our hypothesis we need to estimate the strength of the Coriolis force compared to the other forces that are operative. In the balance of forces, the pressure force is balanced by the outward forces (Coriolis and/or centrifugal). For this reason, it is sufficient to estimate the strength of the Coriolis force compared to the centrifugal force. We derive in Section 12.2 that the Coriolis force is given by $\mathbf{F}_{Cor} = -2\Omega \times \mathbf{v}$. Since we are only estimating orders of magnitude, this means that the Coriolis force is of the order

$$F_{Cor} \sim \Omega v. \tag{11.1}$$

The strength of the centrifugal force is given by (Goldstein, 1980):

$$F_{cent} = \frac{v^2}{r}. \tag{11.2}$$

Problem a Use Figure 11.1 to deduce that $v/r = \omega$ and then use this result to show that the ratio of the Coriolis force to the centrifugal force is approximately given by

$$\frac{F_{Cor}}{F_{cent}} \sim \frac{\Omega}{\omega}. \tag{11.3}$$

The ratio of the Coriolis force to the centrifugal force is thus of the order of the ratio of the rotation rate of the Earth to the rotation rate of the vortex in the bathtub.

Problem b Assuming that the vortex rotates once a second, use the previous expression to show that

$$\frac{F_{Cor}}{F_{cent}} \sim 10^{-5}. \tag{11.4}$$

This means that the Coriolis force is much smaller than all other forces that are operating on the fluid parcel. The asymmetry in the balance of forces created by the Coriolis force is in practice also much smaller than the asymmetry in the shape of the bathtub and the asymmetry in the initial conditions of the fluid motion before you drain the bathtub. We can conclude from this that the Earth's rotation is negligible in the dynamics of the bathtub vortex; hence, the sense of rotation of a vortex of that scale does not depend on the geographic location. This example shows that one can sometimes learn more from simple physical arguments and estimates of orders of magnitude than from very complex numerical calculations.

It follows from (11.3) that the centrifugal force and the Coriolis force are of the same order of magnitude when the rotation rate of the fluid in the bathtub is of the same order of magnitude as the Earth's rotation.

Problem c   Estimate the ratio of the Coriolis force to the centrifugal force for the
   atmospheric motion around a depression.

You will have found that for the atmosphere the Coriolis force *is* one of the dominant forces. This is the reason that in the northern hemisphere the air moves in the counterclockwise direction around a low-pressure area and in the clockwise direction around a high-pressure center, and vice versa on the southern hemisphere. This is called the *cyclonic direction*. We encountered an example of the vortex around a low-pressure area in Figure 7.3, which should remind you of the vortex in a bathtub. The analogy is in fact the cause of the misconception that the rotational direction of the vortex in a bathtub depends on geographic location. The major difference is – as we now learned with a scale analysis – that the relative strengths of the forces that are operative in the atmosphere and in a bathtub are completely different.

There is, of course, always the risk of oversimplifying a problem. To give an example, the theory of this section is equally applicable to tornadoes because the rate of rotation of the flow in a tornado is much higher than the rate of rotation of the Earth. Yet tornadoes do mostly rotate in a cyclonic direction. As discussed by Davies-Jones (1984), the direction of rotation of tornadoes depends on the pattern with which the winds veer with height. This change in the wind direction with height *is* usually determined by the large-scale flow, which does depend on the rotation of the Earth.

11.2 Three ways to estimate a derivative

In this section we estimate the derivative of a function $f(x)$ in three different ways. The *first* way is to realize that the derivative is merely the slope of the function $f(x)$. Consider Figure 11.2 in which the function $f(x)$ is assumed to be known at neighboring points x and $x + h$.

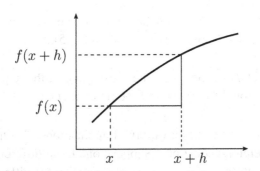

Figure 11.2   The slope of a function $f(x)$ that is known at positions x and $x+h$.

Problem a   Deduce from the geometry of this figure that the slope of the function at x is approximately given by $[f(x+h) - f(x)]/h$.

Since the slope is the derivative, this means that the derivative of the function is approximately given by

$$\frac{df}{dx} \approx \frac{f(x+h) - f(x)}{h}. \tag{11.5}$$

The *second* way to derive the same result is to realize that the derivative is defined by the following limit:

$$\frac{df}{dx} \equiv \lim_{h\to 0} \frac{f(x+h) - f(x)}{h}. \tag{11.6}$$

Note that without the limit $h \to 0$, the right-hand side is not necessarily equal to the derivative; this is why expression (11.5) is an approximation. The accuracy of this approximation depends on the degree to which the function f behaves like a linear function on the scale of h.

The problem with estimating the derivative of $f(x)$ in the previous ways is that although we obtain an estimate of the derivative, we do not know how good these estimates are. We know that if $f(x)$ was a straight line, which has a constant slope, the estimate (11.5) would be exact. Hence, it is the deviation of $f(x)$ from a straight line that makes (11.5) only an approximation. This means that it is the curvature of $f(x)$ that accounts for the error in the approximation (11.5). The *third* way of estimating the derivative provides this error estimate as well.

Problem b   Consider the Taylor series (3.17). Truncate this series after the second-order term and solve the resulting expression for df/dx to derive that

$$\frac{df}{dx} = \frac{f(x+h) - f(x)}{h} - \frac{1}{2}\frac{d^2 f}{dx^2}h + \cdots, \tag{11.7}$$

where the dots indicate terms of order h^2 and higher.

In the limit $h \to 0$ the last term vanishes and (11.6) is obtained. Ignoring the last term in (11.7) for finite h leads to approximation (11.5).

Problem c Use (11.7) to show that the error made in the approximation (11.5) indeed depends on the *curvature* of the function $f(x)$.

The approximation (11.5) has a variety of applications. The first is the numerical solution of differential equations. Suppose one has a differential equation that cannot be solved in closed form. For example, consider the differential equation

$$\frac{df}{dx} = G(f(x), x), \tag{11.8}$$

with initial value

$$f(0) = f_0. \tag{11.9}$$

If this equation cannot be solved in closed form, one can solve it numerically by evaluating the function $f(x)$ not for every value of x, but only at a finite number of x-values that are separated by a distance h. These points x_n are given by $x_n = nh$, and the function $f(x)$ at location x_n is denoted by f_n:

$$f_n \equiv f(x_n). \tag{11.10}$$

Problem d Show that the derivative df/dx at location x_n can be approximated by:

$$\frac{df}{dx}(x_n) = \frac{1}{h}(f_{n+1} - f_n). \tag{11.11}$$

Problem e Insert this result into the differential equation (11.8) and solve the resulting expression for f_{n+1} to show that:

$$f_{n+1} = f_n + hG(f_n, x_n). \tag{11.12}$$

This is all we need to numerically solve the differential equation (11.8) with the boundary condition (11.9). Once f_n is known, (11.12) can be used to compute f_{n+1}. This means that the function can be computed at all values of the grid points x_n recursively. To start this process, one uses the boundary condition (11.9) that gives the value of the function at location $x_0 = 0$. This technique for estimating the derivative of a function can be extended to higher-order derivatives as well so that second-order differential equations can also be solved numerically.

In practice, one has to pay serious attention to the *stability* of the numerical solution. The requirements of stability and numerical efficiency have led to many refinements of the numerical methods for solving differential equations. The interested reader can consult Press et al. (1992) for an introduction and many practical algorithms.

The estimate (11.5) has a second important application, because it allows us to estimate the order of magnitude of a derivative. Suppose a function $f(x)$ varies over a characteristic range of values F and that this variation takes place over a characteristic distance L. It follows from (11.5) that the derivative of $f(x)$ is of the order of the ratio of the variation of the function $f(x)$ divided by the length scale over which the function varies. In other words:

$$\left| \frac{df}{dx} \right| \approx \frac{\text{variation of the function } f(x)}{\text{length scale of the variation}} \sim \frac{F}{L}. \tag{11.13}$$

In this expression the term $\sim (F/L)$ indicates that the derivative is of the *order* F/L. Note that this is not in general an accurate estimate of the precise value of the derivative of $f(x)$, it only provides us with an estimate of the order of magnitude of a derivative. However, this is all we need to carry out a scale analysis.

Problem f Suppose $f(x)$ is a sinusoidal wave with amplitude A and wavelength λ:

$$f(x) = A \sin \left(\frac{2\pi x}{\lambda} \right). \tag{11.14}$$

Show that (11.13) implies that the order of magnitude of the derivative of this function is given by $|df/dx| \sim (A/\lambda)$. Compare this estimate of the order of magnitude with the true value of the derivative and pay attention both to the numerical value as well as to the spatial variation.

From the previous estimate we can learn two things. First, the estimate (11.13) is only a rough estimate that *locally* can be very poor. One should always be aware that (11.13) may break down at certain points and that this can cause errors in the subsequent scale analysis. Second, (11.13) differs by a factor 2π from the true derivative at $x = 0$. However, $2\pi \approx 6.28$, which is not a small number (compared to 1). Therefore, be aware that hidden numerical factors may enter scaling arguments.

11.3 Advective terms in the equation of motion

As a first example of scale analysis we consider the role of advective terms in the equation of motion. As shown in (25.12), the equation of motion for a continuous medium is given by

$$\frac{\partial \mathbf{v}}{\partial t} + \mathbf{v} \cdot \nabla \mathbf{v} = \frac{1}{\rho} \mathbf{F}. \tag{11.15}$$

Note that we have divided by the density compared to the original expression (25.12). This equation can describe the propagation of acoustic waves when **F** is the pressure force, and it accounts for elastic waves when **F** is given by the elastic forces in the medium. We consider the situation in which waves with a wavelength λ and a period T propagate through the medium.

The advective terms $\mathbf{v} \cdot \nabla \mathbf{v}$ often pose a problem in solving this equation. This is because the partial time derivative $\partial \mathbf{v}/\partial t$ is *linear* in the velocity **v** but the advective terms $\mathbf{v} \cdot \nabla \mathbf{v}$ are *nonlinear* in the velocity **v**. Since linear equations are in general much easier to solve than nonlinear equations, it is useful to know under which conditions the advective terms $\mathbf{v} \cdot \nabla \mathbf{v}$ can be ignored compared with the partial derivative $\partial \mathbf{v}/\partial t$.

Problem a   Let the velocity of the continuous medium have a characteristic value V. Show that $|\partial \mathbf{v}/\partial t| \sim V/T$ and that $|\mathbf{v} \cdot \nabla \mathbf{v}| \sim V^2/\lambda$.

Problem b   Show that this means that the ratio of the advective terms to the partial time derivative is given by

$$\frac{|\mathbf{v} \cdot \nabla \mathbf{v}|}{|\partial \mathbf{v}/\partial t|} \sim \frac{V}{c}, \tag{11.16}$$

where $c = \lambda/T$ is the velocity with which the waves propagate through the medium.

This result implies that the advective terms can be ignored when the velocity of the medium itself is much less than the velocity of the waves propagating through the medium:

$$V \ll c. \tag{11.17}$$

In other words, when the amplitude of the wave motion is so small that the velocity of the particle motion in the medium is much less than the phase velocity of the waves, one can ignore the advective terms in the equation of motion.

Problem c   Suppose an earthquake causes a ground displacement of 1 mm at a frequency of 1 Hz at a large distance. The wave velocity of seismic P-waves is of the order of 5 km/s near the surface. Show that in that case $V/c \sim 10^{-6}$.

The small value of V/c implies that for the propagation of elastic waves due to earthquakes one can ignore advective terms in the equation of motion. Note, however, that this is not necessarily true near the earthquake where the motion is much more violent and where the associated velocity of the rocks is not necessarily much smaller than the wave velocity.

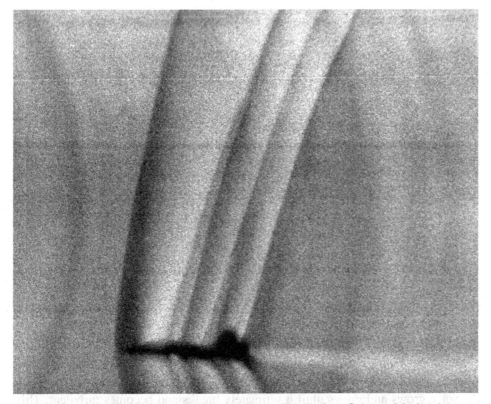

Figure 11.3 The shock waves generated by a T38 airplane flying at Mach 1.1 (a speed of 1.1 times the speed of sound) made visible with the schlieren method.

There are a number of physical phenomena that are intimately related to the presence of the advective terms in the equation of motion. One important phenomenon is the occurrence of shock waves when the motion of the medium exceeds the wave velocity. A prime example of shock waves is the sonic boom made by an aircraft that moves at a velocity greater than the speed of sound (Kermode et al., 1996; Thompson and Beavers, 1972). A spectacular example can be seen in Figure 11.3 where the shock waves generated by a T38 airplane flying at a speed of Mach 1.1 at an altitude of 13,700 ft can be seen. These shock waves are visualized using the schlieren method (Lauterborn and Kurz, 2003) which is an optical technique to convert phase differences of light waves into amplitude differences.

Another example of shock waves is the formation of the *hydraulic jump*. You may not know what a hydraulic jump is, but you likely have seen one! Consider water flowing down a channel such as a mountain stream as shown in Figure 11.4. The flow velocity is denoted by v. In this channel a rock disrupts the flow, generating water waves that propagate with a velocity c compared to the

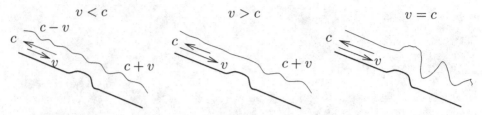

Figure 11.4   Water flowing with a speed v encounters a rock, creating secondary waves with a speed c. The three panels are scenarios for varying relative values of v and c.

moving water. When the flow velocity is less than the wave speed ($v < c$ in the left-hand panel of Figure 11.4), the waves propagate upstream with an absolute velocity $c - v$ and propagate downstream with an absolute velocity $c + v$. When the flow velocity is larger than the wave velocity ($v > c$ in the middle panel of Figure 11.4), the waves move downstream only because the wave velocity is not sufficiently large to move the waves against the current. The most interesting case is when the flow velocity equals the wave velocity ($v = c$ in the right-hand panel of Figure 11.4). In that case, the waves that move upstream have an absolute velocity given by $c - v = 0$. In other words, these waves do not move with respect to the rock that generates the waves. This wave is continuously excited by the rock, and through a process similar to an oscillator that is driven at its resonance frequency the wave grows and grows until it ultimately breaks and becomes turbulent. This is the reason why one sees strong turbulent waves over boulders and other irregularities in streams. For further details on channel flow and hydraulic jumps, read Chapter 9 of Whitaker (1968). In general, the advective terms play a crucial role in the steepening and breaking of waves and in the formation of shock waves. This is described in much detail by Whitham (2011).

11.4 Geometric ray theory

Geometric ray theory is an approximation that accounts for the propagation of waves along lines through space. The theory finds its conceptual roots in optics, where for a long time one has observed that a light beam propagates along a well-defined trajectory through lenses and other optical devices. Mathematically, this behavior of waves is accounted for in geometric ray theory, or more briefly "ray theory."

Ray theory is derived here for the acoustic wave equation rather than for the propagation of light because pressure waves are described by a scalar equation rather than the vector equation that governs the propagation of electromagnetic waves. The starting point is the acoustic wave equation in the frequency domain (8.8):

$$\rho \nabla \cdot \left(\frac{1}{\rho} \nabla p \right) + \frac{\omega^2}{c^2} p = 0. \tag{11.18}$$

For simplicity, the source term on the right-hand side has been set to zero. In addition, the relation $c^2 = \kappa / \rho$ has been used to eliminate the bulk modulus κ in favor of the wave velocity c. Both the density ρ and the wave velocity are arbitrary functions of space.

In general it is not possible to solve this differential equation in closed form. Instead we seek an approximation by writing the pressure as:

$$p(\mathbf{r}, \omega) = A(\mathbf{r}, \omega) e^{i\psi(\mathbf{r}, \omega)}, \tag{11.19}$$

with A and ψ real functions. Any function $p(\mathbf{r}, \omega)$ can be written in this way.

Problem a   Insert the solution (11.19) in the acoustic wave equation (11.18), separate the real and imaginary parts of the resulting equation to deduce that (11.18) is equivalent to the following equations:

$$\underbrace{\nabla^2 A}_{(1)} - \underbrace{A\,|\nabla\psi|^2}_{(2)} - \underbrace{\frac{1}{\rho}(\nabla\rho \cdot \nabla A)}_{(3)} + \underbrace{\frac{\omega^2}{c^2} A}_{(4)} = 0, \tag{11.20}$$

and

$$2\,(\nabla A \cdot \nabla\psi) + A\nabla^2\psi - \frac{1}{\rho}(\nabla\rho \cdot \nabla\psi)\, A = 0. \tag{11.21}$$

These equations are even harder to solve than the acoustic wave equation because they are nonlinear in the unknown functions A and ψ, whereas the acoustic wave equation is linear in the pressure p. However, (11.20) and (11.21) form a good starting point for making the ray-geometric approximation. First we analyze (11.20).

Assume that the density varies on a length scale L_ρ, and that the amplitude A of the wave field varies on a characteristic length scale L_A. Furthermore, the wavelength of the waves is denoted by λ.

Problem b   Explain that the wavelength is the length scale over which the phase ψ of the waves varies.

Problem c   Use the results of Section 11.2 to obtain the following estimates of the order of magnitude of the terms (1)–(4) in equation (11.20):

$$\left.
\begin{array}{ll}
|\nabla^2 A| \sim \dfrac{A}{L_A^2}, & A\,|\nabla\psi|^2 \sim \dfrac{A}{\lambda^2}, \\[2ex]
\left| \dfrac{1}{\rho}(\nabla\rho \cdot \nabla A) \right| \sim \dfrac{A}{L_A L_\rho}, & \dfrac{\omega^2}{c^2} A \sim \dfrac{A}{\lambda^2}.
\end{array}
\right\} \tag{11.22}$$

Next, we assume that the length scales of both the density variations and the amplitude variations are much longer than a wavelength: $\lambda \ll L_A$ and $\lambda \ll L_\rho$.

Problem d Show that under this assumption terms (1) and (3) in equation (11.20) are much smaller than terms (2) and (4).

Problem e Convince yourself that ignoring terms (1) and (3) in (11.20) gives the following (approximate) expression:

$$|\nabla \psi|^2 = \frac{\omega^2}{c^2}. \tag{11.23}$$

Problem f The approximation (11.23) was obtained under the premise that $|\nabla \psi| \sim 1/\lambda$. Show that this assumption is satisfied by the function ψ in (11.23).

Whenever one makes approximations by deleting terms that scale analysis predicts to be small, one has to check that the final solution is consistent with the scale analysis that is used to derive the approximation.

Note that the original equation (11.20) contains both the amplitude A and the phase ψ but that (11.23) contains the phase only. The approximation that we have made has thus decoupled the phase from the amplitude; this simplifies the problem considerably. The frequency enters the right-hand side of this equation only through a simple multiplication with ω^2. The frequency dependence of ψ can be found by substituting

$$\psi(\mathbf{r}, \omega) = \omega \tau(\mathbf{r}). \tag{11.24}$$

Problem g Show that after this substitution (11.23) and (11.21) are given by

$$|\nabla \tau(\mathbf{r})|^2 = \frac{1}{c^2}, \tag{11.25}$$

and

$$2(\nabla A \cdot \nabla \tau) + A \nabla^2 \tau - \frac{1}{\rho}(\nabla \rho \cdot \nabla \tau) A = 0. \tag{11.26}$$

According to (11.25) the function $\tau(\mathbf{r})$ does not depend on frequency. Note that (11.26) for the amplitude does not contain any frequency dependence either. This means that the amplitude also does not depend on frequency: $A = A(\mathbf{r})$. This has important consequences for the shape of the wave field in the ray-geometric approximation. Suppose that the wave field is excited by a source function $s(t)$ in the time domain that is represented in the frequency domain by a complex function $S(\omega)$.

(The forward and backward Fourier transforms are defined in expressions (14.42) and (14.43).) In the frequency domain the response is given by (11.19) multiplied with the source function $S(\omega)$. (We show in Section 14.6 that this multiplication corresponds to a convolution with the source function $s(t)$ in the time domain.) Using that A and τ do not depend on frequency, the pressure in the time domain can be written as:

$$p(\mathbf{r}, t) = \frac{1}{2\pi} \int_{-\infty}^{\infty} A(\mathbf{r}) e^{i\omega\tau(\mathbf{r})} e^{-i\omega t} S(\omega) \, d\omega. \tag{11.27}$$

Problem h Use this expression to show that the pressure in the time domain can be written as:

$$p(\mathbf{r}, t) = A(\mathbf{r}) s(t - \tau(\mathbf{r})). \tag{11.28}$$

This is an important result, because it implies that the time dependence of the wave field is everywhere given by the same source-time function $s(t)$. In a ray-geometric approximation the shape of the waveforms is everywhere the same. There are no frequency-dependent effects in a ray-geometric approximation.

Problem i Explain why this means that geometric ray theory cannot be used to explain why the sky is blue.

The absence of any frequency-dependent wave propagation effects is both the strength and the weakness of ray theory. It is a strength because the wave fields can be computed in a simple way once $\tau(\mathbf{r})$ and $A(\mathbf{r})$ are known. The theory also tells us that this is an adequate description of the wave field as long as the frequency is sufficiently high that $\lambda \ll L_A$ and $\lambda \ll L_\rho$. However, many wave propagation phenomena are in practice frequency-dependent, and it is the weakness of ray theory that it cannot account for these phenomena.

According to (11.28) the function $\tau(\mathbf{r})$ accounts for the time delay of the waves to travel to the point $\mathbf{r}$. Therefore, $\tau(\mathbf{r})$ is the *travel time* of the wave field. The travel time is described by the differential equation (11.25); this equation is called the *eikonal equation*.

Problem j Show that it follows from the eikonal equation that $\nabla\tau$ can be written as:

$$\nabla\tau = \hat{\mathbf{n}}/c, \tag{11.29}$$

where $\hat{\mathbf{n}}$ is a unit vector. Show also that $\hat{\mathbf{n}}$ is perpendicular to the surface $\tau = const$.

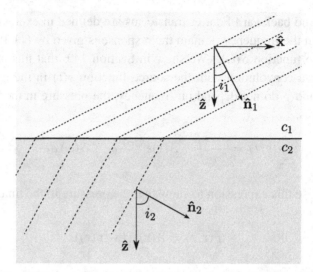

Figure 11.5   Geometry of a plane incoming wave that propagates from a half-space $z < 0$ with constant velocity c_1 that moves into a half-space $z > 0$ with constant velocity c_2. Wavefronts are indicated with dashed lines.

The vector $\hat{\mathbf{n}}$ defines the direction of the *rays* along which the wave energy propagates through the medium. Taking suitable derivatives of (11.29) one can derive the *equation of kinematic ray-tracing*. This is a second-order differential equation for the position of the rays, details are given by Aki and Richards (2002).

As an example of the use of expression (11.29), we consider the case shown in Figure 11.5 of a half space $z < 0$ with constant velocity c_1 that is joined to a half space $z > 0$ with a constant velocity c_2. In this example we assume that $c_2 > c_1$. A plane wave comes in from the upper half space. The words *plane wave* refer to the fact that the travel time $\tau(x, z)$ is a linear function of the space coordinates. The curves where $\tau(x, z) = constant$ are indicated by dashed lines, these are the wavefronts of the wave. The normal vector to the wavefronts in the upper and lower half spaces is indicated by $\hat{\mathbf{n}}_1$ and $\hat{\mathbf{n}}_2$, respectively. These vectors give the direction of propagation of the waves, and they have an *angle of incidence* i_1 and i_2, respectively, with the normal vector to the interface.

Problem k   Use the geometry of Figure 11.5 and the fact that the vectors $\hat{\mathbf{n}}_1$ and $\hat{\mathbf{n}}_2$ are normalized to show that

$$\hat{\mathbf{n}}_1 = \begin{pmatrix} \sin i_1 \\ \cos i_1 \end{pmatrix} \qquad \text{and} \qquad \hat{\mathbf{n}}_2 = \begin{pmatrix} \sin i_2 \\ \cos i_2 \end{pmatrix}. \tag{11.30}$$

Problem l   Use expression (11.29) to show that in the two half spaces

$$\nabla \tau_1 = \frac{1}{c_1} \begin{pmatrix} \sin i_1 \\ \cos i_1 \end{pmatrix} \qquad \text{and} \qquad \nabla \tau_2 = \frac{1}{c_2} \begin{pmatrix} \sin i_2 \\ \cos i_2 \end{pmatrix}. \tag{11.31}$$

At the interface $z = 0$, the travel time is continuous; this means that $\tau(x, z)$ is continuous. This implies that the derivative of $\tau(x, z)$ along the interface is also continuous; in other words, $\partial \tau(x, z)/\partial x$ is continuous.

Problem m   Use the continuity of $\partial \tau(x, z)/\partial x$ at the interface and expression (11.31) to show that

$$\frac{\sin i_1}{c_1} = \frac{\sin i_2}{c_2}. \tag{11.32}$$

This expression is called *Snell's law*, named after the seventeenth-century Dutch scientist Willebrord Snellius.[1] This law relates the direction of propagation of waves through media with different velocities. The reasoning used here can be extended to a number of horizontal layers, and as the wave propagates through these layers, the quantity $\sin i/c$, called the "ray parameter" is always constant. Snell's law can also be used to reflected waves. In that case $c_1 = c_2$ because the incident wave and reflected waves propagate through the same medium. According to Snell's law (11.32), the angles of incidence and of the incoming and reflected waves therefore are the same. As discussed in Section 24.6, the angles of incidence for the incident and reflected waves may be different when a conversion between different wave modes occurs in the reflection process.

We now return to the case of a general velocity. Once $\tau(\mathbf{r})$ is known, one can compute the amplitude $A(\mathbf{r})$ from (11.26). We have not yet applied any scale analysis to this expression. We will not do so, because it can be solved exactly. Let us first simplify this differential equation by considering the dependence on the density ρ in more detail.

Problem n   Insert $A = \rho^\alpha B$ in expression (11.26), and show that this results in the following differential equation for $B(\mathbf{r})$:

$$(2\alpha - 1)(\nabla\rho \cdot \nabla\tau) B + 2\rho(\nabla B \cdot \nabla\tau) + \rho B \nabla^2 \tau = 0. \tag{11.33}$$

Choose the constant α in such a way that the gradient of the density disappears from the equation and show that the remaining terms can be written as $\nabla \cdot (B^2 \nabla\tau) = 0$. Finally show using (11.29) that this implies the following differential equation for the amplitude:

$$\nabla \cdot \left(\frac{1}{\rho c} A^2 \hat{\mathbf{n}}\right) = 0. \tag{11.34}$$

Equation (11.34) states that the divergence of the vector $(A^2/\rho c)\,\hat{\mathbf{n}}$ vanishes; hence, the flux of this vector through any closed surface that does not contain the source

---

[1] The French refer to this principle as the *Lois de Descartes*.

of the wave field vanishes, see Section 8.1. This is not surprising, because the vector $(A^2/\rho c)\,\hat{\mathbf{n}}$ accounts for the energy flux of acoustic waves. Expression (11.34) implies that the net flux of this vector through any closed surface is equal to zero. This means that all the energy that flows into the surface must also flow out through the surface again. The transport equation in the form (11.34) is therefore a statement of energy conservation. Aki and Richards (2002) show how one can compute this amplitude A once the location of the rays is known.

An interesting complication arises when the energy is focused at a point, or on a surface. Such an area of focusing is called a *caustic*. A familiar example of a caustic is the rainbow. At a caustic, the ray-geometric approximation leads to an infinite amplitude of the wave field.

Problem o   Show that when the amplitude becomes infinite in a finite region of space, the condition $\lambda \ll L_A$ must be violated.

This means that ray theory is not valid in or near a caustic. A clear account of the physics of caustics can be found in Berry and Upstill (1980) and Kravtsov (1988). The former reference contains many beautiful images of caustics.

11.5 Is the Earth's mantle convecting?

The Earth is a body that continuously loses heat to outer space. This heat is partly a remnant of the heat that has been converted from the gravitational energy during the Earth's formation and partly, and more importantly, this heat is generated by the decay of unstable isotopes in the Earth. This heat is transported to the Earth's surface; but is the heat transported by conduction or by convection?

If the material in the Earth does not flow, heat can only be transported by conduction as shown in the left-hand panel of Figure 11.6. This means that it is the average transfer of the molecular motion from warm regions to cold regions that is responsible for the transport of heat. On the other hand, if the material in the Earth does flow, heat can be carried by the flow as shown in the right-hand panel of Figure 11.6. This process is called convection.

The starting point of the analysis is the heat equation (25.30). In the absence of source terms, this equation, for the special case of a constant heat conduction coefficient κ, may be written as:

$$\frac{\partial T}{\partial t} + \nabla \cdot (\mathbf{v}T) = \kappa \nabla^2 T. \tag{11.35}$$

The term $\nabla \cdot (\mathbf{v}T)$ describes the convective heat transport, while the term $\kappa \nabla^2 T$ accounts for the conductive heat transport.

Conduction   Convection

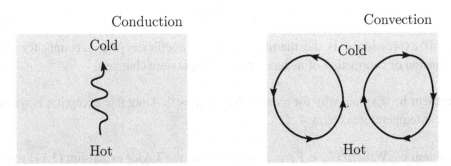

Figure 11.6 Two alternatives for the heat transport in the Earth. In the left-hand panel the material does not move and heat is transported by conduction. In the right-hand panel the material flows and heat is transported by convection.

Problem a Let the characteristic velocity be denoted by V, the characteristic length scale by L, and the characteristic temperature perturbation by T. Show that the ratio of the convective heat transport to the conductive heat transport is of the following order:

$$\frac{\text{convective heat transport}}{\text{conductive heat transport}} \sim \frac{VL}{\kappa}. \tag{11.36}$$

This estimate gives the ratio of the two modes of heat transport. This does not help us much yet, however, because we do not know the order of magnitude V of the flow velocity. This quantity can be obtained from the Navier–Stokes equation of Section 25.5:

$$\frac{\partial(\rho\mathbf{v})}{\partial t} + \nabla \cdot (\rho\mathbf{v}\mathbf{v}) = \mu\nabla^2\mathbf{v} + \mathbf{F}. \tag{25.48}$$

The force (per unit volume) $\mathbf{F}$ on the right-hand side is the buoyancy force that is associated with the flow, while the term $\mu\nabla^2\mathbf{v}$ accounts for the viscosity of the flow with viscosity coefficient μ. The mantle of the Earth is extremely viscous and mantle convection (if it exists at all) is a slow process. We therefore assume that the inertia term $\partial(\rho\mathbf{v})/\partial t$ and the advection term $\nabla \cdot (\rho\mathbf{v}\mathbf{v})$ are small compared to the viscous term $\mu\nabla^2\mathbf{v}$. (This assumption would have to be supported by a proper scale analysis.) Under this assumption, the mantle flow is predominantly governed by a balance between the viscous force and the buoyancy force:

$$\mu\nabla^2\mathbf{v} = -\mathbf{F}. \tag{11.37}$$

The next step is to relate the buoyancy to the temperature perturbation T. According to equation (7.99) of Fowler (2005), a temperature perturbation T from a reference temperature T_0 leads to a density perturbation ρ from the reference density ρ_0 given by:

$$\rho = -\alpha T. \tag{11.38}$$

In this expression α is the thermal expansion coefficient that accounts for the expansion or contraction of material due to temperature changes.

Problem b Explain why for most materials $\alpha > 0$. A notable exception is water at temperatures below 4 °C.

Problem c Write $\rho (T_0 + T) = \rho_0 + \rho$ and use the Taylor expansion (3.11) truncated after the first-order term to show that the expansion coefficient is given by $\alpha = -\partial\rho/\partial T$.

Problem d The buoyancy force is given by Archimedes's law, which states that this force equals the weight of the displaced fluid. Use this result, (11.37), and (11.38) in a scale analysis to show that the velocity is of the following order:

$$V \sim \frac{g\alpha TL^2}{\mu}, \tag{11.39}$$

where g is the acceleration of gravity.

Problem e Use this to derive that the ratio of the convective heat transport to the conductive heat transport is given by

$$\frac{\text{convective heat transport}}{\text{conductive heat transport}} \sim \frac{g\alpha TL^3}{\mu\kappa}. \tag{11.40}$$

The right-hand side of this expression is dimensionless, and is called the *Rayleigh number*, which is denoted by Ra:

$$Ra \equiv \frac{g\alpha T L^3}{\mu\kappa}. \tag{11.41}$$

The Rayleigh number is an indicator for the mode of heat transport. When $Ra \gg 1$, heat is predominantly transported by convection. When the thermal expansion coefficient α is large and when the viscosity μ and the heat conduction coefficient κ are small, the Rayleigh number is large and heat is transported by convection.

Problem f Explain physically why a large value of α and small values of μ and κ lead to convective heat transport rather than conductive heat transport.

discussion of the Rayleigh number and other dimensionless diagnostics such as the Prandtl number and the Grashof number can be found in Section 14.2 of Tritton (1988). The implications of the different values of the Rayleigh number on the character of convection in the Earth's mantle is discussed in Olson (1989) and Turcotte and Schubert (2002). Of course, if one wants to use a scale analysis, one must know the values of the physical properties involved. For the Earth's mantle, the thermal expansion coefficient α is not very well known because of the complications involved in laboratory measurements of the thermal expansion under the extremely high ambient pressure of the Earth's mantle (Chopelas, 1996). Of course, any theory or numerical model is only as good as the numbers that go into it. Seismic tomography has made it possible to make detailed models of the structure of Earth's mantle, such models suggest convective patterns deep into the mantle (e.g., French et al., 2013).

11.6 Making an equation dimensionless

Usually the terms in equations that one wants to analyze have a physical dimension such as temperature and velocity. It can sometimes be useful to rescale all the variables in the equation in such a way that the rescaled variables are dimensionless. This is convenient when setting up numerical solutions of the equations, and it also introduces dimensionless numbers that govern the physics of the problem in a natural way. As an example we will apply this technique to the heat equation (11.35).

Any variable can be made dimensionless by dividing out a constant that has the dimension of the variable. As an example, let the characteristic temperature variation be denoted by T_0, then the dimensional temperature perturbation can be written as:

$$T = T_0 T'. \tag{11.42}$$

The quantity T' is dimensionless. Note that in this section T is the absolute temperature, whereas in the previous section T denoted the temperature perturbation.

In this section, dimensionless variables are denoted by a prime. For example, let the characteristic time used to scale the time variable be denoted by τ, then:

$$t = \tau t'. \tag{11.43}$$

We can still leave τ open and later choose a value that simplifies the equations as much as possible. Of course, when we want to express the heat equation (11.35) in the new time variable, we need to specify how the dimensional time derivative $\partial/\partial t$ is related to the dimensionless time derivative $\partial/\partial t'$.

Problem a   Use the chain rule for differentiation to show that

$$\frac{\partial}{\partial t} = \frac{1}{\tau}\frac{\partial}{\partial t'}. \tag{11.44}$$

Problem b   Let the velocity be scaled with the characteristic velocity (11.39):

$$\mathbf{v} = \frac{g\alpha T_0 L^2}{\mu}\mathbf{v}', \tag{11.45}$$

and let the position vector be scaled with the characteristic length L of the system: $\mathbf{r} = L\mathbf{r}'$. Use a result similar to (11.44) to convert the spatial derivatives to the new space coordinate and rescale all terms in the heat equation (11.35) to derive the following dimensionless form of this equation

$$\underbrace{\frac{1}{\tau}\frac{\partial T'}{\partial t'}}_{(1)} + \underbrace{\frac{g\alpha T_0 L}{\mu}\nabla'\cdot\left(\mathbf{v}'T'\right)}_{(2)} = \underbrace{\frac{\kappa}{L^2}\nabla'^2 T'}_{(3)}, \tag{11.46}$$

where ∇' is the gradient operator with respect to the dimensionless coordinates $\mathbf{r}'$.

At this point we have not yet specified the time scale τ for the scaling of the time variable. There are three possible situations that determine τ. First, when term (2) is much larger than term (3), the time derivative (1) balances term (2). Second, when term (3) is much larger than term (2), the time derivative balances term (3). The third possibility is that terms (2) and (3) are of the same order of magnitude, in which case the balance of terms is more complex.

Problem c   Show that the ratio of term (2) to term (3) is given by the Rayleigh number, which is defined in (11.41).

In this expression all the primed terms are (by definition) of order one. At this point we assume that the convective heat transport dominates the conductive heat transport, that is, that $Ra \gg 1$. This means that term (2) is much larger than term (3); hence, the time derivative in term (1) must be of the same order as the convective term (2). Terms (1) and (2) can therefore balance only when

$$\frac{1}{\tau} = \frac{g\alpha T_0 L}{\mu}. \tag{11.47}$$

This condition determines the time scale of the evolution of the convecting system.

Problem d   Show that with this choice of τ the dimensionless heat equation is given by:

$$\frac{\partial T'}{\partial t'} + \nabla'\cdot\left(\mathbf{v}'T'\right) = \frac{1}{Ra}\nabla'^2 T', \tag{11.48}$$

where Ra is the Rayleigh number.

The advantage of this dimensionless equation over the original heat equation is that (11.48) contains only a single constant Ra, whereas the dimensional heat equation (11.35) depends on a large number of constants. In addition, the scaling of the heat equation has led in a natural way to the key role of the Rayleigh number in the mode of heat transport in a fluid. Since the Rayleigh number is assumed to be large, $1/Ra \ll 1$ and the last term can be seen as a small perturbation. One can either delete this term, or treat it with perturbation theory as described in Chapter 23. The last term in (11.48) contains the highest spatial derivatives of that equation because it contains second-order space derivatives, whereas the other terms either have no space derivatives (term (1)) or only first-order space derivatives (term (2)). Deleting this term turns (11.48) from a second-order differential equation (in the space variables) into a first-order differential equation. This entails a change in the number of boundary conditions that need to be imposed. As shown in Section 23.7, the last term in (11.48) constitutes a *singular perturbation*, and one should be careful in deleting this (small) term altogether.

Transforming dimensional equations into dimensionless equations is often used to derive the relevant dimensionless physical constants of the system as well as to set up algorithms for solving equations numerically. The basic rationale behind this approach is that the physical units that are used are completely arbitrary. It is immaterial whether we express length in meters or in inches, but of course the numerical value of a given length changes when we change from meters to inches. Making the system dimensionless removes all physical units from the system, because all the resulting terms in the equation are dimensionless. This has the drawback that dimensional analysis cannot be used to check for errors, as shown in Section 2.2.

12

Linear algebra

In this chapter several elements of linear algebra are treated that have important applications in physics or that serve to illustrate methodologies used in other areas of mathematical physics. For example, linear algebra provides a foundation for the inverse theory presented in Chapter 22.

12.1 Projections and the completeness relation

In mathematical physics, projections play an important role. This is true not only in linear algebra, but also in the analysis of linear systems such as linear filters in data processing (Section 14.10), and in the analysis of vibrating systems such as the normal modes of the Earth (Section 19.7). Let us consider a vector $\mathbf{v}$ that we want to project along a unit vector $\hat{\mathbf{n}}$ (Figure 12.1). In the examples in this section we work in a three-dimensional space, but the arguments presented here can be generalized to any number of dimensions.

We denote the projection of $\mathbf{v}$ along $\hat{\mathbf{n}}$ as $\mathbf{Pv}$, where $\mathbf{P}$ stands for the projection operator. In a three-dimensional space this operator can be represented by a 3×3 matrix. It is our goal to find the operator $\mathbf{P}$ in terms of the unit vector $\hat{\mathbf{n}}$, as well as the matrix form of this operator. By definition the projection of $\mathbf{v}$ is directed along $\hat{\mathbf{n}}$; hence,

$$\mathbf{Pv} = C\hat{\mathbf{n}}. \tag{12.1}$$

This means that we know the projection operator once the constant C is known.

Problem a Show that with the variables defined in Figure 12.1 the length of the vector $\mathbf{Pv}$ is $|\mathbf{Pv}| = |\mathbf{v}| \cos \varphi$. Use $(\hat{\mathbf{n}} \cdot \mathbf{v}) \equiv |\hat{\mathbf{n}}| \, |\mathbf{v}| \cos \varphi = |\mathbf{v}| \cos \varphi$ to show that $C = (\hat{\mathbf{n}} \cdot \mathbf{v})$.

166

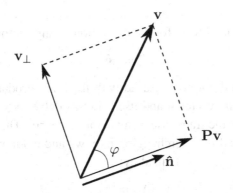

Figure 12.1   Definition of the geometric variables for the projection of a vector **v**.

Inserting this expression for the constant C in (12.1) leads to an expression for the projection

$$\mathbf{Pv} = \hat{\mathbf{n}}\left(\hat{\mathbf{n}} \cdot \mathbf{v}\right).\qquad(12.2)$$

Problem b   Show that the component $\mathbf{v}_\perp$ perpendicular to $\hat{\mathbf{n}}$ as defined in Figure 12.1 is given by:

$$\mathbf{v}_\perp = \mathbf{v} - \hat{\mathbf{n}}\left(\hat{\mathbf{n}} \cdot \mathbf{v}\right).\qquad(12.3)$$

Problem c   As an example, consider the projection along the unit vector along the x-axis: $\hat{\mathbf{n}} = \hat{\mathbf{x}}$. Show using (12.2) and (12.3) that in that case:

$$\mathbf{Pv} = \begin{pmatrix} v_x \\ 0 \\ 0 \end{pmatrix} \quad \text{and} \quad \mathbf{v}_\perp = \begin{pmatrix} 0 \\ v_y \\ v_z \end{pmatrix}.$$

Problem d   When we project the vector **Pv** once more along the same unit vector $\hat{\mathbf{n}}$, the vector will not change. We therefore expect that $\mathbf{P(Pv)} = \mathbf{Pv}$. Show, using (12.2), that this is indeed the case. Since this property holds for any vector **v**, we can also write it as:

$$\mathbf{P}^2 = \mathbf{P}.\qquad(12.4)$$

An operator with this property is called *idempotent*.

Problem e   If **P** were a scalar, the expression above would imply that **P** is the identity operator $\mathbf{P} = 1$ or that $\mathbf{P} = 0$. Explain why for vectors in expression (12.4) this does *not* imply that **P** is either the identity operator or equal to zero.

In (12.2) we derived the action of the projection operator on a vector $\mathbf{v}$. We will now show that the general form for the projection of any vector $\mathbf{v}$ can be written as

$$\mathbf{P} = \hat{\mathbf{n}}\hat{\mathbf{n}}^T. \tag{12.5}$$

This expression should not be confused with the inner product $(\hat{\mathbf{n}} \cdot \hat{\mathbf{n}})$; instead it denotes the dyad of the vector $\hat{\mathbf{n}}$ and itself. Let us delve into what a dyad is. The superscript T denotes the transpose of a vector or matrix. The transpose of a vector (or matrix) is found by interchanging its rows and columns. For example, the transpose $\mathbf{A}^T$ of a matrix $\mathbf{A}$ is defined by:

$$A_{ij}^T = A_{ji}, \tag{12.6}$$

and the transpose of the vector $\mathbf{u}$ is defined by:

$$\mathbf{u}^T = (u_x, u_y, u_z) \quad \text{when} \quad \mathbf{u} = \begin{pmatrix} u_x \\ u_y \\ u_z \end{pmatrix}. \tag{12.7}$$

Taking the transpose converts a column vector into a row vector. As discussed, the projection operator $\mathbf{P}$ in (12.5) is a 3×3 matrix, or tensor. This tensor is called a dyad, because it is created from the dyadic product of two vectors. In general, the dyad $\mathbf{T}$ of two vectors $\mathbf{u}$ and $\mathbf{v}$ is defined as

$$\mathbf{T} = \mathbf{u}\mathbf{v}^T. \tag{12.8}$$

This expression simply means that the components T_{ij} of the dyad are defined by

$$T_{ij} = u_i v_j, \tag{12.9}$$

where u_i is the i-component of $\mathbf{u}$ and v_j is the j-component of $\mathbf{v}$. Contrast this with the dot product (or inner product) of two vectors. In the literature you will find different notations for the dot product of two vectors $\mathbf{u}$ and $\mathbf{v}$:

$$(\mathbf{u} \cdot \mathbf{v}) = \mathbf{u}^T \mathbf{v}. \tag{12.10}$$

In components, the dot product is $\sum_i u_i v_i$ and you can see this results in a scalar. This is why "scalar product" is another name for the dot product.

Problem f   Show that with the component notation of a dyad of (12.9) inserted into expression (12.5), you get the projection operator (12.2).

Problem g   Show that the operator for the projection along the unit vector

$$\hat{\mathbf{n}} = \frac{1}{\sqrt{14}} \begin{pmatrix} 1 \\ 2 \\ 3 \end{pmatrix},$$

is given by

$$\mathbf{P} = \frac{1}{14} \begin{pmatrix} 1 & 2 & 3 \\ 2 & 4 & 6 \\ 3 & 6 & 9 \end{pmatrix}.$$

Verify explicitly that for this example $\mathbf{P}\hat{\mathbf{n}} = \hat{\mathbf{n}}$, and explain this result.

Up to this point we have projected the vector $\mathbf{v}$ along a single unit vector $\hat{\mathbf{n}}$. Suppose we have a set of mutually orthogonal unit vectors $\hat{\mathbf{n}}_i$, which we call orthonormal. The fact that these unit vectors are orthonormal means that the different unit vectors are perpendicular to each other: $(\hat{\mathbf{n}}_i \cdot \hat{\mathbf{n}}_j) = 0$ when $i \neq j$. We can project $\mathbf{v}$ on each of these unit vectors and add these projections. This gives us the projection of $\mathbf{v}$ on the subspace spanned by the unit vectors $\hat{\mathbf{n}}_i$:

$$\mathbf{P}\mathbf{v} = \sum_i \hat{\mathbf{n}}_i \left(\hat{\mathbf{n}}_i \cdot \mathbf{v} \right). \tag{12.11}$$

When the unit vectors $\hat{\mathbf{n}}_i$ span the full space we are working in, the projected vector is identical to the original vector. To see this, consider for example a three-dimensional space. Any vector can be decomposed into its components along the x-, y-, and z-axes, and this can be written as:

$$\mathbf{v} = v_x \hat{\mathbf{x}} + v_y \hat{\mathbf{y}} + v_z \hat{\mathbf{z}} = \hat{\mathbf{x}} \left(\hat{\mathbf{x}} \cdot \mathbf{v} \right) + \hat{\mathbf{y}} \left(\hat{\mathbf{y}} \cdot \mathbf{v} \right) + \hat{\mathbf{z}} \left(\hat{\mathbf{z}} \cdot \mathbf{v} \right). \tag{12.12}$$

Note that this expression has the same form as (12.11). This implies that when in (12.11) we sum over a set of unit vectors that completely spans the space we are working in; the right-hand side of (12.11) is identical to the original vector $\mathbf{v}$; that is, $\sum_i \hat{\mathbf{n}}_i \left(\hat{\mathbf{n}}_i \cdot \mathbf{v} \right) = \mathbf{v}$. The operator of the left-hand side of this equality is therefore identical to the identity operator $\mathbf{I}$:

$$\sum_{i=1}^{N} \hat{\mathbf{n}}_i \hat{\mathbf{n}}_i^T = \mathbf{I}. \tag{12.13}$$

The dimension of the space we are working in is N; if we sum over a smaller number of unit vectors, we project on a subspace of the N-dimensional space. Expression (12.13) expresses that the vectors $\hat{\mathbf{n}}_i$ (with $i = 1, \ldots, N$) can be used to give a complete representation of any vector. Such a set of vectors is called a *complete set*, and expression (12.13) is called the *closure relation*.

Problem h Verify explicitly that when the unit vectors $\hat{\mathbf{n}}_i$ are chosen to be the unit vectors $\hat{\mathbf{x}}$, $\hat{\mathbf{y}}$, and $\hat{\mathbf{z}}$ along the x-, y-, and z-axes, the right-hand side of (12.13) is given by the 3×3 identity matrix.

There are, of course, many different ways of choosing a set of three orthogonal unit vectors in three dimensions. Expression (12.13) should hold for every choice of a complete set of unit vectors.

Problem i   Verify explicitly that when the unit vectors $\hat{\mathbf{n}}_i$ are chosen to be the unit vectors $\hat{\mathbf{r}}$, $\hat{\boldsymbol{\theta}}$, and $\hat{\boldsymbol{\varphi}}$ defined in (4.7) for a system of spherical coordinates, the right-hand side of (12.13) is given by the 3×3 identity matrix.

12.2 Coriolis and centrifugal force

As an example of working with the cross-product of vectors, we consider the inertia forces that occur in the mechanics of rotating coordinate systems. This is of importance in the Earth sciences, because the rotation of the Earth plays a crucial role in the motion of wind in the atmosphere and currents in the ocean. In addition, the Earth's rotation is essential for the generation of the magnetic field of the Earth in the outer core.

To describe the motion of a particle in a rotating coordinate system, we need to characterize the rotation somehow. This can be achieved by introducing a vector $\boldsymbol{\Omega}$ that is aligned with the rotation axis and whose length is given by the *rate* of rotation expressed in radians per seconds (Section 7.3).

Problem a   What is the direction of $\boldsymbol{\Omega}$ and the length $\Omega = |\boldsymbol{\Omega}|$ for the Earth's rotation?

Let us assume we are considering a vector $\mathbf{q}$ that is constant in the rotating coordinate system. In a nonrotating system this vector changes with time because it corotates with the rotating system. The vector $\mathbf{q}$ can be decomposed into a component $\mathbf{q}_{\parallel}$ parallel to rotation vector $\boldsymbol{\Omega}$ and a component $\mathbf{q}_{\perp}$ perpendicular to $\boldsymbol{\Omega}$. In addition, a vector $\mathbf{b}$ is defined in Figure 12.2 that is perpendicular to both $\mathbf{q}_{\perp}$ and $\boldsymbol{\Omega}$ in such a way that $\boldsymbol{\Omega}$, $\mathbf{q}_{\perp}$, and $\mathbf{b}$ form a right-handed orthogonal system.

Problem b   Show that:

$$\left.\begin{aligned}
\mathbf{q}_{\parallel} &= \hat{\boldsymbol{\Omega}}\left(\hat{\boldsymbol{\Omega}} \cdot \mathbf{q}\right), \\
\mathbf{q}_{\perp} &= \mathbf{q} - \hat{\boldsymbol{\Omega}}\left(\hat{\boldsymbol{\Omega}} \cdot \mathbf{q}\right), \\
\mathbf{b} &= \hat{\boldsymbol{\Omega}} \times \mathbf{q}.
\end{aligned}\right\} \tag{12.14}$$

The last identity follows by inserting the second equation into the identity $\mathbf{b} = \hat{\boldsymbol{\Omega}} \times \mathbf{q}_{\perp}$.

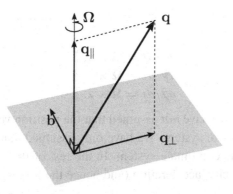

Figure 12.2 Decomposition of a vector **q** in a rotating coordinate system.

In a fixed (nonrotating) coordinate system, the vector **q** rotates; hence, its position is time dependent: $\mathbf{q} = \mathbf{q}(t)$. How does the vector change over a time interval Δt? Since the component $\mathbf{q}_\parallel$ is at all times directed along the rotation vector $\mathbf{\Omega}$, it is constant in time.

Problem c Over a time interval Δt the coordinate system rotates over an angle $\Omega \Delta t$. Use this to show that the component of **q** perpendicular to the rotation vector satisfies:

$$\mathbf{q}_\perp(t + \Delta t) = \cos{(\Omega \Delta t)}\, \mathbf{q}_\perp(t) + \sin{(\Omega \Delta t)}\, \mathbf{b}, \qquad (12.15)$$

and that time evolution of **q** is therefore given by

$$\mathbf{q}(t + \Delta t) = \mathbf{q}(t) + [\cos{(\Omega \Delta t)} - 1]\, \mathbf{q}_\perp(t) + \sin{(\Omega \Delta t)}\, \mathbf{b}. \qquad (12.16)$$

The goal is to obtain the time derivative of the vector **q**.

Problem d Use that $d\mathbf{q}/dt = \lim_{\Delta t \to 0}[\mathbf{q}(t + \Delta t) - \mathbf{q}(t)]/\Delta t$ and expression (12.16) to show that

$$\dot{\mathbf{q}} = \Omega \mathbf{b}, \qquad (12.17)$$

where the dot denotes the time derivative. Use (12.14) to show that the time derivative of the vector **q** is given by

$$\dot{\mathbf{q}} = \mathbf{\Omega} \times \mathbf{q}. \qquad (12.18)$$

At this point, the vector **q** can be any vector that corotates with the rotating coordinate system. In this rotating coordinate system, three Cartesian basis vectors $\hat{\mathbf{x}}$, $\hat{\mathbf{y}}$, and $\hat{\mathbf{z}}$ can be used as a basis to decompose the position vector:

$$\mathbf{r}_{rot} = x\hat{\mathbf{x}} + y\hat{\mathbf{y}} + z\hat{\mathbf{z}}. \qquad (12.19)$$

Since these basis vectors are constant in the rotating coordinate system, they satisfy (12.18), so that

$$\left.\begin{array}{l} d\hat{\mathbf{x}}/dt = \boldsymbol{\Omega} \times \hat{\mathbf{x}}, \\ d\hat{\mathbf{y}}/dt = \boldsymbol{\Omega} \times \hat{\mathbf{y}}, \\ d\hat{\mathbf{z}}/dt = \boldsymbol{\Omega} \times \hat{\mathbf{z}}. \end{array}\right\} \qquad (12.20)$$

It should be noted that we have *not* assumed that the position vector $\mathbf{r}_{rot}$ in (12.19) rotates with the coordinate system. We have only assumed that the unit vectors $\hat{\mathbf{x}}$, $\hat{\mathbf{y}}$, and $\hat{\mathbf{z}}$ rotate with the coordinate system. In the rest of this section, we explore the imprint on the velocity, acceleration (and hence the forces) on particles in this rotating system.

In general, the velocity and the acceleration follow by differentiating (12.19) with time. If the unit vectors $\hat{\mathbf{x}}$, $\hat{\mathbf{y}}$, and $\hat{\mathbf{z}}$ were fixed, they would not contribute to the time derivative. However, the unit vectors $\hat{\mathbf{x}}$, $\hat{\mathbf{y}}$, and $\hat{\mathbf{z}}$ rotate with the coordinate system and the associated time derivative is given by (12.20).

Problem e   Differentiate the position vector in (12.19) with respect to time and show that the velocity vector $\mathbf{v}$ is given by:

$$\mathbf{v} = \dot{x}\hat{\mathbf{x}} + \dot{y}\hat{\mathbf{y}} + \dot{z}\hat{\mathbf{z}} + \boldsymbol{\Omega} \times \mathbf{r}_{rot}. \qquad (12.21)$$

The terms $\dot{x}\hat{\mathbf{x}} + \dot{y}\hat{\mathbf{y}} + \dot{z}\hat{\mathbf{z}}$ denote the velocity as seen in the rotating coordinate system; this velocity is denoted by $\mathbf{v}_{rot}$. The velocity vector can therefore be written as:

$$\mathbf{v} = \mathbf{v}_{rot} + \boldsymbol{\Omega} \times \mathbf{r}_{rot}. \qquad (12.22)$$

Problem f   Give an interpretation of the last term in this expression.

Problem g   The acceleration follows by differentiating (12.21) for the velocity once more with respect to time. Show that the acceleration is given by

$$\mathbf{a} = \ddot{x}\hat{\mathbf{x}} + \ddot{y}\hat{\mathbf{y}} + \ddot{z}\hat{\mathbf{z}} + 2\boldsymbol{\Omega} \times \left(\dot{x}\hat{\mathbf{x}} + \dot{y}\hat{\mathbf{y}} + \dot{z}\hat{\mathbf{z}}\right) + \boldsymbol{\Omega} \times \left(\boldsymbol{\Omega} \times \mathbf{r}_{rot}\right). \qquad (12.23)$$

The terms $\ddot{x}\hat{\mathbf{x}} + \ddot{y}\hat{\mathbf{y}} + \ddot{z}\hat{\mathbf{z}}$ on the right-hand side denote the acceleration as seen in the rotating coordinate system; this quantity will be denoted by $\mathbf{a}_{rot}$. The terms $\dot{x}\hat{\mathbf{x}} + \dot{y}\hat{\mathbf{y}} + \dot{z}\hat{\mathbf{z}}$ again denote the velocity $\mathbf{v}_{rot}$ as seen in the rotating coordinate system. The left-hand side is by Newton's law equal to $\mathbf{F}/m$, where $\mathbf{F}$ is the force acting on the particle.

Problem h   Use this to show that in the rotating coordinate system Newton's law is given by:

$$m\mathbf{a}_{rot} = \mathbf{F} - 2m\boldsymbol{\Omega} \times \mathbf{v}_{rot} - m\boldsymbol{\Omega} \times \left(\boldsymbol{\Omega} \times \mathbf{r}_{rot}\right). \qquad (12.24)$$

The rotation manifests itself through two additional forces. The term $-2m\mathbf{\Omega} \times \mathbf{v}_{rot}$ describes the Coriolis force and the term $-m\mathbf{\Omega} \times (\mathbf{\Omega} \times \mathbf{r}_{rot})$ describes the centrifugal force.

Problem i Show that the centrifugal force is perpendicular to the rotation axis and is directed from the rotation axis toward the particle. Note that the minus sign in the centrifugal force in expression (12.24) ensures that the centrifugal force is directed away from the axis of rotation.

Problem j Air flows from high-pressure areas to low-pressure areas. Does the Coriolis force deflect air flow in the northern hemisphere to the left or to the right, when seen from above?

Problem k Compute the magnitude of the centrifugal force and the Coriolis force you experience due to the Earth's rotation when you ride your bicycle at the north pole. Compare this with the force $m\mathbf{g}$ you experience due to the gravitational attraction of the Earth. It suffices to compute the orders of magnitude of the different terms. Have you ever noticed a tilt due to the Coriolis force while riding your bicycle?

In meteorology and oceanography it is often convenient to describe the motion of air or water along the Earth's surface using a Cartesian coordinate system that rotates with the Earth with unit vectors pointing eastwards ($\hat{\mathbf{e}}_1$), northwards ($\hat{\mathbf{e}}_2$), and upwards ($\hat{\mathbf{e}}_3$), as defined in Figure 12.3. These unit vectors can be related to the unit vectors $\hat{\mathbf{r}}$, $\hat{\boldsymbol{\varphi}}$, and $\hat{\boldsymbol{\theta}}$ that are defined in (4.7). Let the velocity in the eastward direction be denoted by u, the velocity in the northward direction by v, and the vertical velocity by w.

Problem l Show that:

$$\hat{\mathbf{e}}_1 = \hat{\boldsymbol{\varphi}}, \quad \hat{\mathbf{e}}_2 = -\hat{\boldsymbol{\theta}}, \quad \hat{\mathbf{e}}_3 = \hat{\mathbf{r}}, \tag{12.25}$$

and that the velocity in this rotating coordinate system is given by

$$\mathbf{v} = u\hat{\mathbf{e}}_1 + v\hat{\mathbf{e}}_2 + w\hat{\mathbf{e}}_3. \tag{12.26}$$

Problem m We assume that the axes of the spherical coordinate system are chosen in such a way that the direction $\theta = 0$ is aligned with the rotation axis. This is a different way of saying that the rotation vector is parallel to the z-axis: $\mathbf{\Omega} = \Omega\hat{\mathbf{z}}$. Use the first two expressions of (4.13) to show that the rotation vector has the following expansion in terms of the unit vectors $\hat{\mathbf{r}}$ and $\hat{\boldsymbol{\theta}}$:

$$\mathbf{\Omega} = \Omega \left(\cos\theta\, \hat{\mathbf{r}} - \sin\theta\, \hat{\boldsymbol{\theta}} \right). \tag{12.27}$$

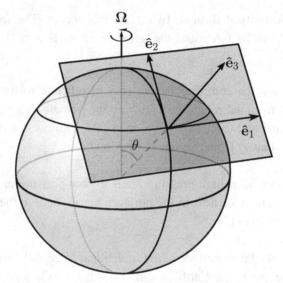

Figure 12.3   Definition of a local Cartesian coordinate system that is aligned with the Earth's surface.

Problem n   In the rotating coordinate system, the Coriolis force is given by $\mathbf{F}_{cor} = -2m\mathbf{\Omega} \times \mathbf{v}$. Use expressions (12.25)–(12.27) and (4.11) for the cross-product of the unit vectors to show that the Coriolis force is given by

$$\mathbf{F}_{cor} = 2m\Omega \sin\theta \, u \, \hat{\mathbf{r}} + 2m\Omega \cos\theta \, u \, \hat{\boldsymbol{\theta}} + 2m\Omega \, (v\cos\theta \, - w\sin\theta) \, \hat{\boldsymbol{\varphi}}.$$
$$(12.28)$$

Problem o   Both the ocean and the atmosphere are shallow in the sense that the vertical length scale (a few kilometers for the ocean and around 10 kilometers for the atmosphere) is much less than the horizontal length scale. This causes the vertical velocity to be much smaller than the horizontal velocity. For this reason the vertical velocity w can be neglected in (12.28). Use this approximation and definition (12.25) to show that the horizontal component $\mathbf{a}_{cor}^{H}$ of the Coriolis acceleration is given in this approach by:

$$\mathbf{a}_{cor}^{H} = -f\hat{\mathbf{e}}_3 \times \mathbf{v},$$
$$(12.29)$$

with

$$f = 2\Omega \cos\theta.$$
$$(12.30)$$

This result is widely used in meteorology and oceanography, because (12.29) states that, in the Cartesian coordinate system aligned with the Earth's surface, the Coriolis force generated by the rotation around the true Earth's axis of rotation is identical to the Coriolis force generated by the rotation around a vertical axis with a rotation rate given by $\Omega \cos \theta$. This rotation rate is largest at the poles, where $\cos \theta = \pm 1$, and vanishes at the equator, where $\cos \theta = 0$. The parameter f in (12.29) acts as a coupling parameter; it is called the *Coriolis parameter*. (In the literature on geophysical fluid dynamics one often uses latitude rather than the colatitude θ that is used here, and for this reason one often sees a sin term rather than a cos term in the definition of the Coriolis parameter.) In many applications one disregards the dependence of f on the colatitude θ; in that approach f is a constant and one speaks of the *f-plane approximation*. However, the dependence of the Coriolis parameter on θ is crucial in explaining a number of atmospheric and oceanographic phenomena such as the propagation of Rossby waves and the formation of the Gulf stream. In a further refinement one linearizes the dependence of the Coriolis parameter with colatitude. This leads to the *β-plane approximation*. Details can be found in the books of Holton and Hakim (2012) and Pedlosky (1982).

12.3 Eigenvalue decomposition of a square matrix

In this section we consider the way in which a square $N \times N$ matrix $\mathbf{A}$ operates on a vector. Since a matrix describes a linear transformation from a vector to a new vector, the action of the matrix $\mathbf{A}$ can be quite complicated. However, suppose the matrix has a set of eigenvectors $\hat{\mathbf{v}}^{(n)}$. We assume these eigenvectors are normalized; hence, a caret is used in the notation. These eigenvectors are useful because the action of $\mathbf{A}$ on an eigenvector $\hat{\mathbf{v}}^{(n)}$ is simple:

$$\mathbf{A}\hat{\mathbf{v}}^{(n)} = \lambda_n \hat{\mathbf{v}}^{(n)}, \tag{12.31}$$

where λ_n is the eigenvalue of the eigenvector $\hat{\mathbf{v}}^{(n)}$. When $\mathbf{A}$ acts on an eigenvector, the resulting vector is parallel to the original vector, and the only effect of $\mathbf{A}$ on this vector is to elongate the vector (when $\lambda_n \geq 1$), compress the vector (when $0 \leq \lambda_n < 1$), or reverse the vector (when $\lambda_n < 0$). We restrict ourselves here to matrices that are real and symmetric.

Problem a   Show that for such matrices the eigenvalues are real and the eigenvectors are orthogonal.

Strang (2003) is an excellent resource for properties like this. The fact that the eigenvectors $\hat{\mathbf{v}}^{(n)}$ are normalized and mutually orthogonal can be expressed as

$$\left(\hat{\mathbf{v}}^{(n)} \cdot \hat{\mathbf{v}}^{(m)} \right) = \delta_{nm}, \tag{12.32}$$

where δ_{nm} is the Kronecker delta, which is defined as

$$\delta_{nm} = \begin{cases} 1 & \text{when} \quad n = m \\ 0 & \text{when} \quad n \neq m. \end{cases} \tag{25.15}$$

The eigenvectors $\hat{\mathbf{v}}^{(n)}$ can be used to define the columns of a matrix $\mathbf{V}$:

$$
\mathbf{V} = \begin{pmatrix} \vdots & \vdots & & \vdots \\ \hat{\mathbf{v}}^{(1)} & \hat{\mathbf{v}}^{(2)} & \cdots & \hat{\mathbf{v}}^{(N)} \\ \vdots & \vdots & & \vdots \end{pmatrix}. \tag{12.33}
$$

This definition implies that

$$
V_{ij} \equiv v_i^{(j)}. \tag{12.34}
$$

Problem b   Use the orthogonality of the eigenvectors $\hat{\mathbf{v}}^{(n)}$ (expression (12.32)) to show that the matrix $\mathbf{V}$ is unitary (see Section 26.23), that is,

$$
\mathbf{V}^T \mathbf{V} = \mathbf{I}, \tag{12.35}
$$

where $\mathbf{I}$ is the identity matrix with elements $I_{kl} = \delta_{kl}$. The superscript T denotes the transpose.

There are N eigenvectors that are mutually orthonormal in an N-dimensional space. These eigenvectors therefore form a complete set and analogously to (12.13) the completeness relation can be expressed as

$$
\mathbf{I} = \sum_{n=1}^{N} \hat{\mathbf{v}}^{(n)} \hat{\mathbf{v}}^{(n)T}. \tag{12.36}
$$

When the terms in this expression operate on an arbitrary vector $\mathbf{p}$, an expansion of $\mathbf{p}$ in the eigenvectors that is analogous to (12.11) is obtained:

$$
\mathbf{p} = \sum_{n=1}^{N} \hat{\mathbf{v}}^{(n)} \hat{\mathbf{v}}^{(n)T} \mathbf{p} = \sum_{n=1}^{N} \hat{\mathbf{v}}^{(n)} \left(\hat{\mathbf{v}}^{(n)} \cdot \mathbf{p} \right). \tag{12.37}
$$

This is a useful expression because it can be used to simplify the effect of the matrix $\mathbf{A}$ on an arbitrary vector $\mathbf{p}$.

Problem c   Let $\mathbf{A}$ act on (12.37) and show that:

$$
\mathbf{A}\mathbf{p} = \sum_{n=1}^{N} \lambda_n \hat{\mathbf{v}}^{(n)} \left(\hat{\mathbf{v}}^{(n)} \cdot \mathbf{p} \right). \tag{12.38}
$$

This expression has an interesting geometric interpretation. When $\mathbf{A}$ acts on $\mathbf{p}$, the vector $\mathbf{p}$ is projected on each of the eigenvectors; this is described by the term $(\hat{\mathbf{v}}^{(n)} \cdot \mathbf{p})$. The corresponding eigenvector $\hat{\mathbf{v}}^{(n)}$ is multiplied by the eigenvalue $\hat{\mathbf{v}}^{(n)} \rightarrow \lambda_n \hat{\mathbf{v}}^{(n)}$, and the result is summed over all the eigenvectors. The action of $\mathbf{A}$ can thus be reduced to a projection on eigenvectors, a multiplication with the corresponding eigenvalue and a summation over all eigenvectors. The eigenvalue λ_n can be seen as the sensitivity of the eigenvector $\hat{\mathbf{v}}^{(n)}$ to the matrix $\mathbf{A}$.

Problem d Expression (12.38) holds for every vector $\mathbf{p}$. Use this to show that $\mathbf{A}$ can be written as:

$$\mathbf{A} = \sum_{n=1}^{N} \lambda_n \hat{\mathbf{v}}^{(n)} \hat{\mathbf{v}}^{(n)T}. \tag{12.39}$$

Problem e Show that with the definition (12.33) this result can also be written as:

$$\mathbf{A} = \mathbf{V}\Sigma\mathbf{V}^T, \tag{12.40}$$

where Σ is a matrix that has the eigenvalues on its diagonal and whose other elements are equal to zero:

$$\Sigma = \begin{pmatrix} \lambda_1 & 0 & \cdots & 0 \\ 0 & \lambda_2 & \cdots & 0 \\ \vdots & \vdots & \ddots & \vdots \\ 0 & 0 & & \lambda_N \end{pmatrix}. \tag{12.41}$$

Hint: Let (12.40) act on an arbitrary vector; use definition (12.34) and see what happens.

12.4 Computing a function of a matrix

Expansion (12.39) (or equivalently (12.40)) is useful because it provides a way to compute the inverse of a matrix and to compute functions of a matrix such as the exponential of a matrix. Let us first use (12.39) to compute the inverse $\mathbf{A}^{-1}$ of the matrix. In order to do this we must know the effect of $\mathbf{A}^{-1}$ on the eigenvectors $\hat{\mathbf{v}}^{(n)}$.

Problem a Use the identities $\hat{\mathbf{v}}^{(n)} = \mathbf{I}\hat{\mathbf{v}}^{(n)} = \mathbf{A}^{-1}\mathbf{A}\hat{\mathbf{v}}^{(n)} = \mathbf{A}^{-1}\lambda_n\hat{\mathbf{v}}^{(n)}$ to show that $\hat{\mathbf{v}}^{(n)}$ is also an eigenvector of the inverse $\mathbf{A}^{-1}$ with eigenvalue $1/\lambda_n$:

$$\mathbf{A}^{-1}\hat{\mathbf{v}}^{(n)} = \frac{1}{\lambda_n}\hat{\mathbf{v}}^{(n)}. \tag{12.42}$$

Problem b Use this result and the eigenvector decomposition (12.37) to show that the effect of $\mathbf{A}^{-1}$ on a vector $\mathbf{p}$ can be written as

$$\mathbf{A}^{-1}\mathbf{p} = \sum_{n=1}^{N} \frac{1}{\lambda_n} \hat{\mathbf{v}}^{(n)} \left(\hat{\mathbf{v}}^{(n)} \cdot \mathbf{p} \right). \tag{12.43}$$

Also show that this implies that $\mathbf{A}^{-1}$ can also be written as:

$$\mathbf{A}^{-1} = \mathbf{V}\Sigma^{-1}\mathbf{V}^T, \tag{12.44}$$

with

$$\Sigma^{-1} = \begin{pmatrix} 1/\lambda_1 & 0 & \cdots & 0 \\ 0 & 1/\lambda_2 & \cdots & 0 \\ \vdots & \vdots & \ddots & \vdots \\ 0 & 0 & & 1/\lambda_N \end{pmatrix}. \tag{12.45}$$

This is an important result: it means that once we have computed the eigenvectors and eigenvalues of a matrix, we can compute the inverse matrix efficiently. Note that this procedure gives problems when one of the eigenvalues is equal to zero because for such an eigenvalue $1/\lambda_n$ is not defined. This makes sense: when one (or more) of the eigenvalues vanishes, the matrix is singular and the inverse does not exist. Also when one of the eigenvalues is nonzero but close to zero, the corresponding term $1/\lambda_n$ is large, and in practice this gives rise to numerical instabilities. In this situation the inverse of the matrix exists, but the result is very sensitive to computational (and other) errors. Such a matrix is called *poorly conditioned*. More aspects on the inverse of a matrix are discussed in Chapter 22.

In general, a function of a matrix, such as the exponent of a matrix, is not defined. However, suppose we have a function $f(z)$ that operates on a scalar z and that this function can be written as a power series:

$$f(z) = \sum_p a_p z^p. \tag{12.46}$$

For example, when $f(z) = \exp(z)$, then according to (3.14):

$f(z) = \sum_{p=0}^{\infty} (1/p!)z^p$. On replacing the scalar z by the matrix $\mathbf{A}$, the power series expansion can be used to *define* the effect of the function f when it operates on the matrix $\mathbf{A}$:

$$f(\mathbf{A}) \equiv \sum_p a_p \mathbf{A}^p. \tag{12.47}$$

Although this may seem to be a simple rule for computing $f(\mathbf{A})$, it is actually not so useful because in many applications the summation (12.47) consists of infinitely many terms and the computation of $\mathbf{A}^p$ can be computationally very demanding. Again, the eigenvalue decomposition (12.39) or (12.40) allows us to simplify the evaluation of $f(\mathbf{A})$.

Problem c Show that $\hat{\mathbf{v}}^{(n)}$ is also an eigenvector of $\mathbf{A}^p$ with eigenvalue $(\lambda_n)^p$, that is, show that

$$\mathbf{A}^p\hat{\mathbf{v}}^{(n)} = (\lambda_n)^p\,\hat{\mathbf{v}}^{(n)}. \tag{12.48}$$

Hint: First compute $\mathbf{A}^2\hat{\mathbf{v}}^{(n)} = \mathbf{A}\left(\mathbf{A}\hat{\mathbf{v}}^{(n)}\right)$, then $\mathbf{A}^3\hat{\mathbf{v}}^{(n)}$, etc.

Problem d Use this result to show that (12.40) can be generalized to:

$$\mathbf{A}^p = \mathbf{V}\Sigma^p\mathbf{V}^T, \tag{12.49}$$

with Σ^p given by

$$\Sigma^p = \begin{pmatrix} \lambda_1^p & 0 & \cdots & 0 \\ 0 & \lambda_2^p & \cdots & 0 \\ \vdots & \vdots & \ddots & \vdots \\ 0 & 0 & & \lambda_N^p \end{pmatrix}. \tag{12.50}$$

Problem e Finally use (12.46) and (12.47) to show that $f(\mathbf{A})$ can be written as:

$$f(\mathbf{A}) = \mathbf{V}f(\Sigma)\mathbf{V}^T, \tag{12.51}$$

with $f(\Sigma)$ given by

$$f(\Sigma) = \begin{pmatrix} f(\lambda_1) & 0 & \cdots & 0 \\ 0 & f(\lambda_2) & \cdots & 0 \\ \vdots & \vdots & \ddots & \vdots \\ 0 & 0 & & f(\lambda_N) \end{pmatrix}. \tag{12.52}$$

Problem f In order to revert to an explicit eigenvector expansion, show that (12.51) can be written as:

$$f(\mathbf{A}) = \sum_{n=1}^{N} f(\lambda_n)\,\hat{\mathbf{v}}^{(n)}\hat{\mathbf{v}}^{(n)T}. \tag{12.53}$$

With this expression (or the equivalent expression (12.51)) the evaluation of $f(\mathbf{A})$ is simple once the eigenvectors and eigenvalues of $\mathbf{A}$ are known, because in (12.53) the function f only acts on the eigenvalues, but not on the matrix. Since the function f normally acts on a scalar (such as the eigenvalues), the eigenvector decomposition has obviated the need for computing higher powers of the matrix $\mathbf{A}$. From a numerical point of view, however, computing functions of matrices can be a tricky issue. For example, Moler and Van Loan (1978) give nineteen dubious ways to compute the exponential of a matrix.

12.5 Normal modes of a vibrating system

An eigenvector decomposition is not only useful for computing the inverse of a matrix and other functions of a matrix, it also provides a way to analyze characteristics of dynamical systems. As an example, a simple model for the oscillations of a vibrating molecule is discussed here. This system is the prototype of a vibrating system that has different modes of vibration. The natural modes of vibration are usually called the *normal modes* of that system. Consider the mechanical system shown in Figure 12.4. Three particles with mass m are coupled by two springs with spring constant k. It is assumed that the three masses are constrained to move along a line. The displacements of the masses from their equilibrium positions are denoted with x_1, x_2, and x_3. This mechanical model can be considered a grossly oversimplified model of a tri-atomic molecule such as CO_2.

Each of the masses can experience an external force F_i, where the subscript i denotes the mass under consideration. The equations of motion for the three masses are given by:

$$\left.\begin{array}{l} m\ddot{x}_1 = k(x_2 - x_1) + F_1, \\ m\ddot{x}_2 = -k(x_2 - x_1) + k(x_3 - x_2) + F_2, \\ m\ddot{x}_3 = -k(x_3 - x_2) + F_3. \end{array}\right\} \tag{12.54}$$

For the moment we consider harmonic oscillations; that is, we assume that both the driving forces F_i and the displacements x_i vary with time as $e^{-i\omega t}$. The displacements x_1, x_2, and x_3 can be used to form a vector $\mathbf{x}$, and similarly a vector $\mathbf{F}$ can be formed from the three forces F_1, F_2, and F_3 that act on the three masses.

Problem a   Show that for a harmonic motion with frequency ω the equations of motion can be written in vector form as:

$$\left(\mathbf{A} - \frac{m\omega^2}{k}\mathbf{I}\right)\mathbf{x} = \frac{1}{k}\mathbf{F}, \tag{12.55}$$

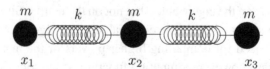

Figure 12.4   Three masses m connected via two springs with constant k vibrate around their equilibrium positions with amplitudes x_1, x_2, x_3, respectively.

with the matrix **A** given by

$$\mathbf{A} = \begin{pmatrix} 1 & -1 & 0 \\ -1 & 2 & -1 \\ 0 & -1 & 1 \end{pmatrix}. \tag{12.56}$$

The normal modes of the system are given by the patterns of oscillations of the system when there is no driving force. For this reason, we set the driving force **F** on the right-hand side of (12.55) momentarily to zero. Equation (12.55) then reduces to a homogeneous system of linear equations; such a system of equations can only have nonzero solutions when the determinant of the matrix vanishes (Strang, 2003). Since the matrix **A** has three eigenvalues, the system can only oscillate freely at three discrete eigenfrequencies. The system can only oscillate at other frequencies when it is driven by the force **F** at such a frequency.

Problem b Show that the eigenfrequencies ω_i of the vibrating system are given by

$$\omega_i = \sqrt{\frac{k\lambda_i}{m}}, \tag{12.57}$$

where λ_i are the eigenvalues of the matrix **A**.

Problem c Show that the eigenfrequencies of the system are given by:

$$\omega_1 = 0, \quad \omega_2 = \sqrt{\frac{k}{m}}, \quad \omega_3 = \sqrt{\frac{3k}{m}}. \tag{12.58}$$

Problem d The frequencies do not give the vibrations of each of the three particles respectively. Instead these frequencies give the eigenfrequencies of the three modes of oscillation of the system. The eigenvector that corresponds to each eigenvalue gives the displacement of each particle for that mode of oscillation. Show that these eigenvectors are given by:

$$\hat{\mathbf{v}}^{(1)} = \frac{1}{\sqrt{3}} \begin{pmatrix} 1 \\ 1 \\ 1 \end{pmatrix}, \quad \hat{\mathbf{v}}^{(2)} = \frac{1}{\sqrt{2}} \begin{pmatrix} 1 \\ 0 \\ -1 \end{pmatrix}, \quad \hat{\mathbf{v}}^{(3)} = \frac{1}{\sqrt{6}} \begin{pmatrix} 1 \\ -2 \\ 1 \end{pmatrix}. \tag{12.59}$$

Remember that the eigenvectors can be multiplied by an arbitrary constant, and this constant is chosen in such a way that each eigenvector has length 1.

Problem e Show that these eigenvectors satisfy the requirement (12.32).

Problem f   Sketch the motion of the three masses for each normal mode. Explain physically why the third mode $\hat{\mathbf{v}}^{(3)}$ has a higher eigenfrequency than the second mode $\hat{\mathbf{v}}^{(2)}$.

Problem g   Explain why the second mode has an eigenfrequency $\omega_2 = \sqrt{k/m}$ that is identical to the frequency of a single mass m that is suspended by a spring with spring constant k.

Problem h   What type of motion does the first mode with eigenfrequency ω_1 describe? Explain why this frequency is independent of the spring constant k and the mass m.

Now that we know the normal modes of the system, we consider the case in which the system is driven by a force $\mathbf{F}$ that varies in time as $e^{-i\omega t}$. For simplicity we assume that the frequency ω of the driving force differs from the eigenfrequencies of the system: $\omega \neq \omega_i$. The eigenvectors $\hat{\mathbf{v}}^{(n)}$ defined in (12.59) form a complete orthonormal set; hence, both the driving force $\mathbf{F}$ and the displacement $\mathbf{x}$ can be expanded in this set. Using (12.37) the driving force can be expanded as

$$\mathbf{F} = \sum_{n=1}^{3} \hat{\mathbf{v}}^{(n)}(\hat{\mathbf{v}}^{(n)} \cdot \mathbf{F}). \tag{12.60}$$

Problem i   Write the displacement vector as a superposition of the normal mode displacements: $\mathbf{x} = \sum_{n=1}^{3} c_n \hat{\mathbf{v}}^{(n)}$, use expansion (12.60) for the driving force, and insert these equations in the equation of motion (12.55) to solve for the unknown coefficients c_n. Eliminate the eigenvalues with (12.57) in favor of the eigenfrequencies ω_n, and show that the displacement is given by:

$$\mathbf{x} = \frac{1}{m} \sum_{n=1}^{3} \frac{\hat{\mathbf{v}}^{(n)}(\hat{\mathbf{v}}^{(n)} \cdot \mathbf{F})}{\left(\omega_n^2 - \omega^2\right)}. \tag{12.61}$$

This expression has a nice physical interpretation. Expression (12.61) states that the total response of the system can be written as a superposition of the different normal modes (the $\sum_{n=1}^{3} \hat{\mathbf{v}}^{(n)}$ terms). The effect that the force has on each normal mode is given by the inner product $(\hat{\mathbf{v}}^{(n)} \cdot \mathbf{F})$. This is nothing but the component of the force $\mathbf{F}$ along the eigenvector $\hat{\mathbf{v}}^{(n)}$, see (12.2). The term $1/\left(\omega_n^2 - \omega^2\right)$ gives the sensitivity of each mode to a driving force with frequency ω; this term can be called a sensitivity term. When the driving force is close to one of the eigenfrequencies of the nth mode, $1/\left(\omega_n^2 - \omega^2\right)$ is large. In that case the system is close to resonance and the resulting displacement is large. On the other hand, when the frequency of the driving force is far from the eigenfrequencies of the system, $1/\left(\omega_n^2 - \omega^2\right)$ is small and the system gives a small response. The total response can be seen as a combination of three basic operations: eigenvector expansion, projection, and multiplication with a response function. Note that the same operations were used in the explanation of the action of a matrix $\mathbf{A}$ given below equation (12.38). Chapter 19 has examples of other systems with normal modes, such as the Earth's response to large earthquakes.

12.6 Singular value decomposition

In Section 12.3 the decomposition of a square matrix in terms of eigenvectors was treated. In many practical applications, such as inverse problems, one encounters a system of equations that is not square:

$$\underbrace{\mathbf{A}}_{\substack{M \times N \\ matrix}} \underbrace{\mathbf{x}}_{\substack{N \\ rows}} = \underbrace{\mathbf{y}}_{\substack{M \\ rows}}. \tag{12.62}$$

Consider the example in which the vector $\mathbf{x}$ has N components and there are M equations. In that case the vector $\mathbf{y}$ has M components and the matrix $\mathbf{A}$ has M rows and N columns; that is, it is an $M \times N$ matrix. A relation such as (12.31), which states that $\mathbf{A}\hat{\mathbf{v}}^{(n)} = \lambda_n \hat{\mathbf{v}}^{(n)}$, cannot possibly hold because when the matrix $\mathbf{A}$ acts on an N-vector, it produces an M-vector, whereas in (12.31) the vector on the right-hand side has the same number of components as the vector on the left-hand side. It is clear that the theory of Section 12.3 cannot be applied when the matrix is not square. However, it is possible to generalize the theory of Section 12.3 when $\mathbf{A}$ is not square. For simplicity it is assumed that $\mathbf{A}$ is a real matrix.

In Section 12.3 a single set of orthonormal eigenvectors $\hat{\mathbf{v}}^{(n)}$ was used to analyze the problem. Since the vectors $\mathbf{x}$ and $\mathbf{y}$ in (12.62) have different dimensions, it is necessary to expand the vector $\mathbf{x}$ in a set of N orthogonal vectors $\hat{\mathbf{v}}^{(n)}$ that each have N components and to expand $\mathbf{y}$ in a different set of M orthogonal vectors $\hat{\mathbf{u}}^{(m)}$ that each have M components. Suppose we have chosen a set $\hat{\mathbf{v}}^{(n)}$, let us define vectors $\hat{\mathbf{u}}^{(n)}$ by the following relation:

$$\mathbf{A}\hat{\mathbf{v}}^{(n)} = \lambda_n \hat{\mathbf{u}}^{(n)}. \tag{12.63}$$

The constant λ_n should not be confused with an eigenvalue; this constant follows from the requirement that $\hat{\mathbf{v}}^{(n)}$ and $\hat{\mathbf{u}}^{(n)}$ are both vectors of unit length. At this point, the choice of $\hat{\mathbf{v}}^{(n)}$ is still open. The vectors $\hat{\mathbf{v}}^{(n)}$ will now be required to satisfy in addition to (12.63) the following condition:

$$\mathbf{A}^T \hat{\mathbf{u}}^{(n)} = \mu_n \hat{\mathbf{v}}^{(n)}, \tag{12.64}$$

where $\mathbf{A}^T$ is the transpose of $\mathbf{A}$.

Problem a   In order to find the vectors $\hat{\mathbf{v}}^{(n)}$ and $\hat{\mathbf{u}}^{(n)}$ that satisfy both (12.63) and (12.64), multiply (12.63) by $\mathbf{A}^T$ and use (12.64) to eliminate $\hat{\mathbf{u}}^{(n)}$. Do this to show that $\hat{\mathbf{v}}^{(n)}$ satisfies:

$$\left(\mathbf{A}^T \mathbf{A}\right) \hat{\mathbf{v}}^{(n)} = \lambda_n \mu_n \hat{\mathbf{v}}^{(n)}. \tag{12.65}$$

Use similar steps to show that $\hat{\mathbf{u}}^{(n)}$ satisfies:

$$\left(\mathbf{A}\mathbf{A}^T\right) \hat{\mathbf{u}}^{(n)} = \lambda_n \mu_n \hat{\mathbf{u}}^{(n)}. \tag{12.66}$$

These equations state that the $\hat{\mathbf{v}}^{(n)}$ are the eigenvectors of $\mathbf{A}^T \mathbf{A}$ and that the $\hat{\mathbf{u}}^{(n)}$ are the eigenvectors of $\mathbf{A}\mathbf{A}^T$.

Problem b   Show that both $\mathbf{A}^T \mathbf{A}$ and $\mathbf{A}\mathbf{A}^T$ are real symmetric matrices and that this implies that the basis vectors $\hat{\mathbf{v}}^{(n)}$ $(n = 1, \ldots, N)$ and $\hat{\mathbf{u}}^{(m)}$ $(m = 1, \ldots, M)$ are both orthonormal:

$$\left(\hat{\mathbf{v}}^{(n)} \cdot \hat{\mathbf{v}}^{(m)}\right) = \left(\hat{\mathbf{u}}^{(n)} \cdot \hat{\mathbf{u}}^{(m)}\right) = \delta_{nm}. \tag{12.67}$$

Although (12.65) and (12.66) can be used to find the basis vectors $\hat{\mathbf{v}}^{(n)}$ and $\hat{\mathbf{u}}^{(n)}$, these expressions cannot be used to find the constants λ_n and μ_n, because they state that the *product* $\lambda_n \mu_n$ is equal to the eigenvalues of $\mathbf{A}^T \mathbf{A}$ and $\mathbf{A}\mathbf{A}^T$. This implies that only the product of λ_n and μ_n is defined.

Problem c   In order to find the relation between λ_n and μ_n, take the inner product of (12.63) with $\hat{\mathbf{u}}^{(n)}$ and use the orthogonality relation (12.67) to show that:

$$\lambda_n = \left(\hat{\mathbf{u}}^{(n)} \cdot \mathbf{A}\hat{\mathbf{v}}^{(n)}\right). \tag{12.68}$$

Problem d   Show that for arbitrary vectors $\mathbf{p}$ and $\mathbf{q}$

$$(\mathbf{p} \cdot \mathbf{A}\mathbf{q}) = \left(\mathbf{A}^T \mathbf{p} \cdot \mathbf{q}\right). \tag{12.69}$$

Problem e   Apply this relation to (12.68) and use (12.64) to show that

$$\lambda_n = \mu_n. \tag{12.70}$$

This is all the information we need to find both λ_n and μ_n. Since these quantities are equal, and since by virtue of (12.65) these eigenvectors are equal to the eigenvectors of $\mathbf{A}^T \mathbf{A}$, it follows that both λ_n and μ_n are given by the square root of the eigenvalues of $\mathbf{A}^T \mathbf{A}$. Note that it follows from (12.66) that the product $\lambda_n \mu_n$ also equals the eigenvalues of $\mathbf{A}\mathbf{A}^T$. This can only be the case when $\mathbf{A}^T \mathbf{A}$ and $\mathbf{A}\mathbf{A}^T$ have the same eigenvalues. Before we proceed, let us show that this is indeed true. Let the eigenvalues of $\mathbf{A}^T\mathbf{A}$ be denoted by Λ_n and the eigenvalues of $\mathbf{A}\mathbf{A}^T$ by Υ_n, that is,

$$\mathbf{A}^T\mathbf{A}\hat{\mathbf{v}}^{(n)} = \Lambda_n\hat{\mathbf{v}}^{(n)} \tag{12.71}$$

and

$$\mathbf{A}\mathbf{A}^T\hat{\mathbf{u}}^{(n)} = \Upsilon_n\hat{\mathbf{u}}^{(n)}. \tag{12.72}$$

Problem f Take the inner product of (12.71) with $\hat{\mathbf{v}}^{(n)}$ to show that $\Lambda_n = (\hat{\mathbf{v}}^{(n)} \cdot \mathbf{A}^T A\hat{\mathbf{v}}^{(n)})$, use the properties (12.69) and $\mathbf{A}^{TT} = \mathbf{A}$ and (12.63) to show that $\lambda_n^2 = \Lambda_n$. Use similar steps to show that $\mu_n^2 = \Upsilon_n$. With (12.70) this implies that $\mathbf{A}\mathbf{A}^T$ and $\mathbf{A}^T\mathbf{A}$ have the same eigenvalues.

The proof that $\mathbf{A}\mathbf{A}^T$ and $\mathbf{A}^T\mathbf{A}$ have the same eigenvalues was not only given as a check of the consistency of the theory; the fact that $\mathbf{A}\mathbf{A}^T$ and $\mathbf{A}^T\mathbf{A}$ have the same eigenvalues has important implications. Since $\mathbf{A}\mathbf{A}^T$ is an $M \times M$ matrix, it has M eigenvalues, and since $\mathbf{A}^T\mathbf{A}$ is an $N \times N$ matrix, it has N eigenvalues. The only way for these matrices to have the same eigenvalues, but to have a different number of eigenvalues, is for the number of nonzero eigenvalues to be given by the minimum of N and M. In practice, some of the eigenvalues of $\mathbf{A}\mathbf{A}^T$ may be zero; hence, the number of nonzero eigenvalues of $\mathbf{A}\mathbf{A}^T$ can be less than M. By the same token, the number of nonzero eigenvalues of $\mathbf{A}^T\mathbf{A}$ can be less than N. The number of nonzero eigenvalues will be denoted by P. It is not known a priori how many nonzero eigenvalues there are, but it follows from the arguments above that P is smaller than or equal to M and N. This implies that

$$P \leq \min(N, M), \tag{12.73}$$

where $\min(N, M)$ denotes the minimum of N and M. Therefore, whenever a summation over eigenvalues occurs, we need to take only P eigenvalues into account. Since the ordering of the eigenvalues is arbitrary, we assume in the following that the eigenvectors are ordered in decreasing size: $\lambda_1 \geq \lambda_2 \geq \cdots \geq \lambda_P$. In this ordering the eigenvalues for $n > P$ are equal to zero so that the summation over eigenvalues runs from 1 to P.

Problem g The matrices $\mathbf{A}\mathbf{A}^T$ and $\mathbf{A}^T\mathbf{A}$ have the same eigenvalues. When you need the eigenvalues and eigenvectors, from the point of view of computational efficiency would it be more efficient to compute the eigenvalues and eigenvectors of $\mathbf{A}^T\mathbf{A}$ or of $\mathbf{A}\mathbf{A}^T$? Consider the situations $M > N$ and $M < N$ separately.

Let us now return to the task of making an eigenvalue decomposition of the matrix $\mathbf{A}$. The vectors $\hat{\mathbf{v}}^{(n)}$ form a basis in N-dimensional space. Since the vector $\mathbf{x}$ is N-dimensional, every vector $\mathbf{x}$ can be decomposed according to (12.37): $\mathbf{x} = \sum_{n=1}^{N} \hat{\mathbf{v}}^{(n)} (\hat{\mathbf{v}}^{(n)} \cdot \mathbf{x})$.

Problem h Let the matrix $\mathbf{A}$ act on this expression and use (12.63) to show that:

$$\mathbf{A}\mathbf{x} = \sum_{n=1}^{P} \lambda_n \hat{\mathbf{u}}^{(n)} \left(\hat{\mathbf{v}}^{(n)} \cdot \mathbf{x} \right). \tag{12.74}$$

Problem i This expression must hold for any vector $\mathbf{x}$. Use this property to deduce that:

$$\mathbf{A} = \sum_{n=1}^{P} \lambda_n \hat{\mathbf{u}}^{(n)} \hat{\mathbf{v}}^{(n)T}. \tag{12.75}$$

Problem j The eigenvectors $\hat{\mathbf{v}}^{(n)}$ can be arranged in an $N \times N$ matrix $\mathbf{V}$, defined in (12.33). Similarly the eigenvectors $\hat{\mathbf{u}}^{(n)}$ can be used to form the columns of an $M \times M$ matrix $\mathbf{U}$:

$$\mathbf{U} = \begin{pmatrix} \vdots & \vdots & & \vdots \\ \hat{\mathbf{u}}^{(1)} & \hat{\mathbf{u}}^{(2)} & \cdots & \hat{\mathbf{u}}^{(M)} \\ \vdots & \vdots & & \vdots \end{pmatrix}. \tag{12.76}$$

Show that $\mathbf{A}$ can also be written as:

$$\mathbf{A} = \mathbf{U}\mathbf{\Sigma}\mathbf{V}^{T}, \tag{12.77}$$

with the diagonal matrix $\mathbf{\Sigma}$ defined in (12.41).

This decomposition of $\mathbf{A}$ in terms of eigenvectors is called the *singular value decomposition* of the matrix. This is frequently abbreviated to SVD.

Problem k You may have noticed the similarity between (12.75) and (12.39) for a square matrix and (12.77) and (12.40). Show that for the special case $M = N$ the theory of this section is identical to the eigenvalue decomposition for a square matrix presented in Section 12.3. Hint: What are the vectors $\hat{\mathbf{u}}^{(n)}$ when $M = N$?

Let us now solve the original system of linear equations (12.62) for the unknown vector $\mathbf{x}$. In order to do this, expand the vector $\mathbf{y}$ in the vectors $\hat{\mathbf{u}}^{(n)}$ that span the

M-dimensional space: $\mathbf{y} = \sum_{m=1}^{M} \hat{\mathbf{u}}^{(m)}(\hat{\mathbf{u}}^{(m)} \cdot \mathbf{y})$, and expand the vector $\mathbf{x}$ in the vectors $\hat{\mathbf{v}}^{(n)}$ that span the N-dimensional space:

$$\mathbf{x} = \sum_{n=1}^{N} c_n \hat{\mathbf{v}}^{(n)}. \tag{12.78}$$

Problem 1 At this point the coefficients c_n are unknown. Insert the expansions for $\mathbf{y}$ and $\mathbf{x}$ and the expansion (12.75) for the matrix $\mathbf{A}$ in the linear system (12.62) and use the orthogonality properties of the eigenvectors to show that $c_n = (\hat{\mathbf{u}}^{(n)} \cdot \mathbf{y})/\lambda_n$, so that

$$\mathbf{x} = \sum_{n=1}^{P} \frac{1}{\lambda_n} \left(\hat{\mathbf{u}}^{(n)} \cdot \mathbf{y} \right) \hat{\mathbf{v}}^{(n)}. \tag{12.79}$$

Note that although in the original expansion (12.78) a summation is carried out over all N basis vectors, in solution (12.79) a summation is carried out over the first P basis vectors only. The reason for this is that the remaining eigenvectors have eigenvalues that are equal to zero so that they can be left out of the expansion (12.75) of the matrix $\mathbf{A}$. Indeed, these eigenvalues would give rise to problems because if they were retained they would lead to infinite contributions $1/\lambda \rightarrow \infty$ in solution (12.79). In practice, some eigenvalues may be nonzero, but close to zero, so that the term $1/\lambda$ gives rise to numerical instabilities. Therefore, one also often leaves out nonzero but small eigenvalues in summation (12.79).

This may appear to be an objective procedure for defining solutions for linear problems that are undetermined or for problems that are otherwise ill-conditioned, but there is a price one pays for leaving out basis vectors in the construction of the solution. The vector $\mathbf{x}$ is N-dimensional; hence, one needs N basis vectors to construct an arbitrary vector $\mathbf{x}$, see (12.78). The solution vector given in (12.79) is built by superposing only P basis vectors. This implies that the solution vector is constrained to be within the P-dimensional subspace spanned by the first P eigenvectors. Therefore, it is not clear that the solution vector in (12.79) is identical to the true vector $\mathbf{x}$. However, the point of using the singular value decomposition is that the solution is only constrained by the linear system of (12.62) within the subspace spanned by the first P basis vectors $\hat{\mathbf{v}}^{(n)}$. Solution (12.79) ensures that only the components of $\mathbf{x}$ within that subspace are affected by the right-hand side vector $\mathbf{y}$. This technique is useful in the analysis of linear inverse problems as discussed in Chapter 22 and by Parker (1994).

13

Dirac delta function

The Dirac delta function is not a function in the sense that one cannot write down the values of this "function." It is an example of a *generalized function*, or *distribution*. These are function-like objects that are defined only within an integral. We will not go into the intricacies of distributions, but instead give enough of a flavor so that we can apply its most important quantities when we get to point charges and masses, and the impulse response of systems in Chapters 17 and 18. The book by Strichartz (1994) treats distributions in a very clear way.

13.1 Introduction of the delta function

In linear algebra, the identity matrix $\mathbf{I}$ plays a central role. This operator maps any vector $\mathbf{v}$ onto itself:

$$\mathbf{I}\mathbf{v} = \mathbf{v}. \tag{13.1}$$

This expression can also be written in component form as

$$\sum_j I_{ij} v_j = v_i. \tag{13.2}$$

The identity matrix has diagonal elements that are equal to unity and its off-diagonal elements are equal to zero. This means that the elements of the identity matrix are equal to the Kronecker delta $I_{ij} = \delta_{ij}$, which we earlier defined as:

$$\delta_{ij} = \begin{cases} 1 & \text{when} \quad i = j \\ 0 & \text{when} \quad i \neq j. \end{cases} \tag{25.15}$$

Expression (13.2) shows that when the identity matrix I_{ij} acts on all the components v_j of a vector, it selects the component v_i. The question we address in this chapter is: How can this idea be generalized to continuous functions instead of vectors? In other words, can we find a function $I(x_0, x)$ such that when it is integrated with a function $f(x)$ it selects that function as the location x_0:

$$\int I(x_0, x) f(x) dx = f(x_0)? \qquad (13.3)$$

Note the resemblance between this expression and (13.2) for a vector. The vector $\mathbf{v}$ is replaced by the function f, the summation over j is changed into the integration over x, and the index i in (13.2) corresponds to the value x_0 in (13.3).

As a first guess for the operator $I(x_0, x)$, let us consider the following generalization of the definition (25.15) of the Kronecker delta to continuous functions:

$$I(x_0, x) \overset{?}{\equiv} \begin{cases} 1 & \text{for} \quad x_0 = x \\ 0 & \text{for} \quad x_0 \neq x. \end{cases} \qquad (13.4)$$

This is not a very good guess; when $I(x_0, x)$ is viewed as a function of x, $I(x_0, x)$ is almost everywhere equal to zero except when $x = x_0$. When we integrate $I(x_0, x)$ over x, the integrand is zero except at the point $x = x_0$, but this point gives a vanishing contribution to the integral because the "width" of this point is equal to zero. (Mathematically one would say that a point has zero measure.) This means that the integral of $I(x_0, x)$ can only give a finite result when $I(x_0, x)$ is infinite at $x = x_0$.

As an improved guess let us therefore try the definition

$$I(x_0, x) \overset{?}{\equiv} \begin{cases} \infty & \text{for} \quad x_0 = x \\ 0 & \text{for} \quad x_0 \neq x. \end{cases} \qquad (13.5)$$

This is not a very precise definition because it is not clear what we mean by "∞." However, we can learn something from this naive guess because it shows that $I(x_0, x)$ is not a well-behaved function. It is discontinuous at $x = x_0$ and its value is infinite at that point. It is clear from this that whatever definition of $I(x_0, x)$ we use, it will not lead to a well-behaved function.

A useful definition of $I(x_0, x)$ can be obtained from the boxcar function $B_a(x)$, which is defined as

$$B_a(x) \equiv \begin{cases} \dfrac{1}{2a} & \text{for} \quad |x| \leq a \\ 0 & \text{for} \quad |x| > a. \end{cases} \qquad (13.6)$$

This function is shown for three values of the parameter a in Figure 13.1.[1]

Problem a Show that

$$\int_{-\infty}^{\infty} B_a(x) dx = 1. \qquad (13.7)$$

---

[1] The boxcar function gets its name from the rectangular-shaped boxcar railroad car that carries freight.

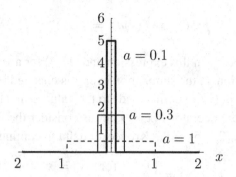

Figure 13.1   The boxcar function $B_a(x)$ for three values of a.

Let us now center the boxcar at a location x_0, multiply it by $f(x)$, and integrate over x.

Problem b   Use the definition of the boxcar function to derive that

$$\int_{-\infty}^{\infty} B_a(x - x_0) f(x) dx = \frac{1}{2a} \int_{x_0-a}^{x_0+a} f(x)\, dx. \qquad (13.8)$$

Pay attention to the limits of integration on the right-hand side.

The integral on the right-hand side is nothing but the average of the function $f(x)$ over the interval $(x_0 - a, x_0 + a)$ because the pre-factor $1/2a$ corrects for the width of that interval. This is a useful expression; as the parameter a goes to zero, the integral (13.8) gives the function at location x_0 because the limit $a \downarrow 0$ gives the mean value of $f(x)$ over the interval $(x_0 - 0, x_0 + 0)$. This means that

$$\lim_{a\downarrow 0} \int_{-\infty}^{\infty} B_a(x - x_0) f(x)\, dx = f(x_0). \qquad (13.9)$$

This means that this limit has the desired properties of the identity operator $I(x_0, x)$ for continuous functions. It is customary to denote this operator as $\delta(x - x_0)$ and to call it the *Dirac delta function*. Usually one refers to this function simply as the "delta function." This means that the delta function satisfies the following property:

$$\int_{-\infty}^{\infty} \delta(x - x_0) f(x) dx = f(x_0). \qquad (13.10)$$

This quality of the delta function is called the *sifting property*, as the integral sifts out the value of $f(x)$, where the delta function is nonzero.

A comparison of this expression with (13.9) suggests that

$$\delta(x - x_0) \text{"} = \text{"} \lim_{a\downarrow 0} B_a(x - x_0). \qquad (13.11)$$

Consider Figure 13.1 again. It can be seen from that figure and definition (13.6) that as a goes to zero, the value of the boxcar function becomes infinite, and the width of the boxcar goes to zero. In this sense the limit on the right-hand side does not exist, and for this reason the $=$ sign is placed between quotes. The word "delta function" is really a misnomer because this "function" is not a function at all. Its value is not defined and it is only nonzero in a region with measure zero. However, within the integral (13.10) the action of the delta function is well defined; it is an operator that selects the function value $f(x_0)$ at position x_0. The delta function is an example of a *distribution*. This is a mathematical object that is only defined within an integral. Note that the limit $a \downarrow 0$ of the integral (13.9) *is* well defined. This means that the properties of the delta function can only be meaningfully stated when the delta function is used in an integral.

Let us first give a formal proof that definition (13.11) indeed leads to a delta function with the desired property (13.10). It is clear from (13.8) that we only need to consider the function $f(x)$ in the interval $(x_0 - a, x_0 + a)$ for small values of a. Therefore, it is useful to represent the function $f(x)$ by a Taylor series around the point x_0.

Problem c   Use (3.17) to show that $f(x)$ can be represented by the following Taylor series around the point x_0:

$$f(x) = \sum_{n=0}^{\infty} \frac{1}{n!} \frac{d^n f}{dx^n}(x_0)(x - x_0)^n \tag{13.12}$$

$$= f(x_0) + \frac{df}{dx}(x_0)(x - x_0) + \frac{1}{2}\frac{d^2 f}{dx^2}(x_0)(x - x_0)^2 + \cdots .$$

Problem d   When this is inserted in (13.8) each term gives a contribution proportional to $\int_{x_0-a}^{x_0+a}(x - x_0)^n dx$. Show that for odd values of n this integral is equal to zero and that for even powers of n it is given by

$$\int_{x_0-a}^{x_0+a} (x - x_0)^n dx = \frac{2}{n+1} a^{n+1}. \tag{13.13}$$

Problem e   Use these results to show that

$$\int_{-\infty}^{\infty} B_a(x - x_0) f(x)\, dx = \sum_{n \text{ even}} \frac{1}{(n+1)!} \frac{d^n f}{dx^n}(x_0)\, a^n =$$

$$f(x_0) + \frac{1}{6}\frac{d^2 f}{dx^2}(x_0)\, a^2 + \frac{1}{120}\frac{d^4 f}{dx^4}(x_0)\, a^4 + \cdots . \tag{13.14}$$

In the limit $a \downarrow 0$ all the terms in this series vanish except the first term. This means that expression (13.9) is indeed satisfied.

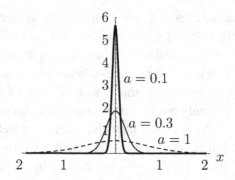

Figure 13.2   The Gaussian function $g_a(x)$ for three values of a.

It should be noted that it is not necessary to define the delta function as the limit $a \downarrow 0$ of the boxcar function. One can also define the delta function using a Gaussian function with width a:

$$g_a(x) = \frac{1}{a\sqrt{\pi}} e^{-x^2/a^2}. \qquad (13.15)$$

This function is shown for various values of a in Figure 13.2. A comparison of this figure with Figure 13.1 shows that for small values of a these functions have similar properties. One can indeed formally define the delta function using the Gaussian function (13.15) instead of the boxcar function. When the analysis of Problem d is applied to this function, the first term is again given by $f(x_0)$. The higher-order terms have different coefficients because they follow from the integral $\int_{-\infty}^{\infty}(x-x_0)^n e^{-x^2/a^2} dx$ rather than the integral $\int_{x_0-a}^{x_0+a}(x-x_0)^n dx$. However, in the limit $a \downarrow 0$ these higher-order terms do not contribute. This example shows that the delta function can be defined as the limit of either boxcar functions or Gaussian functions. In fact the delta function can be defined as the limit of other types of functions as well.

13.2 Properties of the delta function

In the previous section the delta function was formally introduced. In this section we derive some properties of the delta function that are useful in a variety of applications.

Problem a   Apply the identity (13.10) to the function $f(x) = 1$ to derive that

$$\int_{-\infty}^{\infty} \delta(x-x_0) dx = 1. \qquad (13.16)$$

This expression states that the "surface area" under the delta function is equal to one. As the width of this function goes to zero, the value of the function must become infinite to ensure that (13.16) is satisfied. Note that according to (13.7) the boxcar function $B_a(x)$ that we used to define the delta function indeed satisfies this property.

Problem b Show that the integral of the Gaussian function defined in (13.15) is also equal to unity:

$$\int_{-\infty}^{\infty} g_a(x)dx = 1. \tag{13.17}$$

In this derivation you can use (4.49). This means that when the delta function is defined as the limit $a \downarrow 0$ of the Gaussian function $g_a(x)$, property (13.16) is indeed satisfied.

For the next property we consider the delta function $\delta\left(c(x - x_0)\right)$, where c is a constant. Let us first consider the case in which c is positive.

Problem c Make the substitution $y = cx$ to derive the following identity:

$$\int_{-\infty}^{\infty} \delta\left(c\left(x - x_0\right)\right) f(x)\, dx = \int_{-\infty}^{\infty} \delta\left(y - cx_0\right) f(y/c)\frac{1}{c}dy. \tag{13.18}$$

Problem d Carry out the y-integration using property (13.10) of the delta function to show that

$$\int_{-\infty}^{\infty} \delta\left(c\left(x - x_0\right)\right) f(x)\, dx = \frac{1}{c} f(x_0) \qquad \text{(positive } c\text{)}. \tag{13.19}$$

Problem e Carry out the same analysis for negative values of the constant c, and show that (13.18) in that case is given by

$$\int_{-\infty}^{\infty} \delta\left(c\left(x - x_0\right)\right) f(x)\, dx = \int_{+\infty}^{-\infty} \delta\left(y - cx_0\right) f(y/c)\frac{1}{c}dy. \tag{13.20}$$

Explain why the integration runs from $+\infty$ to $-\infty$.

Problem f When the integration limits in the last integral are reversed, the integral obtains an additional minus sign. Carry out the y-integration and derive that

$$\int_{-\infty}^{\infty} \delta\left(c\left(x - x_0\right)\right) f(x)\, dx = -\frac{1}{c} f(x_0) \qquad \text{(negative } c\text{)}. \qquad (13.21)$$

For negative values of c, one can use that $-c = |c|$. For positive values of c, obviously $c = |c|$. This means that (13.19) and (13.20) can be combined in the single property

$$\int_{-\infty}^{\infty} \delta\left(c\left(x - x_0\right)\right) f(x) dx = \frac{1}{|c|} f(x_0). \qquad (13.22)$$

Following (13.10) we also know that the right-hand side of this expression can also be written as $(1/|c|) \int_{-\infty}^{\infty} \delta\left(x - x_0\right) f(x) dx$. A comparison with the left-hand side of (13.22) implies that

$$\delta\left(c\left(x - x_0\right)\right) = \frac{1}{|c|} \delta\left(x - x_0\right). \qquad (13.23)$$

This property also confirms something you may have noticed about the delta function; it is symmetric: $\delta(x - x_0) = \delta(x_0 - x)$.

Problem g   Confirm this symmetry by using $c = -1$ in (13.23).

There is a tricky property of the delta function that is important in dimensional analysis, as treated in Chapter 2. It follows from equation (13.16) that the delta function $\delta(x - x_0)$ has dimension 1/length:

$$\delta(x) \sim [L^{-1}]. \qquad (13.24)$$

The reason for this is that according to expression (13.16) this delta function is equal to 1 upon integration over distance x; hence, this delta function has physical dimension $1/distance$. Using the same reasoning, the delta function $\delta(t)$ has dimension

$$\delta(t) \sim [T^{-1}]. \qquad (13.25)$$

The important conclusion is that the delta function is not dimensionless and that its dimension is the opposite of the dimension of its argument. We return to this issue in Section 18.4.

13.3 The Dirac delta function of a function

In some applications one arrives at an integral of the delta function in which the argument of the delta function is a function $g(x)$ rather than the integration variable x. This means that one needs to evaluate the integral $\int \delta\left(g(x)\right) f(x) dx$. We will encounter such an integral in (18.57). The delta function $\delta(x - x_0)$ only gives a

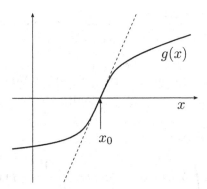

Figure 13.3 The function $g(x)$ (thick solid line) and the tangent line (dashed) at the zero-crossing x_0.

nonzero contribution when x is close to x_0. Therefore, the delta function $\delta(g(x))$ only needs to be evaluated near the value of x where $g(x) = 0$. Let us denote this value by x_0, so that

$$g(x_0) = 0. \tag{13.26}$$

This point x_0 is shown in Figure 13.3. Near the point x_0, the function $g(x)$ can be represented by a Taylor series:

$$g(x) = g(x_0) + \frac{dg}{dx}(x_0)(x - x_0) + \frac{1}{2}\frac{d^2g}{dx^2}(x_0)(x - x_0)^2 + \cdots. \tag{13.27}$$

Because x_0 gives the zero-crossing of $g(x)$, according to (13.26), the zero-order term is equal to zero. Ignoring the second-order term and all higher-order terms then gives the following first-order approximation for the function $g(x)$ near its zero-crossing:

$$g(x) = \frac{dg}{dx}(x_0)(x - x_0). \tag{13.28}$$

Problem a   Show that the right-hand side of (13.28) describes the straight line that is tangent to the curve $g(x)$ at the zero-crossing x_0. This tangent line is shown as the dashed line in Figure 13.3.

Problem b   Using this relation one finds that

$$\int \delta(g(x)) f(x)\, dx = \int \delta\left(\frac{dg}{dx}(x_0)(x - x_0)\right) f(x)\, dx. \tag{13.29}$$

The derivative dg/dx at the point x_0 can be considered to be a constant. Use the results of the previous section to derive that

$$\int \delta\left(g(x)\right) f(x) \, dx = \int \frac{1}{\left|\dfrac{dg}{dx}(x_0)\right|} \delta(x - x_0) f(x) \, dx, \qquad (13.30)$$

and show that this gives

$$\int \delta\left(g(x)\right) f(x) \, dx = \frac{1}{\left|\dfrac{dg}{dx}(x_0)\right|} f(x_0). \qquad (13.31)$$

Problem c   Insert (13.10) into the right-hand side of (13.31) to derive the following property of the delta function

$$\delta\left(g(x)\right) = \frac{1}{\left|\dfrac{dg}{dx}(x_0)\right|} \delta(x - x_0), \qquad (13.32)$$

where it must be kept in mind that x_0 denotes the zero-crossing of $g(x)$.

When the function $g(x)$ has more than one zero-crossing, the analysis of this section can be applied to each of these zero-crossings. The contributions of all the different zero-crossings must be added because all the points where $g(x) = 0$ give a contribution to the integral. This means that when the zero-crossings are denoted by x_i (so that $g(x_i) = 0$):

$$\delta\left(g(x)\right) = \sum_i \frac{1}{\left|\dfrac{dg}{dx}(x_i)\right|} \delta(x - x_i). \qquad (13.33)$$

13.4 The Dirac delta function in more than one dimension

So far, the delta function has been defined for functions of a single variable. Its definition can be extended to functions of more variables. As an example we consider here the delta function in three dimensions. The delta function is then defined as the product of the delta functions for each of the coordinates:

$$\delta(\mathbf{r} - \mathbf{r}_0) \equiv \delta(x - x_0)\,\delta(y - y_0)\,\delta(z - z_0). \qquad (13.34)$$

This definition of the delta function can be used in the integral $\int \delta(\mathbf{r} - \mathbf{r}_0) f(\mathbf{r})\, dV$.

Problem a Write the volume integral as $dV = dxdydz$, insert (13.34) and carry out the integration over x to show that

$$\int \delta(\mathbf{r} - \mathbf{r}_0) f(\mathbf{r}) \, dV = \int \delta(y - y_0) \delta(z - z_0) f(x_0, y, z) \, dydz, \quad (13.35)$$

paying attention to the arguments of the function f.

Problem b Carry out the y-integration and then the z-integration to derive that

$$\int \delta(\mathbf{r} - \mathbf{r}_0) f(\mathbf{r}) \, dV = f(x_0, y_0, z_0). \quad (13.36)$$

The right-hand side of (13.36) is the function f at location $\mathbf{r}_0$. This means that the multidimensional delta function defined in (13.34) satisfies the following property

$$\int \delta(\mathbf{r} - \mathbf{r}_0) f(\mathbf{r}) \, dV = f(\mathbf{r}_0). \quad (13.37)$$

This expression generalizes (13.10) to more dimensions.

13.5 The Dirac delta function on the sphere

Up to this point we have analyzed the delta function in Cartesian coordinates. When the delta function is used in a coordinate system that is not Cartesian, additional terms appear in the definition of the delta function. This is illustrated in this section with the delta function on a sphere. Suppose we define a function $f(\theta, \varphi)$ on the unit sphere. Using definition (13.10) the action of the delta function on the sphere as expressed in the angles θ and φ can be written as

$$\int_0^{2\pi} \int_0^{\pi} \delta(\theta - \theta_0) \delta(\varphi - \varphi_0) f(\theta, \varphi) \, d\theta d\varphi = f(\theta_0, \varphi_0). \quad (13.38)$$

Every point on the unit sphere can be characterized by the unit vector $\hat{\mathbf{r}}$ that points from the origin to that point. It follows from the first identity of (4.7) that for given angles θ and φ this unit vector is given by

$$\hat{\mathbf{r}} = \begin{pmatrix} \sin\theta \cos\varphi \\ \sin\theta \sin\varphi \\ \cos\theta \end{pmatrix}; \quad (13.39)$$

a similar definition holds for the unit vector $\hat{\mathbf{r}}_0$ that corresponds to the angles θ_0 and φ_0.

In this section we rewrite (13.38) as an integration over the unit sphere. According to (4.36), the surface element on the sphere is given by $dS = r^2 \sin\theta d\theta d\varphi$.

On the unit sphere the radius is, by definition, given by $r = 1$, so that on the unit sphere $dS = \sin\theta \, d\theta \, d\varphi$.

Problem a   Show that (13.38) can be written as

$$\oint \frac{1}{\sin\theta} \delta(\theta - \theta_0) \delta(\varphi - \varphi_0) f(\theta, \varphi) \, dS = f(\theta_0, \varphi_0), \qquad (13.40)$$

where the symbol $\oint \cdots dS$ denotes the integration over the area of the unit sphere.

We can consider the function f to be a function of the angles θ and φ, but we can also see f as a function of the unit vector $\hat{\mathbf{r}}$, which means that $f = f(\hat{\mathbf{r}})$. Therefore, we can write the previous expression as

$$\oint \frac{1}{\sin\theta} \delta(\theta - \theta_0) \delta(\varphi - \varphi_0) f(\hat{\mathbf{r}}) \, dS = f(\hat{\mathbf{r}}_0). \qquad (13.41)$$

Formally this can also be written as

$$\oint \delta(\hat{\mathbf{r}} - \hat{\mathbf{r}}_0) f(\hat{\mathbf{r}}) \, dS = f(\hat{\mathbf{r}}_0), \qquad (13.42)$$

where $\delta(\hat{\mathbf{r}} - \hat{\mathbf{r}}_0)$ defines the delta function on the unit sphere.

Problem b   Show that

$$\delta(\hat{\mathbf{r}} - \hat{\mathbf{r}}_0) = \frac{1}{\sin\theta} \delta(\theta - \theta_0) \delta(\varphi - \varphi_0). \qquad (13.43)$$

This expression shows that when the delta function is computed in a non-Cartesian coordinate system (such as spherical coordinates defined on the unit sphere), additional terms appear in the delta function that account for the curvilinear character of the coordinate system (Section 4.4). The terms that appear in the delta function compensate the terms in the Jacobian that account for the curvilinear character of the employed coordinate system.

13.6 Self energy of the electron

Up to this point the delta function has been treated as a mathematical tool. However, it is often used as a description of the concept of a point charge or point mass. Physically, the idea of a point mass is that a finite mass M is concentrated at a certain point $\mathbf{r}_0$. The associated mass-density $\rho(\mathbf{r})$ is then equal to zero everywhere except at $\mathbf{r} = \mathbf{r}_0$. The integral of the mass-density must be equal to the total mass: $\int \rho(\mathbf{r}) d^3\mathbf{r} = M$, where $\int \cdots d^3\mathbf{r}$ denotes the three-dimensional volume integral.

Problem a   Verify that these properties are satisfied by the mass-density

$$\rho(\mathbf{r}) = M\delta(\mathbf{r} - \mathbf{r}_0).\qquad(13.44)$$

For the moment we restrict our attention to the concept of a point mass and the associated gravitational field, but because of the equivalence of the laws for the gravitational field and the electrostatic field, the results can be applied to a stationary electric field as well.

It was shown in Section 8.2 that the gravitational field generated by a spherically symmetric body depends outside the body on the total mass only and not on the mass distribution within the body. For a mass centered at the origin ($\mathbf{r}_0 = 0$), the gravitational field is given by

$$\mathbf{g}(r) = -\frac{GM}{r^2}\hat{\mathbf{r}}.\qquad(8.6)$$

This gravitational field is associated with a gravitational potential $V(\mathbf{r})$ that is related to the gravity field by the expression

$$\mathbf{g}(\mathbf{r}) = -\nabla V(\mathbf{r}).\qquad(13.45)$$

Problem b   Use equation (8.6) to show that the gravitational potential for a point mass M located in the origin is given by

$$V(\mathbf{r}) = -\frac{GM}{r}.\qquad(13.46)$$

In the integration you encounter one integration constant. This integration constant follows from the requirement that the potential energy vanishes at infinity.

In this section the point mass serves as the source of the gravitational field. The response of any linear system to a source function plays an important role in mathematical physics because a general source of the field can always be written as a superposition of delta functions. The response to a delta function excitation is called the *Green's function*: this concept is treated in Chapters 17 and 18. We have seen in Section 13.1 that the delta function is singular, and it is in fact so singular that it cannot be considered to be a function. Very often, the response to such a singular source function is also singular at the location of that source function. This means that the Green's function is usually singular at the point of excitation.

At this point we have computed the gravitational field and its potential energy for a point mass, and it appears that these can be computed without any problems.

However, there is a complication. As shown in expression (1.53) of Jackson (1998), the energy E of a charge-density $\rho(\mathbf{r})$ that is placed in a potential $V(\mathbf{r})$ is given by

$$E = \frac{1}{2} \int \rho(\mathbf{r}) V(\mathbf{r}) \, d^3\mathbf{r}. \qquad (13.47)$$

In Jackson (1998) this is derived for electrostatic energy, but the same expression holds for gravitational energy.

Problem c Use (13.44) and (13.46) to show that the gravitational energy of a point charge placed in the origin is infinite.

This means that a point mass has infinite energy, which indicates that the concept of a point mass is not without complications when the energy is concerned.

Let us consider what the energy is of a spherically symmetric mass distribution when the mass M is homogeneously distributed in a sphere of radius R. The concept of the boxcar function as defined in (13.6) can easily be generalized to three dimensions by the following definition:

$$B_R(\mathbf{r}) \equiv \begin{cases} \dfrac{3}{4\pi R^3} & \text{for} \quad |\mathbf{r}| \leq R \\[2mm] 0 & \text{for} \quad |\mathbf{r}| > R. \end{cases} \qquad (13.48)$$

Problem d Show that $\int B_R(\mathbf{r}) \, d^3\mathbf{r} = 1$ and use this to show that the mass-density of such a mass distribution is given by

$$\rho(\mathbf{r}) = M B_R(\mathbf{r}). \qquad (13.49)$$

Outside the distribution of mass, the gravitational field is given by (8.6) and the associated potential energy is derived in (13.46). Inside the mass distribution the gravitational field is given by (8.7).

Problem e Use (8.7), (13.45), and the requirement that the potential is continuous everywhere to derive that the potential energy is given by

$$V(\mathbf{r}) \equiv \begin{cases} -\dfrac{GM}{2R^3}(3R^2 - r^2) & \text{for} \quad |\mathbf{r}| \leq R \\[3mm] -\dfrac{GM}{r} & \text{for} \quad |\mathbf{r}| > R. \end{cases} \qquad (13.50)$$

Problem f   Use this expression and (13.47) to show that the gravitational energy
is given by

$$E = -\frac{3GM^2}{5R}.$$ (13.51)

Note that when the radius R of the mass goes to zero, the gravitational energy
becomes infinite; this is what we derived in Problem c.

Let us consider this homogeneous mass distribution for the moment as a simplified model of the mass distribution of the Earth. It follows from (13.51) that
the gravitational energy decreases without bound when the radius of the Earth is
decreased. Since physical systems tend to minimize their energy, we can raise the
question of why the radius of the Earth has not become smaller than its present
value of about 6370 km. The fact that the potential energy is negative corresponds
to the fact that mass always attracts itself. The decrease of the gravitational energy
with decreasing radius thus corresponds to a gravitational collapse of the body. So
we can phrase our question in a different way: Why does the Earth not collapse to
a black hole?

There are two effects that need to be considered. First, the gravitational force in
the Earth leads to a compression of the material within the Earth. This compression
induces elastic forces within the Earth. The radius of the Earth is dictated by the
balance between the gravitational force and the elastic reactive force. If energy were
used to describe this balance, one would state that the positive potential energy of
the elastic deformation balances the unbounded negative growth of the gravitational
energy as the radius is decreased. The second factor to take into account is that the
gravitational fields that are treated in this section follow from Newton's law of
gravitation. However, this law only holds for weak fields, and for very strong fields
it should be replaced by the laws of general relativity (Ohanian and Ruffini, 1976).
It follows from (8.6) that the gravitational field of a point mass grows without
bounds as one moves closer to the point mass. This means that ultimately Newton's
law of gravitation ceases to be a good description of the field close to the point
mass. This implies that although the concept of a point mass as described by a
delta function is appealing, it is physically not without complications.

You may think that this is a purely academic issue. However, the same reasoning
applies to electric point charges. By analogy with (13.51), it follows that the energy
as the radius R goes to zero of a homogeneous charge distribution within a sphere
of radius R and a total charge q is given by

$$E = +\frac{3q^2}{20\pi \epsilon_0 R}.$$ (13.52)

The energy becomes infinite as the radius R tends to zero. Note that the minus sign in (13.51) is replaced by a plus sign. This corresponds to the fact that equal masses attract each other while equal charges repel each other. Let us now consider the electron as a homogeneous charge q with radius R. Expression (13.52) then states that an electron has an infinite energy.

Problem g   Show that the energy of the electron is minimized when the radius R goes to infinity.

This means that an electron would grow beyond bounds in order to minimize its energy. Just as with the previous discussion of the gravitational energy of the Earth, there are a number of reasons why this is not a physically accurate description. By analogy with the elastic forces in the Earth, one could argue there may be other forces acting within the electron that keep the charge together. However, the concept of an electron as a homogeneous charge distribution is physically not accurate. The charge of an electron is quantized, and the laws of electrostatics are not applicable to the description of the internal structure of the electron.

Nevertheless, the self energy of the electron is a long-standing problem (Weisskopf, 1939). The more advanced quantum field theory of the electron also predicts an infinite self energy of the electron, but in this case the singularity is logarithmic rather than algebraic (Weisskopf, 1939). This issue has both mathematical and philosophical aspects. Since the electron has a fixed quantized charge, one cannot consider an electron separately from its field. In fact, through quantum fluctuations the electron can interact with virtual particles that are generated in its field. This means that the energy of an electron that we observe is always a mixture of the energy the electron would have in the absence of its interaction with its field and the interaction with its field (and all virtual particles that may be generated in that field). This has led to the concept of renormalization (Sakurai, 1978; Zee, 2005) in which one accounts for the fact that the mass of the electron in the absence of electromagnetic fields is different from the mass that we observe.

14

Fourier analysis

Fourier analysis is concerned with the decomposition of signals into sine and cosine waves. This technique is of obvious relevance for spectral analysis where one decomposes a time signal into its different frequency components. As an example, the spectrum of a low C on a soprano saxophone is shown in Figure 14.1. The low C resonates at ~ 262 Hz, complimented by its overtones at higher frequencies (Section 19.1). However, the use of Fourier analysis goes far beyond this application, because Fourier analysis can also be used to find solutions of differential equations and a large number of other applications. In this chapter the real Fourier transform on a finite interval is used as a starting point. From this, the complex Fourier transform and the Fourier transform on an infinite interval are derived. At several stages of the analysis, the similarity between Fourier analysis and linear algebra is treated.

14.1 Real Fourier series on a finite interval

Consider a function $f(x)$ that is defined on the interval $-L < x \leq L$. This interval is of length $2L$, and let us assume that $f(x)$ is periodic with period $2L$. This means that if one translates this function over a distance $2L$, the value does not change:

$$f(x + 2L) = f(x). \tag{14.1}$$

We want to expand this function into a set of basis functions. Since $f(x)$ is periodic with period $2L$, these basis functions must be periodic with the same period.

Problem a Show that the functions $\cos(n\pi x/L)$ and $\sin(n\pi x/L)$ with integer n are periodic with period $2L$, that is, show that these functions satisfy (14.1).

203

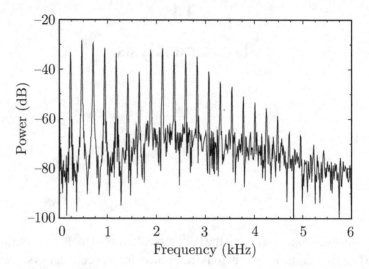

Figure 14.1   The power of the sound of a low C on a soprano saxophone as a function of frequency. The unit used for the horizontal axis is the hertz (the number of oscillations per second), the unit on the vertical axis is decibels (a logarithmic measure of power).

The main statement of Fourier analysis is that one can write $f(x)$ as a superposition of these periodic sine and cosine waves:

$$f(x) = \frac{1}{2}a_0 + \sum_{n=1}^{\infty} a_n \cos(n\pi x/L) + \sum_{n=1}^{\infty} b_n \sin(n\pi x/L). \qquad (14.2)$$

The factor $1/2$ with the coefficient a_0 has no special significance, and is used to simplify subsequent expressions. To show that (14.2) is actually true is not trivial. Providing this proof essentially amounts to showing that the functions $\cos(n\pi x/L)$ and $\sin(n\pi x/L)$ actually contain enough "degrees of freedom" to describe $f(x)$. However, since $f(x)$ is a function of a continuous variable x, this function has infinitely many degrees of freedom and since there are infinitely many coefficients a_n and b_n, counting the number of degrees of freedom does not work. Mathematically one would say that one needs to show that the set of functions $\cos(n\pi x/L)$ and $\sin(n\pi x/L)$ is a "complete set." We will not concern ourselves here with this proof, and simply start working with the Fourier series (14.2).

At this point it is not yet clear what the coefficients a_n and b_n are. To derive these coefficients one needs to use the following integrals:

$$\int_{-L}^{L} \cos^2(n\pi x/L)\,dx = \int_{-L}^{L} \sin^2(n\pi x/L)\,dx = L \qquad \text{if } n \geq 1; \qquad (14.3)$$

$$\int_{-L}^{L} \cos{(n\pi x/L)} \cos{(m\pi x/L)}\, dx = 0 \qquad \text{if } n \neq m; \qquad (14.4)$$

$$\int_{-L}^{L} \sin{(n\pi x/L)} \sin{(m\pi x/L)}\, dx = 0 \qquad \text{if } n \neq m; \qquad (14.5)$$

$$\int_{-L}^{L} \cos{(n\pi x/L)} \sin{(m\pi x/L)}\, dx = 0 \qquad \text{all } n, m. \qquad (14.6)$$

Problem b Derive these identities, using trigonometric identities such as $\cos\alpha \cos\beta = [\cos(\alpha+\beta) + \cos(\alpha-\beta)]/2$. If you have difficulty deriving these identities, you can consult, for example, Boas (2006).

Problem c To find the coefficient b_m, multiply the Fourier expansion (14.2) by $\sin{(m\pi x/L)}$, integrate the result from $-L$ to L and use (14.3)–(14.6) to evaluate the integrals. Show that this gives:

$$b_n = \frac{1}{L} \int_{-L}^{L} f(x) \sin{(n\pi x/L)}\, dx. \qquad (14.7)$$

Problem d Use a similar analysis to show that:

$$a_n = \frac{1}{L} \int_{-L}^{L} f(x) \cos{(n\pi x/L)}\, dx. \qquad (14.8)$$

In deriving this result, treat the cases $n \neq 0$ and $n = 0$ separately.

It is now clear why the factor $1/2$ was introduced in the a_0 term of (14.2); without this factor expression (14.8) would have an additional factor 2 for $n = 0$.

There is a close relation between the Fourier series (14.2) with the coefficients given in the expressions above and the projection of a vector on a number of basis vectors in linear algebra as shown in Section 12.1. To see this relation, we restrict ourselves for simplicity to functions $f(x)$ that are odd functions of x: $f(-x) = -f(x)$, but this restriction is by no means essential. For these functions all coefficients a_n are equal to zero. As an analog of a basis vector in linear algebra let us define the following basis function $u_n(x)$:

$$u_n(x) \equiv \frac{1}{\sqrt{L}} \sin{(n\pi x/L)}. \qquad (14.9)$$

An essential ingredient in the projection operators of Section 12.1 is the inner product between vectors. It is also possible to define an inner product for functions, and for the present example the inner product of two functions $f(x)$ and $g(x)$ is defined as:

$$(f \cdot g) \equiv \int_{-L}^{L} f(x)g(x)dx. \tag{14.10}$$

Problem e   The basis functions $u_n(x)$ defined in (14.9) are the analog of a set of orthonormal basis vectors. To see this, use (14.3) and (14.5) to show that

$$(u_n \cdot u_m) = \delta_{nm}, \tag{14.11}$$

where δ_{nm} is the Kronecker delta (12.33).

This expression implies that the basis functions $u_n(x)$ are mutually orthogonal. If the norm of such a basis function is defined as $\|u_n\| \equiv \sqrt{(u_n \cdot u_n)}$, (14.11) implies that the basis functions are normalized (i.e., have norm 1). These functions are the generalization of orthogonal unit vectors to a function space. The (odd) function $f(x)$ can be written as a sum of the basis functions $u_n(x)$:

$$f(x) = \sum_{n=1}^{\infty} c_n u_n(x). \tag{14.12}$$

Problem f   Take the inner product of (14.12) with $u_m(x)$ and show that $c_m = (u_m \cdot f)$. Use this to show that the Fourier expansion of $f(x)$ can be written as $f(x) = \sum_{n=1}^{\infty} u_n(x) (u_n \cdot f)$, and that on leaving out the explicit dependence on the variable x, the result is

$$f = \sum_{n=1}^{\infty} u_n (u_n \cdot f). \tag{14.13}$$

This equation bears close resemblance to the expression derived in Section 12.1 for the projection of vectors. The projection of a vector $\mathbf{v}$ along a unit vector $\hat{\mathbf{n}}$ was shown to be

$$\mathbf{Pv} = \hat{\mathbf{n}} \left(\hat{\mathbf{n}} \cdot \mathbf{v} \right). \tag{12.2}$$

A comparison with (14.13) shows that $u_n(x) (u_n \cdot f)$ can be interpreted as the projection of the function $f(x)$ on the function $u_n(x)$. To reconstruct the function, one must sum over the projections along all basis functions, hence, the summation in (14.13). It is shown in (12.12) that to find the projection of the vector $\mathbf{v}$ onto the subspace spanned by a finite number of orthonormal basis vectors, one simply has to sum the projections of the vector $\mathbf{v}$ on all the basis vectors that span the subspace: $\mathbf{Pv} = \sum_i \hat{\mathbf{n}}_i \left(\hat{\mathbf{n}}_i \cdot \mathbf{v} \right)$. In a similar way, one can sum the Fourier series (14.13) over only a limited number of basis functions to obtain the projection of $f(x)$ on a limited number of basis functions:

$$f_{filtered} = \sum_{n=n_1}^{n_2} u_n (u_n \cdot f). \tag{14.14}$$

In this expression it was assumed that only values $n_1 \leq n \leq n_2$ have been used. The projected function is called $f_{filtered}$, because this projection is a filtering operation.

Problem g Show that the functions $u_n(x)$ are sinusoidal waves with wavelength $\lambda = 2L/n$.

This means that restricting the n-values in the sum (14.14) amounts to using only wavelengths between $2L/n_2$ and $2L/n_1$ for the projected function. Since only certain wavelengths are used, this projection really acts as a filter that allows only certain wavelengths in the filtered function.

It is the filtering property that makes the Fourier transform so useful for filtering data sets by excluding wavelengths that are unwanted. In fact, the Fourier transform forms the basis of digital filtering techniques that have many applications in science and engineering; see, for example, the books of Claerbout (1976) and Robinson and Treitel (1980).

14.2 Complex Fourier series on a finite interval

In the theory of the preceding section, there is no reason why the function $f(x)$ should be real. Although the basis functions $\cos(n\pi x/L)$ and $\sin(n\pi x/L)$ are real, the Fourier sum (14.2) can be complex because the coefficients a_n and b_n can be complex. Complex numbers can be represented by Euler's formula $\exp(i\varphi) = \cos\varphi + i\sin\varphi$ (Figure 14.2). The equation of de Moivre (Euler's formula with $\varphi = n\pi x/L$) gives the relation between these basis functions and complex exponential functions:

$$e^{in\pi x/L} = \cos(n\pi x/L) + i\sin(n\pi x/L). \tag{14.15}$$

This expression can be used to rewrite the Fourier series (14.2) using the basis functions $e^{in\pi x/L}$ rather than sines and cosines.

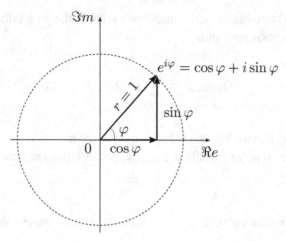

Figure 14.2   The unit circle in the complex plane and Euler's formula.

Problem a   Replace n by $-n$ in (14.15) to show that:

$$\left.\begin{aligned}
\cos(n\pi x/L) &= \frac{1}{2}\left(e^{in\pi x/L} + e^{-in\pi x/L}\right), \\
\sin(n\pi x/L) &= \frac{1}{2i}\left(e^{in\pi x/L} - e^{-in\pi x/L}\right).
\end{aligned}\right\} \tag{14.16}$$

Problem b   Insert this relation in the Fourier series (14.2) to show that this Fourier series can also be written as:

$$f(x) = \sum_{n=-\infty}^{\infty} c_n\, e^{in\pi x/L}, \tag{14.17}$$

with the coefficients c_n given by:

$$\left.\begin{aligned}
c_n &= (a_n - ib_n)/2 &&\text{for } n > 0, \\
c_n &= (a_{|n|} + ib_{|n|})/2 &&\text{for } n < 0, \\
c_0 &= a_0/2.
\end{aligned}\right\} \tag{14.18}$$

Note that the absolute value $|n|$ is used for $n < 0$.

Problem c   Explain why the n-summation in (14.17) extends from $-\infty$ to ∞ rather than from 0 to ∞.

Problem d   Relations (14.7) and (14.8) can be used to express the coefficients c_n in the function $f(x)$. Treat the cases $n > 0$, $n < 0$, and $n = 0$ separately to show that for all values of n the coefficient c_n is given by:

$$c_n = \frac{1}{2L}\int_{-L}^{L} f(x)e^{-in\pi x/L}dx. \tag{14.19}$$

The sum (14.17) with (14.19) constitutes the complex Fourier transform over a finite interval. Again, there is a close analogy with the projections of vectors shown in Section 12.1. Before we can explore this analogy, the inner product between two complex functions $f(x)$ and $g(x)$ needs to be defined. This inner product is not given by $(f \cdot g) = \int f(x)g(x)dx$. The reason for this is that the length of a vector is defined by $\|\mathbf{v}\|^2 = (\mathbf{v} \cdot \mathbf{v})$, and a straightforward generalization of this expression to functions using the inner product given above would give for the norm of the function: $\|f\|^2 = (f \cdot f) = \int f^2(x)dx$. However, when $f(x)$ is purely imaginary, this would lead to a negative norm. This can be avoided by defining the inner product of two complex functions by:

$$(f \cdot g) \equiv \int_{-L}^{L} f^*(x)g(x)dx, \tag{14.20}$$

where the asterisk denotes the complex conjugate.

Problem e Show that with this definition the norm of $f(x)$ is given by $\|f\|^2 = (f \cdot f) = \int |f(x)|^2 dx$.

With this inner product the norm of the function is guaranteed to be positive. Now that we have an inner product, the analogy with the projections in linear algebra can be explored. To do this, define the following basis functions:

$$u_n(x) \equiv \frac{1}{\sqrt{2L}} e^{in\pi x/L}. \tag{14.21}$$

Problem f Show that these functions are orthonormal with respect to the inner product (14.20), that is, show that:

$$(u_n \cdot u_m) = \delta_{nm}. \tag{14.22}$$

Pay special attention to the normalization of these functions, that is, when $n = m$.

Problem g Expand the function $f(x)$ in these basis functions, $f(x) = \sum_{n=-\infty}^{\infty} \gamma_n u_n(x)$, and show that $f(x)$ can be written as:

$$f = \sum_{n=-\infty}^{\infty} u_n (u_n \cdot f). \tag{14.23}$$

Problem h Make the comparison between this expression and the expressions for the projections of vectors in Section 12.1.

14.3 Fourier transform on an infinite interval

In several applications, one wants to compute the Fourier transform of a function that is defined on an infinite interval. This amounts to taking the limit $L \to \infty$. However, a simple inspection of (14.19) shows that one cannot simply take the limit $L \to \infty$ of the expressions of the previous section because in that limit c_n is a possible infinite integral divided by infinity, which is poorly defined. In order to define the Fourier transform for an infinite interval, define the variable k by:

$$k \equiv \frac{n\pi}{L}. \tag{14.24}$$

An increment Δn corresponds to an increment Δk given by $\Delta k = \pi \Delta n / L$. In the summation over n in the Fourier expansion (14.17), n is incremented by unity: $\Delta n = 1$. This corresponds to an increment $\Delta k = \pi / L$ of the variable k. In the limit $L \to \infty$, this increment goes to zero, which implies that the summation over n should be replaced by an integration over k:

$$\sum_{n=-\infty}^{\infty} (\cdots) \to \frac{\Delta n}{\Delta k} \int_{-\infty}^{\infty} (\cdots) \, dk = \frac{L}{\pi} \int_{-\infty}^{\infty} (\cdots) \, dk \quad \text{as } L \to \infty. \tag{14.25}$$

Problem a   Explain the presence of the factor $\Delta n / \Delta k$ and prove the last identity.

This is not enough to generalize the Fourier transform of the previous section to an infinite interval. As noted earlier, the coefficients c_n are poorly defined in the limit $L \to \infty$. Also note that the integral on the right-hand side of (14.25) is multiplied by L/π, and this coefficient is infinite in the limit $L \to \infty$. Both complications can be solved by defining the following function:

$$F(k) \equiv \frac{L}{\pi} c_n, \tag{14.26}$$

where the relation between k and n is given by (14.24).

Problem b   Show that with the replacements (14.25) and (14.26) the limit $L \to \infty$ of the complex Fourier transform (14.17) and (14.19) can be taken and that the result can be written as:

$$f(x) = \int_{-\infty}^{\infty} F(k) e^{ikx} \, dk, \tag{14.27}$$

$$F(k) = \frac{1}{2\pi} \int_{-\infty}^{\infty} f(x) e^{-ikx} \, dx. \tag{14.28}$$

14.4 Fourier transform and the delta function

In this section the Fourier transform of the delta function is treated. This is not only useful in a variety of applications, but it will also establish the relation between the Fourier transform and the closure relation introduced in Section 12.1. Consider the delta function centered at $x = x_0$:

$$f(x) = \delta(x - x_0). \tag{14.29}$$

Problem a Show that the Fourier transform $F(k)$ of $f(x)$ as defined in expression (14.28) is given by:

$$F(k) = \frac{1}{2\pi} e^{-ikx_0}. \tag{14.30}$$

Problem b Show that this implies that the Fourier transform of the delta function $\delta(x)$ centered at $x = 0$ is a constant. Determine this constant.

Problem c Use (14.27) to show that

$$\delta(x - x_0) = \frac{1}{2\pi} \int_{-\infty}^{\infty} e^{ik(x-x_0)} dk. \tag{14.31}$$

Problem d Use a similar analysis to derive that

$$\delta(k - k_0) = \frac{1}{2\pi} \int_{-\infty}^{\infty} e^{-i(k-k_0)x} dx. \tag{14.32}$$

We have shown in Problem g of Section 13.2 that the delta function is symmetric, that is, that $\delta(k - k_0) = \delta(k_0 - k)$. This implies that expression (14.32) also holds with the opposite sign in the exponent. Expressions (14.31) and (14.32) are useful in a number of applications. Again, there is a close analogy between these expressions and the projection of vectors introduced in Section 12.1. To establish this connection, let us define the following basis functions:

$$u_k(x) \equiv \frac{1}{\sqrt{2\pi}} e^{ikx}, \tag{14.33}$$

and use the inner product defined in (14.20) with the integration limits extending from $-\infty$ to ∞.

Problem e Show that (14.32) implies that

$$\left(u_k \cdot u_{k_0} \right) = \delta(k - k_0). \tag{14.34}$$

This implies that the functions $u_k(x)$ form an orthonormal set, because this relation generalizes (14.22) to a continuous basis of functions.

Problem f   Use (14.31) to derive that

$$\int_{-\infty}^{\infty} u_k(x)u_k^*(x_0)\, dk = \delta\,(x - x_0).\tag{14.35}$$

This expression is the counterpart of the closure relation (12.13), introduced in Section 12.1 for finite-dimensional vector spaces. Note that the delta function $\delta\,(x - x_0)$ plays the role of the identity operator $\mathbf{I}$ with components $I_{ij} = \delta_{ij}$ in (12.13) and that the summation $\sum_{i=1}^{N}$ over the basis vectors is replaced by an integration $\int_{-\infty}^{\infty} dk$ over the basis functions. Both differences are due to the fact that we are dealing in this section with an infinite-dimensional function space rather than a finite-dimensional vector space. Also note that in (14.35) the complex conjugate of $u_k(x_0)$ is taken. The reason for this is that for complex unit vectors $\hat{\mathbf{n}}$ the transpose in the completeness relation (12.13) should be replaced by the Hermitian conjugate, which is defined as the complex conjugate of the transpose.

14.5 Changing the sign and scale factor

In the Fourier transformation (14.27) from the wavenumber domain (k) to the position domain (x), the exponent has a plus sign, e^{+ikx}, and the coefficient multiplying the integral is given by 1. In other texts on Fourier transforms you may encounter a different sign in the exponent and different scale factors are sometimes used. For example, the exponent in the Fourier transformation from the wavenumber domain to the position domain may have a minus sign, e^{-ikx}, and there may be a scale factor such as $1/\sqrt{2\pi}$. It turns out that there is a freedom in choosing the sign of the exponential as well as in the scaling of the Fourier transform. We first study the effect of a scaling parameter on the Fourier transform.

Problem a   Let the function $F(k)$ defined in (14.28) be related to a new function $\tilde{F}(k)$ by a scaling with a scale factor $C : F(k) = C\tilde{F}(k)$. Use (14.27) and (14.28) to show that:

$$f(x) = C\int_{-\infty}^{\infty} \tilde{F}(k)e^{ikx}\,dk,\tag{14.36}$$

$$\tilde{F}(k) = \frac{1}{2\pi C}\int_{-\infty}^{\infty} f(x)e^{-ikx}\,dx.\tag{14.37}$$

These expressions are equivalent to the original Fourier transform pair (14.27) and (14.28). The constant C is completely arbitrary. This implies that one may take any multiplication constant for the Fourier transform; the only restriction is that the product of the coefficients for Fourier transform and the backward transform is equal to $1/2\pi$.

Problem b Prove this last statement.

In the literature, notably in quantum mechanics, one often encounters the Fourier transform pair defined by the value $C = 1/\sqrt{2\pi}$. This leads to the Fourier transform pair:

$$f(x) = \frac{1}{\sqrt{2\pi}} \int_{-\infty}^{\infty} \tilde{F}(k)e^{ikx}dk, \qquad (14.38)$$

$$\tilde{F}(k) = \frac{1}{\sqrt{2\pi}} \int_{-\infty}^{\infty} f(x)e^{-ikx}dx. \qquad (14.39)$$

This normalization not only has the advantage that the multiplication factors for the forward and backward transformations are identical $(1/\sqrt{2\pi})$, but the constants are also identical to the constant used in (14.33) to create a set of orthonormal functions.

Next, we investigate a change in the sign of the exponent in the Fourier transform, using the function $\tilde{F}(k)$ defined by $\tilde{F}(k) = F(-k)$.

Problem c Change the integration variable k in (14.27) to $-k$ and show that the Fourier transform pair (14.27) and (14.28) is equivalent to:

$$f(x) = \int_{-\infty}^{\infty} \tilde{F}(k)e^{-ikx}dk, \qquad (14.40)$$

$$\tilde{F}(k) = \frac{1}{2\pi} \int_{-\infty}^{\infty} f(x)e^{ikx}dx. \qquad (14.41)$$

Note that these expressions differ from the earlier expressions only by the sign of the exponents. This means that there is a freedom in the choice of this sign. It does not matter which sign convention you use. Any choice of the sign and the multiplication constant for the Fourier transform can be used as long as:

1.   the product of the constants for the forward and backward transform is equal to $1/2\pi$, and

2.   the sign of the exponent for the forward and the backward transformation is opposite.

In this book, the Fourier transform pair (14.27) and (14.28) is generally used for the Fourier transform from the space (x) domain to the wavenumber (k) domain.

Of course, the Fourier transform can also be used to transform a function from the time (t) domain to the frequency (ω) domain. Perhaps illogically the following convention is used in this book for this Fourier transform pair:

$$f(t) = \frac{1}{2\pi} \int_{-\infty}^{\infty} F(\omega)e^{-i\omega t}d\omega, \qquad (14.42)$$

$$F(\omega) = \int_{-\infty}^{\infty} f(t)e^{i\omega t}dt. \qquad (14.43)$$

This choice has implications for the combined Fourier transform from the (x, t)-domain to the (k, ω)-domain.

Problem d   Use (14.28) and (14.43) to derive the inverse of the double Fourier transform:

$$f(x, t) = \frac{1}{2\pi} \iint_{-\infty}^{\infty} F(k, \omega)e^{i(kx-\omega t)}dkd\omega. \qquad (14.44)$$

The function $e^{i(kx-\omega t)}$ in this integral describes a wave that moves for positive values of k and ω in the direction of increasing values of x. To see this, let us assume that we are at a crest of this wave and that we follow the motion of the crest over a time Δt and we want to find the distance Δx that the crest has moved in that time interval. If we follow a wave crest, the phase of the wave is constant, and hence, $kx - \omega t$ is constant.

Problem e   Show that this implies that $\Delta x = c\Delta t$, with c given by $c = \omega/k$. Why does this imply that the wave moves with velocity c?

In Section 18.3 we treat the properties of waves, their direction and speed, in more detail.

The exponential in the double Fourier transform (14.44) therefore describes for positive values of ω and k a wave traveling in the positive direction with velocity $c = \omega/k$. However, this is no hard reason that we should use the Fourier transform (14.44), as opposed to a transform with a different choice of the sign in the exponent. In fact, one should realize that the Fourier transform (14.44) involves the integration over *all* values of ω and k so that negative values of ω and k contribute to the integral as well.

14.6 Convolution theorem

There are different ways in which one can combine signals to create a new signal. Before we discuss cross correlation in Section 14.11, we first treat the convolution of two signals here. For the sake of argument the signals are taken to be functions of time, and the Fourier transform pair (14.42) and (14.43) is used for the forward and inverse Fourier transforms. Suppose a function $f(t)$ has a Fourier transform $F(\omega)$ defined by (14.42) and another function $h(t)$ has a similar Fourier transform $H(\omega)$:

$$h(t) = \frac{1}{2\pi} \int_{-\infty}^{\infty} H(\omega)e^{-i\omega t}d\omega. \tag{14.45}$$

The two Fourier transforms $F(\omega)$ and $H(\omega)$ can be multiplied in the frequency domain, and we want to find out what the Fourier transform of the product $F(\omega)H(\omega)$ is in the time domain.

Problem a Show that:

$$F(\omega)H(\omega) = \int\!\!\!\int_{-\infty}^{\infty} f(t_1)h(t_2)e^{i\omega(t_1+t_2)}dt_1dt_2. \tag{14.46}$$

Problem b Show that after a Fourier transformation this function corresponds in the time domain to:

$$\int_{-\infty}^{\infty} F(\omega)H(\omega)e^{-i\omega t}d\omega = \frac{1}{2\pi} \int\!\!\!\int\!\!\!\int_{-\infty}^{\infty} f(t_1)h(t_2)e^{i\omega(t_1+t_2-t)}dt_1dt_2d\omega. \tag{14.47}$$

Problem c Use the representation (14.31) of the delta function to carry out the integration over ω and show that this gives:

$$\int_{-\infty}^{\infty} F(\omega)H(\omega)e^{-i\omega t}d\omega = \int\!\!\!\int_{-\infty}^{\infty} f(t_1)h(t_2)\delta(t_1 + t_2 - t)\, dt_1dt_2. \tag{14.48}$$

Problem d Carry out the integration over t_1 with the sifting property of the delta function (13.10), and show that after renaming the variable t_2 as τ the result can be written as:

$$\int_{-\infty}^{\infty} F(\omega)H(\omega)e^{-i\omega t}d\omega = \int_{-\infty}^{\infty} f(t-\tau)h(\tau)d\tau = (f * h)(t). \tag{14.49}$$

The τ-integral in the middle term is called the *convolution* of the functions f and h; this operation is denoted by the symbol $(f * h)$. Equation (14.49) states that a multiplication of the spectra of two functions in the frequency domain corresponds to the convolution of these functions in the time domain. For this reason, (14.49) is called the *convolution theorem*. This theorem is schematically indicated in the following diagram:

$$f(t) \longleftrightarrow F(\omega),$$
$$h(t) \longleftrightarrow H(\omega),$$
$$(f * h)(t) \longleftrightarrow F(\omega)H(\omega). \tag{14.50}$$

The convolution of two time series plays an important role in exploration geophysics. Imagine a seismic experiment where dynamite or a vibroseis truck generates waves that propagate through the Earth. Let the source signal in the time domain be given by $s(t)$. The waves reflect off layers in the Earth and are recorded by geophones at or near the surface. Suppose first that a single reflector with reflection coefficient r is located in the Earth, and that the travel time from the source to the reflector and back to the geophone is denoted by τ. In that case the recorded signal is given by $r \times s(t - \tau)$. In practice, more reflectors are present. We label the reflectors with an index i, and each reflector has reflection coefficient r_i and travel time τ_i. In that case the recorded data are given by

$$d(t) = \sum_i r_i s(t - \tau_i). \tag{14.51}$$

If the reflectors are continuous, we replace the sum over the index i by an integral over the travel time τ:

$$d(t) = \int r(\tau)s(t - \tau)d\tau. \tag{14.52}$$

A comparison with expression (14.49) shows that the recorded data are the convolution of the Earth's reflectivity with the source signal. The goal of the seismic survey is to determine the reflectivity $r(\tau)$ of the subsurface. One thus needs to undo the convolution (14.52) to remove the imprint of the source signal. This process is called *deconvolution*.

Deconvolution can, in principle, be carried out in the frequency domain. According to the convolution theorem (14.50), the counterpart of equation (14.52) is, in the frequency domain given by

$$D(\omega) = R(\omega)S(\omega). \tag{14.53}$$

From this expression, it follows that $R(\omega) = D(\omega)/S(\omega)$; hence, deconvolution corresponds to spectral division. An inverse Fourier transform then gives $r(t)$ in the time domain. Carrying out the deconvolution thus seems trivial in the frequency domain. The problem is that in practice the source spectrum $S(\omega)$ is not well known. This makes seismic deconvolution a difficult process (see the collection of articles compiled by Webster, 1981). It has been strongly argued by Ziolkowski (1991) that the seismic industry should make a larger effort to record the source signal accurately.

14.7 Linear filters

One can consider the previous discussion of convolution as the Earth filtering an input seismic signal. In this section we give a general description of linear filters. Consider a linear system that has an output signal $o(t)$ when it is given an input signal $i(t)$ (Figure 14.3). There are numerous examples of this kind of system. As an example, consider a damped harmonic oscillator that is driven by a force; this system is described by the differential equation $\ddot{x} + 2\beta\dot{x} + \omega_0^2 x = F/m$, where the dots denote a time derivative. The force $F(t)$ can be seen as the input signal, and the response $x(t)$ of the oscillator can be seen as the output signal. The relation between the input signal and the output signal is governed by the characteristics of the system under consideration; in this example it is the physics of the damped harmonic oscillator that determines the relation between the input signal $F(t)$ and the output signal $x(t)$.

Note that we have not defined yet what a *linear* filter is. A filter is linear when an input $c_1 i_1(t) + c_2 i_2(t)$ leads to an output $c_1 o_1(t) + c_2 o_2(t)$, in which $o_1(t)$ is the output corresponding to the input $i_1(t)$ and $o_2(t)$ is the output corresponding to the input $i_2(t)$.

Problem a Can you think of another example of a linear filter?

Problem b Can you think of a system that has one input signal and one output signal, in which these signals are related through a nonlinear relation? This would be an example of a nonlinear filter; the theory of this section would not apply to such a filter.

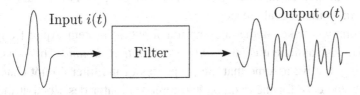

Figure 14.3 Schematic representation of a linear filter.

It is possible to determine the output $o(t)$ for any input $i(t)$ if the output to a delta function input is known. Consider the special input signal $\delta(t - \tau)$ that consists of a delta function centered at $t = \tau$. Since a delta function has "zero-width" (if it has a width at all), such an input function is impulsive. Let the output for this particular input be denoted by $g(t, \tau)$. Since this function is the response at time t to an impulsive input at time τ, this function is called the *impulse response*:

> The impulse response function $g(t, \tau)$ is the output of the system at time t due to an impulsive input at time τ.

How can the impulse response be used to find the response to an arbitrary input function? Any input function can be written as:

$$i(t) = \int_{-\infty}^{\infty} \delta(t - \tau) i(\tau) d\tau. \tag{14.54}$$

This identity follows from the sifting property of the delta function (13.10). However, we can also look at this expression from a different point of view. The integral on the right-hand side of (14.54) can be seen as a superposition of infinitely many delta functions $\delta(t - \tau)$. Each delta function when considered as a function of t is centered at time τ. Since we integrate over τ, these different delta functions are superposed to construct the input signal $i(t)$. Each of the delta functions in the integral (14.54) is multiplied by $i(\tau)$. This term plays the role of a coefficient that gives a weight to the delta function $\delta(t - \tau)$.

At this point it is crucial to use that the filter is linear. Since the response to the input $\delta(t - \tau)$ is the impulse response $g(t, \tau)$, and since the input can be written as the superposition (14.54) of delta function input signals $\delta(t - \tau)$, the output can be written as the same superposition of impulse response signals $g(t, \tau)$:

$$o(t) = \int_{-\infty}^{\infty} g(t, \tau) i(\tau) d\tau. \tag{14.55}$$

Problem c   Carefully compare (14.54) and (14.55). Note the similarity and make sure you understand the reasoning that has led to (14.55).

You may find this "derivation" of (14.55) rather vague. The notion of the impulse response will be treated in much greater detail in Chapter 17, because it plays a crucial role in mathematical physics.

At this point we make another assumption about the system. Apart from its linearity we will also assume it is *invariant for translations in time*. This is a complex way of saying that we assume that the properties of the filter do not change with time. This is the case for the damped harmonic oscillator described at the beginning of this section. However, this oscillator would not be invariant for translations in time if the damping parameter were a function of time as well: $\beta = \beta(t)$. In that case, the system would give different responses when the same input is used at different times.

When the properties of the filter do not depend on time, the impulse response $g(t, \tau)$ depends only on the *difference* $t - \tau$. To see this, consider the damped harmonic oscillator again. The response at a certain time depends only on the time that has lapsed between the excitation at time τ and the time of observation t. Therefore, for a time-invariant filter:

$$g(t, \tau) = g(t - \tau). \tag{14.56}$$

Inserting this in (14.55) shows that for a linear time-invariant filter the output is given by the convolution of the input with the impulse response:

$$o(t) = \int_{-\infty}^{\infty} g(t - \tau)i(\tau)d\tau = (g * i)(t). \tag{14.57}$$

Problem d Let the Fourier transform of $i(t)$ be given by $I(\omega)$ according to (14.42), the Fourier transform of $o(t)$ by $O(\omega)$ and the Fourier transform of $g(t)$ by $G(\omega)$. Use (14.57) to show that these Fourier transforms are related by:

$$O(\omega) = G(\omega)I(\omega). \tag{14.58}$$

Expressions (14.57) and (14.58) are key results in the theory in linear time-invariant filters. The first of these states that one only needs to know the response $g(t)$ to a single impulse to compute the output of the filter to *any* input signal $i(t)$. Equation (14.58) has two important consequences. First, if one knows the Fourier transform $G(\omega)$ of the impulse response, one can compute the Fourier transform $O(\omega)$ of the output. An inverse Fourier transform then gives the output $o(t)$ in the time domain.

Problem e Show that for a fixed value of ω that $G(\omega)e^{-i\omega t}$ is the response of the system to the input signal $e^{-i\omega t}$.

This means that if one knows the response of the filter to the harmonic signal $e^{-i\omega t}$ at any frequency, one knows $G(\omega)$ and the response to any input signal can be determined.

The second important consequence of (14.58) is that the output at frequency ω depends only on the input and impulse response at the same frequency ω, and not on other frequencies. This last property does not hold for nonlinear systems, because for these different frequencies components of the input signal are mixed by the nonlinearity of the system. An example of this phenomenon is the Earth's climate with variations that contain frequency components that cannot be explained by periodic variations in the orbital parameters in the Earth, but that are due to the nonlinear character of the climate response to the amount of energy received by the Sun (Snieder, 1985).

The fact that a filter can be used either by specifying its Fourier transform $G(\omega)$ (or equivalently the response to an harmonic input $e^{-i\omega t}$) or by prescribing the impulse response $g(t)$ implies that a filter can be designed either in the frequency domain or in the time domain. In Section 14.8 the action of a filter in the time domain is described. A Fourier transform then leads to a compact description of the filter response in the frequency domain. In Section 14.9 the converse route is taken; the filter is designed in the frequency domain, and a Fourier transform is used to derive an expression for the filter in the time domain.

As a last reminder it should be mentioned that although the theory of linear filters is introduced here for filters that act in the time domain, the theory is equally valid for filters in the spatial domain. In the latter case, the wavenumber k plays the role that the angular frequency played in this section. Since there may be more than one spatial dimension, the theory for the spatial domain must be generalized to include higher-dimensional spatial Fourier transforms. However, this does not change the principles involved. Examples can be found in Sections 20.2 and 20.3 for the upward continuation of the Earth's gravity field.

14.8 Dereverberation filter

In this section we derive a filter in the time domain to subdue reverberations in the water column in marine seismics. Seismic surveys at sea commonly involve a ship towing a string of hydrophones just below sea level to record the pressure variations from seismic reflections from deep inside the Earth (Figure 14.4). Since the pressure vanishes at the surface of the water, the water surface totally reflects pressure waves. In other words, the reflection coefficient for reflection at the water surface is equal to -1.

Let the reflection coefficient for waves reflecting upwards from the sea bed be denoted by r. Since the impedance in the solid earth is, in general, much larger than it is in water, the reflection coefficient is positive: $r > 0$. Because of energy conservation, the reflected wave must be weaker than the incoming wave, so that $r \leq 1$. In this situation, seismic pressure waves bounce back and forth repeatedly between the water surface and the sea floor. These reverberations are an unwanted artifact in seismic experiments, because a wave that has bounced back and forth in the water layer can be misinterpreted on a seismic section as a reflector in the Earth. For this reason one wants to eliminate these reverberations from seismic data.

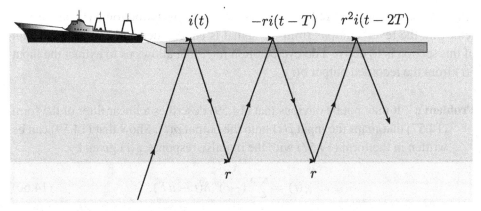

Figure 14.4 Pressure-wave recordings of seismic signals from the solid earth ($i(t)$), and their reverberations in the water column.

Suppose the wave field recorded by the hydrophones in the absence of reverberations is denoted by $i(t)$. We assume that the wave comes in at a specific angle of incidence and denote the time it takes for that wave to travel from the water surface to the sea bed and back to the surface by T; this time is called the two-way travel time.

Problem a Show that a wave that has bounced back and forth once is given by $-ri(t-T)$. Hint: Determine the amplitude of this wave from the reflection coefficients it encounters on its path and account for the time delay due to the bouncing up and down once in the water layer.

Problem b Generalize this result to a wave that bounces back and forth n-times in the water layer and show that the signal $o(t)$ recorded by the hydrophones is given by:

$$o(t) = i(t) - ri(t-T) + r^2i(t-2T) + \cdots$$

or

$$o(t) = \sum_{n=0}^{\infty} (-r)^n \, i(t - nT). \tag{14.59}$$

The notation $i(t)$ and $o(t)$ that was introduced in the previous section is deliberately used here. The action of the reverberation in the water layer is seen as a linear filter. The input of the filter $i(t)$ is the wavefield that would have been recorded if the waves did not bounce back and forth in the water layer. The output is the wave field that results from the reverberations in the water layer. In a marine seismic experiment one records the wave field $o(t)$ while one would like to know the signal

$i(t)$ that contains just the reflections from below the water bottom. The process of removing the reverberations from the signal is called "dereverberation." The aim of this section is to derive a dereverberation filter that allows us to extract the input $i(t)$ from the recorded output $o(t)$.

Problem c   It may not be obvious that (14.59) describes a linear filter of the form (14.57) that maps the input $i(t)$ onto the output $o(t)$. Show that (14.59) can be written in the form (14.57) with the impulse response $g(t)$ given by:

$$g(t) = \sum_{n=0}^{\infty} (-r)^n \, \delta(t - nT),$$ (14.60)

with $\delta(t)$ the Dirac delta function.

Problem d   Show that $g(t)$ is indeed the impulse response. In other words, show that if a delta function is incident as a primary arrival at the water surface, the reverberations within the water layer lead to the signal (14.60).

The deconvolution can be carried out easiest in the frequency domain. Let the Fourier transforms of $i(t)$ and $o(t)$, as defined by the transform (14.43), be denoted by $I(\omega)$ and $O(\omega)$, respectively. It follows from (14.59) that one needs to find the Fourier transform of $i(t - nT)$.

Problem e   According to the definition (14.43), the Fourier transform of $i(t - \tau)$ is given by $\int_{-\infty}^{\infty} i(t - \tau)e^{i\omega t} dt$. Use a change of the integration variable to show that the Fourier transform of $i(t - \tau)$ is given by $I(\omega)e^{i\omega\tau}$.

What you have derived here is the *shift property* of the Fourier transform: a translation of a function over a time τ corresponds in the frequency domain to a multiplication by $e^{i\omega\tau}$

$$\left.\begin{array}{ccc} i(t) & \longleftrightarrow & I(\omega), \\ i(t - \tau) & \longleftrightarrow & I(\omega)e^{i\omega\tau}. \end{array}\right\}$$ (14.61)

Problem f   Apply a Fourier transform to (14.59) for the output, use the shift property (14.61) for each term and show that the output in the frequency domain is related to the Fourier transform of the input by the following expression:

$$O(\omega) = \sum_{n=0}^{\infty} (-r)^n \, e^{i\omega nT} I(\omega).$$ (14.62)

Problem g Use the theory of Section 14.7 to show that the filter that describes the generation of reverberations is given in the frequency domain by:

$$G(\omega) = \sum_{n=0}^{\infty} (-r)^n e^{i\omega nT}. \qquad (14.63)$$

Problem h Because of energy conservation, the reflection coefficient r is less than or equal to unity; therefore, this series is guaranteed to converge. Sum this series to show that

$$G(\omega) = \frac{1}{1 + re^{i\omega T}}. \qquad (14.64)$$

Hint: Use that $(-r)^n e^{i\omega nT} = \left(-re^{i\omega T}\right)^n$, and use expression (3.15).

This is a useful result because it implies that the output and the input in the frequency domain are related by

$$O(\omega) = \frac{1}{1 + re^{i\omega T}} I(\omega). \qquad (14.65)$$

Note that the action of the reverberation leads in the frequency domain to a simple division by $(1 + re^{i\omega T})$. Note also that (14.65) has a form similar to (3.40), which accounts for the reverberation of waves between two stacks of reflectors. This resemblance is no coincidence because the physics of waves bouncing back and forth between two reflectors is similar.

Problem i The goal of this section was to derive a dereverberation filter that produces $i(t)$ when $o(t)$ is given. Use (14.65) to derive a dereverberation filter in the frequency domain.

The dereverberation filter you have just derived is simple in the frequency domain; it only involves a multiplication of every frequency component $O(\omega)$ by a scalar. Since multiplication is a simple and efficient procedure, it is attractive to carry out dereverberation in the frequency domain. The dereverberation filter you have just derived was developed originally by Backus (1959).

The simplicity of the dereverberation filter hides a complication. If the reflection coefficient r and the two-way travel time T are known exactly and if the sea bed is exactly horizontal, there is no problem with the dereverberation filter. However, in practice one only has *estimates* of these quantities. Let these estimates be denoted by r' and T', respectively. The reverberations lead in the frequency domain to a division by $1 + re^{i\omega T}$, while the dereverberation filter based on the estimated parameters leads to a multiplication with $1 + r'e^{i\omega T'}$. The net effect of the generation of the reverberations and the subsequent dereverberation is thus given in the frequency domain by a multiplication by

$$\frac{1 + r' e^{i\omega T'}}{1 + r e^{i\omega T}}.$$

Problem j Show that when the reflection coefficients are close to unity and when the estimate of the travel time is not accurate ($T' \neq T$), the term given above differs appreciably from unity. Explain why this implies that the dereverberation does not work well.

In practice one faces not only the problem that the estimates of the reflection coefficients and the two-way travel time may be inaccurate, but in addition the sea bed may not be exactly flat and there may be variations in the reflection coefficient along the sea bed. In which case the performance of the dereverberation filter can be significantly degraded.

14.9 Design of frequency filters

In this section we consider the problem in which a time series $i(t)$ is recorded and is contaminated with high-frequency noise. The aim of this section is to derive a filter in the time domain that removes the frequency components with a frequency greater than a cut-off frequency ω_0 from the time series. Such a filter is called a low-pass filter because only frequency components lower than the threshold ω_0 pass the filter.

Problem a Show that this filter is given in the frequency domain by:

$$G(\omega) = \begin{cases} 1 & \text{if } |\omega| \leq \omega_0 \\ 0 & \text{if } |\omega| > \omega_0 \end{cases}. \tag{14.66}$$

Problem b Explain why the absolute value of the frequency should be used in this expression.

Problem c Show that this filter is given in the time domain by

$$g(t) = \int_{-\omega_0}^{\omega_0} e^{-i\omega t} d\omega. \tag{14.67}$$

Problem d Carry out the integration over frequency to derive that the filter is, in the time domain, given by

$$g(t) = 2\omega_0 \, \text{sinc}\,(\omega_0 t), \tag{14.68}$$

where the sinc function is defined by

$$\operatorname{sinc} x \equiv \frac{\sin x}{x}. \tag{14.69}$$

Problem e Sketch the impulse response (14.68) of the low-pass filter as a function of time. Determine the behavior of the filter for $t = 0$ and show that the first zero-crossing of the filter is at time $t = \pm\pi/\omega_0$.

The zero-crossing of the filter is of fundamental importance. It implies that the width of the impulse response in the time domain is given by $2\pi/\omega_0$.

Problem f Use expression (14.66) to show that the width of the filter in the frequency domain is given by $2\omega_0$.

This means that when the cut-off frequency ω_0 is increased, the width of the filter in the frequency domain increases but the width of the filter in the time domain decreases. A large width of the filter in the frequency domain corresponds to a small width of the filter in the time domain, and vice versa.

Problem g Show that the product of the width of the filter in the time domain and the width of the same filter in the frequency domain is given by 4π.

The significance of this result is that this product is independent of the frequency ω_0. This implies that the filter cannot be arbitrarily peaked in both the time domain and the frequency domain. This effect has pronounced consequences since it is the essence of the *uncertainty relation of Heisenberg*, which states that the position and momentum of a particle can never be known exactly; more details can be found in the book of Merzbacher (1961).

The filter (14.68) does not actually have very desirable properties; it has two basic problems. The first problem is that the filter decays slowly with time as $1/t$. This means that the filter in the time domain is long, so that the convolution of a time series with the filter is numerically inefficient. This can be solved by making the cut-off of the filter in the frequency domain more gradual than the frequency cut-off defined in (14.66), for example, by using the filter $G(\omega) = (1 + |\omega|/\omega_0)^{-n}$ with n a positive integer.

Problem h Does this filter have the steepest cut-off for low values of n or for high values of n? Hint: Make a plot of $G(\omega)$.

The second problem is that filter (14.68) is not *causal* in the sense that $g(t)$ is nonzero for $t < 0$. This means that when a signal is convolved with the filter, the output signal depends on the value of the input at later times.

Problem i   Show that this is the case, and that the output depends on the input of earlier times only when $g(t) = 0$ for $t < 0$.

A causal filter can be designed by using the theory of analytic functions described in Chapter 15. The design of filters is quite an art; details can be found for example in the books of Robinson and Treitel (1980) or Claerbout (1976).

14.10 Linear filters and linear algebra

There is a close analogy between the theory of linear filters of Section 14.7 and the eigenvector decomposition of a matrix in linear algebra as treated in Section 12.3. To see this, we will use the same notation as in Section 14.7 and use the Fourier transform (14.45) to write the output of the filter in the time domain as:

$$o(t) = \frac{1}{2\pi} \int_{-\infty}^{\infty} O(\omega)e^{-i\omega t}d\omega. \tag{14.70}$$

Problem a   Use expression (14.58) to show that this can be written as

$$o(t) = \frac{1}{2\pi} \int_{-\infty}^{\infty} G(\omega)I(\omega)e^{-i\omega t}d\omega, \tag{14.71}$$

and carry out an inverse Fourier transformation of $I(\omega)$ to derive that

$$o(t) = \frac{1}{2\pi} \iint_{-\infty}^{\infty} G(\omega)e^{-i\omega t}e^{i\omega \tau}i(\tau)d\omega d\tau. \tag{14.72}$$

To establish the connection with linear algebra, we introduce by analogy with (14.33) the following basis functions:

$$u_{\omega}(t) \equiv \frac{1}{\sqrt{2\pi}}e^{-i\omega t}, \tag{14.73}$$

and the inner product

$$(f \cdot g) \equiv \int_{-\infty}^{\infty} f^*(t)g(t)dt. \tag{14.74}$$

Problem b   Use the results of Section 14.4 to show that these basis functions are orthonormal for this inner product in the sense that

$$(u_{\omega} \cdot u_{\omega'}) = \delta(\omega - \omega'). \tag{14.75}$$

Problem c These functions play the same role as the eigenvectors in Section 12.3. To which expression in Section 12.3 does the above expression correspond?

Problem d Show that (14.72) can be written as

$$o(t) = \int_{-\infty}^{\infty} G(\omega)u_\omega(t)\,(u_\omega \cdot i)\,d\omega. \tag{14.76}$$

This expression should be compared with

$$\mathbf{Ap} = \sum_{n=1}^{N} \lambda_n \hat{\mathbf{v}}^{(n)} \left(\hat{\mathbf{v}}^{(n)} \cdot \mathbf{p} \right). \tag{12.38}$$

The integration over frequency plays the same role as the summation over eigenvectors in (12.38). Expression (14.76) can be seen as a description for the operator $g(t)$ in the time domain that maps the input function $i(t)$ onto the output $o(t)$.

Problem e Use (14.57), (14.74), and (14.76) to show that:

$$g(t - \tau) = \int_{-\infty}^{\infty} G(\omega)u_\omega(t)u_\omega^*(\tau)d\omega. \tag{14.77}$$

There is a close analogy between this expression and the dyadic decomposition of a matrix into its eigenvectors and eigenvalues derived in Section 12.3.

Problem f To see this connection, show that (12.39) can be written in component form as:

$$A_{ij} = \sum_{n=1}^{N} \lambda_n \hat{v}_i^{(n)} \hat{v}_j^{(n)T}. \tag{14.78}$$

The sum over eigenvalues in (14.78) corresponds to the integration over frequency in (14.77). In Section 12.3 linear algebra in a finite-dimensional vector space was treated. In such a space there is a finite number of eigenvalues. In this section, a function space with infinitely many degrees of freedom is analyzed; for this reason the sum over a finite number of eigenvalues should be replaced by an integration over the continuous variable ω. The index i in (14.78) corresponds to the variable t in (14.77), while the index j corresponds to the variable τ.

Problem g Establish the connection between *all* variables in (14.77) and (14.78). Show specifically that $G(\omega)$ plays the role of eigenvalue λ_n and u_ω plays the role of eigenvector. Which operation in (14.77) corresponds to the transpose that is taken of the second eigenvector in (14.78)?

You may wonder why the function $u_\omega(t) = e^{-i\omega t}/\sqrt{2\pi}$, defined in (14.73), and not some other function plays the role of the eigenvector of the impulse response operator $g(t - \tau)$. To see why this is so, we have to understand what a linear filter actually does. Let us first consider the example of the reverberation filter of Section 14.8. According to (14.59) the output of the reverberation filter is

$$o(t) = i(t) - ri(t - T) + r^2i(t - 2T) + \cdots . \quad (14.59)$$

It follows from this expression that what the filter really does is to take the input $i(t)$, translate it over a time nT to a new function $i(t - nT)$, multiply each term with $(-r)^n$, and sum over all values of n. This means that the filter is a combination of three operations: (i) translation in time, (ii) multiplication, and (iii) summation over n. The same conclusion holds for any general time-invariant linear filter.

Problem h   Use a change of the integration variable to show that the action of a time-invariant linear filter as given in (14.57) can be written as

$$o(t) = \int_{-\infty}^{\infty} g(\tau)i(t - \tau)d\tau. \quad (14.79)$$

The function $i(t - \tau)$ is the function $i(t)$ translated over a time τ. This translated function is multiplied by $g(\tau)$ and an integration over all values of τ is carried out. This means that in general the action of a linear filter can be seen as a combination of translation in time, multiplication, and integration over all translations τ. How can this be used to explain that the correct eigenfunctions to be used are $u_\omega(t) = e^{-i\omega t}/\sqrt{2\pi}$? The answer does not lie in the multiplication because *any* function is an eigenfunction of the operator that carries out multiplication by a constant; that is, $af(t) = \lambda f(t)$ for every function $f(t)$ with the eigenvalue $\lambda = a$. This means that the translation operator is the reason that the eigenfunctions are $u_\omega(t) = e^{-i\omega t}/\sqrt{2\pi}$. Let the operator that carries out a translation over a time τ be denoted by T_τ:

$$T_\tau f(t) \equiv f(t - \tau). \quad (14.80)$$

Problem i   Show that the functions $u_\omega(t)$ defined in (14.73) are the eigenfunctions of the translation operator T_τ, that is, show that $T_\tau u_\omega(t) = \lambda u_\omega(t)$. Express the eigenvalue λ of the translation operator in terms of the translation time τ.

Problem j   Compare this result with the shift property of the Fourier transform that was derived in (14.61).

This means that the functions $u_\omega(t)$ are the eigenfunctions to be used for the eigenfunction decomposition of a linear time-invariant filter, because these functions are eigenfunctions of the translation operator.

Problem k You identified in Problem e the eigenvalues of the filter with $G(\omega)$. Show that this interpretation is correct: in other words, show that when the filter g acts on the function $u_\omega(t)$, the result can be written as $G(\omega)u_\omega(t)$. Hint: Go back to Problem e of Section 14.7.

This analysis shows that the Fourier transform, which uses the functions $e^{-i\omega t}$, is so useful because these functions are the eigenfunctions of the translation operator. However, this also points to a limitation of the Fourier transform. Consider a linear filter that is not time invariant; that is, a filter in which the output does not depend only on the *difference* between the input time τ and the output time t. Such a filter satisfies the general equation (14.55) rather than the convolution integral (14.57). The action of a filter that is not time-invariant *cannot* in general be written as a combination of multiplication, translation, and integration. This means that for such a filter the functions $e^{-i\omega t}$ that form the basis of the Fourier transform are not the appropriate eigenfunctions. The upshot of this is that in practice the Fourier transform is only useful for systems that are time-invariant, or in general that are translationally invariant in the coordinate that is used.

14.11 Correlation of two signals

The convolution of two signals $f(t)$ and $h(t)$ was obtained via Fourier transformation to the frequency domain, taking the product $F(\omega)H(\omega)$, and carrying out a Fourier transformation back to the time domain. The same steps can be taken by multiplying $F(\omega)$ by the complex conjugate $H^*(\omega)$ and applying a Fourier transformation back to the time domain.

Problem a Take steps similar to those in the derivation of expression (14.49) to show that

$$\int_{-\infty}^{\infty} F(\omega)H^*(\omega)e^{-i\omega t}d\omega = \int_{-\infty}^{\infty} f(t+\tau)h^*(\tau)d\tau. \tag{14.81}$$

The right-hand side of this expression is called the *correlation* of the functions $f(t)$ and $h^*(t)$. Note that this expression is similar to the convolution theorem (14.49). This result implies that the Fourier transform of the product of a function and the complex conjugate in the frequency domain corresponds to the correlation in the time domain.

Measured time series are, in general, real functions, so that the correlation in that case reduces to $C(t) = \int_{-\infty}^{\infty} f(t+\tau)h(\tau)d\tau$. The function $f(t+\tau)$ is just a shifted version of the function $f(\tau)$. The integral $\int_{-\infty}^{\infty} f(t+\tau)h(\tau)d\tau$ is largest when the functions $f(t+\tau)$ and $h(\tau)$ are most similar to each other. The correlation $C(t)$ thus measures the degree to which functions resemble each other after shifting one of them compared to the other.

Problem b   Set $t = 0$ in (14.81) and let the function $h(t)$ be equal to $f(t)$ to show that

$$\int_{-\infty}^{\infty} |F(\omega)|^2 \, d\omega = \int_{-\infty}^{\infty} |f(t)|^2 \, dt. \qquad (14.82)$$

This is known as *Parseval's theorem*. To see its significance, note that $\int_{-\infty}^{\infty} |f(t)|^2 \, dt = (f \cdot f)$, with the inner product of (14.20) with t as the integration variable and with the integration extending from $-\infty$ to ∞. Since $\sqrt{(f \cdot f)}$ is the norm of f measured in the time domain, and since $\int_{-\infty}^{\infty} |F(\omega)|^2 \, d\omega$ is the square of the norm of F measured in the frequency domain, Parseval's theorem states that with this definition of the norm, the norm of a function in the time domain is equal to the norm in the frequency domain.

The Fourier transform $F(\omega)$ is the *complex spectrum*, and $|F(\omega)|^2$ is called the *power spectrum*. The power spectrum integrated over frequency is the total energy. Similarly, the power of the signal in the time domain is $|f(t)|^2$ and after integration over time this gives the total energy in the signal. Parseval's theorem (14.82) thus states that the energy in a signal is the same regardless of whether one computes it in the time domain or in the frequency domain.

14.12 Extracting the system response from noise

In many experiments in physics or in the earth sciences one uses a controlled source to excite waves that probe a physical system. That source can be a laser in optical experiments, or an explosion or air gun in seismic experiments. Sometimes, one cannot use an active source for practical reasons. In this section we show a simple example that one can retrieve waves propagating between two sensors by cross-correlating noise recorded at these two sensors. As a simplest prototype of this idea we consider a one-dimensional homogeneous medium where waves propagate with a constant velocity c. Waves are recorded at two sensors A and B at locations $x = L$ and $x = 0$, respectively (Figure 14.5).

We assume that right-going noise with frequency spectrum $S_R(\omega)$ is incident on both sensors from the left, and left-going noise $S_L(\omega)$ comes in from the right side. The noise propagating to the right is, in the frequency domain, given by $S_R(\omega)e^{ikx}$,

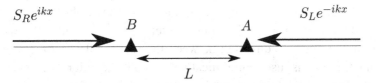

Figure 14.5   Geometry of experiment where noise recorded at two sensors A and B is used to find the ways that propagate between these sensors.

where the wavenumber is given by $k = \omega/c$. The noise propagating to left is given by $S_L(\omega)e^{-ikx}$, so that the total wave field is

$$U(x, \omega) = S_R(\omega)e^{ikx} + S_L(\omega)e^{-ikx}. \tag{14.83}$$

The noise is statistical, and we characterize it with its expectation value, which is indicated by angled brackets $\langle \cdots \rangle$. We assume that on average, the power spectrum of the noise is identical, so that

$$\langle |S_R(\omega)|^2 \rangle = \langle |S_L(\omega)|^2 \rangle = \langle |S(\omega)|^2 \rangle. \tag{14.84}$$

We also assume that the noise coming in from the left is, on average, uncorrelated from the noise coming in from the right. In the frequency domain this implies that

$$\langle S_R(\omega)S_L^*(\omega) \rangle = 0. \tag{14.85}$$

Problem a   Use equation (14.81) to show that expression (14.85) implies that

$$\int \langle s_R(t + \tau)s_L(\tau) \rangle d\tau = 0, \tag{14.86}$$

where $s_R(t)$ and $s_L(t)$ are the right-going and left-going noise in the time domain.

When the noise is uncorrelated, the product of $s_R(t_1)$ and $s_L(t_2)$ vanishes on average for all times t_1 and t_2; hence, expression (14.86) is satisfied, and hence, equation (14.85) also holds.

We next consider the cross-correlation of the noise recorded at the receivers A and B and denote the expectation of this cross-correlation in the frequency domain as

$$C_{AB}(\omega) = \langle U(x = L, \omega)U^*(x = 0, \omega) \rangle. \tag{14.87}$$

Problem b   Use the previous result and expressions (14.83)–(14.85) to show that

$$C_{AB}(\omega) = \langle |S(\omega)|^2 \rangle \left(e^{ikL} + e^{-ikL} \right). \tag{14.88}$$

The term e^{ikL} describes waves that propagate toward the right from the sensor at $x = 0$ to the sensor at $x = L$, while the term e^{-ikL} describes the waves that propagate in the opposite direction between the sensors. The term $\langle |S(\omega)|^2 \rangle$ is a positive number that is just a multiplicative constant. Equation (14.88) thus states the important result that one can obtain the superposition of waves propagating in opposite directions between two sensors from the cross-correlation of random noise recorded at these two sensors.

The property that the system response can be extracted from the cross-correlation of noise is shown here for a one-dimensional homogeneous wave system, but this property holds for waves in any number of dimensions and for arbitrarily inhomogeneous media (Curtis et al., 2006; Larose et al., 2006; Snieder and Larose, 2013). Figure 14.6 shows the result of ambient noise correlations in North America of the seismological station indicated with star with noise recordings at all other stations marked by triangles. The cross-correlation is shown in the

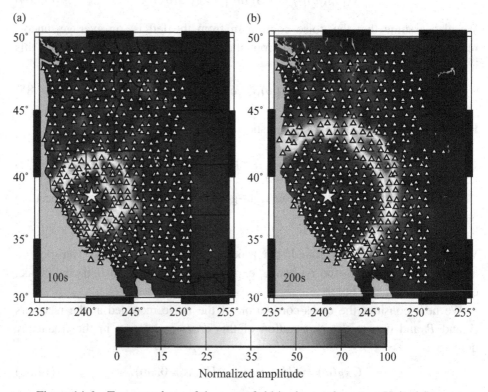

Figure 14.6   Two snapshots of the wave field in the southwestern United States from Lin et al. (2009). These are obtained by cross-correlating noise recorded at a master station (indicated by a star) with noise recorded at a network of stations (triangles).

time domain at two different lag times, 100 s (left panel) and 200 s (right panel). The cross-correlation thus obtained clearly behaves as if a wave is radiated from the station shown with the star, as though a source was present at that location (Lin et al., 2009). In reality, no such source is present, and for this reason one speaks of the *virtual source method* (Bakulin and Calvert, 2006). This technique is also referred to as *seismic interferometry* (Schuster, 2009). In the example of Figure 14.6, the waves are Rayleigh waves, these are elastic waves that propagate along Earth's surface (Aki and Richards, 2002), and the ambient noise is generated by wind-generated waves on the surface of the Pacific Ocean. The property that one can extract the response of a system from ambient noise is not restricted to transmitted scalar waves; the principle is also valid for a large class of phenomena that include electromagnetic waves, refracted waves, and diffusion (Wapenaar et al., 2006; Snieder et al., 2007; Mikesell and van Wijk, 2011).

15

Analytic functions

In this chapter we consider complex functions in the complex plane. The reason for doing this is that the requirement that a complex function "behaves well" (what that exactly means is defined shortly) imposes remarkable constraints on such functions. Since these constraints coincide with some of the laws of physics, the theory of complex functions has a number of important applications in mathematical physics. Some are explored in the later sections of this chapter. In addition, we lay the foundation for (contour) integration of complex functions that is treated in Chapter 16.

15.1 Theorem of Cauchy–Riemann

Before we get to complex functions, let us first consider a real function $F(x)$ of a real variable x. The derivative of such a function is defined by the rule

$$\frac{dF}{dx} = \lim_{\Delta x \to 0} \frac{F(x + \Delta x) - F(x)}{\Delta x}. \tag{15.1}$$

If the function is not continuous, the difference $F(x + \Delta x) - F(x)$ remains finite as $\Delta x \to 0$ and the derivative is infinite. Jumps in the function thus lead to infinite derivatives. But even when the function in continuous, there may be issues with defining the derivative. In general there are two ways in which Δx can approach zero: from above and from below. For a function that is differentiable it does not matter whether Δx approaches zero from above or from below. If the limits $\Delta x \downarrow 0$ and $\Delta x \uparrow 0$ do give a different result, it is a sign that the function does not behave well, it has a kink, and the derivative is not unambiguously defined; see Figure 15.1.

Complex functions $h(z)$ are decomposed into real and imaginary parts:

$$h(z) = f(z) + ig(z); \tag{15.2}$$

where, the functions $f(z)$ and $g(z)$ are real. The argument of these functions is complex: $z = x + iy$, so that $x = \Re e(z)$ and $y = \Im m(z)$, where $\Re e$ and $\Im m$ denote the real and imaginary part, respectively.

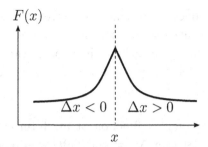

Figure 15.1   A function $F(x)$ that is not differentiable at x.

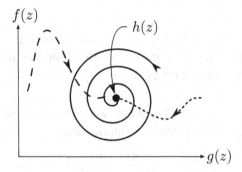

Figure 15.2   Three paths along which the limit $\Delta z \to 0$ can be taken in the complex plane.

For complex functions the derivative is defined in the same way as in (15.2) for real functions:

$$\frac{dh}{dz} = \lim_{\Delta z \to 0} \frac{h(z + \Delta z) - h(z)}{\Delta z}. \tag{15.3}$$

For real functions, Δx could approach zero in two ways: from below and from above. However, the limit $\Delta z \to 0$ in (15.3) can be taken in infinitely many ways. As an example, see Figure 15.2 where some different paths that one can use to let Δz approach zero are sketched. These do not necessarily give the same result.

Problem a   Consider the function $h(z) = e^{1/z}$. Using the definition (15.3), compute dh/dz at the point $z = 0$ when Δz approaches zero: (i) from the positive real axis, (ii) from the negative real axis, (iii) from the positive imaginary axis, and (iv) from the negative imaginary axis.

You have discovered that for some functions the result of the limit Δz depends critically on the path that one uses in the limit process. The derivative of such a function is not defined unambiguously. However, for many functions the value of the derivative does *not* depend on the way that Δz approaches zero. When these functions and their derivatives are also finite, they are called *analytic functions*. The requirement that the derivative does not depend on the way in which Δz approaches zero imposes a constraint on the real and imaginary parts of the complex function. To see this, we will let Δz approach zero along the real axis and along the imaginary axis.

Problem b   Consider a complex function of the form (15.2) and compute the derivative dh/dz by setting $\Delta z = \Delta x$ with Δx a real number. (Hence, Δz approaches zero along the real axis.) Show that the derivative is given by $dh/dz = \partial f/\partial x + i \partial g/\partial x$.

Problem c   Compute the derivative dh/dz also by setting $\Delta z = i\Delta y$ with Δy a real number. (Hence, Δz approaches zero along the imaginary axis.) Show that the derivative is given by $dh/dz = \partial g/\partial y - i \partial f/\partial y$.

Problem d   When $h(z)$ is analytic, these two expressions for the derivative are, by definition, equal. Show that this implies that:

$$\frac{\partial f}{\partial x} = \frac{\partial g}{\partial y}, \tag{15.4}$$

$$\frac{\partial g}{\partial x} = -\frac{\partial f}{\partial y}. \tag{15.5}$$

These are puzzling expressions, because they imply that the real and imaginary parts of an analytic complex function $h(z)$ are not independent. Instead, they are coupled by the constraints imposed by equations (15.4) and (15.5), called the *Cauchy–Riemann relations*.

Problem e   Use these relations to show that both $f(x, y)$ and $g(x, y)$ are harmonic functions. These are functions for which the Laplacian vanishes:

$$\nabla^2 f = \nabla^2 g = 0. \tag{15.6}$$

In addition to the coupling between f and g, we now found that the functions f and g must be harmonic functions, as well. This is exactly the reason why this theory is so useful in mathematical physics because harmonic functions arise in several applications, as we will see in the coming sections. However, we have not found all the properties of analytic functions yet.

Problem f   Show that:

$$(\nabla f \cdot \nabla g) = 0. \tag{15.7}$$

Since the gradient of a function is perpendicular to the lines where the function is constant, this implies that the curves where f is constant and where g is constant intersect each other at a fixed angle.

Problem g   Determine this angle.

Problem h   Verify the properties (15.4)–(15.7) explicitly for the function $h(z) = z^2$. Also sketch the lines in the complex plane where $f = \Re e(h)$ and $g = \Im m(h)$ are constant.

We have still not fully explored all the properties of analytic functions. Let us consider a line integral $\oint_C h(z)dz$ along a closed contour C in the complex plane.

Problem i   Use the property $dz = dx + idy$ to deduce that:

$$\oint_C h(z)dz = \oint_C \mathbf{v} \cdot d\mathbf{r} + i \oint_C \mathbf{w} \cdot d\mathbf{r}, \tag{15.8}$$

where $d\mathbf{r} = \begin{pmatrix} dx \\ dy \end{pmatrix}$ and the vectors $\mathbf{v}$ and $\mathbf{w}$ are defined by:

$$\mathbf{v} = \begin{pmatrix} f \\ -g \end{pmatrix} \quad \mathbf{w} = \begin{pmatrix} g \\ f \end{pmatrix}. \tag{15.9}$$

Note that we are now using x and y in a dual role as the real and imaginary parts of a complex number, as well as the Cartesian coordinates in a plane. In the following problem we will (perhaps confusingly) use the notation z both for a complex number in the complex plane, as well as for the familiar z-coordinate in a three-dimensional Cartesian coordinate system.

Problem j   Show that the Cauchy–Riemann relations (15.4)–(15.5) imply that the z-component of the *curl* of $\mathbf{v}$ and $\mathbf{w}$ vanishes: $(\nabla \times \mathbf{v})_z = (\nabla \times \mathbf{w})_z = 0$, and use (15.8) and the theorem of Stokes (9.2) to show that when $h(z)$ is analytic everywhere within the contour C:

$$\oint_C h(z)\, dz = 0, \quad \text{where } h(z) \text{ is analytic within } C. \tag{15.10}$$

This means that the line integral of a complex functions along *any* contour that encloses a region of the complex plane where that function is analytic is equal to zero. We will make extensive use of this property in Chapter 16 where we treat integration in the complex plane.

15.2 Electric potential

Analytic functions are often useful in the determination of the electric field and the potential for two-dimensional problems. The electric field satisfies the field equation

$$(\nabla \cdot \mathbf{E}) = \rho(\mathbf{r})/\epsilon_0. \qquad (6.24)$$

In free space the charge-density vanishes; hence, $(\nabla \cdot \mathbf{E}) = 0$. The electric field is related to the potential V through the relation

$$\mathbf{E} = -\nabla V. \qquad (15.11)$$

Problem a   Show that in free space the potential V is described by the Laplace equation (10.21):

$$\nabla^2 V(x, y) = 0. \qquad (15.12)$$

We can exploit the theory of analytic functions by noting that the real and imaginary parts of analytic functions both satisfy (15.12). This implies that if we take $V(x, y)$ to be the real part of a complex analytic function $h(x + iy)$, equation (15.12) is automatically satisfied.

Problem b   It follows from (15.11) that the electric field is perpendicular to the lines where V is constant. Show that this implies that the electric field lines are also perpendicular to the lines $V = const$. Use the theory of the previous section to argue that the field lines are the lines where the imaginary part of $h(x + iy)$ is constant.

This means that we receive a bonus by expressing the potential V as the real part of a complex analytic function, because the field lines simply follow from the requirement that $\Im m(h) = const$.

Suppose we want to know the potential in the half-space $y \geq 0$ when we have specified the potential on the x-axis. (Mathematically this means that we want to solve the equation $\nabla^2 V = 0$ for $y \geq 0$ when $V(x, y = 0)$ is given.) If we can find an analytic function $h(x + iy)$ such that on the x-axis (where $y = 0$) the real part of h is equal to the potential, we have solved our problem because the real part of h by definition satisfies both the required boundary condition and the field equation (15.12).

Problem c   Consider a potential that is given on the x-axis by

$$V(x, y = 0) = V_0\, e^{-x^2/a^2}. \qquad (15.13)$$

Show that on the x-axis this function can be written as $V = \Re e(h)$ with

$$h(z) = V_0\, e^{-z^2/a^2}. \qquad (15.14)$$

Problem d With this expression for $h(z)$ we can determine the potential and the field lines throughout the half-plane $y \geq 0$. Insert the relation $z = x + iy$ into expression (15.14) and take the real part to show that the potential is given by

$$V(x, y) = V_0 \, e^{(y^2 - x^2)/a^2} \cos\left(\frac{2xy}{a^2}\right). \qquad (15.15)$$

Problem e Verify explicitly that this solution satisfies the boundary condition at the x-axis and that it satisfies the field equation (15.12).

Problem f Show that the field lines are given by the relation

$$e^{(y^2 - x^2)/a^2} \sin\left(\frac{2xy}{a^2}\right) = const. \qquad (15.16)$$

Problem g Sketch the field lines and the lines where the potential is constant in the half-space $y \geq 0$.

In this derivation we have extended the solution $V(x, y)$ into the upper half-plane by identifying it with an analytic function in the half-plane, which has on the x-axis a real part that equals the potential. Note that we found the solution to this problem without explicitly solving the partial differential equation (15.12) which governs the potential. The approach we have taken is called *analytic continuation* since we continue an analytic function from one region (the x-axis) into the upper half-plane. Analytic continuation turns out to be an unstable process. This can be verified explicitly for this example.

Problem h Sketch the potential $V(x, y)$ as a function of x for the values $y = 0$, $y = a$, and $y = 10a$. What is the wavelength of the oscillations in the x-direction of the potential $V(x, y)$ for these values of y? Show that when we slightly perturb the constant a, the perturbation of the potential increases for increasing values of y. This implies that when we slightly perturb the boundary condition, the solution is more perturbed as we move further away from that boundary.

15.3 Fluid flow and analytic functions

As a second application of the theory of analytic functions we consider fluid flow. At the end of Section 6.2 we saw that the streamlines of fluid flow from a source can be determined by solving the differential equations

$$d\mathbf{r}/dt = \mathbf{v}(\mathbf{r}) \tag{6.19}$$

for the velocity field (6.17)–(6.18). This requires the solution of a system of nonlinear differential equations, which is difficult. Here the theory of analytic functions is used to solve this problem in a simple way. We consider once again a fluid that is incompressible ($\nabla \cdot \mathbf{v} = 0$) and consider the special case in which the vorticity of the flow vanishes:

$$\nabla \times \mathbf{v} = 0. \tag{15.17}$$

Such a flow is called *irrotational*, because it does not rotate (see Sections 7.3 and 7.4). The requirement (15.17) is automatically satisfied when the velocity is the gradient of a scalar function f:

$$\mathbf{v} = \nabla f. \tag{15.18}$$

Problem a   Show this by taking the curl of the previous expression.

The function f plays the same role for the velocity field $\mathbf{v}$ as the electric potential V does for the electric field $\mathbf{E}$. For this reason, flow with a vorticity equal to zero is called *potential flow*.

Problem b   Show that the requirement that the flow is incompressible implies that

$$\nabla^2 f = 0. \tag{15.19}$$

For an incompressible and irrotational flow in two dimensions, we can again use the theory of analytic functions to describe this flow field. Once we can identify the potential $f(x, y)$ with the real part of an analytic function $h(x+iy)$, we know that (15.19) must be satisfied.

Problem c   Consider the velocity field (6.14) due to a point source at $\mathbf{r} = 0$. Show that this flow field follows from the potential

$$f(x, y) = \frac{\dot{V}}{2\pi} \ln r, \tag{15.20}$$

where $r = \sqrt{x^2 + y^2}$.

Problem d   Verify explicitly that this potential satisfies (15.19), except for $r = 0$. For what physical reason is (15.19) not satisfied at $r = 0$?

We now want to identify the potential $f(x, y)$ with the real part of an analytic function $h(x+iy)$. We know already that the flow follows from the requirement that the velocity is the gradient of the potential; hence, it follows by taking the gradient of the real part of h. The curves $f = const.$ are perpendicular to the flow because $\mathbf{v} = \nabla f$ is perpendicular to these curves. However, it was shown in Section 15.1 that the curves $g = \Im m(h) = const.$ are perpendicular to the curves $f = \Re e(h) = const.$ This means that the flow lines are given by the curves $g = \Im m(h) = const.$ To use this, we first have to find an analytic function with a real part given by (15.20). The simplest guess is to replace r in (15.20) by the complex variable z:

$$h(z) = \frac{\dot{V}}{2\pi} \ln z. \tag{15.21}$$

Problem e Verify that for complex z the real part of this function indeed satisfies (15.20) and that the imaginary part of this function is

$$g = \frac{\dot{V}}{2\pi} \varphi, \tag{15.22}$$

where $\varphi = \arctan(y/x)$ is the argument of the complex number. Hint: Use the representation $z = r \exp(i\varphi)$ for the complex numbers.

Problem f Sketch the lines $g(x, y) = const.$ and verify that these lines indeed represent the flow lines in this flow.

Now we consider the more complicated problem in Section 6.2, where the flow has a source at $\mathbf{r}_+ = (L, 0)$ and a sink at $\mathbf{r}_- = (-L, 0)$. The velocity field is given by (6.17) and (6.18) and our goal is again to determine the flow lines without solving the differential equation (6.19). We translate the coordinates $\mathbf{r}$ of the flow to the complex plane $z = x + iy$, and the source $\mathbf{r}_+$ and sink $\mathbf{r}_-$ are placed on the real axis: $z_+ = L + i0$ and $z_- = -L + i0$, respectively. For the source at $\mathbf{r}_+$ flow is represented by the potential $(\dot{V}/2\pi) \ln |z - z_+|$; this follows from the solution (15.21) for a single source by moving this source to $z = z_+$.

Problem g Using a similar reasoning, determine the contribution to the potential f by the sink at $\mathbf{r}_-$. Show that the potential of the total flow is given by:

$$f(z) = \frac{\dot{V}}{2\pi} \ln \left(\frac{|z - z_+|}{|z - z_-|} \right). \tag{15.23}$$

Problem h Express this potential in x and y, compute the gradient, and verify that this potential indeed gives the flow of (6.17)–(6.18). You may find Figure 15.3 helpful.

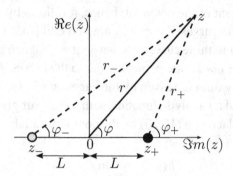

Figure 15.3   Definition of the geometric variables for the fluid flow with a source and a sink.

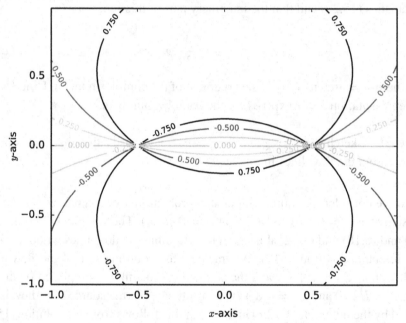

Figure 15.4   Streamlines for a source at $L = 0.5$ and a sink at $L = -0.5$ from the contours of expression (15.26).

We have found that the potential f is the real part of the complex function

$$h(z) = \frac{\dot{V}}{2\pi} \ln \left(\frac{z - z_+}{z - z_-} \right).$$

(15.24)

Problem i   Write $z - z_\pm = r_\pm e^{i\varphi_\pm}$ (with $r_\pm$ and $\varphi_\pm$ defined in Figure 15.3), take the imaginary part of (15.24), and show that $g = \Im m(h)$ is given by:

$$g = \frac{\dot{V}}{2\pi} (\varphi_+ - \varphi_-).$$

(15.25)

Problem j Show that the streamlines of the flow are given by the relation

$$\arctan\left(\frac{y}{x-L}\right) - \arctan\left(\frac{y}{x+L}\right) = const. \qquad (15.26)$$

This treatment is simpler than solving the differential equation (6.19). The streamlines of Figure 15.4 are the contours of the left-hand side of expression (15.26). If you compare these to Figure 6.3, you can see the connection between the flow velocity arrows and the streamlines.

16

Complex integration

In Chapter 15 the properties of analytic functions in the complex plane were treated. One of the key results is that the contour integral of a complex function is equal to zero when the function is analytic everywhere in the area of the complex plane enclosed by that contour, equation (15.10). From this result it follows that the integral of a complex function along a closed contour is nonzero only when the function is not analytic in the area enclosed by the contour. Functions that are not analytic come in different types. In this chapter complex functions are considered that are not analytic only at isolated points, these points are called the *poles* of the function.

16.1 Nonanalytic functions

When a complex function is analytic at a point z_0, it can be expanded in a Taylor series around that point, analogous to the treatment of real functions in Section 3.1. This implies that within a certain region around z_0 the function can be written as:

$$h(z) = \sum_{n=0}^{\infty} a_n (z - z_0)^n. \tag{16.1}$$

Note that in this sum only positive integer powers of $(z - z_0)$ appear.

Problem a   Show that the function $h(z) = \sin(z)/z$ can be written as a Taylor series around the point $z_0 = 0$ of the form (16.1) and determine the coefficients a_n.

Not all functions can be represented in a series of the form (16.1). As an example consider the function $h(z) = e^{1/z}$. The function is not analytic at the point $z = 0$, as we saw in Problem a of Section 15.1. Expanding the exponential leads to the expansion

$$h(z) = e^{1/z} = 1 + \frac{1}{z} + \frac{1}{2!}\frac{1}{z^2} + \cdots = \sum_{n=0}^{\infty} \frac{1}{n!}\frac{1}{z^n}. \tag{16.2}$$

In this case an expansion in *negative* powers of z is needed to represent this function. Each of the terms $1/z^n$ is for $n \geq 1$ singular at $z = 0$; this reflects the fact that the function $e^{1/z}$ has a pole at $z = 0$. In this section we consider complex functions that can be expanded around a point z_0 as an infinite sum of all *integer* powers of $(z - z_0)$:

$$h(z) = \sum_{n=-\infty}^{\infty} a_n (z - z_0)^n. \tag{16.3}$$

This type of series is called the *Laurent series*. Such a series does resemble a Taylor series, such as expression (16.1), but the difference is that in the Taylor series only powers $n \geq 0$ are present, while the Laurent series include negative powers n as well.

However, not *every* complex function can be represented as such a sum. Take for example the function $h(z) = \sqrt{z}$; it can be written as $h(z) = z^{1/2}$. When viewed as a sum of powers of z^n, the term $z^{1/2}$ is the only term that contributes. Since $n = \frac{1}{2}$ is not an integer, this function cannot be written in the form (16.3).

16.2 Residue theorem

It was argued in expression (15.10) that the integral of a complex function around a closed contour in the complex plane is only nonzero when the function is not analytic at some point in the area enclosed by the contour. Here we derive the value of the contour integral in the complex plane that encloses a pole of the function at the point z_0, as in the left-hand panel of Figure 16.1. Note that the integration is carried out in the counterclockwise direction along the path C. It is assumed that around the point z_0 the function $h(z)$ can be expanded in a power series of the form (16.3).

Our goal is to determine the value of the contour integral $\oint_C h(z)dz$. To do so, we first deviate the path C by making a small break in the path (C_- and C_+) with at the end a circular path with radius ϵ around the pole at z_0 (C_ϵ). This new path is called C^*. The first thing is to recognize that in the gray area defined by C^* the function $h(z)$ is analytic, because we have assumed that $h(z)$ is only nonanalytic at the point z_0. By virtue of the identity (15.10), this implies that

$$\oint_{C^*} h(z)dz = 0. \tag{16.4}$$

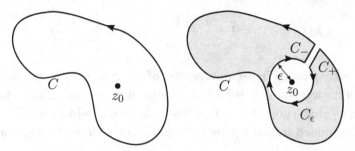

Figure 16.1   Left: A contour C in the complex plane, enclosing a single pole at z_0. Right: The contour C is adjusted to exclude the only pole at z_0.

Problem a   Show that the integrals along C^+ and C^- do not give a net contribution to the total integral:

$$\int_{C^+} h(z)dz + \int_{C^-} h(z)dz = 0. \tag{16.5}$$

Hint: Note the opposite directions of integration along the paths C^+ and C^-.

Problem b   Use this result and (16.4) to show that the integral along the original contour C is identical to the integral along the small circle C_ϵ around the point where $h(z)$ is not analytic:

$$\oint_C h(z)dz = \oint_{C_\epsilon} h(z)dz. \tag{16.6}$$

Problem c   The integration along C is in the counterclockwise direction. Show that the integration along C_ϵ in the expression above should also be taken in the counterclockwise direction.

Expression (16.6) is useful because the integral along the small circle can be evaluated by using that close to z_0 the function $h(z)$ can be written as the series (16.3). When one does this, the integration path C_ϵ needs to be parameterized. This can be achieved by writing the points on the path C_ϵ as

$$z = z_0 + \epsilon e^{i\varphi}, \tag{16.7}$$

with φ running from 0 to 2π since C_ϵ forms a complete circle.

Problem d   Use (16.3), (16.6), and (16.7) to derive that

$$\oint_C h(z)\, dz = \sum_{n=-\infty}^{\infty} i a_n \epsilon^{(n+1)} \int_0^{2\pi} e^{i(n+1)\varphi} d\varphi. \tag{16.8}$$

This expresses the contour integral in the coefficients a_n of the expansion (16.3).

Problem e Show by direct integration that:

$$\int_0^{2\pi} e^{im\varphi}d\varphi = \begin{cases} 0 & \text{for} \quad m \neq 0 \\ 2\pi & \text{for} \quad m = 0 \end{cases}. \tag{16.9}$$

Problem f Use this result to derive that only the term $n = -1$ contributes to the sum on the right-hand side of (16.8) and that

$$\oint_C h(z)dz = 2\pi i a_{-1}. \tag{16.10}$$

It may seem surprising that only the term $n = -1$ contributes to the sum on the right-hand side of (16.8). However, we could have anticipated this result because we had already discovered that the contour integral does not depend on the precise choice of the integration path. It can be seen that in the sum (16.8) each term is proportional to $\epsilon^{(n+1)}$. Since we know that the integral does not depend on the choice of the integration path, and hence, on the size of the circle C_ϵ, one would expect that only terms that do not depend on the radius ϵ contribute. This is only the case when $n + 1 = 0$; hence, only for the term $n = -1$ is the contribution independent of the size of the circle. It is indeed only this term that gives a nonzero contribution to the contour integral.

The coefficient a_{-1} is usually called the *residue* and is denoted by the symbol Res $h(z_0)$ rather than a_{-1}. However, remember that there is nothing mysterious about the residue, it is simply defined as

$$\text{Res } h(z_0) \equiv a_{-1}. \tag{16.11}$$

With this definition the result (16.10) can trivially be written as

$$\oint_C h(z)dz = 2\pi i \text{Res } h(z_0) \quad \text{(counterclockwise direction)}. \tag{16.12}$$

This may appear to be a rather uninformative rewrite of (16.10) but it is the form (16.12) that you will find in the literature and is called the *residue theorem*.

Of course, the residue theorem is only useful when one can determine the coefficient a_{-1} in the expansion (16.3). You can find in Section 2.12 of Butkov (1968) an overview of methods for computing the residue. Here we present the two most widely used ones. The first method is to determine the power series expansion (16.3) of the function explicitly.

Problem g Determine the power series expansion of the function $h(z) = \sin(z)/z^4$ around $z = 0$ and use this expansion to find the residue.

Unfortunately, this method does not always work. For some special functions, other tricks can be used. Here we consider functions with a simple pole; these are functions where the terms for $n < -1$ do not contribute in the expansion (16.3):

$$h(z) = \sum_{n=-1}^{\infty} a_n (z - z_0)^n. \tag{16.13}$$

An example of such a function is $h(z) = \cos(z)/z = z^{-1} - (1/2)z + \cdots$. The residue at the point $z_0 = 0$ follows by "extracting" the coefficient a_{-1} from (16.13).

Problem h Multiply (16.13) by $(z - z_0)$, and take the limit $z \to z_0$ to show that:

$$\text{Res } h(z_0) = \lim_{z \to z_0} (z - z_0) h(z) \quad \text{(simple pole).} \tag{16.14}$$

However, remember that this recipe works only for functions with a simple pole; it gives the wrong answer (infinity) when applied to a function that has nonzero coefficients a_n for $n < -1$ in (16.3).

In the treatment in this section we have considered an integration in the counterclockwise direction around the pole z_0.

Problem i Repeat the derivation of this section for a contour integration in the *clockwise* direction around the pole z_0 and show that in that case

$$\oint_C h(z)dz = -2\pi i \, \text{Res } h(z_0) \quad \text{(clockwise direction).} \tag{16.15}$$

Find out in which step of the derivation the minus sign is picked up!

Problem j It may happen that a contour encloses not a single pole but a number of poles at points z_j. Find for this case a contour similar to the contour C^* in the right-hand panel of Figure 16.1 to show that the contour integral is equal to the *sum* of the contour integrals around the individual poles z_j. Use this to show that for this situation:

$$\oint_C h(z)dz = 2\pi i \sum_j \text{Res } h(z_j) \quad \text{(counterclockwise direction).} \tag{16.16}$$

16.3 Solving integrals without knowing the antiderivative

The residue theorem has important implications for integrals that do not contain a complex variable at all! For example, consider the integral

$$I = \int_{-\infty}^{\infty} \frac{1}{1+x^2} dx, \tag{16.17}$$

where x is a real number. If you know that $1/(1 + x^2)$ is the derivative of $\arctan x$, it is not difficult to solve this integral:

$$I = [\arctan x]_{-\infty}^{\infty} = \frac{\pi}{2} - \left(-\frac{\pi}{2}\right) = \pi. \tag{16.18}$$

Now suppose you did not know that $\arctan x$ is the antiderivative of $1/(1 + x^2)$. In that case you would be unable to see that the integral (16.17) is equal to π. Complex integration offers a way to obtain the value of this integral without explicit calculation of the antiderivative.

First, note that the path of integration in (16.17) can be viewed as the real axis in the complex plane. Nothing prevents us from viewing the real function $1/(1 + x^2)$ as a complex function $1/(1 + z^2)$, because on the real axis z is equal to x. This means that

$$I = \int_{C_{real}} \frac{1}{1+z^2} dz, \tag{16.19}$$

where C_{real} denotes the real axis as integration path. At this point we cannot yet apply the residue theorem, because the integration is not over a closed contour in the complex plane. Let us close the integration path using the circular path C_R in the upper half-plane with radius R, as in Figure 16.2. At the end of the calculation, we will let R go to infinity so that the integration path over the semicircle moves to infinity.

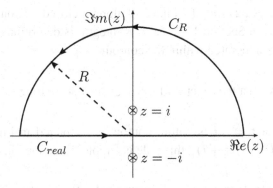

Figure 16.2   The paths C_{real} and C_R form a closed contour in the complex plane. Poles are at $z = \pm i$.

Problem a   Show that

$$I = \int_{C_{real}} \frac{1}{1+z^2} dz = \oint_C \frac{1}{1+z^2} dz - \int_{C_R} \frac{1}{1+z^2} dz. \qquad (16.20)$$

The first integral on the right-hand side is over the closed contour in Figure 16.2. What we have done is first close the contour and then we subtract the integral over the semicircle that we added to solve the integral along the real axis. This is the general approach: adding segments to the integration path, we obtain an integral over a closed contour in the complex plane. We apply the residue theorem to this contour integral. Then, we correct for the segments added to the original integration path. Obviously, this is only useful when the integral over the segment that one has added can be computed easily. In this example the integral over the semicircle vanishes as $R \to \infty$. This can be seen from the following estimations:

$$\left| \int_{C_R} \frac{1}{1+z^2} dz \right| \leq \int_{C_R} \left| \frac{1}{1+z^2} \right| |dz| \leq \int_{C_R} \frac{1}{|z|^2 - 1} |dz| =$$

$$\frac{\pi R}{R^2 - 1} \to 0, \quad \text{as } R \to \infty. \qquad (16.21)$$

If these estimates look complicated, there is a more intuitive way to show the same result. As $R \to \infty$, the absolute value of the integrand behaves as $|1/(1+z^2)| \to 1/R^2$. The length of the integration path is πR; hence, the integral is of the order $\pi R/R^2 = \pi/R$, which goes to zero as $R \to \infty$.

The estimate (16.21) implies that the last integral in (16.20) vanishes in the limit $R \to \infty$; this means that

$$I = \oint_C \frac{1}{1+z^2} dz. \qquad (16.22)$$

Now we are in the position to apply the residue theorem because we have reduced the problem to the evaluation of an integral along a closed contour in the complex plane. We know from Section 16.2 that this integral is determined by the poles of the function that is integrated within the contour.

Problem b   Show that the function $1/(1+z^2)$ has poles for $z = +i$ and $z = -i$.

Only the pole at $z = +i$ is within contour C, depicted in Figure 16.2. Since $1/(1+z^2) = 1/[(z-i)(z+i)]$, this pole is simple, as defined in Section 16.2.

Problem c   Use (16.14) to show that for the pole at $z = i$ the residue is given by: Res $= 1/2i$.

Problem d   Use the residue theorem (16.12) to deduce that

$$I = \int_{-\infty}^{\infty} \frac{1}{1+x^2} dx = \pi. \qquad (16.23)$$

This value is identical to the value obtained at the beginning of this section by using that the antiderivative of $1/(1+x^2)$ is equal to $\arctan x$. Note that the analysis is systematic and that we did not need to "know" the antiderivative.

In the treatment of this problem there is no reason why the contour should be closed in the upper half-plane. The estimate (16.21) holds equally well for a semicircle in the lower half-plane.

Problem e   Repeat the analysis of this section with the contour closed in the lower half-plane. Use that now the pole at $z = -i$ contributes and take into account that the sense of integration is in the clockwise rather than the anticlockwise direction. Show that this leads to the same result (16.23) that was obtained by closing the contour in the upper half-plane.

In the evaluation of the integral (16.17) there was freedom whether to close the contour in the upper half-plane or in the lower half-plane. This is not always the case. To see this, consider the integral

$$J = \int_{-\infty}^{\infty} \frac{\cos (x - \pi/4)}{1 + x^2} dx. \qquad (16.24)$$

Since $e^{ix} = \cos x + i \sin x$, this integral can be written as

$$J = \Re e \left(\int_{-\infty}^{\infty} \frac{e^{i(x-\pi/4)}}{1+x^2} dx \right), \qquad (16.25)$$

where $\Re e(\cdots)$ again denotes the real part. We want to evaluate this integral by closing this integration path with a semicircle either in the upper half-plane or in the lower half-plane. Due to the term e^{ix} in the integral we now have no real choice in this issue. The decision whether to close the integral in the upper half-plane or in the lower half-plane is dictated by the requirement that the integral over the semicircle vanishes as $R \to \infty$. The integral over a semicircle is approximately equal to the integrand on the semicircle times the length of the semicircle. This length is given by πR. This means that for the integral over the semicircle to vanish, the integrand times πR must go to zero as $R \to \infty$. This means that the integrand must vanish faster than $1/R$ as $R \to \infty$.

We next apply this reasoning to the evaluation of the integral (16.25). Let z be a point in the complex plane on the semicircle C_R that we use for closing the contour.

On the semicircle z can be written as $z = Re^{i\varphi}$, where in the upper half-plane $0 \le \varphi < \pi$ and in the lower half-plane $\pi \le \varphi < 2\pi$.

Problem f   Use this representation of z to show that

$$\left| e^{iz} \right| = e^{-R \sin \varphi}. \tag{16.26}$$

Problem g   Show that in the limit $R \to \infty$ the absolute value of the integral (16.25) over the semicircle is of the order

$$\left| \int_{C_R} \frac{e^{i(z-\pi/4)}}{1+z^2} dz \right| < \int \frac{e^{-R \sin \varphi}}{R^2} \pi R d\varphi = \pi \int \frac{e^{-R \sin \varphi}}{R} d\varphi. \tag{16.27}$$

This integral vanishes in the limit $R \to \infty$ when $\sin \varphi > 0$. This condition is satisfied in the upper half-plane where $0 \le \varphi < \pi$. This means that we must close the integral along the real axis with a semicircle in the upper half-plane if we want the integral along the semicircle to vanish at $R \to \infty$.

Problem h   Take exactly the same steps as in the derivation of (16.23) and show that

$$\int_{-\infty}^{\infty} \frac{\cos(x - \pi/4)}{1 + x^2} dx = \frac{\pi}{\sqrt{2}e}. \tag{16.28}$$

Note that this integral is equal to a combination of the three irrational numbers π, $\sqrt{2}$, and e!

Problem i   Determine the integral $\int_{-\infty}^{\infty} [\sin(x - \pi/4)/(1 + x^2)] dx$ without doing any additional calculations. Hint: Look carefully at (16.25) and spot $\sin(x - \pi/4)$.

It is important to note that in finding the integrals in this section we did not need to know the antiderivative of the function that we integrated. This is the reason why complex integration is such a powerful technique; it allows us to solve many integrals without knowing their antiderivative. The structure of the singularities of the integrand in the complex plane and the behavior of the function on a bounding contour carry sufficient information to integrate the function. This rather miraculous property is due to the fact that according to (15.6) in regions where the integrand is analytic it satisfies Laplace's equation: $\nabla^2 f = \nabla^2 g = 0$ (where f and g are the real and imaginary parts of the integrand, respectively). The singularities of the integrand act as sources or sinks for f and g. We have learned already that the specification of the sources and sinks (ρ) plus boundary conditions is sufficient to solve Poisson's equation: $\nabla^2 V = \rho$. In the same way, the specification of the singularities (the sources and sinks) of a complex function plus its values on a bounding contour (the boundary conditions) are sufficient to compute the integral of this complex function.

16.4 Response of a particle in syrup

Up to this point, complex integration has been applied to mathematical problems. However, this technique does have important applications in physical problems. In this section we consider a particle with mass m on which a force $f(t)$ is acting. The particle is suspended in syrup, which damps the velocity of the particle, and it is assumed that this damping force is proportional to the velocity $v(t)$ of the particle. The equation of motion of the particle is given by

$$m\frac{dv}{dt} + \beta v = f(t), \tag{16.29}$$

where β is a parameter that determines the strength of the damping of the motion by the fluid. The question we want to solve is: What is the velocity $v(t)$ for a given force $f(t)$?

We solve this problem using a Fourier transform technique. The Fourier transform of $v(t)$ is denoted by $V(\omega)$ and the force $f(t)$ in the frequency domain is denoted by $F(\omega)$.

Problem a Use the definition (14.42) of the Fourier transform to show that the equation of motion (16.29) is given in the frequency domain by

$$-i\omega m V(\omega) + \beta V(\omega) = F(\omega). \tag{16.30}$$

Comparing this with the original equation (16.29), we can see why the Fourier transform is so useful. The original expression (16.29) is a differential equation, while (16.30) is an algebraic equation. Since algebraic equations are much easier to solve than differential equations, we have made considerable progress.

Problem b Solve the algebraic equation (16.30) for $V(\omega)$ and use the Fourier transform (14.42) to derive that

$$v(t) = \frac{i}{2\pi m} \int_{-\infty}^{\infty} \frac{F(\omega)e^{-i\omega t}}{\left(\omega + \frac{i\beta}{m}\right)} d\omega. \tag{16.31}$$

Now we have an explicit relation between the velocity $v(t)$ in the time domain and the force $F(\omega)$ in the frequency domain. This is not quite what we want. We want to find the relation between the velocity and the force $f(t)$, both in the time domain.

Problem c   Use the inverse Fourier transform (14.43) for the force to show that

$$v(t) = \frac{i}{2\pi m} \int_{-\infty}^{\infty} f(t') \int_{-\infty}^{\infty} \frac{e^{-i\omega(t-t')}}{\left(\omega + \frac{i\beta}{m}\right)} d\omega dt'. \qquad (16.32)$$

This equation looks messy, but we can simplify it by writing it as

$$v(t) = \frac{i}{2\pi m} \int_{-\infty}^{\infty} f(t') I(t - t') dt', \qquad (16.33)$$

with

$$I(t - t') = \int_{-\infty}^{\infty} \frac{e^{-i\omega(t-t')}}{\left(\omega + \frac{i\beta}{m}\right)} d\omega. \qquad (16.34)$$

A comparison of (16.33) with expression (14.57) shows that $I(t - t')$ acts as a linear filter that converts the input $f(t)$ into the output $v(t)$.

The problem we now face is to evaluate this last integral. For this we use complex integration. The integration variable is now called ω rather than the x variable we used before, but this does not change the principles. We close the contour by adding a semicircle either in the upper half-plane or in the lower half-plane to the integral (16.34) along the real axis. On a semicircle with radius R the complex number ω can be written as $\omega = Re^{i\varphi}$.

Problem d   Show that $\left| e^{-i\omega(t-t')} \right| = e^{-R(t'-t)\sin\varphi}$.

Problem e   The integral along the semicircle should vanish in the limit $R \to \infty$. Use the result of Problem d to show that the contour should be closed in the upper half-plane for $t < t'$ and in the lower half-plane for $t > t'$, as in Figure 16.3.

Problem f   Show that the integrand in (16.34) has one pole at the negative imaginary axis at $\omega = -i\beta/m$ and that the residue at this pole is given by

$$Res = e^{-(\beta/m)(t-t')}. \qquad (16.35)$$

Problem g   Use these results and the theorems derived in Section 16.2 to show that:

$$I(t - t') = \begin{cases} 0 & \text{for} \quad t < t' \\ -2\pi i e^{-(\beta/m)(t-t')} & \text{for} \quad t > t' \end{cases}. \qquad (16.36)$$

Hint: Treat the cases $t < t'$ and $t > t'$ separately.

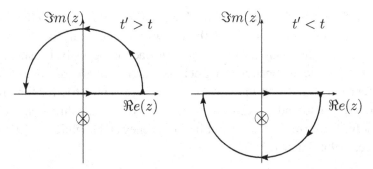

Figure 16.3   A pole in the complex plane and the closure of contours for $t' > t$ (left) and $t' < t$ (right).

Let us first consider (16.36). It follows from (16.33) that $I(t - t')$ is a function that describes the effect of a force acting at time t' on the velocity at time t. Expression (16.36) tells us that this effect is zero when $t < t'$. In other words, this expression tells us that the force $f(t')$ has no effect on the velocity $v(t)$ when $t < t'$. This is equivalent to saying that the force only affects the velocity at *later* times. In this sense, (16.36) can be seen as an expression of *causality*; the cause $f(t')$ only influences the effect $v(t)$ for later times.

Problem h   Insert (16.36) and (16.34) in (16.32) to show that

$$v(t) = \frac{1}{m} \int_{-\infty}^{t} e^{-(\beta/m)(t-t')} f(t') dt'. \tag{16.37}$$

Pay particular attention to the limits of integration.

Problem i   Convince yourself that this expression states that the force (the "cause") only has an influence on the velocity (the "effect") for *later* times.

It may appear that for this problem we have managed to give a proof of the causality principle. However, there is a problem hidden in the analysis. Suppose we switch off the damping parameter β, that is, we remove the syrup from the problem. One can see that setting $\beta = 0$ in the final result (16.37) poses no problem. However, suppose that we had set $\beta = 0$ at the start of the problem.

Problem j   Show that in that case the pole in Figure 16.3 is located on the *real* axis.

This implies that it is not clear how this pole affects the response. In particular, it is not clear whether in the absence of damping this pole gives a nonzero contribution for $t < t'$ (as it would if we consider it to lie in the upper half-plane) or for $t > t'$

(as it would if we consider it to lie in the lower half-plane). This is a disconcerting result, since it implies that causality only follows from the analysis when the problem contains some dissipation. This is not an artifact of the analysis using complex integration. What we are encountering here is a manifestation of the problem that in the absence of dissipation the laws of physics are symmetric for time reversal, whereas the world around us seems to move in only one time direction. This is the poorly resolved issue of the "arrow of time" (Coveney and Highfield, 1991; Popper, 1956; Price, 1996; Snieder, 2002).

17

Green's functions: principles

Green's functions play an important role in mathematical physics, similar to that of the impulse response for linear filters (Section 14.7). The general idea is that if one knows the response of a system to a delta-function input, the response of the system to any input can be reconstructed by superposing the response to the delta-function input in an appropriate manner. However, the use of Green's functions suffers from the same limitation as the use of the impulse response for linear filters: since the superposition principle underlies the use of Green's functions they are only useful for systems that are linear. Excellent treatments of Green's functions can be found in Barton (1989), which is completely devoted to Green's functions, but also in Butkov (1968).

17.1 Girl on a swing

To become familiar with Green's functions, let us consider a girl on a swing who is pushed by her mother (Figure 17.1). When the amplitude of the swing is not too large, the motion of the swing is described by the equation of a harmonic oscillator that is driven by a time-dependent external force $F(t)$:

$$\ddot{x} + \omega_0^2 x = F(t)/m. \tag{17.1}$$

The length of the swing and gravity determine the resonant frequency of the oscillator, denoted by ω_0. It is not trivial to solve this equation for a general driving force. For simplicity, we first solve (17.1) for the special case that the mother gives a single push to her daughter. The push is given at time $t = 0$, has duration Δ, and the magnitude of the force is denoted by F_0. This means that:

$$F(t) = \begin{cases} 0 & \text{for} \quad t < 0 \\ F_0 & \text{for} \quad 0 \leq t < \Delta \\ 0 & \text{for} \quad \Delta \leq t. \end{cases} \tag{17.2}$$

We will look here for a *causal* solution. This is another way of saying that we are looking for a solution in which the cause (the driving force) precedes the effect (the

Figure 17.1   The girl on a swing.

motion of the oscillator). This means that we require that the oscillator does not move for times $t < 0$.

When $t < 0$ and $t \geq \Delta$, the function $x(t)$ satisfies the differential equation $\ddot{x} + \omega_0^2 x = 0$. This differential equation has general solutions of the form $x(t) = A \cos(\omega_0 t) + B \sin(\omega_0 t)$. For $t < 0$, there is no displacement; hence, the constants A and B vanish. For $0 \leq t < \Delta$, the displacement satisfies the differential equation $\ddot{x} + \omega_0^2 x = F_0/m$. The solution can be found by writing $x(t) = (F_0/m\omega_0^2) + y(t)$. The function $y(t)$ then satisfies the equation $\ddot{y} + \omega_0^2 y = 0$, which has the general solution $y(t) = C \cos(\omega_0 t) + D \sin(\omega_0 t)$. Summarizing, the general solution $x(t)$ of the oscillator is given by

$$x(t) = \begin{cases} 0 & \text{for} \quad t < 0 \\ (F_0/m\omega_0^2) + C \cos(\omega_0 t) + D \sin(\omega_0 t) & \text{for} \quad 0 \leq t < \Delta \quad (17.3) \\ A \cos(\omega_0 t) + B \sin(\omega_0 t) & \text{for} \quad \Delta \leq t. \end{cases}$$

The integration constants A, B, C, and D follow from the requirement that the displacement $x(t)$ and velocity $\dot{x}(t)$ of the oscillator are continuous at all times. The latter condition follows from the consideration that when the force is finite, the acceleration is finite, and the velocity therefore is continuous. These two requirements applied at $t = 0$ and $t = \Delta$ lead to the following equations:

$$\left(F_0/m\omega_0^2\right) + C = 0,$$
$$\omega_0 D = 0,$$
$$\left(F_0/m\omega_0^2\right) + C \cos(\omega_0\Delta) + D \sin(\omega_0\Delta) = A \cos(\omega_0\Delta) + B \sin(\omega_0\Delta),$$
$$-C \sin(\omega_0\Delta) + D \cos(\omega_0\Delta) = -A \sin(\omega_0\Delta) + B \cos(\omega_0\Delta). \quad (17.4)$$

These are four linear equations for the four unknown integration constants A, B, C, and D. The upper two equations can be solved directly for the constants C and D to give the values $C = -\left(F_0/m\omega_0^2\right)$ and $D = 0$.

Problem a Insert these values for C and D into the lower two equations, to show that $A = -\left(F_0/m\omega_0^2\right)(1 - \cos(\omega_0\Delta))$ and $B = \left(F_0/m\omega_0^2\right) \sin(\omega_0\Delta)$.

Inserting these values of the constants into (17.3) shows that the motion of the oscillator is given by:

$$x(t) = \begin{cases} 0 & \text{for} \quad t < 0 \\ \left(F_0/m\omega_0^2\right)[1 - \cos(\omega_0 t)] & \text{for} \quad 0 \le t < \Delta \\ \left(F_0/m\omega_0^2\right)\{\cos(\omega_0 t)\cos(\omega_0\Delta) + \\ \quad \sin(\omega_0 t)\sin(\omega_0\Delta) - \cos(\omega_0 t)\} & \text{for} \quad \Delta \le t. \end{cases} \quad (17.5)$$

This is the solution for a push of duration Δ starting at time $t = 0$. Suppose now that the push is short compared to the period of the oscillator, then $\omega_0\Delta \ll 1$. In that case one can use a Taylor expansion in $\omega_0\Delta$ in (17.5). This can be achieved by using the Taylor expansions (3.12) and (3.13) for $\sin(\omega_0\Delta)$ and $\cos(\omega_0\Delta)$. Retaining only the term of order $(\omega_0\Delta)$ shows that for an impulsive push $(\omega_0\Delta \ll 1)$, the solution is given by:

$$x(t) = \begin{cases} 0 & \text{for} \quad t < 0 \\ \left(F_0/m\omega_0^2\right)(\omega_0\Delta)\sin(\omega_0 t) & \text{for} \quad t > \Delta. \end{cases} \quad (17.6)$$

We will not bother anymore with the solution between $0 \le t < \Delta$ because in the limit $\Delta \to 0$ this interval is of vanishing duration.

At this point we have all the ingredients needed to determine the response of the oscillator for a general driving force $F(t)$. Suppose we divide the time-axis in intervals of duration Δ. In the ith interval, the force is given by $F_i = F(t_i)$, where t_i is the time of the ith interval. We know from (17.6) the response to a force of duration Δ at time $t = 0$. More generally, the response to a force F_i at a time t_i follows by replacing F_0 by F_i and by replacing t by $t - t_i$. With these substitutions, it follows that the response to a force F_i delivered over a time interval Δ at time t_i is

$$x(t) = \begin{cases} 0 & \text{for} \quad t < t_i \\ \dfrac{1}{m\omega_0} \sin\left[\omega_0 \left(t - t_i\right)\right] \, F(t_i)\Delta & \text{for} \quad t > t_i. \end{cases} \tag{17.7}$$

This is the response due to the force acting at time t_i only. To obtain the response to the full force $F(t)$, one should sum over the forces delivered at all times t_i. In the language of the girl on the swing, one would say that (17.7) gives the motion of the swing for a single impulsive push, and that we want to use this result to find the displacement caused by a number of pushes. Since the differential equation (17.1) is linear, we can use the superposition principle, which states that the response to the superposition of two pushes is the sum of the response to the individual pushes. In the language of Section 14.7 we would say that the swing is a linear system. This means that when the swing receives a number of pushes at different times t_i, the response can be written as the sum of the responses to every individual push.

Problem b   Use (17.7) to show that

$$x(t) = \sum_{t_i < t} \frac{1}{m\omega_0} \sin\left[\omega_0 \left(t - t_i\right)\right] \, F(t_i)\Delta. \tag{17.8}$$

Note that in (17.7) the response at time t only depends on the pushes at *earlier* times, because the pushes at later times give a vanishing contribution for causal systems. For this reason the summation is limited to times $t \geq t_i$.

   Suppose now that the swing is not given a finite number of impulse pushes but instead that the driving force is a continuous function. This case can be handled by taking the limit $\Delta \to 0$. The summation in (17.8) then needs to be replaced by an integration. This can naturally be achieved because the duration Δ is then equal to the infinitesimal interval dt used in the integration. What we are really doing here is replacing the continuous function $F(t)$ by a function that is constant within every interval Δ at times t_i, as depicted in Figure 17.2. Then we take the limit where the width of the intervals goes to zero; that is, $\Delta \to 0$. A similar treatment may be familiar to you from the theory of integration. When the limit $\Delta \to 0$ is taken the summation over t_i can be replaced by an integration: $\sum_{t_i} (\cdots)\Delta \to \int (\cdots)\,d\tau$. The integration variable τ plays the role of the summation variable t_i and the time interval Δ is replaced $d\tau$. The response of the oscillator to a continuous force $F(t)$ is then given by

$$x(t) = \int_{-\infty}^{t} \frac{1}{m\omega_0} \sin\left[\omega_0 \left(t - \tau\right)\right] F(\tau)\,d\tau. \tag{17.9}$$

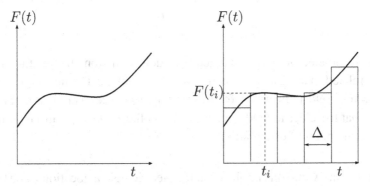

Figure 17.2   A continuous function (left) and an approximation to this function that is constant within finite intervals Δ (right).

The integration is only carried out over times $\tau < t$ because the summation (17.8) extends only over the times $t_i < t$.

With a slight change in notation, this result can be written as:

$$x(t) = \int_{-\infty}^{\infty} G(t, \tau) F(\tau) \, d\tau, \tag{17.10}$$

with

$$G(t, \tau) = \begin{cases} 0 & \text{for} \quad t < \tau \\ \dfrac{1}{m\omega_0} \sin\left[\omega_0 \left(t - \tau\right)\right] & \text{for} \quad t > \tau. \end{cases} \tag{17.11}$$

The function $G(t, \tau)$ in (17.11) is called the *Green's function* of the harmonic oscillator. Note that (17.10) is similar to the response of a linear filter described by (14.55). This is not surprising; in both examples the response of a linear system to an impulsive input was determined, and it will be no surprise that the results are identical. In fact, the Green's function is defined as the response of a linear system to a delta-function input. Although Green's functions are often presented in a rather abstract way, one should remember that:

The Green's function of a system is nothing but the impulse response of the system: it is the response of the system to a delta-function excitation.

17.2 You have seen Green's functions before!

Although the concept of a Green's function may appear to be new, you have already seen several examples of Green's functions. One example is the electric field generated by a point charge q at the origin, which was treated in Section 6.1:

A Guided Tour of Mathematical Methods for the Physical Sciences

$$E(\mathbf{r}) = \frac{q\hat{\mathbf{r}}}{4\pi\epsilon_0 r^2}. \tag{6.2}$$

Since this is the electric field generated by a delta-function charge at the origin, it is closely related to the Green's function for this problem. The field equation (6.24) of the electric field is invariant for translations in space. This is a complex way of saying that the electric field depends only on the *relative* positions of the point charge and the point of observation.

Problem a   Show that this implies that the electric field at location $\mathbf{r}$ due to a point charge at location $\mathbf{r}'$ is given by

$$E(\mathbf{r}) = \frac{q}{4\pi\epsilon_0} \frac{(\mathbf{r} - \mathbf{r}')}{|\mathbf{r} - \mathbf{r}'|^3}. \tag{17.12}$$

Now suppose that we do not have a single point charge, but instead a system of point charges q_i at locations $\mathbf{r}_i$. Since the field equation is linear, the electric field generated by a sum of point charges is the sum of the fields generated by each point charge:

$$E(\mathbf{r}) = \sum_i \frac{q_i}{4\pi\epsilon_0} \frac{(\mathbf{r} - \mathbf{r}_i)}{|\mathbf{r} - \mathbf{r}_i|^3}. \tag{17.13}$$

Problem b   To which expression of the previous section does this equation correspond?

Just as in the previous section we now make the transition from a finite number of discrete inputs (either pushes of the swing or point charges) to an input function that is a continuous function (either the applied force to the oscillator as a function of time or a continuous electric charge). Let the electric charge per unit volume be denoted by $\rho(\mathbf{r})$. This means that the electric charge in a volume dV is given by $\rho(\mathbf{r})dV$.

Problem c   Replace the sum in (17.13) by an integration over the volume, and use the appropriate charge for each volume element dV to show that the electric field for a continuous charge distribution is given by:

$$E(\mathbf{r}) = \iiint \frac{\rho(\mathbf{r}')}{4\pi\epsilon_0} \frac{(\mathbf{r} - \mathbf{r}')}{|\mathbf{r} - \mathbf{r}'|^3} dV', \tag{17.14}$$

where the volume integration is over $\mathbf{r}'$.

Problem d   Show that this implies that the electric field can be written as

$$\mathbf{E}(\mathbf{r}) = \iiint \mathbf{G}(\mathbf{r}, \mathbf{r}')\rho(\mathbf{r}')dV', \qquad (17.15)$$

with the Green's function given by

$$\mathbf{G}(\mathbf{r}, \mathbf{r}') = \frac{1}{4\pi\epsilon_0}\frac{(\mathbf{r} - \mathbf{r}')}{|\mathbf{r} - \mathbf{r}'|^3}. \qquad (17.16)$$

Note that this Green's function has the same form as the electric field for a point charge given in (17.12), and that the Green's function is only a function of the relative distance $\mathbf{r} - \mathbf{r}'$.

Problem e   Explain why the integral (17.15) can be seen as a three-dimensional convolution: $\mathbf{E}(\mathbf{r}) = \iiint \mathbf{G}(\mathbf{r} - \mathbf{r}')\rho(\mathbf{r}')dV'$.

The main purpose of this section is not to show you that you had seen an example of a Green's function before. Rather, it provides an example that the Green's function is not necessarily a function of time and that it is not necessarily a scalar function, either; the Green's function (17.16) depends on the position but not on time, and it describes a vector field rather than a scalar. The most important thing to remember is that the Green's function is the impulse response of a linear system.

Problem f   You have seen another Green's function before if you worked through Section 16.4 where the response of a particle in syrup was treated. Find the Green's function in that section and spot the expressions equivalent to (17.10) and (17.11).

17.3 Green's function as an impulse response

You may have found the derivation of the Green's function in Section 17.1 rather complicated. The reason for this is that in (17.3) the motion of the swing was determined before the push ($t < 0$), during the push ($0 < t < \Delta$) and after the push ($t > \Delta$). The requirement that the displacement x and the velocity $\dot{x}$ were continuous then led to the system of equations (17.4) with four unknowns. However, in the end we took the limit $\Delta \to 0$ and did not use the solution for time $0 < t < \Delta$. This suggests that this method of solution is unnecessarily complicated. This is indeed the case. In this section an alternative derivation of the Green's function (17.11) is given that is based on the idea that the Green's function $G(t, \tau)$ describes the motion of the oscillator due to a Dirac delta-function force at time τ:

$$\ddot{G}(t, \tau) + \omega_0^2 G(t, \tau) = \frac{1}{m}\delta(t - \tau). \qquad (17.17)$$

Problem a For $t \neq \tau$ the delta function vanishes and the right-hand side of this expression is equal to zero. We are looking for the *causal* Green's function, that is, the solution where the cause (the force) precedes the effect (the motion of the oscillator). Show that these conditions imply that for $t \neq \tau$ the Green's function is given by

$$G(t, \tau) = \begin{cases} 0 & \text{for} \quad t < \tau \\ A \cos[\omega_0(t - \tau)] + B \sin[\omega_0(t - \tau)] & \text{for} \quad t > \tau, \end{cases} \quad (17.18)$$

where A and B are unknown integration constants.

The integration constants follow from the conditions at $t = \tau$. Since we have two unknown parameters, we need to impose two conditions. The first condition is that the displacement of the oscillator is continuous at $t = \tau$. If this were not the case, the velocity of the oscillator would be infinite at that moment.

Problem b Show that the requirement of continuity of the Green's function at $t = \tau$ implies that $A = 0$.

The second condition requires more care. We derive the second condition first mathematically and then explore its physical meaning. The second condition follows by integrating (17.17) over t from $\tau - \epsilon$ to $\tau + \epsilon$ and by taking the limit $\epsilon \downarrow 0$. Integrating (17.17) in this way gives:

$$\int_{\tau-\epsilon}^{\tau+\epsilon} \ddot{G}(t, \tau)\, dt + \omega_0^2 \int_{\tau-\epsilon}^{\tau+\epsilon} G(t, \tau)\, dt = \frac{1}{m} \int_{\tau-\epsilon}^{\tau+\epsilon} \delta(t - \tau)\, dt. \quad (17.19)$$

Problem c Show that the right-hand side of (17.19) is equal to $1/m$, regardless of the value of ϵ.

Problem d Show that the absolute value of the second term on the left-hand side of (17.19) is smaller than $2\epsilon\omega_0^2 \max(G)$, where $\max(G)$ is the maximum of G over the integration interval. Since the Green's function is finite, this means that the middle term vanishes in the limit $\epsilon \downarrow 0$.

Problem e Show that the first term on the left-hand side of (17.19) is equal to $\dot{G}(t = \tau + \epsilon, \tau) - \dot{G}(t = \tau - \epsilon, \tau)$. This quantity will be denoted by $[\dot{G}(t, \tau)]_{t=\tau-\epsilon}^{t=\tau+\epsilon}$.

Problem f Use this result to show that in the limit $\epsilon \downarrow 0$ expression (17.19) gives:

$$\left[\dot{G}(t, \tau)\right]_{t=\tau-\epsilon}^{t=\tau+\epsilon} = \frac{1}{m}. \quad (17.20)$$

Problem g Show that this condition together with the continuity of G implies that the integration constants in (17.18) have the values $A = 0$ and $B = 1/m\omega_0$, that is that the Green's function is given by:

$$G(t, \tau) = \begin{cases} 0 & \text{for} \quad t < \tau \\ \dfrac{1}{m\omega_0} \sin\left[\omega_0 (t - \tau)\right] & \text{for} \quad t > \tau. \end{cases} \tag{17.21}$$

A comparison with (17.11) shows that the Green's function derived in this section is identical to the Green's function derived in Section 17.1. Note that the solution was obtained here without invoking the motion of the oscillator during the moment of excitation. This would have been difficult because the duration of the excitation (a delta function) is equal to zero, if it can be defined at all.

There is, however, something strange about the derivation in this section. In Section 17.1 the solution was found by requiring that the displacement x and its first derivative $\dot{x}$ were continuous at all times. As used in Problem b the first condition is also met by the solution (17.21). However, the derivative $\dot{G}$ is *not* continuous at $t = \tau$.

Problem h Which of the equations that you derived above states that the first derivative is not continuous?

Problem i $G(t, \tau)$ denotes the displacement of the oscillator. Show that expression (17.20) states that the velocity of the oscillator changes discontinuously at $t = \tau$.

Problem j Give a physical reason why the velocity of the oscillator was continuous in the first part of Section 17.1 and why the velocity is discontinuous for the Green's function derived in this section. Hint: How large is the force needed to produce a finite jump in the velocity of a particle when the force is applied over a time-interval of length zero (the width of the delta-function excitation).

How can we reconcile this result with the solution obtained in Section 17.1?

Problem k Show that the change in the velocity in the solution $x(t)$ in (17.7) is proportional to $F(t_i)\Delta$, that is, that

$$[\dot{x}]_{t_i - \epsilon}^{t_i + \epsilon} = \frac{1}{m} F(t_i)\Delta. \tag{17.22}$$

This means that the change in the velocity depends on the strength of the force times the duration of the force. The physical reason for this is that the change in the velocity depends on the integral of the force over time divided by the mass of the particle.

Problem l Derive this last statement directly from Newton's law ($F = ma$).

When the force is finite and when $\Delta \to 0$, the jump in the velocity is zero and the velocity is continuous. However, when the force is infinite (as is the case for a delta function), the jump in the velocity is nonzero and the velocity is discontinuous.

In many applications the Green's function is the solution of a differential equation with a delta function as the excitation. This implies that some derivative, or combination of derivatives, of the Green's function is equal to a delta function at the point (or time) of excitation. This usually has the effect that the Green's function, or its derivative, is not a continuous function. The delta function in the differential equation usually leads to a singularity in the Green's function or its derivative.

17.4 Green's function for a general problem

In this section, the theory of Green's functions is treated in a more abstract fashion. Every linear differential equation for a function u with a source term F can be written symbolically as:

$$Lu = F. \tag{17.23}$$

For example, in (17.1) for the girl on the swing, u is the displacement $x(t)$, while L is a differential operator given by

$$L = m\frac{d^2}{dt^2} + m\omega_0^2, \tag{17.24}$$

where it is understood that a differential operator acts term by term on the function to the right of the operator.

Problem a Find the differential operator L and the source term F for the electric field treated in Section 17.2 from the field equation (6.24).

In the notation used in this section, the Green's function depends on the position vector $\mathbf{r}$, but the results derived here are equally valid for a Green's function that depends only on time or on position and time. In general, the differential equation (17.23) must be supplemented with boundary conditions to give a unique solution. In this section the position of the boundary is denoted by $\mathbf{r}_B$ and it is assumed that the function u has the value u_B at the boundary:

$$u(\mathbf{r}_B) = u_B. \qquad (17.25)$$

This boundary condition, where the function value is specified, is called a *Dirichlet boundary condition*.

Let us first find a single solution to the differential equation (17.23) without bothering about boundary conditions. We follow the same treatment as in Section 14.7 where in (14.54) the input of a linear function was written as a superposition of delta functions. In the same way, the source function can be written as:

$$F(\mathbf{r}) = \int \delta(\mathbf{r} - \mathbf{r}') F(\mathbf{r}') \, dV'. \qquad (17.26)$$

This expression follows from the sifting property of the delta function (13.10). One can interpret this expression as an expansion of the function $F(\mathbf{r})$ in delta functions, because the integral (17.26) describes a superposition of delta functions $\delta(\mathbf{r} - \mathbf{r}')$ centered at $\mathbf{r} = \mathbf{r}'$; each of these delta functions is given a weight $F(\mathbf{r}')$. We want to use a Green's function to construct a solution. The Green's function $G(\mathbf{r}, \mathbf{r}')$ is the response at location $\mathbf{r}$ due to a delta-function source at location $\mathbf{r}'$, that is, the Green's function satisfies:

$$LG(\mathbf{r}, \mathbf{r}') = \delta(\mathbf{r} - \mathbf{r}'). \qquad (17.27)$$

The response to the input $\delta(\mathbf{r} - \mathbf{r}')$ is given by $G(\mathbf{r}, \mathbf{r}')$, and the source functions can be written as a superposition of these delta functions with weight $F(\mathbf{r}')$. This suggests that a solution of (17.23) is given by a superposition of Green's functions $G(\mathbf{r}, \mathbf{r}')$, where each Green's function has the same weight factor as the delta function $\delta(\mathbf{r} - \mathbf{r}')$ in the expansion (17.26) of $F(\mathbf{r})$ in delta functions. This means that a solution of (17.23) is given by:

$$u_P(\mathbf{r}) = \int G(\mathbf{r}, \mathbf{r}') F(\mathbf{r}') dV'. \qquad (17.28)$$

Problem b If you worked through Section 14.7, discuss the relation between this expression and (14.55) for the output of a linear function.

It is crucial to understand at this point that we have used three steps to arrive at (17.28): (i) the source function is written as a superposition of delta functions, (ii) the response of the system to each delta-function input is defined, and (iii) the solution is written as the same superposition of Green's function as was used in the expansion of the source function in delta functions:

$$\delta(\mathbf{r} - \mathbf{r}') \quad \leftrightarrow \quad F(\mathbf{r}) \overset{(i)}{=} \int \delta(\mathbf{r} - \mathbf{r}') F(\mathbf{r}') \, dV'$$

$$\Downarrow (ii) \qquad\qquad\qquad\qquad \Downarrow \qquad\qquad\qquad (17.29)$$

$$G(\mathbf{r}, \mathbf{r}') \quad \leftrightarrow \quad u_P(\mathbf{r}) \overset{(iii)}{=} \int G(\mathbf{r}, \mathbf{r}') F(\mathbf{r}') \, dV'.$$

Although this reasoning may sound plausible, we have not proved that $u_P(\mathbf{r})$ in (17.28) actually is a solution of the differential equation (17.23).

Problem c   Give a proof that this is indeed the case by letting the operator L act on (17.28) and by using (17.27) for the Green's function. Hint: The operator L acts on $\mathbf{r}$ while the integration is over $\mathbf{r}'$; the operator can thus be taken inside the integral.

It should be noted that we have not solved our problem yet, because u_P does not necessarily satisfy the boundary conditions. In fact, (17.28) is just one of the many possible solutions to (17.23). It is a particular solution of the inhomogeneous equation (17.23), and this is the reason why the subscript P is used. Equation (17.23) is called an *inhomogeneous equation* because the right-hand side is nonzero. When the right-hand side is zero, one speaks of the *homogeneous equation*. This implies that a solution u_H of the homogeneous equation satisfies

$$Lu_H = 0. \qquad\qquad (17.30)$$

In general one can add a solution of the homogeneous equation (17.30) to a particular solution, and the result still satisfies the inhomogeneous equation (17.23).

Problem d   Give a proof of the previous statement by showing that the function $u = u_P + u_H$ is a solution of (17.23). In other words, show that the general solution of (17.23) is given by

$$u(\mathbf{r}) = u_H(\mathbf{r}) + \int G(\mathbf{r}, \mathbf{r}') F(\mathbf{r}') \, dV'. \qquad\qquad (17.31)$$

The problem is that we still need to enforce the boundary conditions (17.25). This can be achieved by requiring that the solution u_H satisfies specific boundary conditions at $\mathbf{r}_B$.

Problem e   Insert (17.31) into the boundary conditions (17.25) to show that the required solution u_H of the homogeneous equation must satisfy the following boundary conditions:

$$u_H(\mathbf{r}_B) = u(\mathbf{r}_B) - \int G(\mathbf{r}_B, \mathbf{r}') F(\mathbf{r}') \, dV'. \qquad\qquad (17.32)$$

This is all we need to solve the problem. What we have shown is that:

> *The total solution (17.31) is given by the sum of the particular solu-*
> *tion (17.28) plus a solution of the homogeneous equation (17.30), such*
> *that this sum satisfies the boundary condition (17.32).*

This construction may appear to be very complicated. However, you should realize that the main complexity is the treatment of the boundary condition. In many
problems, the boundary condition dictates that the function vanishes at the boundary ($u_B = 0$) and the Green's function also vanishes at the boundary. It follows
from (17.31) that in that case the boundary condition for the homogeneous solution
is $u_H(\mathbf{r}_B) = 0$. This boundary condition is satisfied by the solution $u_H(\mathbf{r}) = 0$,
which implies that in that case one can dispense with the addition of u_H to the
particular solution $u_P(\mathbf{r})$.

Problem f Suppose that the boundary conditions do not prescribe the value of the
solution at the boundary but that instead of (17.25) the normal derivative of
the solution is prescribed by the boundary conditions:

$$\frac{\partial u}{\partial n}(\mathbf{r}_B) = \hat{\mathbf{n}} \cdot \nabla u(\mathbf{r}_B) = w_B, \qquad (17.33)$$

where $\hat{\mathbf{n}}$ is the unit vector perpendicular to the boundary. How should the
theory in this section be modified to accommodate this boundary condition?

The boundary condition (17.33) where the normal derivative is specified is called a
Neumann boundary condition. In some problems one specifies the relation between
the normal derivative and the function itself:

$$\frac{\partial u}{\partial n}(\mathbf{r}_B) = v_B u(\mathbf{r}_B), \qquad (17.34)$$

this is called a *Cauchy boundary condition*. This boundary condition is common
in wave propagation problems; see, for example, equation (18.38), where it is
sometimes called the radiation boundary condition.

The theory in this section is rather abstract. To make the issues at stake
more explicit, the theory is applied in the next section to the calculation of the
temperature in the Earth.

17.5 Radiogenic heating and the Earth's temperature

As an application of the use of the Green's function, we investigate the temperature
structure in the Earth's crust in the presence of the decay of radioactive elements.

Several radioactive elements such as U_{235} do not fit well in the lattice of mantle rocks. For this reason, these elements are expelled from the Earth's mantle and have accumulated in the crust. Radioactive decay of these elements then leads to the production of heat, where these elements have accumulated.

We first briefly derive the partial differential equation that governs the distribution of heat, but the reasoning applied here applies to any diffusive process. Variations in temperature cause a heat flow $\mathbf{J}$. The heat flows from high temperature to low temperature, while according to Section 5.1 the gradient ∇T points from low temperature to high temperature. The heat flow is therefore given by Fick's law, which states that $\mathbf{J} = -\kappa \nabla T$, where κ is the heat conduction coefficient that states how strong the heat flow is for a given temperature gradient. We next need to specify how this heat flow changes the temperature. The rate of change of the temperature depends on the net heat flow through any volume element. Following Section 6.1, the divergence of $\mathbf{J}$ is the net outward flux of heat per unit volume. Therefore, the rate of the change of the temperature satisfies $\partial T / \partial t = -(\nabla \cdot \mathbf{J})$, where the minus sign is present because an outward heat flux lowers the temperature. Combining these expressions gives $\partial T / \partial t = (\nabla \cdot \kappa \nabla T)$. Assuming that κ is constant, and adding a source term S to the right-hand side gives the heat equation, which is also called the diffusion equation:

$$\frac{\partial T}{\partial t} = \kappa \nabla^2 T + S. \tag{17.35}$$

This expression is derived with more rigor in Section 25.4.

In the simplified example of this problem we assume that the temperature T and the radiogenic heating source S depend on depth only, and that we can ignore curvature of the Earth's surface. We assume that the radiogenic heating does not depend on time and that we consider only the equilibrium temperature.

Problem a   Show that these assumptions imply that the temperature is only a function of the z-coordinate: $T = T(z)$.

Problem b   Use this expression to show that for the problem in this section the temperature field satisfies

$$\frac{d^2 T}{dz^2} = -\frac{S(z)}{\kappa}. \tag{17.36}$$

This equation can be solved when the boundary conditions are specified. The thickness of the crust is denoted by H, see Figure 17.3. The temperature is assumed

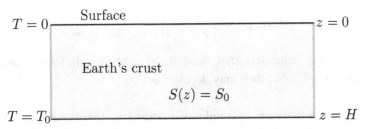

$T = 0$ — Surface — $z = 0$

Earth's crust

$S(z) = S_0$

$T = T_0$ — $z = H$

Figure 17.3 Definition of the geometric variables and boundary conditions for the temperature in the Earth's crust.

to vanish at the Earth's surface. In addition, it is assumed that at the base of the crust the temperature has a fixed value T_0.[1] This implies that the boundary conditions are:

$$T(z = 0) = 0, \qquad T(z = H) = T_0. \tag{17.37}$$

In this section we solve the differential equation (17.36) with the boundary conditions (17.37) using the Green's function technique described in the previous section. Analogously to (17.28) we first determine a particular solution T_P of the differential equation (17.36) and worry about the boundary conditions later. The Green's function $G(z, z')$ to be used is the temperature at depth z due to delta-function heating at depth z':

$$\frac{d^2 G(z, z')}{dz^2} = \delta(z - z'). \tag{17.38}$$

Problem c Use the theory of the previous section to show that the following function satisfies the heat equation (17.36):

$$T_P(z) = -\frac{1}{\kappa} \int_0^H G(z, z') S(z') \, dz'. \tag{17.39}$$

Before further progress can be made, it is necessary to find the Green's function. This means we need to solve the differential equation (17.38), but for that the boundary conditions need to be specified. In this example we will use a Green's function that vanishes at the endpoints of the depth interval:

$$G(z = 0, z') = G(z = H, z') = 0. \tag{17.40}$$

Problem d Use (17.38) to show that for $z \neq z'$ the Green's function satisfies the differential equation $d^2 G(z, z')/dz^2 = 0$ and use this to show that the Green's function, which satisfies the boundary conditions (17.40), must be of the form

[1] Geophysically this is an oversimplified boundary condition because in reality the temperature in the Earth is determined by the radiogenic heating everywhere in the Earth and by the heat that was formed during the Earth's formation.

$$G(z, z') = \begin{cases} \beta z & \text{for } z < z' \\ \gamma(z - H) & \text{for } z > z', \end{cases} \tag{17.41}$$

with β and γ constants that need to be determined. These constants are independent of z, but they may depend on z'.

Problem e Since there are two unknown constants, two conditions are needed. The first condition is that the Green's function is continuous for $z = z'$. Use the theory of Section 17.3 and the differential equation (17.38) to show that the second requirement is

$$\lim_{\epsilon \downarrow 0} \left[\frac{dG(z, z')}{dz} \right]_{z=z'-\epsilon}^{z=z'+\epsilon} = 1. \tag{17.42}$$

Note that this expression gives a jump condition similar to the jump condition (17.20) in the treatment of the Green's function for the harmonic oscillator. It should not be a surprise that a jump condition occurs, the Green's function is the response to a singular forcing – a delta function – and such a response is likely to be singular as well. One can show that the Green's function for an ordinary differential equation of order N has a jump in the $(N - 1)$-th of size a_N, where a_N is the coefficient that multiplies the Nth derivative (Barton, 1989).

Problem f Apply these two conditions to the solution (17.41) to determine the constants β and γ to show that the Green's function is

$$G(z, z') = \begin{cases} -\dfrac{1}{H}(H - z')z & \text{for } z < z' \\ -\dfrac{1}{H}(H - z)z' & \text{for } z > z'. \end{cases} \tag{17.43}$$

In this notation the two regions $z < z'$ and $z > z'$ are separated. Note, however, that the solution in the two regions has a highly symmetric form. In the literature you will find that a solution such as (17.43) is often rewritten by defining $z_>$ to be the maximum of z and z' and $z_<$ to be the minimum of z and z':

$$\left. \begin{array}{l} z_> \equiv \max(z, z'), \\ z_< \equiv \min(z, z'). \end{array} \right\} \tag{17.44}$$

Problem g Show that in this notation the Green's function (17.43) can be written as

$$G(z, z') = -\frac{1}{H}(H - z_>)z_<. \tag{17.45}$$

For simplicity we assume that the radiogenic heating is constant in the crust:

$$S(z) = S_0 \qquad \text{for} \quad 0 < z < H. \tag{17.46}$$

Problem h Show that the particular solution (17.39) for this heating function is given by

$$T_P(z) = \frac{S_0 H^2}{2\kappa} \frac{z}{H}\left(1 - \frac{z}{H}\right). \tag{17.47}$$

Problem i Show that this particular solution satisfies the boundary conditions

$$T_P(z = 0) = T_P(z = H) = 0. \tag{17.48}$$

Problem j This means that this solution does not satisfy the boundary conditions (17.37) of our problem. Use the theory in Section 17.4 to show that to obtain this solution we must add a solution $T_H(z)$ of the homogeneous equation $d^2 T_H / dz^2 = 0$, which satisfies the boundary conditions $T_H(z = 0) = 0$ and $T_H(z = H) = T_0$.

Problem k Show that the solution of the homogeneous equation is given by $T_H(z) = T_0 z / H$ and that the total solution is given by

$$T(z) = T_0 \frac{z}{H} + \frac{S_0 H^2}{2\kappa} \frac{z}{H}\left(1 - \frac{z}{H}\right). \tag{17.49}$$

Problem l Verify explicitly that this solution satisfies the differential equation (17.36) with the boundary conditions (17.37).

As shown in expression (25.29), the conductive heat flow is given by $\mathbf{J} = -\kappa \nabla T$. Since our problem is one-dimensional, the (vertical) heat flow is

$$J = -\kappa \frac{dT}{dz}. \tag{17.50}$$

Problem m Compute the heat flow at the top ($z = 0$) and at the bottom ($z = H$) of the crust. Assuming that T_0 and S_0 are both positive, does the heat flow at these locations increase or decrease because of the radiogenic heating S_0? Give a physical interpretation of this result.

The derivation in this section used a Green's function that satisfies the boundary conditions (17.40) rather than the boundary conditions (17.37) of the temperature field. However, there is no particular reason why one should use these boundary conditions for the Green's function. That is to say, one might think one could avoid the step of adding a solution $T_H(z)$ of the homogeneous equation by using a Green's function $\tilde{G}$ that satisfies the differential equation (17.41) and an inhomogeneous boundary condition at the base of the crust:

$$\tilde{G}(z = 0, z') = 0, \qquad \tilde{G}(z = H, z') = H. \tag{17.51}$$

The boundary value at the base of the crust is set equal to H because the Green's function has the dimension of length (see (17.43)) and the crustal thickness H is the only length-scale in the problem.

Problem n   Go through the same steps as you did earlier in this section by constructing the Green's function $\tilde{G}(z, z')$, computing the corresponding particular solution $\tilde{T}_P(z)$, verifying whether the boundary conditions (17.37) are satisfied by this particular solution and if necessary adding a solution of the homogeneous equation in order to satisfy the boundary conditions. Show that this again leads to (17.49).

The lesson to be learned from this section is that usually one needs to add a solution of the homogeneous equation to a particular solution in order to satisfy the boundary conditions. However, suppose that the boundary conditions of the temperature field were also homogeneous: $T = (z = 0) = T(z = H) = 0$. In that case the particular solution (17.47) that was constructed using a Green's function that satisfies the homogeneous boundary conditions (17.40) satisfies the boundary conditions of the full problem as well. This implies that it only pays to use a Green's function that satisfies the boundary conditions of the full problem when these boundary conditions are *homogeneous*, that is, when the function itself vanishes ($T = 0$) or when the normal gradient of the function vanishes ($\partial T / \partial n = 0$) or when a linear combination of these quantities vanishes ($aT + b\partial T / \partial n = 0$). In all other cases one cannot avoid adding a solution of the homogeneous equation in order to satisfy the boundary conditions and the most efficient procedure is usually to use the Green's function that can most easily be computed.

17.6 Nonlinear systems and Green's functions

Up to this point, Green's functions have been applied to linear systems. The definition of a linear system was introduced in Section 14.7. Suppose that a force F_1 leads to a response x_1 and that a force F_2 leads to a response x_2. A system is linear when the response to the linear combination $c_1 F_1 + c_2 F_2$ (with c_1 and c_2 constants) is the superposition response $c_1 x_1 + c_2 x_2$. This definition implies that the response to the input times a constant is given by the response multiplied by the same constant. In other words, for a linear system an input that is twice as large leads to a response that is twice as large.

Problem a   Show that the definition of linearity given above implies that the response to the sum of two force functions is the sum of the responses to the individual force functions.

This last property reflects that a linear system satisfies the *superposition principle*, which states that for a linear system one can superpose the response to a sum of force functions.

Not every system is linear, and we exploit here the extent to which Green's functions are useful for nonlinear systems. As an example we consider the Verhulst equation:

$$\dot{x} = x - x^2 + F(t). \tag{17.52}$$

This equation has been used in mathematical biology to describe the growth of a population. Suppose that only the term x were present on the right-hand side. In that case the solution would be given by $x(t) = Ce^t$. This means that the first term on the right-hand side accounts for the exponential population growth, which is due to the fact that the number of offspring is proportional to the size of the population. However, a population cannot grow indefinitely; when a population is too large, limited resources restrict the growth, and this is accounted for by the $-x^2$ term on the right-hand side. The term $F(t)$ accounts for external influences on the population. For example, a mass extinction could be described by a strongly negative forcing function $F(t)$. We will consider first the solution for the case that $F(t) = 0$. Since the population size is positive, we consider positive solutions $x(t)$ only.

Problem b   Show that for the case $F(t) = 0$, the change of variable $y = 1/x$ leads to the linear equation $\dot{y} = 1 - y$. Solve this equation and show that the general solution of (17.52) (with $F(t) = 0$) is given by:

$$x(t) = \frac{1}{Ae^{-t} + 1}, \tag{17.53}$$

with A an integration constant.

Problem c   Use this solution to show that any solution of the unforced equation goes to 1 for infinite times:

$$\lim_{t \to \infty} x(t) = 1. \tag{17.54}$$

In other words, the population of the unforced Verhulst equation always converges to the same population size. Note that when the force vanishes after a finite time, the solution after that time must satisfy (17.53), which implies that the long-time limit is then also given by (17.54).

Now, consider the response to a delta-function excitation at time t_0 with strength F_0. The associated response $g(t, t_0)$ thus satisfies

$$\dot{g} - g + g^2 = F_0\delta(t - t_0). \tag{17.55}$$

Since this function is the impulse response of the system, the notation g is used in order to bring out the resemblance with the Green's functions used earlier. We consider only causal solutions, that is, we require that $g(t, t_0)$ vanishes for $t < t_0$: $g(t, t_0) = 0$ for $t < t_0$. For $t > t_0$ the solution satisfies the Verhulst equation without force; hence, the general form is given by (17.53). The only task remaining is to find the integration constant A. This constant follows by a treatment similar to the analysis of Section 17.3.

Problem d Integrate (17.55) over t from $t_0 - \epsilon$ to $t_0 + \epsilon$, take the limit $\epsilon \downarrow 0$, and show that this leads to the following requirement for the discontinuity in g:

$$\lim_{\epsilon \downarrow 0} \left[g(t, t_0)\right]_{t_0-\epsilon}^{t_0+\epsilon} = F_0. \tag{17.56}$$

Problem e Use this condition to show that the constant A in the solution (17.53) is given by $A = (1/F_0 - 1)e^{t_0}$ and that the solution is given by

$$g(t, t_0) = \begin{cases} 0 & \text{for } t < t_0 \\ \dfrac{F_0}{(1 - F_0)e^{-(t-t_0)} + F_0} & \text{for } t > t_0. \end{cases} \tag{17.57}$$

At this point we should be suspicious of interpreting $g(t, t_0)$ as a Green's function. An important property of linear systems is that the response is proportional to the force. However, the solution $g(t, t_0)$ in (17.57) is not proportional to the strength F_0 of the force.

Let us now check whether we can use the superposition principle. Suppose the force function is the superposition of a delta-function force F_1 at $t = t_1$ and a delta-function force F_2 at $t = t_2$:

$$F(t) = F_1\delta(t - t_1) + F_2\delta(t - t_2). \tag{17.58}$$

By analogy with (17.10) you might think that a Green's function-type solution is given by:

$$x^{Green}(t) = \frac{F_1}{(1 - F_1)e^{-(t-t_1)} + F_1} + \frac{F_2}{(1 - F_2)e^{-(t-t_2)} + F_2}, \tag{17.59}$$

for times larger than both t_1 and t_2. You can verify by direct substitution that this function is not a solution of the differential equation (17.52). However, this process is rather tedious and there is a simpler way to see that the function $x^{Green}(t)$ violates the differential equation (17.52).

Problem f   To see this, show that the solution $x^{Green}(t)$ has the following long-time behavior:

$$\lim_{t \to \infty} x^{Green}(t) = 2. \qquad (17.60)$$

This limit is at odds with the limit (17.54) that every solution of the differential equation (17.52) should satisfy when the force vanishes after a certain finite time. This proves that $x^{Green}(t)$ is not a solution of the Verhulst equation.

This implies that the Green's function technique introduced in the previous sections cannot be used for a nonlinear equation such as the forced Verhulst equation. The reason for this is that Green's functions are based on the superposition principle; by knowing the response to a delta-function force and by writing a general force as a superposition of delta functions, one can construct a solution by making the corresponding superposition of Green's functions, as in (17.29). However, solutions of a nonlinear equation such as the Verhulst equation do not satisfy the principle of superposition. This implies that Green's function cannot be used effectively to construct the behavior of nonlinear systems, and is the reason why Green's functions are in practice only used for constructing the response of *linear* systems.

18

Green's functions: examples

After the introduction in the previous chapter of the basic theory of Green's function, this chapter consists of examples of Green's functions that are often used in mathematical physics. A common strategy in these examples is to turn a partial differential equation (PDE) into an ordinary differential equation (ODE) via the Fourier transform (Chapter 14). We then solve the ODE, followed by a Fourier back-transform to obtain the solution to the original PDE of interest.

18.1 The heat equation in N dimensions

In this section we consider the distribution of heat due to an impulsive source. The heat equation for a general source S was derived in Section 17.5 and is given by

$$\frac{\partial T}{\partial t} = \kappa \nabla^2 T + S. \tag{17.35}$$

First we construct a Green's function for this equation in N space dimensions. This equation is identical to the diffusion equation. Perhaps surprisingly, we will see that the analysis for N dimensions is just as easy (or difficult) as the analysis for only one spatial dimension.

The heat equation is invariant for translations in both space and time. For this reason the Green's function $G(\mathbf{r}, t; \mathbf{r}_0, t_0)$ that gives the temperature at location $\mathbf{r}$ and time t due to a delta-function heat source at location $\mathbf{r}_0$ and time t_0 depends only on the relative distance $\mathbf{r} - \mathbf{r}_0$ and the relative time $t - t_0$. This implies that $G(\mathbf{r}, t; \mathbf{r}_0, t_0) = G(\mathbf{r} - \mathbf{r}_0, t - t_0)$. Since the Green's function depends only on $\mathbf{r} - \mathbf{r}_0$ and $t - t_0$, it suffices to construct the simplest solution by considering the special case of a source at $\mathbf{r}_0 = 0$ at time $t_0 = 0$. This means that we construct the Green's function $G(\mathbf{r}, t)$ that satisfies:

$$\frac{\partial G(\mathbf{r}, t)}{\partial t} - \kappa \nabla^2 G(\mathbf{r}, t) = \delta(\mathbf{r})\delta(t). \tag{18.1}$$

This Green's function can most easily be constructed by carrying out a spatial Fourier transform. Using the Fourier transform (14.27) for each of the N spatial dimensions, one finds that the Green's function has the following Fourier expansion:

$$G(\mathbf{r}, t) = \int g(\mathbf{k}, t)e^{i\mathbf{k}\cdot\mathbf{r}}d^N k. \qquad (18.2)$$

Note that the Fourier transform is only carried out over the spatial dimensions and not over time, this implies that $g(\mathbf{k}, t)$ is a function of time as well. The differential equation that g satisfies can be obtained by inserting the Fourier representation (18.2) in the differential equation (18.1). In doing this we also need the Fourier representation of $\nabla^2 G(\mathbf{r}, t)$.

Problem a   Show by applying the Laplacian to the Fourier integral (18.2) that:

$$\nabla^2 G(\mathbf{r}, t) = -\int k^2 g(\mathbf{k}, t)e^{i\mathbf{k}\cdot\mathbf{r}}d^N k. \qquad (18.3)$$

Problem b   As a last ingredient we need the Fourier representation of the delta function on the right-hand side of (18.1). This multidimensional delta function is a shorthand notation for $\delta(\mathbf{r}) = \delta(x_1)\delta(x_2)\cdots\delta(x_N)$. Use the Fourier representation (14.31) of the delta function to show that

$$\delta(\mathbf{r}) = \frac{1}{(2\pi)^N}\int e^{i\mathbf{k}\cdot\mathbf{r}}d^N k. \qquad (18.4)$$

Problem c   Insert these results into the differential equation (18.1) of the Green's function to show that $g(\mathbf{k}, t)$ satisfies the differential equation

$$\frac{\partial g(\mathbf{k}, t)}{\partial t} + \kappa k^2 g(\mathbf{k}, t) = \frac{1}{(2\pi)^N}\delta(t). \qquad (18.5)$$

We have made considerable progress. The original equation (18.1) was a partial differential equation, whereas (18.5) is an ordinary differential equation for g with only a time derivative. In fact, you saw this equation before when you read Section 16.4, which dealt with the response of a particle in syrup. Equation (18.5) is equivalent to the equation of motion (16.29) for a particle in syrup when the forcing force equals a delta function.

Problem d   Use the theory of Section 17.3 to show that the causal solution of (18.5) is given by:

$$g(\mathbf{k}, t) = \frac{1}{(2\pi)^N}e^{-\kappa k^2 t}. \qquad (18.6)$$

This solution can be inserted into the Fourier representation (18.2) of the Green's function, so that

$$G(\mathbf{r}, t) = \frac{1}{(2\pi)^N} \int e^{-\kappa k^2 t + i\mathbf{k}\cdot\mathbf{r}} d^N k. \tag{18.7}$$

The Green's function can be found by solving this Fourier integral. Before we do this, let us pause to consider the solution (18.6) for the Green's function in the wavenumber–time domain. The function $g(\mathbf{k}, t)$ gives the coefficient of the plane-wave component $e^{i\mathbf{k}\cdot\mathbf{r}}$ as a function of time. According to (18.6) each Fourier component decays exponentially with time with a characteristic decay time $1/\kappa k^2$.

Problem e   Show that this implies that plane waves with a smaller wavelength decay faster with time than plane waves with a larger wavelength. Explain this result physically.

To find the Green's function, we need to solve the Fourier integral (18.7). The integrations over the different components k_i of the wavenumber integration all have the same form.

Problem f   Show that (18.7) can be written as

$$G(\mathbf{r}, t) = \frac{1}{(2\pi)^N} \left(\int e^{-\kappa k_1^2 t + i k_1 x_1} dk_1 \right) \left(\int e^{-\kappa k_2^2 t + i k_2 x_2} dk_2 \right)$$
$$\times \cdots \times \left(\int e^{-\kappa k_N^2 t + i k_N x_N} dk_N \right). \tag{18.8}$$

You will notice that each of the integrals is of the same form; hence, the Green's function can be written as

$$G(x_1, x_2, \ldots, x_N, t) = I(x_1, t) I(x_2, t) \cdots I(x_N, t)$$

with $I(x, t)$ given by

$$I(x, t) = \frac{1}{2\pi} \int_{-\infty}^{\infty} e^{-\kappa k^2 t + i k x} dk. \tag{18.9}$$

This means that our N-dimensional problem is solved when the one-dimensional Fourier integral (18.9) is evaluated.

To solve this integral, it is important to realize that the exponent in the integral is a quadratic function of the integration variable k. If the integral were of the form $\int_{-\infty}^{\infty} e^{-\alpha k^2} dk$, the problem would not be difficult because (4.49) states that this integral has the value $\sqrt{\pi/\alpha}$. Therefore, the problem can be solved by rewriting the integral (18.9) in the form of the integral $\int_{-\infty}^{\infty} e^{-\alpha k^2} dk$.

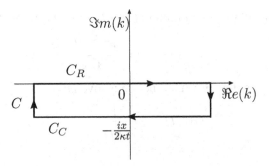

Figure 18.1   The contours C_R, C_C, and C in the complex k-plane.

Problem g   Complete the square of the exponent in (18.9), that is, show that

$$- \kappa k^2 t + ikx = -\kappa t \left(k - \frac{ix}{2\kappa t} \right)^2 - \frac{x^2}{4\kappa t}, \qquad (18.10)$$

and use this result to show that $I(x, t)$ can be written as:

$$I(x, t) = \frac{1}{2\pi} e^{-x^2/4\kappa t} \int_{-\infty-i(x/2\kappa t)}^{\infty-i(x/2\kappa t)} e^{-\kappa k'^2 t} dk'. \qquad (18.11)$$

With these steps we have achieved our goal of having an integrand of the form $e^{-\alpha k^2}$, but have paid a price. In the integral (18.9) the integration was along the real axis C_R. In the transformed integral the integration now takes place along the integration path $-C_C$ in the complex plane that lies below the real axis (Figure 18.1). The minus sign indicates that C_C runs from the right to the left. The integrand is analytic within the closed contour of Figure 18.1; therefore, according to (15.10) the circular integral along the contour vanishes. The contribution to the two intervals on the left and the right vanish as the endpoints go to $\pm\infty$; so that $\int_{C_R} + \int_{C_C} = 0$. Since an integral changes sign as the direction is reversed, $\int_{C_C} = -\int_{-C_C}$; hence, the condition that the closed integral vanishes implies that $\int_{C_R} = \int_{-C_C}$, which is a different way of saying that the integration path in (18.11) can be replaced by an integration along the real axis:

$$I(x, t) = \frac{1}{2\pi} e^{-x^2/4\kappa t} \int_{-\infty}^{\infty} e^{-\kappa k^2 t} dk. \qquad (18.12)$$

Problem h   In Section 16.2 you saw all the material necessary to prove that (18.12) is indeed identical to (18.11). Show this by using that the integral along the closed contour C in Figure 18.1 vanishes.

Problem i   Carry out the integration in (18.12) with (4.49) to find that

$$I(x, t) = \frac{e^{-x^2/4\kappa t}}{\sqrt{4\pi\kappa t}},$$
(18.13)

and that therefore the Green's function is given by

$$G(\mathbf{r}, t) = \frac{1}{(4\pi\kappa t)^{N/2}} e^{-r^2/4\kappa t}.$$
(18.14)

A comparison with (13.15) implies the Green's function in any dimension has the form of the Gaussian.

Problem j   Show that this Gaussian changes shape with time. Is the Gaussian broadest at early times or at late times? What is the shape of the Green's function in the limit $t \downarrow 0$; the time just after the heat force has been applied.

Problem k   Sketch the time behavior of the Green's function for a fixed distance r as a function of time. Does the Green's function decay more rapidly as a function of time in three dimensions than in one dimension? Give a physical interpretation of this result.

In one dimension, the Green's function (18.14) is given by

$$G(x, t) = \frac{1}{\sqrt{4\pi\kappa t}} e^{-x^2/4\kappa t}.$$
(18.15)

It is a remarkable property of the derivation in this section that the Green's function for the heat equation can be derived with a single derivation for any number of dimensions. However, this is not generally the case. In many problems, the behavior of the system depends critically on the number of spatial dimensions. We will see in Section 18.5 that wave propagation in two dimensions is fundamentally different from wave propagation in one or three dimensions. Another example is chaotic behavior of dynamical systems where the occurrence of chaos is intricately linked to the number of dimensions (Tabor, 1989).

Diffusion is a process that cannot be characterized by a velocity. To see this, we compute the expected value of x^2 that is defined by

$$\langle x^2 \rangle = \frac{\int_{-\infty}^{-\infty} x^2 G(x, t) dx}{\int_{-\infty}^{-\infty} G(x, t) dx}.$$
(18.16)

This quantity is similar to the expectation value in statistics that we discuss in more detail in Section 21.2.

Problem 1 Insert equation (18.15) in this expression. Use equation (4.49) to show that $\int_{-\infty}^{-\infty} G(x, t)dx = 1$. Differentiate expression (4.49) with respect to b to derive that $\int_{-\infty}^{-\infty} x^2 \exp(-bx^2)dx = \sqrt{\pi}/2b^{3/2}$, and use these results to show that

$$\langle x^2 \rangle = 2\kappa t. \tag{18.17}$$

If diffusion was a process that could be characterized by a velocity, then $\langle x^2 \rangle = c^2t^2$ with c the velocity of the diffusive process, but according to equation (18.17) $\langle x^2 \rangle$ is proportional to t instead of t^2. Therefore, diffusive processes do not have a characteristic velocity.

The Green's function (18.15) is the solution to a delta-function source at time $t = 0$ at location $x = 0$. A quirk of the diffusion equation is that its solution (18.15) is nonzero for a fixed value to x for every time t, including $t \to 0$. This means that the diffusive field spreads with infinite velocity, which cannot be physical. An alternative to the diffusion equation that takes into account that the diffusive field does not spread with infinite speed is the theory of *radiative transfer* (Özisik, 1973; Paasschens, 1997).

18.2 Schrödinger equation with an impulsive source

In this section we study the Green's function for the Schrödinger equation, which was introduced in Section 8.4:

$$i\hbar\frac{\partial \psi(\mathbf{r}, t)}{\partial t} = -\frac{\hbar^2}{2m}\nabla^2\psi(\mathbf{r}, t) + V(\mathbf{r})\psi(\mathbf{r}, t). \tag{8.14}$$

Solving this equation for a general potential $V(\mathbf{r})$ is a formidable problem, and solutions are known for only very few examples such as the free particle, the harmonic oscillator, and the Coulomb potential. We restrict ourselves here to the simplest case of a free particle: this is the case where the potential vanishes ($V(\mathbf{r}) = 0$). The corresponding Green's function satisfies the following partial differential equation:

$$\frac{\hbar}{i}\frac{\partial G(\mathbf{r}, t)}{\partial t} - \frac{\hbar^2}{2m}\nabla^2 G(\mathbf{r}, t) = \delta(\mathbf{r})\delta(t). \tag{18.18}$$

Before we compute the Green's function for this problem, let us pause to consider the meaning of this Green's function. First, the Green's function is a solution of Schrödinger's equation for $\mathbf{r} \neq 0$ and $t \neq 0$. It was discussed in Section 8.4 that $|G|^2$ represents the probability density of a particle. However, the right-hand side of (18.18) contains a delta-function forcing at time $t = 0$ at location $\mathbf{r} = 0$. This is a source term of G and hence this is a source of the probability of the presence of the particle. One can say that this source term creates the probability for having a particle at the origin at $t = 0$. Of course, this particle will not necessarily remain at the origin; it will move according to the laws of quantum mechanics. This motion is represented by (18.18), which describes the time evolution of matter waves when matter is injected at $t = 0$ at location $\mathbf{r} = 0$.

Problem a   The Green's function $G(\mathbf{r}, t; \mathbf{r}', t')$ gives the wavefunction at location $\mathbf{r}$ and time t for a source of particles at location $\mathbf{r}'$ at time t'. Express the Green's function $G(\mathbf{r}, t; \mathbf{r}', t')$ in terms of the solution $G(\mathbf{r}, t)$ of (18.18), and show how you obtained this result. Is this result also valid for the Green's function for the quantum-mechanical harmonic oscillator (where the potential $V(\mathbf{r})$ depends on position)?

In the previous section the Green's function gave the evolution of the temperature field due to a delta-function injection of heat at the origin at time $t = 0$. Similarly, the Green's function of this section describes the time evolution of the quantum mechanical wave function at position $\mathbf{r}$ and time t that is due to a point source at $\mathbf{r} = 0$ at time $t = 0$. These two Green's functions are not only conceptually very similar, the differential equations (18.1) for the temperature field and (18.18) for the Schrödinger equation are first-order differential equations in time and second-order differential equations in the space coordinate that have a delta-function excitation on the right-hand side. In this section we exploit this similarity and derive the Green's function for Schrödinger's equation from the Green's function for the heat equation derived in the previous section rather than constructing the solution from first principles. This approach is admittedly not very rigorous, but it shows that analogies are useful for making shortcuts.

The principle difference between (18.1) and (18.18) is that the time derivative for Schrödinger's equation is multiplied by $i = \sqrt{-1}$, whereas the heat equation is purely real. We will relate the two equations by introducing the new time variable τ for the Schrödinger equation that is proportional to the original time: $\tau = \gamma t$.

Problem b   Determine the proportionality constant γ, so that (18.18) transforms to

$$\frac{\partial G(\mathbf{r}, \tau)}{\partial \tau} - \frac{\hbar^2}{2m} \nabla^2 G(\mathbf{r}, \tau) = C\delta(\mathbf{r})\delta(\tau). \tag{18.19}$$

The constant C on the right-hand side cannot easily be determined from the change of variables $\tau = \gamma t$ because γ is not necessarily real and it is not clear how a delta function with a complex argument should be interpreted. For this reason we will not bother to specify C.

The key point is that this equation is of exactly the same form as the heat equation (18.1), where $\hbar^2/2m$ plays the role of the heat conductivity κ. The only difference is the constant C on the right-hand side of (18.19). However, since the equation is linear, this term only leads to an overall multiplication by C.

Problem c Show that the Green's function defined in (18.19) for the Schrödinger equation can be obtained from the Green's function (18.14) for the heat equation by making the following substitutions:

$$\left. \begin{array}{rcl} t & \longrightarrow & it/\hbar, \\ \kappa & \longrightarrow & \hbar^2/2m, \\ G & \longrightarrow & C\,G. \end{array} \right\} \tag{18.20}$$

It is interesting to note that the "diffusion constant" κ, which governs the spreading of the waves with time is proportional to the square of Planck's constant. Classical mechanics follows from quantum mechanics by letting Planck's constant go to zero: $\hbar \to 0$.[1] It follows from (18.20) that in that limit the diffusion constant of the matter waves goes to zero. This reflects the fact that in classical mechanics the probability of the presence of a particle does not spread out with time.

Problem d Use the substitutions (18.20) to show that the Green's function for the Schrödinger equation in N dimensions is

$$G(\mathbf{r}, t) = C \frac{1}{(2\pi i \hbar t/m)^{N/2}} e^{imr^2/2\hbar t}. \tag{18.21}$$

This Green's function plays a crucial role in the formulation of the *Feynman path integrals*, which have been a breakthrough in quantum mechanics as well as in other fields. A clear description of the Feynman path integrals is given by Feynman et al. (1965).

Problem e Sketch the real part of $e^{imr^2/2\hbar t}$ in the Green's function for a fixed time as a function of distance r. Does the wavelength of the Green's function increase or decrease with distance?

The Green's function (18.21) actually has an interesting physical meaning based on the fact that it describes the propagation of matter waves injected at $t = 0$ at the

---

[1] Of course, Planck's constant has a fixed value, but the limit $\hbar \to 0$ means in this context that in the Schrödinger equation (18.18) the term with the time derivative $(\hbar/i)\partial G/\partial t$ dominates the spatial derivative $(\hbar^2/2m)\nabla^2 G$.

origin. The Green's function can be written as $G = C \, (2\pi i \hbar t / m)^{-N/2} \, e^{i\Phi}$, where the phase of the Green's function is given by

$$\Phi = \frac{mr^2}{2\hbar t}. \tag{18.22}$$

As noted in Problem e of Section 8.4 the wavenumber of the waves depends on position. For a plane wave $e^{i\mathbf{k}\cdot\mathbf{r}}$ the phase is given by $\Phi = (\mathbf{k} \cdot \mathbf{r})$ and the wavenumber follows by taking the gradient of this function.

Problem f   Show that for a plane wave

$$\mathbf{k} = \nabla\Phi. \tag{18.23}$$

Relation (18.23) has a wider applicability than plane waves. It has been shown by Whitham (2011) that for a general phase function $\Phi(\mathbf{r})$ that varies smoothly with $\mathbf{r}$ the local wavenumber $\mathbf{k}(\mathbf{r})$ is defined by (18.23).

Problem g   Use this to show that for the Green's function of the Schrödinger equation the local wavenumber is given by

$$\mathbf{k} = \frac{m\mathbf{r}}{\hbar t}. \tag{18.24}$$

Problem h   Use the definition $\mathbf{v} = \mathbf{r}/t$ to show that this expression is equivalent to:

$$\mathbf{v} = \frac{\hbar\mathbf{k}}{m}. \tag{8.21}$$

In Problem e you discovered that for a fixed time, the wavelength of the waves decreases when the distance r to the source is increased. This is consistent with (8.21): when a particle has moved further away from the source in a fixed time, its velocity is larger. This corresponds according to (8.21) with a larger wavenumber and hence with a smaller wavelength. This is indeed the behavior that is exhibited by the full wavefunction (18.21).

The analysis in this chapter was not rigorous because the substitution $t \rightarrow (i/\hbar) \, t$ implies that the independent parameter is purely imaginary rather than real. This means that all the arguments used in the previous section for the complex integration should be carefully reexamined. However, a more rigorous analysis shows that (18.21) is indeed the correct Green's function for the Schrödinger equation (Feynman et al., 1965). The approach taken in this section shows that an educated guess can be very useful in deriving new results. One can in fact argue that many innovations in mathematical physics have been obtained using intuition or analogies rather than formal derivations. Of course, a formal derivation should ultimately substantiate results obtained from a more intuitive approach.

18.3 Helmholtz equation in one, two, and three dimensions

The Helmholtz equation plays an important role in mathematical physics because it is closely related to the wave equation. A complete analysis of the Green's function for the wave equation and the Helmholtz equation is given by DeSanto (1992). In this section and the next we consider the Green's function for the Helmholtz and wave equations in one, two, and three spatial dimensions. The wave equation (6.42) for a medium with constant sound speed c and constant density describes the pressure field $p(\mathbf{r}, t)$ that is excited by a source $S(\mathbf{r}, t)$:

$$\nabla^2 p(\mathbf{r}, t) - \frac{1}{c^2} \frac{\partial^2 p(\mathbf{r}, t)}{\partial t^2} = S(\mathbf{r}, t). \tag{18.25}$$

The Green's function for an impulsive source at $(\mathbf{r}_0, t_0)$ for this problem is

$$\nabla^2 G(\mathbf{r}, t; \mathbf{r}_0, t_0) - \frac{1}{c^2} \frac{\partial^2 G(\mathbf{r}, t; \mathbf{r}_0, t_0)}{\partial t^2} = \delta(\mathbf{r} - \mathbf{r}_0)\delta(t - t_0). \tag{18.26}$$

As shown in Section 18.1, the Green's function depends for a constant velocity c only on the relative location $\mathbf{r} - \mathbf{r}_0$ and the relative time $t - t_0$ so that without loss of generality we can take the source at the origin ($\mathbf{r}_0 = 0$) and let the source act at time $t_0 = 0$. In addition, it follows from symmetry considerations that the Green's function depends only on the relative distance $|\mathbf{r} - \mathbf{r}_0|$ and not on the orientation of the vector $\mathbf{r} - \mathbf{r}_0$. This means that the Green's function then satisfies $G(\mathbf{r}, t; \mathbf{r}_0, t_0) = G(|\mathbf{r} - \mathbf{r}_0|, t - t_0)$ and we need to solve the following equation:

$$\nabla^2 G(r, t) - \frac{1}{c^2} \frac{\partial^2 G(r, t)}{\partial t^2} = \delta(\mathbf{r})\delta(t). \tag{18.27}$$

Problem a Under which conditions is this approach justified?

Problem b Use a similar treatment to that in Section 18.1 to show that when the Fourier transform (14.42) is used, the Green's function satisfies the following equation in the frequency domain:

$$\nabla^2 G(r, \omega) + k^2 G(r, \omega) = \delta(\mathbf{r}), \tag{18.28}$$

where the wavenumber k satisfies $k = \omega/c$.

This equation is called the *Helmholtz equation* and is the reformulation of the wave equation in the frequency domain. In the following we suppress the factor ω in the

Green's function but it should be remembered that the Green's function depends on frequency.

Let us first solve (18.28) for one dimension. In that case the Green's function is defined by

$$\frac{d^2G}{dx^2} + k^2G = \delta(x).\tag{18.29}$$

Problem c For $x \neq 0$, the right-hand side of this equation is equal to zero. Use this to show that for $x \neq 0$ the solution is given by

$$G(x) = A\,e^{ikx} + B\,e^{-ikx},\tag{18.30}$$

where A and B are integration constants, which need to be determined for the regions $x < 0$ and $x > 0$ separately.

In the Fourier transformation (14.42), both terms are multiplied by $e^{-i\omega t}$. Using the relation $k = \omega/c$, this means that in the time domain the term e^{ikx} becomes $e^{-i\omega(t-x/c)}$. For increasing time t, the phase of the wave remains constant only when x increases. This means that the term e^{ikx} corresponds in the time domain to a right-going wave. Similarly, the term e^{-ikx} describes a wave in the time domain that moves to the left. We are looking here for a Green's function that describes waves that move *away* from the source. This means that for $x > 0$ only the term e^{ikx} contributes, so in that region we must have $B = 0$, while for $x < 0$ only the term e^{-ikx} contributes; hence, we must take $A = 0$. This means that the Green's function is given by

$$G(x) = \begin{cases} B\,e^{-ikx} & \text{for } x < 0 \\ A\,e^{ikx} & \text{for } x > 0. \end{cases}\tag{18.31}$$

The constants A and B follow from the requirements that $G(x)$ is continuous at $x = 0$ and that the first derivative is discontinuous at that point.

Problem d Use (17.42) to show that for this second-order differential equation, the jump in the first derivative at $x = 0$ is given by

$$\left[\frac{dG}{dx}\right]_{x=0-\epsilon}^{x=0+\epsilon} = 1.\tag{18.32}$$

Problem e Use these two boundary conditions to derive that $A = B = -i/2k$. With these constants, confirm that the Green's function of the Helmholtz equation in one dimension is

$$G^{1D}(x) = \frac{-i}{2k}e^{ik|x|}.\tag{18.33}$$

Now we solve (18.28) for two and three space dimensions. To do this, we consider the case of N dimensions, where N is either 2 or 3. Because the problem is spherically symmetric, we just need to consider a Green's function that depends on radius only.

Problem f Use (10.35) and (10.36) to show that for such a radially symmetric function in N dimensions, the Laplacian is

$$\nabla^2 G(r) = \frac{1}{r^{N-1}} \frac{\partial}{\partial r} \left(r^{N-1} \frac{\partial G}{\partial r} \right). \tag{18.34}$$

The differential equation for the Green's function in N dimensions is thus given by

$$\frac{1}{r^{N-1}} \frac{\partial}{\partial r} \left(r^{N-1} \frac{\partial G}{\partial r} \right) + k^2 G(r, \omega) = \delta(\mathbf{r}). \tag{18.35}$$

This differential equation is not difficult to solve for two or three space dimensions for locations away from the source ($r \neq 0$). However, we need to consider carefully how the source $\delta(\mathbf{r})$ should be coupled to the solution of the differential equation. In one dimension, we obtained in expression (18.32) a boundary condition from the jump across the source. In more than one dimension, there are an infinite number of ways to jump across the source! Instead, the necessary boundary condition can be found by integrating (18.28) over a sphere of radius R centered at the source and letting its radius go to zero.

Problem g Integrate (18.28) over this volume, use Gauss' theorem (8.1) and let the radius R go to zero to show that the Green's function satisfies

$$\oint_{S_R} \frac{\partial G}{\partial r} dS = 1, \tag{18.36}$$

where the surface integral is over a sphere S_R with radius R in the limit $R \downarrow 0$. Show that this can also be written as

$$\lim_{r \downarrow 0} S_r \frac{\partial G}{\partial r} = 1, \tag{18.37}$$

where S_r is the surface of a sphere in N dimensions with radius r.

Note that the surface of the sphere goes to zero as $r \downarrow 0$, which implies that $\partial G/\partial r$ must be infinite in the limit $r \downarrow 0$ in more than one space dimension.

The differential equation (18.35) is a second-order differential equation. Such an equation must be supplemented with two boundary conditions. The first boundary condition is given by (18.37), and specifies how the solution is coupled to the source at $\mathbf{r} = 0$. The second boundary condition that we will use reflects the fact that the waves generated by the source will move away from the source. Mathematically, the solutions that we will find behave for large distances as $e^{\pm ikr}$, but it is not clear whether we should use the upper sign $(+)$ or the lower sign $(-)$.

Problem h   Use the Fourier transform (14.42) and the relation $k = \omega/c$ to show that the integrand in the Fourier transformation to the time domain is proportional to $e^{-i\omega(t\mp r/c)}$. (The convention of the notation $\mp$ is that the upper sign in $e^{-i\omega(t\mp r/c)}$ corresponds to the upper sign in $e^{\pm ikr}$. Similarly, the lower signs in both expressions correspond to each other. This means that the solution $e^{-i\omega(t-r/c)}$ corresponds to e^{+ikr} and that the solution $e^{-i\omega(t+r/c)}$ corresponds to e^{-ikr}.) Show that the waves only move away from the source for the upper sign by showing that for this solution the distance r must increase for increasing time t to keep the phase constant. This means that this boundary condition dictates that the solution behaves in the limit $r \to \infty$ as e^{+ikr}.

The derivative of function e^{+ikr} is given by ike^{+ikr}; that is, the derivative is ik times the original function. When the Green's function behaves for large r as e^{+ikr}, then the derivative of the Green's function must satisfy the same relation as the derivative of e^{+ikr}. This means that the Green's function satisfies for large distance r:

$$\frac{\partial G}{\partial r} = ikG. \tag{18.38}$$

This relation specifies that the energy radiates *away* from the source. For this reason (18.38) is called the *radiation boundary condition*.

Now we are at a point where we can actually construct the solution for each dimension. Before we go to two dimensions, we first solve the Green's function in three dimensions.

Problem i   Make for three dimensions ($N = 3$) the substitution $G(r) = f(r)/r$ and show that (18.35) implies that away from the source the function $f(r)$ satisfies

$$\frac{\partial^2 f}{\partial r^2} + k^2 f = 0. \tag{18.39}$$

This equation has the solution $f(r) = Ce^{\pm ikr}$. According to Problem h the upper sign should be used and the Green's function is given by $G(r) = Ce^{ikr}/r$.

Problem j   Show that condition (18.37) dictates that $C = -1/4\pi$, so that in three dimensions the Green's function is given by

$$G^{3D}(r) = \frac{-1}{4\pi}\frac{e^{ikr}}{r}. \tag{18.40}$$

The problem is actually most difficult in two dimensions, because in that case the Green's function cannot be expressed in the simplest elementary functions.

Problem k Show that in two dimensions ($N = 2$) the differential equation of the Green's function away from the source is given by

$$\frac{\partial^2 G}{\partial r^2} + \frac{1}{r} \frac{\partial G}{\partial r} + k^2 G(r) = 0, \qquad r \neq 0. \tag{18.41}$$

This equation cannot be solved in terms of elementary functions. However, there is a close relation between (18.41) and the Bessel equation that is given by

$$\frac{d^2 F}{dx^2} + \frac{1}{x} \frac{dF}{dx} + \left(1 - \frac{m^2}{x^2}\right) F = 0. \tag{18.42}$$

Problem l Show that $G(kr)$ satisfies the Bessel equation for order $m = 0$.

This implies that the Green's function is given by the solution of the zeroth order Bessel equation with argument kr. The Bessel equation is a second-order differential equation; therefore, there are two independent solutions. The solution that is finite everywhere is denoted by $J_m(x)$ and is called the regular Bessel function. The second solution is singular at the point $x = 0$ and is called the Neumann function and denoted by $N_m(x)$. The Green's function is obviously a linear combination of $J_0(kr)$ and $N_0(kr)$. To determine how this linear combination is constructed, it is crucial to consider the behavior of these functions at the source (i.e., for $kr = 0$) and at infinity (i.e., for $kr \gg 1$). The required asymptotic behavior can be found in Butkov (1968) and Arfken and Weber (2005), but is summarized in Table 18.1.

Problem m Show that neither $J_0(kr)$ nor $N_0(kr)$ behaves as e^{+ikr} for large values of r. Show that in this case the linear combination $J_0(kr) + i N_0(kr)$ does behave as e^{+ikr}.

The Green's function thus is a linear combination of the regular Bessel function and the Neumann function. This particular combination is called the *first Hankel function of degree zero* and is denoted by $H_0^{(1)}(kr)$. The Hankel functions are simply linear combinations of the Bessel function and the Neumann function:

$$\left. \begin{array}{l} H_m^{(1)}(x) \equiv J_m(x) + i N_m(x), \\ H_m^{(2)}(x) \equiv J_m(x) - i N_m(x). \end{array} \right\} \tag{18.43}$$

Problem n Use this definition and Table 18.1 to show that $H_0^{(1)}(kr)$ behaves for large values of r as $e^{+ikr - i\pi/4}/\sqrt{(\pi/2)kr}$ and that in this limit $H_0^{(2)}(kr)$

Table 18.1 *Asymptotic behavior of the Bessel and Neumann functions of order zero.*

	$J_0(x)$	$N_0(x)$
$x \to 0$	$1 - \frac{1}{4}x^2 + \mathcal{O}(x^4)$	$\frac{2}{\pi}\ln(x) + \mathcal{O}(1)$
$x \gg 1$	$\sqrt{\dfrac{2}{\pi x}}\cos\left(x - \dfrac{\pi}{4}\right) + \mathcal{O}(x^{-3/2})$	$\sqrt{\dfrac{2}{\pi x}}\sin\left(x - \dfrac{\pi}{4}\right) + \mathcal{O}(x^{-3/2})$

behaves as $e^{-ikr+i\pi/4}/\sqrt{(\pi/2)kr}$. Use this to argue that the Green's function is given by

$$G(r) = C H_0^{(1)}(kr), \tag{18.44}$$

where the constant C still needs to be determined.

Problem o This constant follows from the requirement (18.37) at the source. Use (18.43) and the asymptotic value of the Bessel function and the Neumann function given in Table 18.1 to derive the asymptotic behavior of the Green's function near the source and use this to show that $C = -i/4$.

This result implies that in two dimensions the Green's function of the Helmholtz equation is given by

$$G^{2D}(r) = \frac{-i}{4} H_0^{(1)}(kr). \tag{18.45}$$

Summarizing these results and reverting to the more general case of a source at location $\mathbf{r}_0$, it follows that in one, two, and three dimensions the Green's functions of the Helmholtz equation are given by

$$\left.\begin{aligned}
G^{1D}(x, x_0) &= \frac{-i}{2k} e^{ik|x-x_0|}, \\[1em]
G^{2D}(\mathbf{r}, \mathbf{r}_0) &= \frac{-i}{4} H_0^{(1)}(k\,|\mathbf{r} - \mathbf{r}_0|), \\[1em]
G^{3D}(\mathbf{r}, \mathbf{r}_0) &= \frac{-1}{4\pi} \frac{e^{ik|\mathbf{r}-\mathbf{r}_0|}}{|\mathbf{r} - \mathbf{r}_0|}.
\end{aligned}\right\} \tag{18.46}$$

Note that in two and three dimensions the Green's function is singular at the source $\mathbf{r}_0$, while in one dimension the first derivative of the Green's function is singular due to the absolute value $|x - x_0|$.

Problem p Show that these singularities are *integrable*: when the Green's function is integrated over a sphere with finite radius around the source, the result is finite.

There is a physical reason why the Green's function in two and three dimensions has an integrable singularity. Suppose one has a source that is not a point source but that the source is spread out over a sphere with radius R centered around the origin. The response p to this source is given by $p(\mathbf{r}) \propto \int_{r'<R} G(\mathbf{r}, \mathbf{r}')dV'$, where the integration over the variable $\mathbf{r}'$ is over a sphere with radius R. (We use the proportionality sign ($\propto$) because the response depends on the source density.) It follows from this expression that the response at the origin is given by

$$p(\mathbf{r} = 0) \propto \int_{r'<R} G(\mathbf{r} = 0, \mathbf{r}')dV'. \tag{18.47}$$

Since the excitation of this field is finite everywhere, the response $p(\mathbf{r} = 0)$ should be finite. This implies that the integral (18.47) should be finite as well, which is a different way of stating that the singularity of the Green's function must be integrable.

18.4 Green's functions and dimensionality

The Green's functions of the Helmholtz equation in 1D, 2D, and 3D are given in expression (18.46). In this section we treat the decay of these Green's functions with distance, as well as their physical dimensions.

Problem a Use equation (18.46) to show that the absolute value of the Green's functions in one and three dimensions is given by

$$|G^{1D}(x, x_0)| = \frac{1}{2k},$$
$$|G^{3D}(\mathbf{r}, \mathbf{r}_0)| = \frac{1}{4\pi |\mathbf{r} - \mathbf{r}_0|}. \tag{18.48}$$

Problem b The absolute value of the Green's function in two dimensions is a little more complicated because according to equation (18.46) it is given by a Hankel function. For this function we only have a closed-form expression for the behavior at large distances when $k|\mathbf{r} - \mathbf{r}_0| \gg 1$. Use equation (18.43) and Table 18.1 to show that for such large distances the Green's function in two dimensions satisfies

$$|G^{2D}(\mathbf{r}, \mathbf{r}_0)| = \frac{1}{\sqrt{8\pi k |\mathbf{r} - \mathbf{r}_0|}} \qquad (k|\mathbf{r} - \mathbf{r}_0|/ \gg 1). \tag{18.49}$$

Equations (18.48) and (18.49) imply that in one dimension the magnitude of the Green's function does not depend on distance, that in two dimensions the Green's function decays as $1/\sqrt{distance}$ and in three dimensions as $1/distance$. This means that in a higher number of dimensions the Green's function decays more rapidly with distance than in a lower number of dimensions. This is due to geometrical spreading. In more dimensions, the wave energy is "spread out" over a larger part of space as the waves move away from the source. Since energy is conserved, this corresponds for a fixed propagation distance to a faster decay in a higher number of dimensions. In one dimension the wave does not "spread out" over space at all, and as a result the wave does not decay with distance as it moves away from the source. We treat the issue of geometrical spreading in more detail in Sections 19.5 and 19.6.

In Chapter 2 we treated dimensional analysis. We return to this issue by considering the physical dimension of the Green's functions.

Problem c   Use expressions (18.48) and (18.49) to derive that the Green's functions have the following dimensions:

$$G^{1D} \sim [L], \quad G^{2D} \sim [1], \quad G^{3D} \sim [L^{-1}]. \tag{18.50}$$

Regardless of the number of dimensions, the Green's functions satisfy the same Helmholtz equation (18.28), so why the difference in physical dimension of the Green's function in 1D, 2D, and 3D? The answer lies in the elusive property that the Dirac delta function is not dimensionless. According to expression (13.16), $\int_{-\infty}^{\infty} \delta(x - x_0)dx = 1$.

Problem d   Use this property to deduce that $\delta(x) \sim [L^{-1}]$.

Problem e   Use expression (13.37) to derive that in three dimensions $\delta(\mathbf{r}) \sim [L^{-3}]$.

This last property actually follows from Problem d because $\delta(\mathbf{r}) = \delta(x)\delta(y)\delta(z)$. We can generalize this by stating that in N dimensions the delta function has physical dimension $\delta(\mathbf{r}) \sim [L^{-N}]$.

Problem f   Use this last result in a dimensional analysis of the Helmholtz equation (18.28) to show that the Green's function in N dimensions satisfies

$$G^N \sim [L^{2-N}]. \tag{18.51}$$

Problem g   Verify that this result is the same as equation (18.50).

The differences in the physical dimension of G^{1D}, G^{2D}, and G^{3D} caused by the fact that the Dirac delta function is not dimensionless!

18.5 Wave equation in one, two, and three dimensions

In this section we consider the Green's function for the wave equation in one, two and three dimensions. This means that we consider solutions to the wave equation with an impulsive source at location $\mathbf{r}_0$ at time t_0:

$$\nabla^2 G(\mathbf{r}, t; \mathbf{r}_0, t_0) - \frac{1}{c^2} \frac{\partial^2 G(\mathbf{r}, t; \mathbf{r}_0, t_0)}{\partial t^2} = \delta(\mathbf{r} - \mathbf{r}_0)\delta(t - t_0). \quad (18.26)$$

It was shown in the previous section that this Green's function depends only on the relative distance $|\mathbf{r} - \mathbf{r}_0|$ and the relative time $t - t_0$. For the case of a source at the origin ($\mathbf{r}_0 = 0$) acting at time zero ($t_0 = 0$), the time domain solution follows by applying a Fourier transform to the solutions $G(\mathbf{r}, \omega)$ of (18.46) in one, two and three dimensions. This Fourier transform is simplest in three dimensions; hence, we will start with this case.

Problem a Apply the Fourier transform (14.42) to the three-dimensional Green's function (18.40) and use the relation $k = \omega/c$ and the properties of the delta function to show that the Green's function is given in the time domain by

$$G^{3D}(\mathbf{r}, t) = -\frac{1}{4\pi r} \delta\left(t - \frac{r}{c}\right). \quad (18.52)$$

Problem b Consider the wave equation (18.25) with a general source term $S(\mathbf{r}, t)$. Use the Green's function (18.52) to show that a solution of this equation is given by

$$p(\mathbf{r}, t) = -\frac{1}{4\pi} \int \frac{S\left(\mathbf{r}', t - \frac{|\mathbf{r}-\mathbf{r}'|}{c}\right)}{|\mathbf{r} - \mathbf{r}'|} dV'. \quad (18.53)$$

Note that since $|\mathbf{r} - \mathbf{r}'|$ is always positive, the response $p(\mathbf{r}, t)$ depends only on the source function at *earlier* times. The solution therefore has a causal behavior and the Green's function (18.52) is called the *retarded Green's function*. However, in several applications one does not want to use a Green's function that depends on excitation at earlier times. An example is reflection seismology, in which one records the wave field at the surface, and from these observations one wants to reconstruct the wave field at *earlier* times while it was being reflected off layers inside the Earth. This is discussed in Section 8.3 and Schneider (1978), for example. A Green's function with waves that propagate toward the source and are then annihilated by the source can be obtained by replacing the radiation condition (18.38) by $\partial G / \partial r = -ikG$. The only difference is the minus sign on the right-hand side, which is equivalent to replacing k by $-k$.

Problem c   Apply the Fourier transform (14.42) to the three-dimensional Green's function (18.40) with k replaced by $-k$ and show that the resulting Green's function is

$$G^{3D,advanced}(\mathbf{r}, t) = -\frac{1}{4\pi r}\delta\left(t + \frac{r}{c}\right),\tag{18.54}$$

and that a solution of the wave equation (18.25) is

$$p(\mathbf{r}, t) = -\frac{1}{4\pi}\int \frac{S\left(t + \frac{|\mathbf{r}-\mathbf{r'}|}{c}\right)}{|\mathbf{r} - \mathbf{r'}|}dV'.\tag{18.55}$$

Note that in this representation the wave field is expressed in terms of the source function at *later* times. For this reason the Green's function (18.54) is called the *advanced Green's function*. The fact that the wave equation has both a retarded and an advanced solution is mathematically due to the wave equation (18.25) being invariant for time-reversal. This means that when one replaces t by $-t$, the equation does not change. Physically this means that the wave equation does not know the "direction of time." In practice one most often works with the retarded Green's function, but keep in mind that in some applications, such as exploration seismology, the advanced Green's functions are crucial. In the remaining part of this section we will focus exclusively on the retarded Green's functions that represent causal solutions.

In order to obtain the Green's function for two dimensions in the time domain one could apply a Fourier transform to the solution (18.45). This involves taking the Fourier transform of a Hankel function, and it is not obvious how this Fourier integral should be solved (although it *can* be solved). Here we follow an alternative route by recognizing that the Green's function in two dimensions is identical to the solution of the wave equation in three dimensions when the source is not a point source but a line source. In other words, we obtain the two-dimensional Green's function, spanned by the (x, y)-plane, by considering the wave field in three dimensions that is generated by a source distributed homogeneously along the z-axis. To separate the distance to the origin from the distance to the z-axis, the variables r and ρ are used, as defined in Figure 18.2.

Problem d   Use this line source to show that

$$G^{2D}(\rho, t) = \int_{-\infty}^{\infty} G^{3D}(r, t)\, dz.\tag{18.56}$$

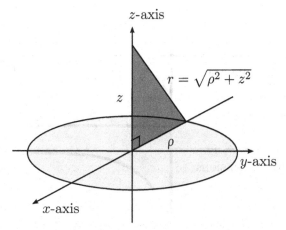

Figure 18.2 The variable r defines the distance from a point on the z-axis to a point in the (x, y)-plane with a distance ρ to the origin.

Problem e Use the Green's function (18.52) and the relation

$$r = \sqrt{\rho^2 + z^2}$$

to show that

$$G^{2D}(\rho, t) = -\frac{1}{2\pi} \int_0^\infty \frac{\delta\left(t - \frac{\sqrt{\rho^2+z^2}}{c}\right)}{\sqrt{\rho^2 + z^2}} \, dz. \qquad (18.57)$$

Note that the integration interval has been changed from $(-\infty, \infty)$ to $(0, \infty)$, and show how this was achieved.

The distance r in three dimensions no longer appears in this expression.

Problem f The integral (18.57) can be solved by introducing the new integration variable $u \equiv \sqrt{\rho^2 + z^2}$, instead of the old integration variable z. Show that with this new variable the integral (18.57) can be written as

$$G^{2D}(\rho, t) = -\frac{1}{2\pi} \int_\rho^\infty \frac{\delta(t - u/c)}{\sqrt{u^2 - \rho^2}} du. \qquad (18.58)$$

Pay attention to the limits of integration!

Problem g Use (13.23) to rewrite this integral and evaluate the result separately for $t < \rho/c$ and $t > \rho/c$. Finally, rename the distance ρ in the two-dimensional (x, y)-plane to r, to show that

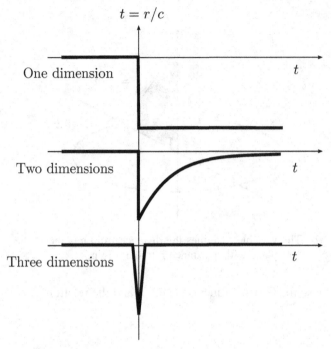

Figure 18.3   The Green's function of the wave equation in one, two, and three dimensions as a function of time.

$$G^{2D}(r, t) = \begin{cases} 0 & \text{for} \quad t < r/c \\ -\dfrac{1}{2\pi} \dfrac{1}{\sqrt{t^2 - r^2/c^2}} & \text{for} \quad t > r/c. \end{cases} \qquad (18.59)$$

This Green's function and the Green's function for the three-dimensional case are sketched in Figure 18.3. There is a fundamental difference between the Green's function for two dimensions and the Green's function (18.52) for three dimensions. In three dimensions the Green's function is a delta function $\delta(t - r/c)$ modulated by the geometrical spreading $-1/4\pi r$. This means that the response to a delta-function source has the same shape as the input function $\delta(t)$ that excites the wave field. An impulsive input leads to an impulsive output with a time delay given by r/c and the solution is only nonzero at the wave front $t = r/c$. However, (18.59) shows that an impulsive input in two dimensions leads to a response that is not impulsive. The response has an infinite duration and decays with time as $1/\sqrt{t^2 - r^2/c^2}$: the solution is not only nonzero *at* the wave front $t = r/c$, it is nonzero everywhere *within* this wave front. This means that in two dimensions an impulsive input leads to a sound response that is of infinite duration. Following one of our mentors at the University of Utrecht, Professor Eckhaus, one can therefore say that:

Any word spoken in two dimensions will reverberate forever (albeit weakly).

The approach we have taken to compute the Green's function in two dimensions is interesting in that we solved the problem first in a higher dimension and retrieved the solution by integrating over one space dimension. Note that for this trick it is not necessary that this higher-dimensional space indeed exists! (Although in this case it does.) Remember that we took this approach because we did not want to evaluate the Fourier transform of a Hankel function. We can also turn this around: the Green's function (18.59) can be used to determine the Fourier transform of the Hankel function.

Problem h Show that the Fourier transform of the Hankel function is given by:

$$
\frac{1}{2\pi} \int_{-\infty}^{\infty} H_0^{(1)}(x) e^{-iqx} dx =
\begin{cases}
0 & \text{for } q < 1 \\
\dfrac{2}{i\pi} \dfrac{1}{\sqrt{q^2 - 1}} & \text{for } q > 1.
\end{cases} \tag{18.60}
$$

Hint: Take the Fourier transform $G^{2D}(\mathbf{r}, \mathbf{r}_0)$ in (18.46) in order to obtain the Green's function for two dimensions in the time domain and compare the result with the corresponding expression (18.59). Make a suitable change of variables to arrive at (18.60).

Let us continue with the Green's function of the wave equation in one dimension in the time domain.

Problem i Use the Green's function (18.33) for the Helmholtz equation to show that in the Green's function for the wave equation is

$$
G^{1D}(x, t) = -\frac{ic}{4\pi} \int_{-\infty}^{\infty} \frac{1}{\omega} e^{-i\omega(t - |x|/c)} d\omega. \tag{18.61}
$$

This integral resembles the integral used for the calculation of the Green's function in three dimensions. The only difference is the term $1/\omega$ in the integrand, and because of this term, we cannot immediately evaluate the integral. However, the $1/\omega$ term can be removed by differentiating (18.61) with respect to time, and the remaining integral can then be evaluated analytically.

Problem j Show that

$$
\frac{\partial G^{1D}(x, t)}{\partial t} = -\frac{c}{2} \delta\left(t - \frac{|x|}{c}\right). \tag{18.62}
$$

Problem k   This expression can be integrated but one condition is needed to specify the integration constant that appears. We will use here that at $t = -\infty$ the
Green's function vanishes. Show that with this condition the Green's function
is given by:

$$G^{1D}(x, t) = \begin{cases} 0 & \text{for} \quad t < |x|/c \\ -c/2 & \text{for} \quad t > |x|/c. \end{cases} \tag{18.63}$$

Just as in two dimensions the solution is nonzero everywhere *within* the expanding
wave front and not only on the wave front $|x| = ct$ as in three dimensions. How-
ever, there is an important difference: in two dimensions the solution varies for all
moments with time, whereas in one dimension the solution is constant except for
$t = |x|/c$. The human ear is only sensitive to pressure variations; it is insensitive
to a static pressure. Therefore, a one-dimensional human will only hear a sound at
$t = |x|/c$ but not at later times.

   In order to appreciate the difference in sound propagation in one, two, and three
space dimensions, the Green's functions for the different dimensions are sketched
in Figure 18.3. Note the dramatic change in the response for different numbers of
dimensions. This change in the properties of the Green's function with change in
dimension has been used somewhat jokingly by Morley (1985) to give "a simple
proof that the world is three dimensional." When you worked through Sections 18.1
and 18.2, you learned that for both the heat equation and the Schrödinger equation,
the solution does not depend fundamentally on the number of dimensions. This is
in stark contrast with the solutions of the wave equation, which depend critically
on the number of dimensions.

18.6 If I can hear you, you can hear me

In this section we treat a property of the Green's function of acoustic waves. We
start with the acoustic wave equation in the frequency domain:

$$\nabla \cdot \left(\frac{1}{\rho}\nabla p\right) + \frac{\omega^2}{\kappa}p = \frac{f}{\rho}. \tag{8.8}$$

The density $\rho(\mathbf{r})$ and the bulk modulus $\kappa(\mathbf{r})$ can be arbitrary functions of position.
Let a solution p_1 be excited by an excitation f_1/ρ and a solution p_2 by an excitation f_2/ρ. As shown in Section 8.3, the two solutions are related by the following
expression

$$\oint_S \frac{1}{\rho}(p_2\nabla p_1 - p_1\nabla p_2) \cdot d\mathbf{S} = \int_V \frac{1}{\rho}(p_2 f_1 - p_1 f_2)\,dV. \tag{8.11}$$

Let us consider a volume V that is either bounded by a free surface where $p = 0$, or that extends to infinity. The parts of the boundary that form a free surface do not contribute to the surface integral in the left-hand side, because at the free surface $p = 0$, so that the integrand vanishes.

Problem a At the parts of the surface S that are placed at infinity, the radiation boundary condition (18.38) applies. Show that with this boundary condition the contributions of the surface S to the integral in the left-hand side of (8.11) vanishes.

This means that for the volume under consideration the left-hand side of (8.11) vanishes. This result holds for a general excitations f_1 and f_2. Let us take $f_1(\mathbf{r})/\rho(\mathbf{r}) = \delta(\mathbf{r} - \mathbf{r}_1)$, a point source at location $\mathbf{r}_1$. The response to this excitation is the Green's function: $p_1(\mathbf{r}) = G(\mathbf{r}, \mathbf{r}_1, \omega)$. Similarly, we take for the excitation of p_2 a delta function placed at location $\mathbf{r}_2$, so that $f_2(\mathbf{r})/\rho(\mathbf{r}) = \delta(\mathbf{r} - \mathbf{r}_2)$. The corresponding pressure field is given by the Green's function $p_2(\mathbf{r}) = G(\mathbf{r}, \mathbf{r}_2, \omega)$.

Problem b Insert these solutions into equation (8.11), use that the left-hand side vanishes, and show that

$$G(\mathbf{r}_1, \mathbf{r}_2, \omega) = G(\mathbf{r}_2, \mathbf{r}_1, \omega). \qquad (18.64)$$

Problem c The analysis up to this point was carried out in the frequency domain. Expression (18.64) holds for every frequency. Apply a Fourier transform to show that reciprocity also holds in the time domain:

$$G(\mathbf{r}_1, \mathbf{r}_2, t) = G(\mathbf{r}_2, \mathbf{r}_1, t). \qquad (18.65)$$

Expressions (18.64) or (18.65) are called the *reciprocity theorem* in the frequency domain or time domain, respectively. The Green's function $G(\mathbf{r}_1, \mathbf{r}_2)$ gives

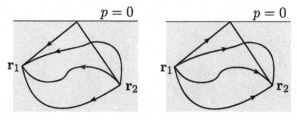

Figure 18.4 Paths that connect the points $\mathbf{r}_1$ and $\mathbf{r}_2$ in opposite directions. The waves that propagate in the two opposite directions are identical, even in the presence of the free surface, where the pressure $p = 0$.

the pressure at location $\mathbf{r}_1$ due to a point source at $\mathbf{r}_2$, while the Green's function $G(\mathbf{r}_2, \mathbf{r}_1)$ gives the pressure at location $\mathbf{r}_2$ due to a point source at $\mathbf{r}_1$. According to expression (18.64), these solutions are identical. This means that the role of the source and receiver can be interchanged, and the recorded pressure field is identical. This situation is sketched in Figure 18.4. Because density and bulk modulus can be arbitrary functions of space, waves may scatter and thus travel along a multitude of paths. In addition, we allowed for the presence of a free surface, so that the pressure waves may also bounce off this free surface. Intuitively, it is clear that the travel time from $\mathbf{r}_1$ to $\mathbf{r}_2$ is the same as the travel time from $\mathbf{r}_2$ to $\mathbf{r}_1$. However, the reciprocity theorem (18.64) states that this holds for the amplitude as well, and that the complete waveforms are identical.

This means that if I can hear you, you can hear me, regardless of how complicated the medium is, and how many many echoes it may produce. (Provided that our sense of hearing is equally good.) This also implies that if one submarine can detect another submarine with its sonar, then the second submarine can detect the first one as well, assuming that their sonar equipment is equally sensitive. This property holds no matter how complicated the distribution of the speed of sound in the ocean may be. The reciprocity theorem does not only hold for acoustic waves. Elastic waves satisfy a similar reciprocity theorem that incorporates the vector character of elastic waves (Aki and Richards, 2002; Fokkema and van den Berg, 1993), as do solutions of the heat equation (18.1) for any spatial distribution of the conductivity $\kappa(\mathbf{r})$.

19

Normal modes

Many physical systems can only oscillate at certain specific frequencies. As a child (and hopefully also as an adult), you observed that a swing in a playground moves only with a specific natural period, and that the force pushing the swing is only effective when the period of the force matches the period of the swing. The patterns of motion at which a system oscillates are called the *normal modes* of the system. A swing has one normal mode (Section 17.1), but you have seen in Section 12.5 that a simple model of a tri-atomic molecule has three normal modes. An example of the normal modes of a system is given in Figure 19.1, which shows the pattern of oscillation of a metal plate driven by an oscillator at six different frequencies. The screw in the middle of the plate shows the point at which the force on the plate is applied, and before any force is applied, sugar is evenly sprinkled on the plate. When an external force is applied at the frequency of one of the normal modes of the plate, oscillations of the plate result in a pattern of motion with nodal lines. These nodal lines define where the motion vanishes and as a result the sugar on the plate collects at these lines.

In this chapter, the normal modes of a variety of systems are analyzed. Normal modes play an important role in many applications, because the frequencies of normal modes provide important information about physical systems. Examples include the spectral lines of the light emitted by atoms, which have led to the advent of quantum mechanics and its description of the structure of atoms. In another example, the normal modes of the Earth provide information about the internal structure of our planet. In addition, normal modes are used in this chapter to introduce some properties of special functions, such as Bessel and Legendre functions. This is achieved by analyzing the normal modes of systems in one, two, and three dimensions in Sections 19.1–19.3.

19.1 Normal modes of a string

In this and the following two sections we consider the motion u of a system governed by the Helmholtz equation

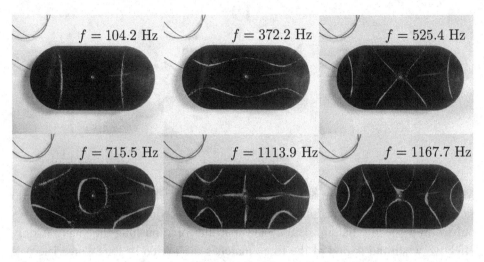

Figure 19.1   A metal plate oscillating at six frequencies that correspond to modes of the plate. For each mode, sugar collects in the nodal lines, where the motion vanishes.

$$\nabla^2 u + k^2 u = 0. \tag{19.1}$$

In this expression the wavenumber k is related to the angular frequency ω by

$$k = \frac{\omega}{c}. \tag{19.2}$$

For simplicity we assume the medium is homogeneous, which means that the velocity c and wavenumber k are constant at angular frequency ω. Note that the word *homogeneous* has more than one meaning. A *homogeneous medium* is a medium whose properties are constant in space. A *homogeneous differential equation* is a differential equation without a forcing term. Since normal modes are unforced oscillations, the corresponding differential equation is, by definition, homogeneous. The endpoints of this string of length $2R$ are fixed so that the boundary conditions are:

$$u(0) = u(2R) = 0. \tag{19.3}$$

In Sections 19.1–19.3 we consider a body of radius R. Since a circle or sphere of radius R has a diameter $2R$, we consider here a string of length $2R$, for consistency.

Problem a   Show that the solutions of (19.1) that satisfy the boundary conditions (19.3) are given by $\sin(k_n x)$ with wavenumber

$$k_n = \frac{n\pi}{2R}, \tag{19.4}$$

where n is an integer.

It is useful to normalize the modes $u_n(x)$, so that they satisfy the condition $\int_0^{2R} u_n^2(x)dx = 1$.

Problem b Show that the normalized modes are given by

$$u_n(x) = \frac{1}{\sqrt{R}} \sin(k_n x).$$ (19.5)

Problem c Sketch the modes for several values of n as a function of the distance along the string x.

Problem d The modes $u_n(x)$ are orthogonal, which means that the inner product $\int_0^{2R} u_n(x)u_m(x)dx$ vanishes when $n \neq m$. Show this by deriving that

$$\int_0^{2R} u_n(x)u_m(x)dx = \delta_{nm}.$$ (19.6)

We conclude from this section that the modes of a string are oscillatory functions with a wavenumber that can only have discrete well-defined values k_n. According to (19.2) this means that the string can only vibrate at discrete frequencies given by

$$\omega_n = \frac{n\pi c}{2R}.$$ (19.7)

This property may be familiar to you because you probably know that a guitar string vibrates only at very specific frequencies, which determines the pitch of the sound that you hear.[1] The results of this section imply that each string oscillates not only at one particular frequency, but at many discrete frequencies. The oscillation with the lowest frequency is given by (19.7) with $n = 1$: this is called the *fundamental mode* or *ground-tone*. This is what the ear perceives as the pitch of the tone. The oscillations corresponding to larger values of n are called the *higher modes* or *overtones*. The particular mix of overtones determines the timbre of the signal. If the higher modes are strongly excited, the ear perceives this sound as metallic, whereas just the fundamental mode is perceived as a smooth sound. The reader who is interested in the theory of musical instruments can consult Rossing et al. (1990).

Discrete modes are not a peculiarity of the string. Most systems of finite extent that involve waves support modes. For example, in Figure 14.1 the spectrum of the sound of a soprano saxophone is shown. This spectrum is characterized by well-defined peaks that correspond to the modes of the air-waves in the instrument. Mechanical systems in general have discrete modes; these modes can be destructive when they are excited at their resonance frequency. The matter waves in atoms

---

[1] It is not quite true that a guitar string always vibrates at fixed discrete frequencies. The more advanced players often bend a string sideways. This increases the tension in the string, and according to expression (19.91) the wave velocity changes accordingly. This in turn leads to variations in pitch that enrich the music.

are organized in modes as well, and this is ultimately the reason why atoms in an excited state emit light only at very specific frequencies, called spectral lines.

19.2 Normal modes of a drum

In the previous section we looked at the modes of a one-dimensional system. Here we derive the modes of a drum, which is a two-dimensional model. We consider a two-dimensional membrane that satisfies the Helmholtz equation (19.1). The membrane is circular and has radius R. At the edge $r = R$, the membrane is fixed to the rim of the drum and cannot move. This means that in cylindrical coordinates the boundary condition for the waves $u(r, \varphi)$ is given by:

$$u(R, \varphi) = 0. \tag{19.8}$$

To find the modes of the drum, we use *separation of variables*, which means that we seek solutions that can be written as a product of a function that depends only on r and a function that depends only on φ:

$$u(r, \varphi) = F(r)G(\varphi). \tag{19.9}$$

Problem a Insert this solution into the Helmholtz equation (19.1), use the expression of the Laplacian in cylindrical coordinates (10.35), and show that the resulting equation can be written as

$$\left[\frac{1}{F(r)} r \frac{\partial}{\partial r} \left(r \frac{\partial F}{\partial r} \right) + k^2 r^2 \right] = -\frac{1}{G(\varphi)} \frac{\partial^2 G}{\partial \varphi^2}. \tag{19.10}$$

The left-hand side of equation (19.10) depends only on the variable r, whereas the right-hand side depends only on the variable φ. These sides can only be equal for all values of r and φ if they depend neither on r nor on φ; that is, when they are a constant.

Problem b Use a constant value μ to show that $F(r)$ and $G(\varphi)$ satisfy the following differential equations:

$$\frac{d^2 F}{dr^2} + \frac{1}{r} \frac{dF}{dr} + \left(k^2 - \frac{\mu}{r^2} \right) F = 0, \tag{19.11}$$

$$\frac{d^2 G}{d\varphi^2} + \mu G = 0. \tag{19.12}$$

These differential equations need to be supplemented with boundary conditions. The boundary conditions for $F(r)$ follow from the requirements that displacement of the membrane – and hence this function – is finite everywhere and that it vanishes at the edge of the drum:

$$F(r) \text{ is finite everywhere, and } F(R) = 0. \tag{19.13}$$

The boundary condition for $G(\varphi)$ follows from the requirement that if we rotate the drum through $360°$, every point on the drum returns to its original position. In other words, the modes satisfy $u(r, \varphi) = u(r, \varphi + 2\pi)$. This implies that $G(\varphi)$ satisfies the *periodic boundary condition*:

$$G(\varphi) = G(\varphi + 2\pi). \tag{19.14}$$

Problem c   The general solution of (19.12) is given by

$$G(\varphi) = e^{\pm i\sqrt{\mu}\varphi}.$$

Show that the boundary condition (19.14) implies that $\mu = m^2$, where m is an integer.

This means that the dependence of the modes on the angle φ is given by:

$$G(\varphi) = e^{im\varphi}. \tag{19.15}$$

With $\mu = m^2$ in (19.11), the resulting equation bears a close resemblance to the Bessel equation:

$$\frac{d^2 J_m}{dx^2} + \frac{1}{x}\frac{d J_m}{dx} + \left(1 - \frac{m^2}{x^2}\right) J_m = 0. \tag{19.16}$$

This equation has two independent solutions: the Bessel function $J_m(x)$, which is finite everywhere, and the Neumann function $N_m(x)$, which is singular at $x = 0$. The Bessel functions $J_m(x)$ for several orders m are shown in Figure 19.2. All the Bessel functions are oscillatory functions, which decay with increasing values of the argument x. Note that there is a phase shift of a quarter cycle between the successive orders m. For small arguments, the Bessel functions $J_m(x) \propto x^m$ (see the right panel of Figure 19.2 and Arfken and Weber, 2005).

Problem d   Show that the general solution of (19.11) can be written as:

$$F(r) = A J_m(kr) + B N_m(kr), \tag{19.17}$$

with integration constants A and B. Hint: Change variables $s = kr$, and $F(r) = F(s/k) = \tilde{F}(s)$.

Problem e   Use the boundary conditions of $F(r)$ to show that $B = 0$ and that the wavenumber k must take a value such that $J_m(kR) = 0$.

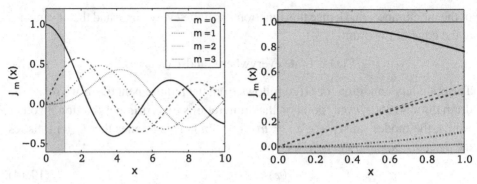

Figure 19.2   The Bessel functions $J_m(x)$ for orders $m = 0, 1, 2$, and 3. The right panel compares these functions to the approximation $J_m(x) \approx \left(\frac{x}{2}\right)^m / m!$ for $x < 1$.

This last condition for the wavenumber is analogous to the condition (19.4) for the one-dimensional string. For both the string and the drum the wavenumber can only take discrete values: these values are dictated by the condition that the displacement vanishes at the outer boundary of the string or the drum. For the drum there are for every value of the angular degree m infinitely many wavenumbers that satisfy the requirement $J_m(kR) = 0$. These wavenumbers are labeled with a subscript n, but since these wavenumbers are different for each value of the angular order m, the allowed wavenumbers carry two indices and are denoted here by $k_n^{(m)}$. They satisfy the condition

$$J_m(k_n^{(m)} R) = 0. \tag{19.18}$$

The zeroes (or roots) of the Bessel function $J_m(x)$ are not known in closed analytical form. However, numerical tables exists for the roots of Bessel functions, see for example, Table 9.4 of Abramowitz and Stegun (1965). Take a look at this reference, which contains a bewildering collection of formulas, graphs, and tables of mathematical functions. The lowest order zeroes of the Bessel functions from $J_0(x)$ to $J_5(x)$ are shown in Table 19.1.

Using the results in this section it follows that the modes of the drum are given by

$$u_{nm}(r, \varphi) = J_m(k_n^{(m)} r) e^{im\varphi}. \tag{19.19}$$

Problem f   Let us first consider the φ-dependence of these modes. Show that when one follows the mode $u_{nm}(r, \varphi)$ along a complete circle around the origin, one encounters exactly $|m|$ oscillations of that mode. We used the absolute value because m can be positive or negative.

Problem g   Find the five modes of the drum with the lowest frequencies and make a sketch of the associated standing waves of the drum. Use (19.18) and

Table 19.1   *The lowest roots of the Bessel function $J_m(x)$, these are the values of*
x for which $J_m(x) = 0$.

n	$m = 0$	$m = 1$	$m = 2$	$m = 3$	$m = 4$	$m = 5$
1	2.40482	3.83171	5.13562	6.38016	7.58834	8.77148
2	5.52007	7.01559	8.41724	9.76102	11.06471	12.33860
3	8.65372	10.17347	11.61984	13.01520	14.37254	15.70017
4	11.79153	13.32369	14.79595	16.22347	17.61597	18.98013
5	14.93091	16.47063	17.95982	19.4092	20.82693	22.21780
6	18.07106	19.61586	21.11700	22.58273	24.01902	25.43034
7	21.21163	22.76008	24.27011	25.74817	27.19909	28.62662

Table 19.1 to determine what the values of n and m are for these five modes. Figure 19.2 gives the dependence of the lowest-order Bessel functions on their argument.

Problem h   Use Table 19.1 to compute the separation between the different zero-crossings for a fixed value of m. To which number does this separation converge for the zero-crossings at large values of x?

The shape of the Bessel function is more difficult to see than the properties of the functions $e^{im\varphi}$. As shown in Section 9.7 of Butkov (1968), these functions satisfy a large number of properties, which include recursion relations and series expansions. However, at this point the following facts are most important:

- The Bessel functions $J_m(x)$ are oscillatory functions that decay with distance: in a sense they behave as decaying standing waves. We will return to this issue in Section 19.5.
- The Bessel functions satisfy an orthogonality relation similar to the orthogonality relation (19.6) for the modes of the string. This orthogonality relation is treated in more detail in Section 19.4.

19.3 Normal modes of a sphere

In this section we consider the normal modes of a spherical surface with radius R. We only consider the modes that are associated with the waves that propagate along the surface; hence, we do not consider wave motion in the *interior* of the sphere. The modes are assumed to satisfy the Helmholtz equation (19.1). Since the waves propagate on the spherical surface, they are only a function of the angles θ and φ that are used in spherical coordinates: $u = u(\theta, \varphi)$. Using the Laplacian expressed in spherical coordinates (10.36), The Helmholtz equation (19.1) is then given by

$$\frac{1}{R^2}\left[\frac{1}{\sin\theta}\frac{\partial}{\partial\theta}\left(\sin\theta\frac{\partial u}{\partial\theta}\right)+\frac{1}{\sin^2\theta}\frac{\partial^2 u}{\partial\varphi^2}\right]+k^2 u = 0. \tag{19.20}$$

Again, we seek a solution by applying separation of variables by writing the solution in a form similar to (19.9):

$$u(\theta,\varphi) = F(\theta)G(\varphi). \tag{19.21}$$

Problem a   Insert this into (19.20) and apply separation of variables to show that $F(\theta)$ satisfies the following differential equation:

$$\sin\theta\frac{d}{d\theta}\left(\sin\theta\frac{dF}{d\theta}\right)+\left(k^2 R^2 \sin^2\theta - \mu\right)F = 0, \tag{19.22}$$

and that $G(\varphi)$ satisfies (19.12), where the unknown constant μ does not depend on θ or φ.

Next, we apply the boundary conditions. Just as with the drum in Section 19.2, the system is invariant when a rotation of the angle φ over 2π is applied: $u(\theta,\varphi) = u(\theta,\varphi+2\pi)$. This means that $G(\varphi)$ satisfies the same differential equation (19.12) as for the case of the drum and the same periodic boundary condition (19.14). The solution is therefore given by $G(\varphi) = e^{im\varphi}$ and the separation constant satisfies $\mu = m^2$, with m an integer. Using this, the differential equation for $F(\theta)$ can be written as:

$$\frac{1}{\sin\theta}\frac{d}{d\theta}\left(\sin\theta\frac{dF}{d\theta}\right)+\left(k^2 R^2 - \frac{m^2}{\sin^2\theta}\right)F = 0. \tag{19.23}$$

Before we continue, let us compare this equation with (19.11) for the modes of the drum, which using the relation $\mu = m^2$ we can rewrite as

$$\frac{1}{r}\frac{d}{dr}\left(r\frac{dF}{dr}\right)+\left(k^2 - \frac{m^2}{r^2}\right)F = 0. \tag{19.24}$$

Note that these equations are identical when we compare r in (19.24) with $\sin\theta$ in (19.23). There is a good reason for this. For a source in the middle of a drum, the variable r measures the distance from a point on the drum to the source. In the case of waves on a spherical surface that are excited by a source at the north pole, $\sin\theta$ is a measure of the distance from a point to the source. The only difference is that $\sin\theta$ enters the equation rather than the true angular distance θ. This is a consequence of the fact that the surface is curved, and this curvature leaves an imprint on the differential equation that the modes satisfy.

Problem b   The differential equation (19.11) was reduced in Section 19.2 to the Bessel equation by changing to a new variable $x = kr$. Define a new variable

$$x = \cos\theta \tag{19.25}$$

and show that the differential equation (19.23) is given by

$$\frac{d}{dx}\left[(1-x^2)\frac{dF}{dx}\right] + \left(k^2R^2 - \frac{m^2}{1-x^2}\right)F = 0. \tag{19.26}$$

The solution of this differential equation is given by the associated Legendre functions $P_l^m(x)$. These functions are described in great detail in Section 9.8 of Butkov (1968). In fact, just like the Bessel equation, the differential equation (19.26) has a solution that is regular as well as a solution $Q_l^m(x)$ that is singular at the point $x = 1$ where $\theta = 0$. However, since the modes are finite everywhere, they are represented by the regular solutions $P_l^m(x)$.

The wavenumber k is related to frequency by the relation $k = \omega/c$. At this point it is not clear what k is; hence, the eigenfrequencies of the spherical surface are not yet known. It is shown in Section 9.8 of Butkov (1968) that:

- The associated Legendre functions are only finite when the wavenumber satisfies

$$k^2R^2 = l(l+1), \tag{19.27}$$

where l is a positive integer. Using this in (19.26) implies that the associated Legendre functions satisfy the following differential equation:

$$\frac{1}{\sin\theta}\frac{d}{d\theta}\left[\sin\theta\frac{dP_l^m(\cos\theta)}{d\theta}\right] + \left[l(l+1) - \frac{m^2}{\sin^2\theta}\right]P_l^m(\cos\theta) = 0. \tag{19.28}$$

Seen as a function of $x = \cos\theta$, this is equivalent to

$$\frac{d}{dx}\left[(1-x^2)\frac{dP_l^m(x)}{dx}\right] + \left[l(l+1) - \frac{m^2}{1-x^2}\right]P_l^m(x) = 0.$$

- The integer l must be larger than or equal to the absolute value of the angular order m.

Problem c Show that the last condition can also be written as:

$$-l \le m \le l. \tag{19.29}$$

Problem d Derive that the eigenfrequencies of the modes are given by

$$\omega_l = \sqrt{l(l+1)}\frac{c}{R}. \tag{19.30}$$

It is interesting to compare this result with the eigenfrequencies (19.7) of the string. The eigenfrequencies of the string all have the same spacing in frequency, but the eigenfrequencies of the spherical surface are not spaced at the same interval. In musical jargon one would say that the overtones of a string are harmonious: this means that the eigenfrequencies of the overtones are multiples of the eigenfrequency of the ground tone. In contrast, the overtones of a spherical surface are not harmonious.

Problem e Show that for large values of l the eigenfrequencies of the spherical surface have an almost equal spacing.

Problem f The eigenfrequency ω_l depends on the order l, but not on the degree m. For each value of l, the angular degree m can according to (19.29) take the values $-l, -l+1, \ldots, l-1, l$. Show that this implies that for every value of l, there are $(2l+1)$ modes with the same eigenfrequency.

When different modes have the same eigenfrequency, one speaks of *degenerate modes*.

The Legendre functions $P_l^0(x)$ for several degrees l are shown in Figure 19.3 as a function of the variable x. The value $x = 1$ corresponds to $\cos\theta = 1$ or $\theta = 0$: this is the north pole of the spherical coordinate system. The value $x = -1$ corresponds to the south pole of the spherical coordinate system. The number of oscillations of these functions increases with the degree l. All Legendre functions $P_l^0(x)$ have the same value at the north pole: $P_l^0(x = 1) = 1$.

The results we obtained imply that the modes on a spherical surface are given by $P_l^m(\cos\theta)e^{im\varphi}$. We used here that the variable x is related to the angle θ through

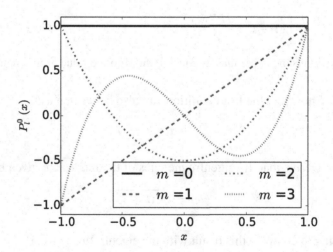

Figure 19.3   The Legendre polynomials $P_l^0(x)$ for $l = 0, 1, 2,$ and 3.

Table 19.2   *The lowest-order spherical harmonics $Y_{lm}(\theta, \varphi)$.*

	$m = -1$	$m = 0$	$m = 1$
$l = 0$		$\dfrac{1}{\sqrt{4\pi}}$	
$l = 1$	$\sqrt{\dfrac{3}{8\pi}} \sin\theta \, e^{-i\varphi}$	$\sqrt{\dfrac{3}{4\pi}} \cos\theta$	$-\sqrt{\dfrac{3}{8\pi}} \sin\theta \, e^{i\varphi}$

(19.25). The modes of the spherical surface are called *spherical harmonics*. These eigenfunctions for $m \geq 0$ are given by:

$$Y_{lm}(\theta, \varphi) = (-1)^m \sqrt{\frac{2l+1}{4\pi} \frac{(l-m)!}{(l+m)!}} P_l^m(\cos\theta)e^{im\varphi} \qquad m \geq 0. \qquad (19.31)$$

For $m < 0$ the spherical harmonics are defined by the relation

$$Y_{lm}(\theta, \varphi) = (-1)^m \, Y_{l,-m}(\theta, \varphi). \qquad (19.32)$$

The lowest-order spherical harmonics are shown in Table 19.2. Note that $Y_{00}(\theta, \varphi) = 1/\sqrt{4\pi}$ does not depend on the angles; hence, it is just a constant.

You may wonder where the square root in front of the associated Legendre function comes from. One can show that with this numerical factor the spherical harmonics are normalized when integrated over the sphere:

$$\iint |Y_{lm}|^2 \, d\Omega = 1, \qquad (19.33)$$

where $\iint \cdots d\Omega$ denotes an integration over the unit sphere. You should be aware different authors use different definitions of the spherical harmonics. For example, one could also define the spherical harmonics as $\tilde{Y}_{lm}(\theta, \varphi) = P_l^m(\cos\theta)e^{im\varphi}$ because the functions also account for the normal modes of a spherical surface.

Problem g   Show that the modes defined in this way satisfy $\iint \left|\tilde{Y}_{lm}\right|^2 d\Omega = 4\pi/(2l+1) \times (l+m)!/(l-m)!$

This means that the modes defined in this way are not normalized when integrated over the sphere. There is no reason why one cannot work with this convention, as long as one accounts for the fact that in this definition the modes are not normalized. Throughout this book we will use the definition (19.31) for the spherical harmonics. In doing so we follow the normalization that is used by Edmonds (1996).

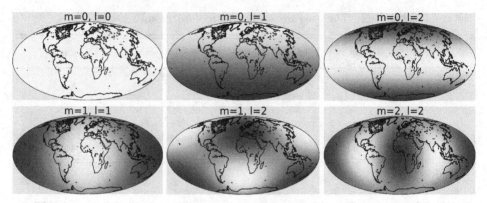

Figure 19.4   The real part of the lowest-order spherical harmonics projected on the Earth in the Mollweide projection. The value of the order l and degree m is shown above each panel. Negative amplitudes are light and positive amplitudes dark.

The real parts of the lowest-order spherical harmonics are shown in Figure 19.4, in which these functions are projected on the Earth's surface. Only those spherical harmonics with $m \geq 0$ are shown, the spherical harmonics for $m < 0$ follow from (19.32). Just as with the Bessel functions, the associated Legendre functions satisfy recursion relations and a large number of other properties that are described in detail in Section 9.8 of Butkov (1968). The most important properties of the spherical harmonics $Y_{lm}(\theta, \varphi)$ are:

- These functions display m oscillations when the angle φ increases by 2π. In other words, there are m oscillations along one circle of constant latitude.
- The associated Legendre functions $P_l^m(\cos\theta)$ behave like Bessel functions in the sense that they behave like standing waves with an amplitude that decays from the pole. We return to this issue in Section 19.6.
- The number of oscillations between the north pole of the sphere and the south pole of the sphere increases with the difference $(l - m)$.
- The spherical harmonics are orthogonal for a suitably chosen inner product; this orthogonality relation is derived in Section 19.4.
- The spherical harmonics are the eigenfunctions of the Laplacian on the sphere.

Problem h   Give a proof of this last property by showing that

$$\nabla_1^2 Y_{lm}(\theta, \varphi) = -l(l+1) Y_{lm}(\theta, \varphi), \tag{19.34}$$

where the Laplacian on the unit sphere is given by

$$\nabla_1^2 = \frac{1}{\sin\theta} \frac{\partial}{\partial\theta} \left(\sin\theta \frac{\partial}{\partial\theta} \right) + \frac{1}{\sin^2\theta} \frac{\partial^2}{\partial\varphi^2}. \tag{19.35}$$

This property is extremely useful in many applications, because the action of the Laplacian on a sphere can be replaced by the much simpler multiplication by the constant $-l\,(l+1)$ when spherical harmonics are concerned. We use this, for instance, in Chapter 20 to determine the gravity field outside the Earth.

19.4 Normal modes and orthogonality relations

The normal modes of a physical system often satisfy orthogonality relations when a suitably chosen inner product for the eigenfunctions is used. In this section this is illustrated by studying once again the normal modes of the Helmholtz equation (19.1) for different geometries. In this section we derive first the general orthogonality relation for these normal modes. This is then applied to the normal modes of the previous sections to derive the orthogonality relations for Bessel functions and associated Legendre functions.

Let us consider two normal modes of the Helmholtz equation (19.1), and let these modes be called u_p and u_q. At this point we leave it open whether the modes are defined on a line, a surface, or in a volume. The integration over the region of space in which the modes are defined is denoted as $\int \cdots d^N x$, where N is the dimension of this space. The wavenumbers of these modes, which are effectively the corresponding eigenvalues of the Helmholtz equation, are defined by k_p and k_q, respectively. In other words, the modes satisfy the equations:

$$\nabla^2 u_p + k_p^2 u_p = 0, \tag{19.36}$$
$$\nabla^2 u_q + k_q^2 u_q = 0. \tag{19.37}$$

The subscript p may stand for a single mode index such as in the index n for the wavenumber k_n for the modes of a string, or it may stand for a number of indices such as the indices n and m that label the eigenfunctions (19.19) of a circular drum or the indices l and m of the spherical harmonics (19.31) on a sphere.

Problem a Multiply (19.36) by u_q^*, take the complex conjugate of (19.37), and multiply the result by u_p. Subtract the resulting equations and integrate this over the region of space for which the modes are defined to show that

$$\int \left(u_q^* \nabla^2 u_p - u_p \nabla^2 u_q^* \right) d^N x + \left(k_p^2 - k_q^{*2} \right) \int u_q^* u_p d^N x = 0. \tag{19.38}$$

Problem b Use the theorem of Gauss (8.1) to derive that

$$\int u_q^* \nabla^2 u_p d^N x = \oint u_q^* \nabla u_p \cdot d\mathbf{S} - \int \left(\nabla u_q^* \cdot \nabla u_p \right) d^N x, \tag{19.39}$$

316   *A Guided Tour of Mathematical Methods for the Physical Sciences*

where the integral $\oint \cdots d\mathbf{S}$ is over the surface that bounds the body. If you have trouble deriving this, you can consult (8.10) where a similar result was used for the derivation of the representation theorem for acoustic waves.

Problem c   Use the last result to show that expression (19.38) can be written as

$$\oint \left(u_q^*\nabla u_p - u_p\nabla u_q^*\right) \cdot d\mathbf{S} + \left(k_p^2 - k_q^{*2}\right) \int u_q^* u_p d^N x = 0. \qquad (19.40)$$

Problem d   The first term is an integral over the boundary of the body. The second term contains a volume integral and this term leads to the orthogonality relation of the modes. Let us assume that on this boundary the modes satisfy one of the three boundary conditions: (i) $u = 0$, (ii) $\hat{\mathbf{n}} \cdot \nabla u = 0$ (where $\hat{\mathbf{n}}$ is the unit vector perpendicular to the surface), or (iii) $\hat{\mathbf{n}} \cdot \nabla u = \alpha u$ (where α is a constant). Show that for all of these boundary conditions the surface integral in (19.40) vanishes.

The last result implies that when the modes satisfy one of these boundary conditions

$$\left(k_p^2 - k_q^{*2}\right) \int u_q^* u_p d^N x = 0. \qquad (19.41)$$

Let us first consider the case in which the modes are equal, that is, in which $p = q$. In that case the integral reduces to $\int |u_p|^2 d^N x$, which is guaranteed to be positive. Equation (19.41) then implies that $k_p^2 = k_p^{*2}$, so that the wavenumbers k_p must be real: $k_p = k_p^*$. For this reason the complex conjugate of the wavenumbers can be dropped and (19.41) can be written as:

$$\left(k_p^2 - k_q^2\right) \int u_q^* u_p d^N x = 0. \qquad (19.42)$$

Now consider the case of two different modes for which the wavenumbers k_p and k_q are different. In that case the term $\left(k_p^2 - k_q^2\right)$ is nonzero; hence, in order to satisfy (19.42) the modes must satisfy

$$\int u_q^* u_p d^N x = 0 \qquad \text{for} \quad k_p \neq k_q. \qquad (19.43)$$

This finally gives the *orthogonality relation* of the modes in the sense that it states that the modes are orthogonal for the following inner product: $f \cdot g \equiv \int f^* g \, d^N x$. Note that the inner product for which the modes are orthogonal follows from the Helmholtz equation (19.1), which defines the modes.

Let us now consider this orthogonality relation for the modes of the string, the drum, and the spherical surface of the previous sections. For the string, the orthogonality relation was derived in Problem d of Section 19.1 and you can see that equation (19.6) is identical to the general orthogonality relation (19.43). For the circular drum the modes are given by (19.19).

Problem e Use (19.19) for the modes of the circular drum to show that the orthogonality relation (19.43) for this case can be written as:

$$\int_0^R \int_0^{2\pi} J_{m_1}(k_{n_1}^{(m_1)}r) J_{m_2}(k_{n_2}^{(m_2)}r) e^{i(m_1-m_2)\varphi} d\varphi r dr = 0, \qquad (19.44)$$

for $k_{n_1}^{(m_1)} \neq k_{n_2}^{(m_2)}$. Explain where the factor r comes from in the integration.

Problem f This integral can be separated into an integral over φ and an integral over r. The φ-integral is given by $\int_0^{2\pi} e^{i(m_1-m_2)\varphi} d\varphi$. Show that this integral vanishes when $m_1 \neq m_2$:

$$\int_0^{2\pi} e^{i(m_1-m_2)\varphi} d\varphi = 0 \quad \text{for} \quad m_1 \neq m_2. \qquad (19.45)$$

Note that you obtained this relation earlier in (16.9) in the derivation of the residue theorem.

Expression (19.45) implies that the modes $u_{n_1 m_1}(r, \varphi)$ and $u_{n_2 m_2}(r, \varphi)$ are orthogonal when $m_1 \neq m_2$ because the φ-integral in (19.44) vanishes when $m_1 \neq m_2$. Let us now consider why the different modes of the drum are orthogonal when m_1 and m_2 are equal to the same integer m. In that case (19.44) implies that

$$\int_0^R J_m(k_{n_1}^{(m)}r) J_m(k_{n_2}^{(m)}r) \, r \, dr = 0 \quad \text{for} \quad n_1 \neq n_2. \qquad (19.46)$$

Problem g Find in your favorite book on mathematical physics an alternative derivation of the orthogonality relation (19.46) of Bessel functions of the same degree m.

In the previous we used that $k_{n_1}^{(m)} \neq k_{n_2}^{(m)}$ when $n_1 \neq n_2$. This integral defines an orthogonality relation for Bessel functions. In this case the Bessel functions in this relation are of the same degree m but the wavenumbers in the argument of the Bessel functions differ. The presence of the term r in the integral (19.46) is due to the cylindrical symmetry of the drum. Note the resemblance between this expression and the orthogonality relation (19.6) of the modes of the string, which can be written as

$$\int_0^{2R} \sin(k_n x) \, \sin(k_m x) \, dx = 0 \quad \text{for} \quad n \neq m. \qquad (19.47)$$

Finally, note that the modes $u_{n_1 m_1}(r, \varphi)$ and $u_{n_2 m_2}(r, \varphi)$ are orthogonal when $m_1 \neq m_2$ because the φ-integral satisfies (19.45), whereas the modes are orthogonal when $n_1 \neq n_2$ but with the same order m because the r-integral (19.46) vanishes in that case. This implies that the eigenfunctions of the drum defined in (19.19) satisfy the following orthogonality relation:

$$\int_0^R \int_0^{2\pi} u_{n_1 m_1}^*(r, \varphi) u_{n_2 m_2}(r, \varphi) \, d\varphi \, r dr = C \delta_{n_1 n_2} \delta_{m_1 m_2}, \qquad (19.48)$$

where δ_{ij} is the Kronecker delta and C is a constant that depends on n_1 and m_1.

A similar analysis can be applied to the spherical harmonics $Y_{lm}(\theta, \varphi)$, which are the eigenfunctions of the Helmholtz equation on a spherical surface. You may wonder in that case what the boundary conditions of these eigenfunctions are because in the step from (19.40) to (19.41) the boundary conditions of the modes have been used. A closed surface, however, has no boundary. This means that the surface integral in (19.40) vanishes. This in turn means that the orthogonality relation (19.43) holds despite the fact that the spherical harmonics do not satisfy one of the boundary conditions that were used in Problem d.

Let us now consider the inner product of two spherical harmonics on the sphere: $\iint Y_{l_1 m_1}^*(\theta, \varphi) Y_{l_2 m_2}(\theta, \varphi) d\Omega$.

Problem h Show that the φ-integral in the integration over the sphere is of the form $\int_0^{2\pi} e^{i(m_2 - m_1)\varphi} d\varphi$ and that this integral is equal to $2\pi \delta_{m_1 m_2}$.

This implies that the spherical harmonics are orthogonal when $m_1 \neq m_2$ because of the φ-integration. We will now continue with the case in which $m_1 = m_2$, and denote this common value with the single index m.

Problem i Use the general orthogonality relation (19.43) to derive that the associated Legendre functions satisfy the following orthogonality relation:

$$\int_0^\pi P_{l_1}^m(\cos \theta) P_{l_2}^m(\cos \theta) \sin \theta d\theta = 0 \quad \text{for} \quad l_1 \neq l_2. \qquad (19.49)$$

Note the common value of the degree m in the two associated Legendre functions. Also show explicitly that the condition $k_{l_1} \neq k_{l_2}$ is equivalent to the condition $l_1 \neq l_2$.

Problem j Use a substitution of variables to show that this orthogonality relation can also be written as

$$\int_{-1}^1 P_{l_1}^m(x) P_{l_2}^m(x) dx = 0 \quad \text{for} \quad l_1 \neq l_2. \qquad (19.50)$$

The result you obtained in Problem h implies that the spherical harmonics are orthogonal when $m_1 \neq m_2$ because of the φ-integration, whereas Problem i implies that the spherical harmonics are orthogonal when $l_1 \neq l_2$ because of the θ-integration. This means that the spherical harmonics satisfy the following orthogonality relation:

$$\iint Y_{l_1 m_1}^*(\theta, \varphi) Y_{l_2 m_2}(\theta, \varphi) \, d\Omega = \delta_{l_1 l_2} \delta_{m_1 m_2}. \tag{19.51}$$

The numerical constant multiplying the delta functions is equal to 1. This is a consequence of the square-root term in (19.31) that scales the associated Legendre functions. Be aware of the fact that when a different convention is used for the normalization of the spherical harmonics, a normalization factor appears on the right-hand side of the orthogonality relation (19.51) of the spherical harmonics.

19.5 Bessel functions behave as decaying cosines

As we have seen in Section 19.2, the modes of the circular drum are given by $J_m(kr)e^{im\varphi}$, where the Bessel function satisfies the differential equation (19.16) and where k is a wavenumber chosen in such a way that the displacement at the edge of the drum vanishes. We show in this section that the waves that propagate through the drum have an approximately constant wavelength, but that their amplitude decays with the distance to the center of the drum. The starting point of the analysis is the Bessel equation

$$\frac{d^2 J_m}{dx^2} + \frac{1}{x} \frac{d J_m}{dx} + \left(1 - \frac{m^2}{x^2}\right) J_m = 0. \tag{19.16}$$

If the terms $(1/x) d J_m/dx$ and m^2/x^2 were absent, this equation would reduce to the differential equation $d^2 F/dx^2 + F = 0$, whose solutions are given by a superposition of $\cos x$ and $\sin x$. Since $1/x$ and $1/x^2$ vanish as $x \to \infty$ we can expect the Bessel functions to display an oscillatory behavior when x, is large.

It follows directly from (19.16) that the term m^2/x^2 is relatively small for large values of x, specifically when $x \gg m$. However, it is not obvious under which conditions the term $(1/x) d J_m/dx$ is relatively small. Fortunately this term can be transformed away.

Problem a   Write $J_m(x) = x^\alpha g_m(x)$, insert this in the Bessel equation (19.16), and show that the term with the first derivative vanishes when $\alpha = -1/2$ and that the resulting differential equation for $g_m(x)$ is given by

$$\frac{d^2 g_m}{dx^2} + \left(1 - \frac{m^2 - 1/4}{x^2}\right) g_m = 0. \tag{19.52}$$

Up to this point we have made no approximations. Although we have transformed the first derivative term out of the Bessel equation, we still cannot solve (19.52). However, when $x \gg m$, the term proportional to $1/x^2$ in this expression is relatively small. This means that for large values of x the function $g_m(x)$ satisfies the approximate differential equation $d^2 g_m/dx^2 + g_m \approx 0$.

Problem b Show that the solution of this equation is given by $g_m(x) \approx A \cos(x + \varphi)$, where A and φ are constants. Also show that this implies that the Bessel function is approximately given by:

$$ J_m(x) \approx A \frac{\cos(x + \varphi)}{\sqrt{x}}. \tag{19.53} $$

This approximation is obtained from a local analysis of the Bessel equation. Since all values of the constants A and φ lead to a solution that approximately satisfies the differential equation (19.52), it is not possible to retrieve the precise values of these constants from the analysis in this section. The constant A follows from the employed normalization of Bessel functions. An analysis based on the asymptotic evaluation of the integral representation of the Bessel function as presented in Section 24.5 or by Bender and Orszag (1999) shows that:

$$ J_m(x) = \sqrt{\frac{2}{\pi x}} \cos\left[x - (2m+1)\frac{\pi}{4}\right] + \mathcal{O}(x^{-3/2}). \tag{19.54} $$

Problem c As a check on the accuracy of this asymptotic expression, let us compare the zeroes of this approximation with the zeroes of the Bessel functions as given in Table 19.1. In Problem h of Section 19.2 you found that the separation of the zero-crossings tends to π for large values of x. Explain this using the approximate expression (19.54). How large must x be for the different values of the degree m so that the difference in the spacing of the zero-crossings with π is less than 0.01?

The asymptotic expression (19.54) is compared in Figure 19.5 with the Bessel functions $J_m(x)$ for the degrees $m = 0$ and $m = 2$, respectively. For large values of x, the approximation (19.54) is very good. Note that for $m = 0$ the approximation is better for smaller values of x than it is for $m = 2$. This is related to the fact that the approximation (19.54) is valid under the condition $(m^2 - 1/4)/x^2 \ll 1$. This requirement is satisfied when $x \gg m$; hence, the approximation (19.54) is for a given value of x better for the smaller degrees m than for larger degrees m.

Physically, (19.54) states that Bessel functions behave like standing waves with a constant wavelength, which decay with distance as $1/\sqrt{kr}$.[2] How can we explain

---

[2] Here it is used that the modes are given by the Bessel functions with argument $x = kr$.

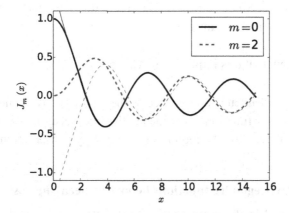

Figure 19.5   The Bessel functions $J_0(x)$ and $J_2(x)$ and their approximations from (19.54) in thinner lines. Note that the approximation for smaller values of x is better for $m = 0$ than for $m = 2$.

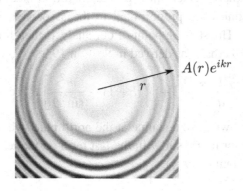

Figure 19.6   An expanding wavefront with radius r on a flat surface. This could be the model for a resonance in a shaken cup of coffee, or a drum struck in the center.

this decay of the amplitude with distance? First, let us note that (19.54) expresses the Bessel function in a cosine; hence, this is a representation of the Bessel function as a standing wave. However, using the relation $\cos kr = \left(e^{ikr} + e^{-ikr}\right)/2$, the Bessel function can be written as two traveling waves that depend on the distance r as $e^{\pm ikr}/\sqrt{kr}$. These waves interfere to give the standing wave pattern of the Bessel function. Now let us consider a propagating wave $A(r)e^{ikr}$ in two dimensions; in this expression $A(r)$ is an amplitude that is at this point unknown. The energy of the wave varies with the square of the wave field, and thus depends on $|A(r)|^2$. The energy current therefore also varies as $|A(r)|^2$. Consider an outgoing wave as shown in Figure 19.6. The total energy flux through a ring of radius r is given by the energy current times the circumference of the ring, which means that the flux is equal to $2\pi r \, |A(r)|^2$. Since energy is conserved, this total energy flux is the same for all values of r, which means that $2\pi r \, |A(r)|^2 = \text{constant}$.

Problem d Show that this implies that $A(r) \propto 1/\sqrt{r}$.

This is the same dependence on distance as the $1/\sqrt{x}$ decay of the approximation (19.54) of the Bessel function. This means that the decay of the Bessel function with distance is dictated by the requirement of energy conservation.

19.6 Legendre functions behave as decaying cosines

The technique used in the previous section for the approximation of the Bessel function can also be applied to spherical harmonics. We show in this section that the spherical harmonics behave asymptotically as standing waves on a sphere with an amplitude decay that is determined by the condition that energy is conserved. The spherical harmonics are proportional to the associated Legendre functions with argument $\cos \theta$. The starting point of our analysis therefore is the differential equation for $P_l^m(\cos \theta)$ that was derived in Section 19.3:

$$\frac{1}{\sin \theta} \frac{d}{d\theta} \left[\sin \theta \frac{d P_l^m(\cos \theta)}{d\theta} \right] + \left[l \, (l+1) - \frac{m^2}{\sin^2 \theta} \right] P_l^m(\cos \theta) = 0. \quad (19.28)$$

Let us assume that we have a source at the north pole, where $\theta = 0$. Far away from the source, the term $m^2/\sin^2 \theta$ in the last term on the left-hand side is much smaller than the constant $l \, (l+1)$.

Problem a Show that the words "far away from the source" stand for the requirement

$$\sin \theta \gg \frac{|m|}{\sqrt{l \, (l+1)}}, \quad (19.55)$$

and show that this implies that the approximation that we derive breaks down near the north pole as well as near the south pole of the employed system of spherical coordinates. In addition, the asymptotic expressions that we derive are most accurate for large values of the angular order l and small values of the degree m.

Problem b Just as in the previous section we can transform the first derivative in the differential equation (19.28) away: here this can be achieved by writing $P_l^m(\cos \theta) = (\sin \theta)^\alpha \, g_l^m(\theta)$. Insert this substitution into the differential equation (19.28) and show that the first derivative $dg_l^m/d\theta$ disappears when $\alpha = -1/2$ and that the resulting differential equation for $g_l^m(\theta)$ is given by:

$$\frac{d^2 g_l^m}{d\theta^2} + \left[\left(l + \frac{1}{2} \right)^2 - \frac{m^2 - 1/4}{\sin^2 \theta} \right] g_l^m(\theta) = 0. \qquad (19.56)$$

It is interesting to note the resemblance of this equation to the corresponding expression (19.52) in the analysis of the Bessel function.

Problem c If the term $(m^2 - 1/4)/\sin^2 \theta$ were absent, this equation would be easy to solve. Show that this term is small compared to the constant $(l + \frac{1}{2})^2$ when the requirement (19.55) is satisfied.

Problem d Show that under this condition the associated Legendre functions satisfy the following approximation:

$$P_l^m(\cos \theta) \approx A \, \frac{\cos \left[\left(l + \frac{1}{2} \right) \theta + \gamma \right]}{\sqrt{\sin \theta}}, \qquad (19.57)$$

where A and γ are constants.

Just as in the previous section the constants A and γ cannot be obtained from this analysis because (19.57) satisfies the approximate differential equation for *any* value of these constants. As shown in (2.5.58) of Edmonds (1996), the asymptotic relation of the associated Legendre functions is given by:

$$P_l^m(\cos \theta) \approx (-l)^m \sqrt{\frac{2}{\pi l \sin \theta}} \cos \left[\left(l + \frac{1}{2} \right) \theta - (2m + 1) \frac{\pi}{4} \right] + \mathcal{O}(l^{-3/2}).$$

$$(19.58)$$

Just like the Bessel functions, the spherical harmonics behave like a standing wave given by a cosine that is multiplied by a factor $1/\sqrt{\sin \theta}$, which modulates the amplitude.

Problem e Use the same reasoning as in Problem d of Section 19.5 to explain that this amplitude decrease follows from the requirement of energy conservation. In doing so you may find Figure 19.7 helpful.

Problem f Deduce from (19.58) that the wavelength of the associated Legendre functions measured in radians is given by $2\pi/(l + \frac{1}{2})$.

This last result can be used to find the number of oscillations in the spherical harmonics when one moves around the globe once. For simplicity we consider here the case of a spherical harmonic $Y_l^0(\theta, \varphi)$ for degree $m = 0$. When one goes from the north pole to the south pole, the angle θ increases from 0 to π. The number of oscillations that fit in this interval is given by $\pi/wavelength$, and according to Problem f this number is equal to $\pi/[2\pi/(l+\frac{1}{2})] = (l+\frac{1}{2})/2$. This is the number of

Figure 19.7   An expanding wavefront on a spherical surface at an angle θ from the source. These wavefronts could describe the propagation of infrasound in the atmosphere, generated by the impact of a meteorite on the north pole.

wavelengths that fit on half the globe. When one returns from the south pole to the north pole, one encounters another $(l + \frac{1}{2})/2$ oscillations. This means that the total number of waves that fit around the globe is given by $(l + \frac{1}{2})$. It may surprise you that the number of oscillations in one loop around the globe is not an integer. One would expect that the requirement of constructive interference would dictate that an integer number of wavelengths should "fit" in this interval. The reason why the total number of oscillations is $(l + \frac{1}{2})$ rather than the integer l is that near the north and south poles the asymptotic approximation (19.58) breaks down; this follows from the requirement (19.55).

The fact that $(l + \frac{1}{2})$ rather than l oscillations fit on the sphere has a profound effect in quantum mechanics. In the first attempts to explain the line spectra of light emitted by atoms, Bohr postulated that an integer number of waves has to fit on a sphere: this can be expressed as $\oint k\,ds = 2\pi n$, where k is the local wavenumber. This condition could not explain the observed spectra of light emitted by atoms. However, the arguments in this section imply that the number of wavelengths that fit on a sphere should be given by the requirement

$$\oint k\,ds = 2\pi \left(n + \frac{1}{2}\right).$$ (19.59)

This is the *Bohr–Sommerfeld quantization rule*, and was the earliest result in quantum mechanics that provided an explanation of the line spectra of light emitted by atoms. More details on this issue and the cause of the factor $\frac{1}{2}$ in the quantization rule are given by Tabor (1989) and by Brack and Bhaduri (1997). The effect of this factor on the Earth's normal modes is discussed in a pictorial way by Dahlen and Henson (1985).

The asymptotic expression (19.58) can give a useful insight into the relation between modes and traveling waves on a sphere. Let us first return to the modes on the string, which according to (19.5) are given by $\sin(k_n x)$. For simplicity, we will leave out normalization constants in the arguments. The wave motion associated with this mode is given by the real part of $\sin(k_n x)e^{-i\omega_n t}$, with $\omega_n = k_n/c$. These modes therefore denote a standing wave. However, using the decomposition $\sin(k_n x) = (e^{ik_n x} - e^{-ik_n x})/2i$, the mode can in the time domain also be seen as a superposition of two waves $e^{i(k_n x - \omega_n t)}$ and $e^{-i(k_n x + \omega_n t)}$. These are two traveling waves that move in opposite directions.

Problem g Suppose we excite a string at the left-hand side at $x = 0$. We know we can account for the motion of the string as a superposition of standing waves $\sin(k_n x)$. However, we can consider these modes to exist also as a superposition of waves $e^{\pm i k_n x}$ that move in opposite directions. The wave $e^{ik_n x}$ moves away from the source at $x = 0$. However, the wave $e^{-ik_n x}$ moves *toward* the source at $x = 0$. Give a physical explanation of why in the string traveling waves also move *toward* the source.

On a sphere the situation is analogous. The modes are according to (19.58) proportional to standing waves on the sphere that are proportional to

$$\cos\left[\left(l + \frac{1}{2}\right)\theta - (2m + 1)\frac{\pi}{4}\right]/\sqrt{\sin\theta}.$$

Using the relation $\cos x = \left(e^{ix} + e^{-ix}\right)/2$, the modes can also be seen as a superposition of traveling waves $e^{i\left(l+\frac{1}{2}\right)\theta}/\sqrt{\sin\theta}$ and $e^{-i\left(l+\frac{1}{2}\right)\theta}/\sqrt{\sin\theta}$ on the sphere.

Problem h Explain why the first wave travels away from the north pole, while the second wave travels toward the north pole.

Problem i Suppose that the waves are excited by a source at the north pole. According to the last problem the motion of the sphere can alternatively be seen as a superposition of standing waves or of traveling waves. The traveling wave $e^{i\left(l+\frac{1}{2}\right)\theta}/\sqrt{\sin\theta}$ moves away from the source. Explain physically why there is also a traveling wave $e^{-i\left(l+\frac{1}{2}\right)\theta}/\sqrt{\sin\theta}$ moving *toward* the source.

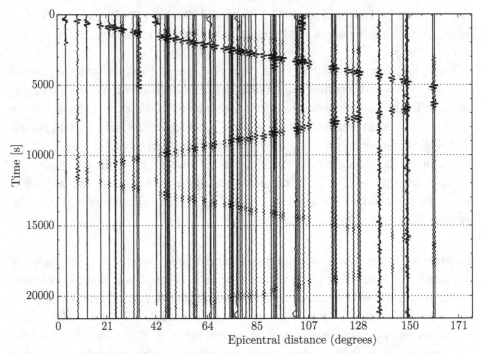

Figure 19.8   Seismograms from stations in the global seismic network IU after the Great Tohoku Earthquake of March 11, 2011. The zig-zagging pattern represents surface waves circling the Earth. Station ANMO, used in the next section, is at an epicentral distance of 83 degrees.

These results imply that the oscillations of the Earth instigated by large earthquakes can be seen as either a superposition of normal modes, or a superposition of waves that travel along the Earth's surface in opposite directions. The seismograms of Figure 19.8 display seismic waves traveling along the Earth's surface (i.e., surface waves), after the Great Tohoku Earthquake (Magnitude 9) on March 11, 2011. The epicenter of the earthquake (epicentral distance equals 0 degrees) was off the east coast of Honsu, Japan. Figure 19.8 shows the superposition of traveling waves that move away from the earthquake, and waves that – after circling the Earth – propagate back to the epicenter. The waves that move away from the earthquake are proportional to $\exp\left(+i\left(l+\frac{1}{2}\right)\theta\right)$ while the waves propagating back to the earthquake are proportional to $\exp\left(-i\left(l+\frac{1}{2}\right)\theta\right)$. The zig-zagging pattern of surface waves in the figure is evidence of these waves going around the world multiple times in the six hours displayed. The relation between normal modes and surface waves is treated in more detail by Dahlen (1979) and Snieder and Nolet (1987), and we also examine this dichotomy further in the next section.

19.7 Normal modes and the Green's function

In Section 12.5 we analyzed the normal modes of a system of three coupled masses. This system had three normal modes, and each mode could be characterized by a vector $\hat{\mathbf{v}}^{(n)}$ with the displacement of the three masses and by an eigenfrequency ω_n. The response of the system to a force $\mathbf{F}$ acting on the three masses with time dependence $e^{-i\omega t}$ was derived to be:

$$\mathbf{x} = \frac{1}{m} \sum_{n=1}^{3} \frac{\hat{\mathbf{v}}^{(n)}(\hat{\mathbf{v}}^{(n)} \cdot \mathbf{F})}{(\omega_n^2 - \omega^2)}. \tag{12.61}$$

This means that the Green's function of this system is given by the following dyad:

$$\mathbf{G} = \frac{1}{m} \sum_{n=1}^{3} \frac{\hat{\mathbf{v}}^{(n)}\hat{\mathbf{v}}^{(n)T}}{(\omega_n^2 - \omega^2)}. \tag{19.60}$$

In this section we derive the Green's function for a general oscillating system that can be continuous. An important example is the Earth, which is a body that has well-defined normal modes and where the displacement is a continuous function of the space coordinates.

We consider a system that satisfies the following equation of motion:

$$\rho \ddot{u} + Hu = F. \tag{19.61}$$

The field u can be either a scalar field or a vector field. The operator H at this point is general, the only requirement that we impose is that this operator is Hermitian, which means that we require that

$$(f \cdot Hg) = (Hf \cdot g), \tag{19.62}$$

where the inner product is defined as the familiar $(f \cdot g) \equiv \int f^* g \ dV$. In the frequency domain, the equation of motion is given by

$$-\rho \omega^2 u + Hu = F(\omega). \tag{19.63}$$

Let the normal modes of the system be denoted by $u^{(n)}$; the normal modes describe the oscillations of the system in the absence of any external force. The normal modes therefore satisfy the following expression

$$Hu^{(n)} = \rho \omega_n^2 u^{(n)}, \tag{19.64}$$

where ω_n is the eigenfrequency of this mode.

Problem a Take the inner product of this expression with a mode $u^{(m)}$, and use the fact that H is Hermitian to derive that

$$\left(\omega_n^2 - \omega_m^{*2}\right)\left(u^{(m)} \cdot \rho u^{(n)}\right) = 0. \tag{19.65}$$

Note the resemblance of this expression to (19.41) for the modes of a system that obeys the Helmholtz equation.

Problem b   Just like in Section 19.4 one can show that the eigenfrequencies are real by setting $m = n$, and one can derive that different modes are orthogonal with respect to the following inner product:

$$\left(u^{(m)} \cdot \rho u^{(n)}\right) = 0 \qquad \text{for} \quad \omega_m \neq \omega_n. \tag{19.66}$$

Use (19.65) to give a proof of this orthogonality relation.

When $\omega_n = \omega_m$, the right-hand side is nonzero with a certain constant value. The modes can always be normalized so that this constant is equal to 1; hence,

$$\left(u^{(m)} \cdot \rho u^{(n)}\right) = \delta_{nm} \qquad \text{for} \quad \omega_m \neq \omega_n. \tag{19.67}$$

Note the presence of the density term ρ in this inner product. In the analysis of this section it was crucial for the operator H to be Hermitian. This property of H has two important implications: (1) the eigenvalues of H are real and (2) the eigenfunctions are orthogonal (Strang, 2003). Therefore, it is crucial to establish whether the operator H is Hermitian or not. In general, the operator of a dynamical system is Hermitian when the system is invariant for time reversal. This means that the equations are invariant when one lets the clock run backward, or mathematically when one replaces t by $-t$. In general, dissipation breaks the symmetry for time reversal. It is shown in detail by Dahlen and Tromp (1998) that attenuation in the Earth makes the eigenfrequencies of the Earth complex and that the normal modes of an attenuating Earth do not satisfy the orthogonality relation (19.67).

Let us now return to the problem (19.63) where an external force $F(\omega)$ is present. Assuming that the normal modes form a complete set, the response to this force can be written as a sum of normal modes:

$$u = \sum_n c_n u^{(n)}, \tag{19.68}$$

where the c_n are unknown coefficients.

Problem c   Find these coefficients by inserting (19.68) into the equation of motion (19.63) and by taking the inner product of the result with a mode $u^{(m)}$ to derive that

$$c_m = \frac{\left(u^{(m)} \cdot F\right)}{\omega_m^2 - \omega^2}. \tag{19.69}$$

This means that the response of the system can be written as:

$$u = \sum_n \frac{u^{(n)} \left(u^{(n)} \cdot F \right)}{\omega_n^2 - \omega^2}. \tag{19.70}$$

Note the resemblance of this expression to (12.61) for a system of three masses. The main difference is that the derivation in this section is also valid for continuous vibrating systems such as the Earth.

It is instructive to rewrite this expression taking the dependence of the space coordinates explicitly into account:

$$u(\mathbf{r}) = \sum_n \frac{u^{(n)}(\mathbf{r}) \int u^{*(n)}(\mathbf{r'}) F(\mathbf{r'}) \, dV'}{\omega_n^2 - \omega^2}. \tag{19.71}$$

It follows from this expression that the Green's function is given by

$$G(\mathbf{r}, \mathbf{r'}, \omega) = \sum_n \frac{u^{(n)}(\mathbf{r}) u^{*(n)}(\mathbf{r'})}{\omega_n^2 - \omega^2}. \tag{19.72}$$

When the mode is a vector, one should take the transpose of the mode $u^{*(n)}(\mathbf{r'})$. Note the similarity between this expression for the Green's function of a continuous medium and the Green's function (19.60) for a discrete system. In this sense, the Earth behaves in the same way as a tri-atomic molecule. For both systems, the dyadic representation of the Green's function provides a compact way to account for the response of the system to external forces.

Note that the response is strongest when the frequency ω of the external force is close to one of the eigenfrequencies ω_n of the system. This implies for example for the Earth that modes with a frequency close to the frequency of the external forcing are most strongly excited. If we jump up and down with a frequency of 1 Hz, we excite the Earth's fundamental mode with a period of about 1 hour only very weakly. In addition, a mode is most effectively excited when the inner product of the forcing $F(\mathbf{r'})$ in (19.71) is maximal. This means that a mode is most strongly excited when the spatial distribution of the force equals the displacement $u^{(n)}(\mathbf{r'})$ of the mode.

Problem d Show that a mode is not excited when the force acts only at one of the nodal lines of that mode.

A strong earthquake – such as the March 11, 2011, Tohoku earthquake – will excite normal modes of the Earth. Figure 19.9 shows the power spectrum of the vertical component of the ground motion after the earthquake. This power spectrum is computed using a Fourier transform of the data at station ANMO in Albuquerque,

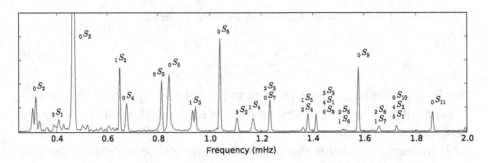

Figure 19.9   Amplitude spectrum of the vertical component of the ground motion at a seismic station ANMO, New Mexico (USA), after the Great Tohoku Earthquake of March 11, 2011. The numbers $_n S_l$ denote the different spheroidal normal modes of the Earth.

New Mexico (USA), depicted as one of the waveforms in Figure 19.9. The frequency is given in units of millihertz (mHz). The peaks in the amplitude spectrum correspond to the normal modes of the Earth. These peaks are described by the function $1/\left(\omega_n^2 - \omega^2\right)$ in (19.72). In reality the detailed structure of these resonances is also affected by the attenuation in the Earth and by a number of factors that perturb the Earth's normal modes. Each of the peaks in the power spectrum is associated with known modes of the Earth. These spheroidal modes are denoted by the capital letter S, and defined as combinations of angular derivatives of the spherical harmonic functions $Y_l^m(\theta, \phi)$ (Stein and Wysession, 2003).

As a next step we consider the Green's function in the time domain. This function follows by applying the Fourier transform (14.42) to the Green's function (19.72).

Problem e   Show that this gives:

$$G(\mathbf{r}, \mathbf{r}', t) = \frac{1}{2\pi} \sum_n u^{(n)}(\mathbf{r}) u^{*(n)}(\mathbf{r}') \int_{-\infty}^{\infty} \frac{e^{-i\omega t}}{\omega_n^2 - \omega^2} d\omega. \qquad (19.73)$$

The integrand is singular at the frequencies $\omega = \pm\omega_n$ of the normal modes. These singularities are located on the integration path, as shown in the left-hand panel of Figure 19.10. Near the singularity at $\omega = \omega_n$ the integrand behaves as $1/(\omega_n^2 - \omega^2) = 1/(\omega_n + \omega)(\omega_n - \omega) \to 1/[2\omega_n(\omega_n - \omega)]$ as $\omega \to \omega_n$. The contribution of these singularities is poorly defined because the integral $\int 1/(\omega - \omega_n)\, d\omega$ is not defined.

This situation is comparable to the treatment in Section 16.4 of the response of a particle in syrup to an external forcing. When this particle was subjected to a damping β, the integrand in the Fourier transform to the time domain had a singularity in

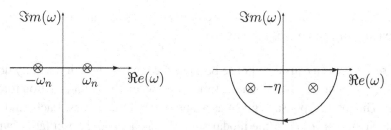

Figure 19.10 The location of the poles and the integration path in the complex ω-plane. In the left-hand panel poles on the real axis are located on the integration path at $\pm\omega_n$. The right-hand panel shows the poles after a slight anelastic damping is introduced.

the lower half-plane. This gave a causal response; as shown in (16.34) the response was different from zero only at times *later* than the time at which the forcing was applied. This suggests that we can obtain a well-defined causal response of the Green's function (19.73) when we introduce a slight damping. This damping breaks the invariance of the problem for time reversal, and is responsible for a causal response. At the end of the calculation we can let the damping parameter go to zero. Damping can be introduced by giving the eigenfrequencies of the normal modes a small negative imaginary component: $\pm\omega_n \to \pm\omega_n - i\eta$, where η is a small positive number.

Problem f The time dependence of the oscillation of a normal mode is given by $e^{-i\omega_n t}$. Show that with this replacement the modes decay with a decay time that is given by

$$\tau = 1/\eta. \qquad (19.74)$$

This last property means that when we ultimately set $\eta = 0$, the decay time becomes infinite: in other words, the modes are not attenuated in that limit.

With the replacement $\pm\omega_n \to \pm\omega_n - i\eta$ the poles that are associated with the normal modes are located in the lower ω-plane: this situation is shown in Figure 19.10. Now that the singularities are moved from the integration path on the real axis, the theory of complex integration can be used to evaluate the resulting integral.

Problem g Use contour integration as treated in Chapter 16 to derive that the Green's function is given in the time domain by:

$$G(\mathbf{r}, \mathbf{r}', t) = \begin{cases} 0 & \text{for } t < 0 \\ \sum_n \dfrac{u^{(n)}(\mathbf{r})u^{*(n)}(\mathbf{r}')}{\omega_n} \sin\omega_n t & \text{for } t > 0. \end{cases} \qquad (19.75)$$

Hint: Use the same steps as in the derivation of the function (16.34) and let the damping parameter η go to zero after integration.

This result gives a causal response because the Green's function is only nonzero at times $t > 0$, which is later than the time $t = 0$ when the delta-function forcing is nonzero. The total response is given as a sum over all the modes. Each mode leads to a time signal $\sin(\omega_n t)$ in the modal sum: this is a periodic oscillation with the frequency ω_n of the mode. The singularities in the integrand of the Green's function (19.73) at the pole positions $\omega = \pm\omega_n$ are thus associated in the time domain with a harmonic oscillation with angular frequency ω_n. Note that the Green's function is continuous at the time $t = 0$ of excitation. It is interesting to compare this Green's function with the Green's function (17.21) for the girl on the swing. The only differences are that in (19.75) modes are present in the Green's function and that a summation over the modes is carried out. This difference is due to the fact that the harmonic oscillator has only one mode and that this mode of oscillation does not have a spatial extent.

Problem h Use the Green's function (19.75) to derive that the response of the system to a force $F(\mathbf{r}, t)$ is given by:

$$u(\mathbf{r}, t) = \sum_n \frac{u^{(n)}(\mathbf{r})}{\omega_n} \int \int_{-\infty}^{t} u^{*(n)}(\mathbf{r}') \sin \omega_n \left(t - t'\right) F(\mathbf{r}', t')dt'dV'.$$

$$(19.76)$$

Justify the integration limit in the t'-integration.

The results in this section imply that the total Green's function of a system is known once the normal modes are known. The total response can then be obtained by summing the contribution of each normal mode to the total response. This technique is called *normal-mode summation*, and is often used to obtain the low-frequency response of the Earth to an excitation (Dahlen and Tromp, 1998; Dziewonski and Woodhouse, 1983). However, in the seismological literature one usually treats a source signal that is given by a step function at $t = 0$ rather than a delta function because this is a more accurate description of the slip on a fault during an earthquake (Aki and Richards, 2002). This leads to a time dependence $(1 - \cos(\omega_n t))$ rather than the time dependence $\sin(\omega_n t)$ in the response (19.75) to a delta-function excitation.

19.8 Guided waves in a low-velocity channel

In this section we treat a system that strictly speaking does not have normal modes, but that can support solutions that behave like traveling waves in one direction and

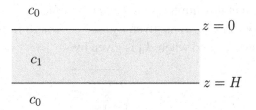

Figure 19.11   A single layer of thickness H and velocity c_1 in between two homogeneous half-spaces with velocity c_0.

as modes in another direction. The waves in such a system propagate as *guided waves*. A beautiful example of guided waves in the ocean is the sound propagation of Baleen whales communicating over long distances in the SOFAR channel (Payne and Webb, 1971).

Consider a system in two dimensions (x and z), where the velocity depends only on the z-coordinate. We assume that the wave field in the frequency domain satisfies the Helmholtz equation (19.1):

$$\nabla^2 u + \frac{\omega^2}{c^2(z)} u = 0. \tag{19.77}$$

In our model a layer with velocity c_1 extends from $z = 0$ to $z = H$ and is embedded in a medium with a constant velocity c_0. The geometry of the problem is shown in Figure 19.11. Since the system is invariant in the x-direction, the problem can be simplified by a Fourier transform over the x-coordinate as defined in (14.27):

$$u(x, z) = \int_{-\infty}^{\infty} U(k, z) e^{ikx} dk. \tag{19.78}$$

Problem a   Use the previous two expressions to show that $U(k, z)$ satisfies the following ordinary differential equation:

$$\frac{d^2 U}{dz^2} + \left[\frac{\omega^2}{c(z)^2} - k^2 \right] U = 0. \tag{19.79}$$

It is important to note at this point that the frequency ω is a fixed constant, and that according to (19.78) the variable k is an integration variable that assumes all values in the integration (19.78). For this reason one should not at this point use the relation $k = \omega/c(z)$.

Now consider the special case of the model shown in Figure 19.11. We require that the waves outside the layer move away from the layer.

Problem b   Show that this implies that the solution for $z < 0$ is given by Ae^{-ik_0z} and the solution for $z > H$ is given by Be^{+ik_0z} where A and B are unknown integration constants and where k_0 is given by

$$k_0 = \sqrt{\frac{\omega^2}{c_0^2} - k^2}. \tag{19.80}$$

Problem c   Show that within the layer the wave field is given by $C\cos(k_1z) + D\sin(k_1z)$ with C and D integration constants and k_1 given by

$$k_1 = \sqrt{\frac{\omega^2}{c_1^2} - k^2}. \tag{19.81}$$

The solution in the three regions of space therefore takes the following form:

$$U(k, z) = \begin{cases} Ae^{-ik_0z} & \text{for } z < 0, \\ C\cos(k_1z) + D\sin(k_1z) & \text{for } 0 < z < H, \\ Be^{+ik_0z} & \text{for } z > H. \end{cases} \tag{19.82}$$

We now have the general form of the solution within the layer and the two half-spaces on either side of the layer. Boundary conditions are needed to find the integration constants A, B, C, and D. For this system both U and dU/dz are continuous at $z = 0$ and $z = H$.

Problem d   Use the results of Problem b and Problem c to show that these requirements impose the following constraints on the integration constants:

$$\left. \begin{array}{l} A - C = 0, \\ ik_0A + k_1D = 0, \\ -Be^{ik_0H} + C\cos k_1H + D\sin k_1H = 0, \\ ik_0Be^{ik_0H} + k_1C\sin k_1H - k_1D\cos k_1H = 0. \end{array} \right\} \tag{19.83}$$

This is a linear system of four equations for the four unknowns A, B, C, and D. Note that this is a homogeneous system of equations, because the right-hand sides vanish. Such a homogeneous system of equations only has nonzero solutions when the determinant of the system of equations vanishes (Strang, 2003).

Problem e   Show that this requirement leads to the following condition:

$$\tan(k_1H) = \frac{-2ik_0k_1}{k_1^2 + k_0^2}. \tag{19.84}$$

This equation – called a dispersion relation[3] – is implicitly an equation for the wavenumber k, because according to (19.80) and (19.81) both k_0 and k_1 are functions of the wavenumber k. Equation (19.84) implies that the system can only support waves when the wavenumber k is such that expression (19.84) is satisfied. The system, strictly speaking, does not have normal modes, because the waves propagate indefinitely in the x-direction. However, in the z-direction the waves only "fit" in the layer for very specific values of the wavenumber k. These waves are called "guided waves" because they propagate along the layer with a well-defined phase velocity that follows from the relation $c(\omega) = \omega/k$. Be careful not to confuse this phase velocity $c(\omega)$ with the velocities c_1 and c_0 in the layer and the half-spaces outside the layer. At this point we do not know yet what the phase velocities of the guided waves are.

The phase velocity of the guided wave follows from expression (19.84) because this expression is implicitly an equation for the wavenumber k. At this point we consider the case of a low-velocity layer, that is, we assume that $c_1 < c_0$. In this case $1/c_0 < 1/c_1$. We look for guided waves with a wavenumber in the following interval: $\omega/c_0 < k < \omega/c_1$.

Problem f Show that in that case k_1 is real and k_0 is purely imaginary.

Problem g Show that the solution decays exponentially away from the low-velocity channel both in the half-space $z < 0$ and the half-space $z > H$.

The fact that the waves decay exponentially with the distance to the low-velocity layer means that the guided waves are trapped near the low-velocity layer. Waves that decay exponentially are called *evanescent waves*.

Inserting equations (19.80) and (19.81) into expression (19.84) gives an equation that contains ω and k. For a fixed value of ω this expression constitutes a constraint on the wavenumber k of the guided waves. Unfortunately, it is not possible to solve this equation for k in closed form. Such an equation is called a *transcendental equation* that must be solved numerically. For each solution $k(\omega)$ the corresponding phase velocity $c = \omega/k(\omega)$ depends, in general, on the frequency ω. This means that these guided waves are *dispersive*, which means that the different frequency components travel with a different phase velocity.

Dispersive waves occur in many different situations. When electromagnetic waves propagate between plates or in a layered structure, guided waves result (Jackson, 1998). The atmosphere, and most importantly the ionosphere, is an excellent waveguide for electromagnetic waves (Guglielmi and Pokhotelov, 1996). This leads to a large variety of electromagnetic guided waves in the upper atmosphere

---

[3] The topic of dispersion is covered in Section 24.4.

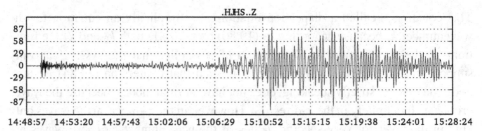

Figure 19.12   Vertical component of the ground motion at Hillside Junior High
School in Boise, Idaho, after an earthquake in Costa Rica. This station is part of a
school network of TC1 seismometers described in Section 9.4. The time is given
in Coordinated Universal Time (in French "Temps Universel Coordonné," UTC),
and the amplitude is in arbitrary units.

with exotic names such as "pearls," "whistlers," "tweaks," "hydromagnetic howl-
ing," and "serpentine emissions"; colorful names associated with the sounds these
phenomena would make if they were audible, or with the patterns they generate
in frequency–time diagrams. These perturbations are excited for example by the
electromagnetic fields generated by lightning. Guided waves play a crucial role
in telecommunication, because light propagates through optical fibers as guided
waves (Lauterborn and Kurz, 2003). The fact that these waves are guided prohibits
the light from propagating out of the fiber, and this allows for the transmission of
light signals over extremely large distances. The topic of dispersion is treated in
more detail in Section 24.4.

   In the Earth the wave velocity increases rapidly with depth. Elastic waves can be
guided near the Earth's surface and the different modes are called "Rayleigh waves"
and "Love waves" (Aki and Richards, 2002). These surface waves in the Earth are
a prime tool for mapping the shear velocity within the Earth (Snieder, 1996).

   Since surface waves in the Earth are trapped near the Earth's surface, they effectively propagate in two dimensions rather than in three dimensions. Surface waves
therefore suffer less from geometrical spreading than body waves that propagate
through the interior of the Earth. For this reason, it is the surface waves that do most
damage after an earthquake. This is illustrated in Figure 19.12, which shows the
vertical displacement at a seismic station at Hillside Junior High in Boise, Idaho,
after an earthquake in Costa Rica. Around 14:50 the primary wave arrives: this is a
compressional wave that travels through the interior of the Earth. The waves with
the largest amplitude between 15:10 and 15:30 are waves guided along the Earth's
surface.

19.9 Leaky modes

The guided waves in the previous section decay exponentially with the distance to
the low-velocity layer. Intuitively, the fact that the waves are confined to a region

near a low-velocity layer can be understood using Snell's Law (11.32). Waves are refracted from regions of high velocity to a region of low velocity. This means that the waves that stray out of the low-velocity channel are refracted back into the channel. Effectively this traps the waves in the vicinity of the channel. This explanation suggests that for a high-velocity channel the waves are refracted *away* from the channel. The resulting wave pattern then corresponds to waves that preferentially move away from the high-velocity layer. For this reason we consider in this section the waves that propagate through the system shown in Figure 19.11 but now we consider the case of a high-velocity layer where $c_1 > c_0$. In this case, $1/c_1 < 1/c_0$, and we consider waves with a wavenumber that is confined to the following interval: $\omega/c_1 < k < \omega/c_0$.[4]

Problem a   Show that in this case the wavenumber k_1 in (19.81) is imaginary and that it can be written as $k_1 = i\kappa_1$, with

$$\kappa_1 = \sqrt{k^2 - \frac{\omega^2}{c_1^2}},\qquad(19.85)$$

and show that the dispersion relation (19.84) is given by:

$$\tan(i\kappa_1 H) = \frac{-2k_0\kappa_1}{\kappa_1^2 - k_0^2}.\qquad(19.86)$$

Problem b   Use the relation $\cos x = \left(e^{ix} + e^{-ix}\right)/2$ and the related expression for $\sin x$ to rewrite the dispersion relation (19.86) in the following form:

$$i\tanh(\kappa_1 H) = \frac{-2k_0\kappa_1}{\kappa_1^2 - k_0^2}.\qquad(19.87)$$

For real values of k, real values of κ_1 and k_0 result in a real right-hand side of equation (19.87). For real values of k, $\tanh(\kappa_1 H)$ is real, as well. However, the factor i makes the left-hand side imaginary, and thus expression (19.87) cannot be satisfied for real values of k. The only reason for this expression to hold therefore is that k must be complex.

We next explore the consequences of the wavenumber k being complex. Suppose that the dispersion relation is satisfied for a complex wavenumber $k = k_r + ik_i$, with k_r and k_i the real and imaginary parts. In the time domain a solution behaves for a fixed frequency as $U(k, z)e^{i(kx-\omega t)}$. This means that for complex values of the wavenumber the solution behaves as $U(k, z)e^{-k_i x}e^{i(k_r x-\omega t)}$. This is a wave that propagates in the x-direction with phase velocity $c = \omega/k_r$ and that decays exponentially with the propagation distance x.

---

[4] We shall discover shortly that in this case k is complex; therefore, it would be more appropriate to use this inequality for the real part of k.

The exponential decay of the wave with the propagation distance x is due to the fact that the wave energy refracts out of the high-velocity layer. A different way of understanding this exponential decay is to consider the character of the wave field outside the layer.

Problem c   Show that in the two half-spaces outside the high-velocity layer the waves propagate away from the layer. Hint: Analyze the wavenumber k_0 in the half-spaces and consider the corresponding solution in these regions.

This means that wave energy is continuously radiated away from the high-velocity layer. The exponential decay of the mode with propagation distance x is thus due to the fact that wave energy continuously leaks out of the layer. For this reason one speaks of *leaky modes* (Watson, 1972). In the Earth a well-observed leaky mode is the S-PL wave. This is a mode in which a transverse propagating wave in the mantle is coupled to a wave that is trapped in the Earth's crust.

In general there is no simple way to find the complex wavenumber k for which the dispersion relation (19.87) is satisfied. However, the presence of leaky modes can be seen in Figure 19.13 where the absolute value of the following function is shown in the complex k-plane:

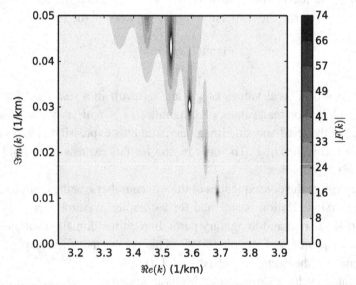

Figure 19.13   Contour plot of the function $|F(k)|$ for a high-velocity layer with velocity $c_1 = 8.4$ km/s and a thickness $H = 15$ km, which is embedded between two half-spaces with velocity $c_0 = 8$ km/s, for waves with a frequency of 5 Hz.

$$F(k) \equiv \left(i \tanh(\kappa_1 H) + \frac{2k_0\kappa_1}{\kappa_1^2 - k_0^2} \right)^{-1}. \qquad (19.88)$$

Problem d Show that this function is infinite for the k-values that correspond to a leaky mode.

The function $F(k)$ in Figure 19.13 is computed for a high-velocity layer with a thickness of 15 km and a velocity of 8.4 km/s that is embedded between two half-spaces with a velocity of 8 km/s. The frequency of the wave is 5 Hz. The leaky modes show up in Figure 19.13 as singularities of the function $F(k)$. The x-dependence of these modes is given by $\exp(ikx) = \exp(-k_ix)\exp(ik_rx)$.

Problem e Measure the imaginary part of the wavenumber of the mode with the lowest velocity from Figure 19.13 and use this to determine the distance over which the amplitude of that mode decays with a factor $1/e$.

Leaky modes have been used by Gubbins and Snieder (1991) to analyze waves that have propagated along a subduction zone.[5] By a fortuitous geometry, compressional waves excited by earthquakes in the Tonga-Kermadec region propagate through the Tonga-Kermadec subduction zone to a seismic station in Wellington, New Zealand. At this station (SNZO), a high-frequency wave arrives before the main compressional wave. This can be seen in Figure 19.14 in which such a seismogram is shown band-pass filtered at different frequencies. Clearly, the waves with a frequency around 6 Hz arrive before waves with a frequency around 1 Hz. This observation can be explained by the propagation of a leaky mode through the subduction zone. The physical reason why the high-frequency components arrive before the lower-frequency components is that the high-frequency waves "fit" in the high-velocity layer in the subducting plate, whereas the lower-frequency components do not fit in the high-velocity layer and are more influenced by the slower material outside the high-velocity layer. Of course, the energy leaks out of the high-velocity layer so that this arrival is very weak. From the data it could be inferred that in the subduction zone a high-velocity layer with a thickness between 6 and 10 km is present (Gubbins and Snieder, 1991).

19.10 Radiation damping

Up to this point we have considered systems that are freely oscillating. When such systems are of finite extent, such a system displays undamped free oscillations. In the previous section leaky modes were introduced. In such a system, energy is

[5] A subduction zone is a tectonic plate in the Earth that slides downward in the mantle.

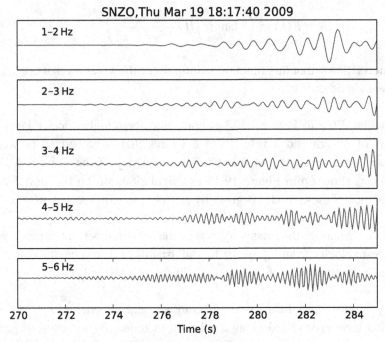

SNZO,Thu Mar 19 18:17:40 2009

Figure 19.14   Seismic waves recorded in Wellington (station SNZO) after an earthquake in the Tonga-Kermadec subduction zone, organized in narrow frequency bands. Note how the waves with a higher frequency appear to arrive faster in Wellington than the ones with a lower frequency.

radiated away, which leads to an exponential decay of waves that propagate through the system. In a similar way, a system that has normal modes when it is isolated from its surroundings can display damped oscillations when it is coupled to the external world.

As a simple prototype of such a system, consider a mass m that can move in the z-direction, which is coupled to a spring with spring constant κ. The mass is attached to a wire that is under a tension T and has a mass ρ per unit length. The system is shown in Figure 19.15. The total force acting on the mass is the sum of the force $-\kappa z$ exerted by the spring and the force F_w that is generated by the wire:

$$m\ddot{z} + \kappa z = F_w, \tag{19.89}$$

where z denotes the vertical displacement of the mass. The motion of the waves that propagate in the wire is given by the one-dimensional version of the wave equation in (6.42):

$$u_{xx} - \frac{1}{c^2}u_{tt} = 0, \tag{19.90}$$

Figure 19.15 Geometry of an oscillating mass that is coupled to a spring and a wire.

where u is the displacement of the wire in the vertical direction and c is given by (Butkov, 1968):

$$c = \sqrt{\frac{T}{\rho}}.$$ (19.91)

Let us first consider the case in which no external force is present and the mass is not coupled to the wire.

Problem a Show that in that case the equation of motion is given by

$$\ddot{z} + \omega_0^2 z = 0,$$ (19.92)

with ω_0 given by

$$\omega_0 = \sqrt{\frac{\kappa}{m}}.$$ (19.93)

The mass that is not coupled to the wire has one free oscillation with angular frequency ω_0. The fact that the system has one free oscillation is a consequence of the fact that this mass can move in the vertical direction only; hence, it has only one degree of freedom.

Before we couple the mass to the wire, let us first analyze the wave motion in the wire in the absence of the mass.

Problem b Show that any function $f(t - x/c)$ satisfies the wave equation (19.90). Show that this function describes a wave that moves in the positive x-direction with velocity c.

Problem c Show that any function $g(t + x/c)$ satisfies the wave equation (19.90) as well. Show that this function describes a wave that moves in the negative x-direction with velocity c.

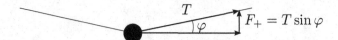

Figure 19.16   Sketch of the force exerted by the wire on the mass.

The general solution is a superposition of the rightward and leftward moving waves:

$$u(x, t) = f\left(t - \frac{x}{c}\right) + g\left(t + \frac{x}{c}\right). \tag{19.94}$$

This general solution is called the *d'Alembert solution*.

Now we want to describe the motion of the coupled system. Let us first assume that the mass oscillates with a prescribed displacement $z(t)$ and find the waves that this displacement generates in the wire. We consider here the situation in which there are no waves moving toward the mass. This means that to the right of the mass the waves can only move rightward and to the left of the mass there are only leftward moving waves.

Problem d   Show that this radiation condition implies that the waves in the wire are given by:

$$u(x, t) = \begin{cases} f\left(t - \dfrac{x}{c}\right) & \text{for } x > 0, \\[2ex] g\left(t + \dfrac{x}{c}\right) & \text{for } x < 0. \end{cases} \tag{19.95}$$

Problem e   At $x = 0$ the displacement of the mass is the same as the displacement of the wire. Show that this implies that $f(t) = g(t) = z(t)$, so that

$$u(x, t) = \begin{cases} z\left(t - \dfrac{x}{c}\right) & \text{for } x > 0, \\[2ex] z\left(t + \dfrac{x}{c}\right) & \text{for } x < 0. \end{cases} \tag{19.96}$$

Now we have solved the problem of finding the wave motion in the wire given the motion of the mass. To complete our description of the system, we also need to specify how the wave motion of the wire affects the mass. In other words, we need to find the force F_w in (19.89) given the motion of the wire. This force can be derived from Figure 19.16. The vertical component F_+ of the force acting on the mass from the right-hand side of the wire is given by $F_+ = T \sin \varphi$, where T is the tension in the wire. When the motion in the wire is sufficiently weak, we can approximate: $F_+ = T \sin \varphi \approx T \varphi \approx T \tan \varphi \approx T u_x(x = 0^+, t)$. In the last identity we used that the derivative $u_x(x = 0^+, t)$ gives the slope of the wire on the right of the point $x = 0$.

Problem f Use a similar reasoning to determine the force acting on the mass from the left-hand part of the wire and show that the net force exerted by the wire on the mass is given by

$$F_w(t) = T\left(u_x(x = 0^+, t) - u_x(x = 0^-, t)\right), \qquad (19.97)$$

where $u_x(x = 0^-, t)$ is the x-derivative of the displacement in the wire just to the left of the mass.

Problem g Show that this expression implies that the net force that acts on the mass is proportional to the *kink* in the spring at the location of the mass.

You may not feel comfortable that we used the approximation of a small angle φ in the derivation of (19.97). However, keep in mind that the wave equation (19.90) for a wire is derived using the same approximation and that this wave equation therefore is only valid for small displacements of the wire (see Section 8.1 of Butkov, 1968). Now we have assembled all the ingredients for solving the coupled problem.

Problem h Use (19.96) and (19.97) to derive that the force exerted by the spring on the mass is given by

$$F_w(t) = -\frac{2T}{c}\,\dot{z}, \qquad (19.98)$$

and that the motion of the mass is therefore given by:

$$\ddot{z} + \frac{2T}{mc}\dot{z} + \omega_0^2 z = 0. \qquad (19.99)$$

It is interesting to compare this expression for the motion of the mass that is coupled to the wire with the equation of motion (19.92) for the mass that is not coupled to the wire. The wire leads to a term $(2T/mc)\,\dot{z}$ in the equation of motion that damps the motion of the mass. How can we explain this damping physically? When the mass moves, the wire moves with it at location $x = 0$. Any motion in the wire at that point excites waves propagating in the wire. This means that the wire radiates wave energy away from the mass whenever the mass moves. Since energy is conserved, the energy that is radiated in the wire must be withdrawn from the energy of the moving mass. This means that the mass loses energy whenever it moves; this effect is described by the damping term in equation (19.99). This damping process is called *radiation damping*, because it is the radiation of waves that damps the motion of the mass.

The system described in this section is extremely simple. However, it does contain the essential physics of radiation damping. Many systems in physics that display normal modes are not quite isolated from their surroundings. Interactions of the systems with their surroundings often lead to the radiation of energy, and hence to a damping of the oscillation of the system. One example of such a system is an atom in an excited state. In the absence of external influences such an atom will not emit any light and will not decay. However, when such an atom can interact with electromagnetic fields, it can emit a photon and subsequently decay.

A second example is a charged particle that moves in a synchrotron. In the absence of external fields, such a particle will continue forever in a circular orbit without any change in its speed. In reality, a charged particle is coupled to electromagnetic fields. This has the effect that a charged particle that is accelerated emits electromagnetic radiation, called synchrotron radiation (Jackson, 1998). The radiated energy corresponds to an energy loss of the particle, so that the particle slows down. This is actually the reason why accelerators such as those used at CERN and Fermilab are so large. The acceleration of a particle in a circular orbit with radius r at the given velocity v is given by v^2/r. This means that for a fixed velocity v the larger is the radius of the orbit, the smaller is the acceleration, and the weaker is the energy loss due to the emission of synchrotron radiation. This is why one needs huge machines to accelerate tiny particles to an extreme energy.

The normal modes of the earth-ocean-atmosphere system are also coupled. The normal modes of the Earth at periods of about 240 s, also called *Earth's hum*, are excited by internal gravity waves in the ocean (Rhie and Romanowicz, 2004, 2006). These internal gravity waves are, in their turn, excited by wind-driven waves at the surface of the ocean. The coupling of different waves is ubiquitous and sometimes exotic; wave motion in the Earth has been shown to be excited by the shock wave generated by the space shuttle while on approach to Edwards Air Force base (Kanamori et al., 1991, 1992). Conversely, atmospheric normal modes are excited by volcanic eruptions (Kanamori and Harkrider, 1994) and earthquakes (Garcia et al., 2013).

Problem i The modes in the plate of Figure 19.1 are also damped because of radiation damping. What form of radiation is emitted by this oscillating plate?

20

Potential field theory

Potential fields play an important role in physics and geophysics because they describe the behavior of gravitational and electric fields as well as a number of other fields. Conversely, measurements of potential fields provide important information about the internal structure of bodies. For example, measurements of the electric potential at the Earth's surface when a current is sent into the Earth give information about the electrical conductivity, while measurements of the Earth's gravity field or geoid provide information about the mass distribution within the Earth.

An example of this can be seen in Figure 20.1 in which the gravity anomaly over the northern part of the Yucatan peninsula in Mexico is shown (Hildebrand et al., 1995). The coast is visible as a thin white line. Note the ring structure in the gravity signal. These rings have led to the discovery of the Chicxulub crater, which was caused by the massive impact of a meteorite. Note that the diameter of the impact crater is about 150 km! This crater is presently hidden by thick layers of sediments: at the surface the only apparent imprint of this crater is the presence of underground water-filled caves called "cenotes" at the outer edge of the crater. It was the measurement of the gravity field that made it possible to find this massive impact crater that caused a mass extinction at the Cretaceous-Tertiary boundary 65 million years ago (Alvarez et al., 1980; Schulte et al., 2010).

The equation that satisfies the gravitational or electric potential satisfies depends critically on the Laplacian of the potential. As shown in Section 6.7 the gravitational field has the mass-density as its source:

$$\nabla \cdot \mathbf{g} = -4\pi G \rho(\mathbf{r}). \tag{6.45}$$

The gravity field $\mathbf{g}$ is (minus) the gradient of the gravitational potential: $\mathbf{g} = -\nabla V$. This means that the gravitational potential satisfies the following partial differential equation:

$$\nabla^2 V(\mathbf{r}) = 4\pi G \rho(\mathbf{r}). \tag{20.1}$$

This equation is called Poisson's equation and is the prototype of the equations that occur in potential field theory. Note that the mathematical structures of the

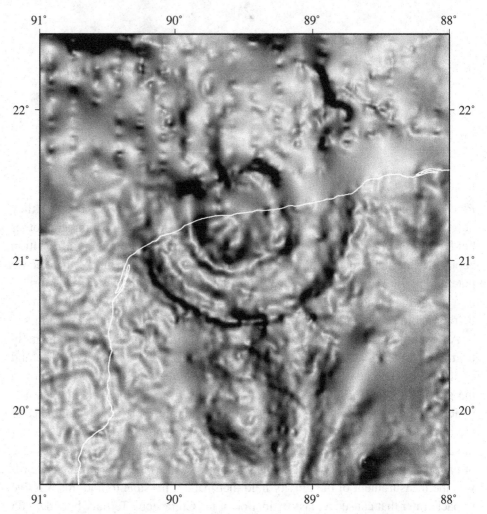

Figure 20.1   Gravity field over the Chicxulub impact crater on the northern coast of Yucatan (Mexico), courtesy of M. Pilkington and A. R. Hildebrand. The coastline is shown by a white line, and the vertical and horizontal axes are latitude and longitude in degrees, respectively. The gray scale represents the magnitude of the horizontal gradient of the Bouguer gravity anomaly (details are given by Hildebrand et al., 1995).

equations of the gravitational field (6.45) and the electrostatic field (6.24) are identical, so that the results derived in this chapter for the gravitational field can be used directly for the electrostatic field as well by replacing the mass-density by the charge-density and by making the following replacement:

$$\underbrace{4\pi G}_{\text{gravity}} \Leftrightarrow \underbrace{-1/\epsilon_0}_{\text{electrostatics}} \ . \tag{20.2}$$

The theory of potential fields is treated in great detail by Blakeley (1995).

20.1 Green's function of the gravitational potential

Poisson's equation (20.1) can be solved using a Green's function technique. In essence the derivation of the Green's function yields the well-known result that the gravitational potential for a point mass m is given by $-Gm/r$. The use of Green's functions was introduced in great detail in Chapter 17. The Green's function $G(\mathbf{r}, \mathbf{r}')$ that describes the gravitational potential at location $\mathbf{r}$ generated by a point mass at location $\mathbf{r}'$ satisfies the following differential equation:

$$\nabla^2 G(\mathbf{r}, \mathbf{r}') = \delta\left(\mathbf{r} - \mathbf{r}'\right). \tag{20.3}$$

Take care not to confuse the Green's function $G(\mathbf{r}, \mathbf{r}')$ with the gravitational constant G.

Problem a Show that the solution of (20.1) is:

$$V(\mathbf{r}) = 4\pi G \int G(\mathbf{r}, \mathbf{r}')\rho(\mathbf{r}')\, dV'. \tag{20.4}$$

Problem b The differential equation (20.3) has translational and rotational invariance. Show that this implies that $G(\mathbf{r}, \mathbf{r}') = G(|\mathbf{r} - \mathbf{r}'|)$. Show by placing the point mass at the origin by setting $\mathbf{r}' = 0$ that $G(r)$ satisfies

$$\nabla^2 G(r) = \delta(\mathbf{r}). \tag{20.5}$$

Problem c Use the expression for the Laplacian in spherical coordinates (10.36) to show that for $r \neq 0$ (20.5) is given by

$$\frac{1}{r^2}\frac{\partial}{\partial r}\left(r^2 \frac{\partial G(r)}{\partial r}\right) = 0. \tag{20.6}$$

Problem d Integrate this equation with respect to r to derive that the solution is given by $G(r) = A/r + B$, where A and B are integration constants.

The constant B in the potential does not contribute to the forces that are associated with this potential because $\nabla B = 0$. For this reason the arbitrary constant B can be set equal to zero. The potential is therefore given by

$$G(r) = \frac{A}{r}. \tag{20.7}$$

Problem e The constant A can be found by integrating (20.5) over a sphere of radius R centered around the origin. Show that Gauss' theorem (8.1) implies that $\int \nabla^2 G(r)dV = \oint \nabla G \cdot d\mathbf{S}$, use equation (20.5) in the left-hand side of this expression, (20.7) in the right-hand side, and show that this gives $A = -1/4\pi$. Note that this result is independent of the radius R that you have used.

Problem f   Show that the Green's function is given by:

$$G(\mathbf{r}, \mathbf{r}') = -\frac{1}{4\pi} \frac{1}{|\mathbf{r} - \mathbf{r}'|}. \tag{20.8}$$

With (20.4) this implies that the gravitational potential is given by:

$$V(\mathbf{r}) = -G \int \frac{\rho(\mathbf{r}')}{|\mathbf{r} - \mathbf{r}'|} dV'. \tag{20.9}$$

This general expression is useful for a variety of different purposes, and we will make extensive use of it. By taking the gradient of this expression, one obtains the gravitational acceleration **g**. This acceleration was also derived in (8.6) for the special case of a spherically symmetric mass distribution. Surprisingly it is a nontrivial calculation to derive (8.6) by taking the gradient of (20.9).

Problem g   If the mass m is centered at the origin, show that $\rho = m\delta(\mathbf{r})$. Use this and the sifting property (13.10) to find the potential:

$$V = -Gm/r. \tag{20.10}$$

20.2 Upward continuation in a flat geometry

Consider a body in two dimensions with a variable mass-density that is only nonzero in the half-space $z < 0$. Here, we determine the gravitational potential V above the half-space when the potential is specified at the plane $z = 0$ that defines the upper boundary of this body. The geometry of this problem is sketched in Figure 20.2. This problem is of relevance for the interpretation of

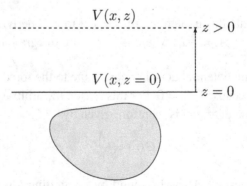

Figure 20.2   Geometry of the upward continuation problem. A mass anomaly (in gray) leaves an imprint on the gravitational potential at $z = 0$. The upward continuation problem states how the potential at the surface $z = 0$ is related to the potential $V(x, z)$ at greater height.

gravity measurements taken above the Earth's surface using aircraft or satellites, because the first step in this interpretation is to relate the values of the potential at the Earth's surface to the measurements taken above the surface. This process is called *upward continuation.*

Mathematically the problem can be stated this way. Suppose one is given the function $V(x, z = 0)$, what is the function $V(x, z)$? When we know that there is no mass above the surface, it follows from (20.1) that the potential satisfies:

$$\nabla^2 V(\mathbf{r}) = 0 \quad \text{for} \quad z > 0. \tag{20.11}$$

It is instructive to solve this problem by making a Fourier expansion of the potential in the variable x, as defined in (14.27):

$$V(x, z) = \int_{-\infty}^{\infty} v(k, z)e^{ikx} dk, \tag{20.12}$$

where k is the wavenumber in the horizontal direction.

Problem a Show that for $z = 0$ the Fourier coefficients can be expressed in terms of the known value of the potential at the edge of the half-space:

$$v(k, z = 0) = \frac{1}{2\pi} \int_{-\infty}^{\infty} V(x, z = 0)e^{-ikx} dx. \tag{20.13}$$

Problem b Use the Laplace equation (20.11) and the Fourier expansion (20.12) to derive that the Fourier components of the potential satisfy for $z > 0$ the following differential equation

$$\frac{\partial^2 v(k, z)}{\partial z^2} - k^2 v(k, z) = 0. \tag{20.14}$$

Problem c Show that the general solution of this differential equation can be written as $v(k, z) = A(k)e^{+|k|z} + B(k)e^{-|k|z}$.

The wavenumber k can be either positive or negative. By using the absolute value, it is explicit that the solution consists of a superposition of an exponentially growing solution (with z), and an exponentially decaying solution.

Since the potential must remain finite at great height ($z \to \infty$), the coefficients $A(k)$ must equal zero. Setting $z = 0$ shows that $B(k) = v(k, z = 0)$, so that the potential is given by:

$$V(x, z) = \int_{-\infty}^{\infty} v(k, z = 0)e^{ikx} e^{-|k|z} dk. \tag{20.15}$$

This expression is interesting because it states that the different Fourier components of the potential decay as $e^{-|k|z}$ with height.

Problem d   Explain that equation (20.15) implies that short horizontal wave-length components in the potential field decay faster with height than the long-wavelength components.

The decrease of the Fourier components with the distance z to the surface is problematic when one wants to infer the mass-density in the body from measurements of the potential or from gravity at a great height above the surface, because the influence of mass perturbations on the gravitational field decays rapidly with height. The measurement of the short-wavelength component of the potential at a great height therefore carries virtually no information about the small-scale details of the horizontal variations in the density distribution within the Earth. This is the reason why gravity measurements from space are preferably carried out using satellites in low orbits rather than in high orbits. Similarly, for gravity surveys at sea, a gravity meter has been developed that is towed far below the sea surface (Zumberge et al., 1997). The idea is that by towing the gravity meter closer to the sea bed, the gravity signal generated at the subsurface for short wavelengths suffers less from the exponential decay due to upward continuation.

Problem e   Take the z-derivative of (20.15) to find the vertical component of the gravity field. Use the resulting expression to show that the gravity field **g** is less sensitive to the loss of short-wavelength features due to upward continuation than the potential V.

This last result is the reason why satellites in low orbits are used to measure the Earth's gravitational potential and satellites in high orbits are used to measure gravity. In fact, the space-borne gradiometer GRACE measures the *gradient* of the gravity vector by monitoring the differential motion between two satellites that are in nearby orbits (Förste et al., 2008). Taking the gradient of the gravity leads to another factor of k in the Fourier expansion so that the decay of short-wavelength features due to upward continuation are further reduced. Figure 20.3 displays the Earth's *geoid*, which is a surface of constant potential V. If you can imagine an Earth covered in oceans, the geoid would be the shape the sea level would take, based on the gravity and rotation of the Earth (Blakeley, 1995). The colorscale is set relative to the best-fitting ellipsoid (model WGS84). Lows and highs in the image indicate deficits and excess mass, respectively, in the subsurface. This image is the superposition of spherical harmonics – such as the ones in Figure 19.4 – up to degree and order 360 from observations of the GRACE and LAGEOS satellites (Förste et al., 2008). Note that the sea level dips about 80 meters south of India due to variations in the mass distribution in the Earth. This large anomaly shows that the word sea *level* really is a misnomer!

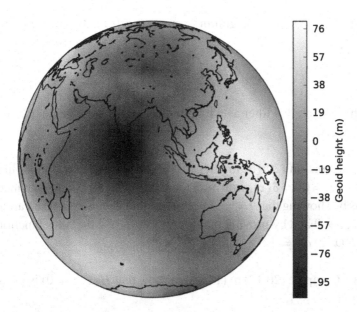

Figure 20.3   Earth's geoid, based on data from the International Centre for Global Earth Models. The greyscale indicates deviations from the Earth's reference ellipsoidal shape.

Let us express the potential at height z to the potential at the surface $z = 0$, explicitly.

Problem f   Insert (20.13) into (20.15) to show that the upward continuation of the potential is given by:

$$V(x, z) = \int_{-\infty}^{\infty} H(x - x', z) V(x', z = 0) \, dx', \qquad (20.16)$$

with

$$H(x, z) = \frac{1}{2\pi} \int_{-\infty}^{\infty} e^{-|k|z} e^{ikx} dk. \qquad (20.17)$$

Note that (20.16) has exactly the same structure as (14.57) for a time-independent linear filter. The only difference is that the variable x now plays the role of the variable t in (14.57). This means that we can consider upward continuation as a linear filtering operation. The convolutional filter $H(x, z)$ maps the potential from the surface $z = 0$ onto the potential at height z.

Problem g   Carry out the integral in expression (20.17) to show that this filter is given by:

$$H(x, z) = \frac{1}{\pi} \frac{z}{z^2 + x^2}. \tag{20.18}$$

Problem h   Sketch this filter as a function of x for a large value of z and a small value of z.

Problem i   Equation (20.16) implies that at the surface $z = 0$ this filter is given by $H(x, z = 0) = \delta(x)$, with $\delta(x)$ the Dirac delta function. Convince yourself of this by showing that $H(x, z)$ becomes more and more peaked around $x = 0$ when $z \rightarrow 0$ and by proving that for all values of z the filter function satisfies $\int_{-\infty}^{\infty} H(x, z)\, dx = 1$.

Problem j   Compare (20.17) to (14.31) to show that $H(x, z = 0)$ is indeed a delta function.

Either way, it makes sense that if we continue upward by zero meters, the potential is unchanged from the one measured at the surface.

20.3 Upward continuation in a flat geometry in three dimensions

The analysis in the previous section is valid for a flat geometry in two dimensions. However, the theory can readily be extended to three dimensions by including another horizontal coordinate y in the derivation.

Problem a   Show that the theory in the previous section up to equation (20.17) can be generalized by carrying out a Fourier transformation over both x and y. Show in particular that in three dimensions:

$$V(x, y, z) = \iint_{-\infty}^{\infty} H^{3D}(x - x', y - y', z) V(x', y', z = 0)\, dx'dy', \tag{20.19}$$

with

$$H^{3D}(x, y, z) = \frac{1}{(2\pi)^2} \iint_{-\infty}^{\infty} e^{-\sqrt{k_x^2 + k_y^2}\, z} e^{i(k_x x + k_y y)}\, dk_x dk_y. \tag{20.20}$$

The only difference with the case in two dimensions is that integral (20.20) leads to a different upward continuation filter than integral (20.17) for the two-dimensional case. The integral can be solved by switching the k-integral to cylindrical coordinates.

Problem b Show that the product $k_x x + k_y y$ can be written as $kr \cos \varphi$, where k and r are the length of the **k**-vector and the position vector in the horizontal plane.

Problem c Use this result to show that H^{3D} can be written as

$$H^{3D}(x, y, z) = \frac{1}{(2\pi)^2} \int_0^\infty \int_0^{2\pi} k e^{-kz} e^{ikr \cos \varphi} d\varphi dk. \tag{20.21}$$

Note the factor k in the integrand.

Problem d As shown in Section 24.5, the Bessel function has the following integral representation:

$$J_0(x) = \frac{1}{2\pi} \int_0^{2\pi} e^{ix \cos \theta} d\theta. \tag{20.22}$$

Use this result to write the upward continuation filter as

$$H^{3D}(x, y, z) = \frac{1}{2\pi} \int_0^\infty e^{-kz} J_0(kr) k \, dk. \tag{20.23}$$

Note that the upward continuation operator in two dimensions in expression (20.17) follows by the replacement $k J_0(kr) \rightarrow e^{ikx}$ and by limiting the k-integral to positive values.

It appears that we have only made the problem more complex, because the integral of the Bessel function is not trivial. Fortunately, books and tables exist with a bewildering collection of integrals. For example, in equation (6.621.1) of Gradshteyn and Ryzhik (1965) you can find an expression for the following integral: $\int_0^\infty e^{-\alpha x} J_\nu(\beta x) x^{\mu-1} dx$.

Problem e What are the values of α, ν, β, and μ if we want to use this integral to solve the integration in (20.23)?

Take a look at the form of the integral in Gradshteyn and Ryzhik (1965). You will probably be discouraged by what you find because the result is expressed in hypergeometric functions, which means that you now have the new problem of finding out what these functions are. There is, however, a way out because (6.611.1) of Gradshteyn and Ryzhik (1965) gives the following integral:

$$\int_0^\infty e^{-\alpha x} J_\nu(\beta x) \, dx = \frac{\beta^{-\nu} \left\{ \sqrt{\alpha^2 + \beta^2} - \alpha \right\}^\nu}{\sqrt{\alpha^2 + \beta^2}}. \tag{20.24}$$

This is not quite the integral that we want because it does not contain a term that corresponds to the factor k in (20.23). However, we can introduce such a factor by differentiating (20.24) with respect to α.

Problem f Do this to show that the upward continuation operator is given by

$$H^{3D}(x, y, z) = \frac{1}{2\pi} \frac{z}{\left(x^2 + y^2 + z^2\right)^{3/2}}. \tag{20.25}$$

Hint: You can make the problem simpler by first inserting the appropriate value of ν in (20.24).

Problem g Compare the upward continuation operator (20.25) for three dimensions with the corresponding operator for two dimensions in (20.18). Which of these operators decays more rapidly as a function of the horizontal distance? Can you explain this difference physically?

Problem h In Section 20.2 you showed that the integral of the upward continuation operator over the horizontal distance is equal to 1. Show that the same holds in three dimensions; that is, show that $\iint_{-\infty}^{\infty} H^{3D}(x, y, z)dxdy = 1$. The integration simplifies by using cylindrical coordinates.

Problem i By the same reasoning we used in Problem i of Section 20.2, it follows that

$$H^{3D}(x, y, z = 0) = \delta(x)\delta(y). \tag{20.26}$$

Use expressions (20.25) and (13.24) to show that the equation above is dimensionally correct.

20.4 Gravity field of the Earth

In this section we obtain an expression for the gravitational potential outside the Earth for an arbitrary distribution of the mass density $\rho(\mathbf{r})$ within the Earth:

$$\nabla^2 V(\mathbf{r}) = 4\pi G\rho(\mathbf{r}). \tag{20.1}$$

This could be done be using the Green's function that is appropriate for Poisson's equation (20.1). As an alternative we solve the problem here by expanding both the mass-density and the potential in spherical harmonics and by using a Green's function technique for every component in the spherical harmonics expansion separately.

When using spherical harmonics, the natural coordinate system is a system of spherical coordinates. For every value of the radius r, both the density and the potential can be expanded in spherical harmonics defined by (19.31) and (19.32):

$$\rho(r, \theta, \varphi) = \sum_{l=0}^{\infty} \sum_{m=-l}^{l} \rho_{lm}(r) Y_{lm}(\theta, \varphi) \tag{20.27}$$

and

$$V(r, \theta, \varphi) = \sum_{l=0}^{\infty} \sum_{m=-l}^{l} V_{lm}(r) Y_{lm}(\theta, \varphi). \tag{20.28}$$

Problem a Use the orthogonality relation (19.51) of spherical harmonics to show that the expansion coefficients for the density are given by:

$$\rho_{lm}(r) = \int Y_{lm}^*(\theta, \varphi) \rho(r, \theta, \varphi) \, d\Omega, \tag{20.29}$$

where $\int (\cdots) \, d\Omega$ denotes an integration over the unit sphere.

Equation (20.1) for the gravitational potential contains the Laplacian. The Laplacian in spherical coordinates can be decomposed into a radial and angular component:

$$\nabla^2 = \frac{1}{r^2} \frac{\partial}{\partial r} \left(r^2 \frac{\partial}{\partial r} \right) + \frac{1}{r^2} \nabla_1^2, \tag{20.30}$$

with ∇_1^2 the Laplacian on the unit sphere:

$$\nabla_1^2 = \frac{1}{\sin \theta} \frac{\partial}{\partial \theta} \left(\sin \theta \frac{\partial}{\partial \theta} \right) + \frac{1}{\sin^2 \theta} \frac{\partial^2}{\partial \varphi^2}. \tag{19.35}$$

The reason why an expansion in spherical harmonics is used for the density and the potential is that the spherical harmonics are the eigenfunctions of the operator ∇_1^2 (see (19.34) or p. 379 of Butkov, 1968):

$$\nabla_1^2 Y_{lm}(\theta, \varphi) = -l(l+1) Y_{lm}(\theta, \varphi). \tag{19.34}$$

Problem b Insert (20.27) and (20.28) into the Laplace equation, and use (19.34) and (19.35) for the Laplacian of the spherical harmonics to show that the expansion coefficients $V_{lm}(r)$ of the potential satisfy the following differential equation:

$$\frac{1}{r^2} \frac{\partial}{\partial r} \left[r^2 \frac{\partial V_{lm}(r)}{\partial r} \right] - \frac{l(l+1)}{r^2} V_{lm}(r) = 4\pi G \rho_{lm}(r). \tag{20.31}$$

What we have gained by making the expansion in spherical harmonics is that (20.31) is an ordinary differential equation in the variable r, whereas the original equation (20.1) is a partial differential equation in the variables r, θ, and φ. The differential equation (20.31) can be solved using the Green's function technique described in Section 17.3. Let us first consider a mass $\delta(r-r')$ located at a radius r'. The response to this mass is the Green's function G_l that satisfies the following differential equation:

$$\frac{1}{r^2}\frac{\partial}{\partial r}\left[r^2\frac{\partial G_l(r,r')}{\partial r}\right] - \frac{l(l+1)}{r^2}G_l(r,r') = \delta(r-r'). \tag{20.32}$$

Note that this equation depends on the angular order l but not on the angular degree m. For this reason the Green's function $G_l(r,r')$ depends on l but not on m.

Problem c   The Green's function can be found by first solving the differential equation (20.32) for $r \neq r'$. Show that when $r \neq r'$, the general solution of the differential equation (20.32) can be written as $G_l = Ar^l + Br^{-(l+1)}$, where the constants A and B do not depend on r.

Problem d   In the regions $r < r'$ and $r > r'$ the constants A and B in general have different values. Show that the requirement that the potential is everywhere finite implies that $B = 0$ for $r < r'$ and that $A = 0$ for $r > r'$. The solution can therefore be written as:

$$G_l(r,r') = \begin{cases} Ar^l & \text{for} \quad r < r', \\ Br^{-(l+1)} & \text{for} \quad r > r'. \end{cases} \tag{20.33}$$

The integration constants follow in the same way as in the analysis of Section 17.3. One constraint on the integration constants follows from the requirement that the Green's function is continuous in the point $r = r'$. The other constraint follows by multiplying (20.32) by r^2 and integrating the resulting equation over r from $r' - \epsilon$ to $r' + \epsilon$.

Problem e   Show by taking the limit $\epsilon \to 0$ that this leads to the requirement

$$\left[r^2\frac{\partial G_l(r,r')}{\partial r}\right]_{r=r'-\epsilon}^{r=r'+\epsilon} = r'^2. \tag{20.34}$$

Problem f Use this condition with the continuity of G_l to find the coefficients A and B and show that the Green's function is given by:

$$G_l(r, r') = \begin{cases} -\dfrac{1}{(2l+1)} \dfrac{r^l}{r'^{(l-1)}} & \text{for} \quad r < r', \\[2ex] -\dfrac{1}{(2l+1)} \dfrac{r'^{(l+2)}}{r^{(l+1)}} & \text{for} \quad r > r'. \end{cases} \tag{20.35}$$

Problem g Use this result to derive that the solution of (20.31) is:

$$\begin{aligned} V_{lm}(r) = &-\frac{4\pi G}{(2l+1)} \frac{1}{r^{l+1}} \int_0^r \rho_{lm}(r') r'^{(l+2)} dr' \\ &-\frac{4\pi G}{(2l+1)} r^l \int_r^\infty \rho_{lm}(r') \frac{1}{r'^{(l-1)}} dr'. \end{aligned} \tag{20.36}$$

Hint: Split the integration over r' into the interval $0 < r' < r$ and the interval $r' > r$.

Problem h Let us now consider the potential outside the Earth. The radius of the Earth is denoted by the symbol a. Use the above expression and (20.28) to show that the potential outside the Earth is given by

$$V(r, \theta, \varphi) = -\sum_{l=0}^\infty \sum_{m=-l}^l \frac{4\pi G}{(2l+1)} \frac{1}{r^{l+1}} \int_0^a \rho_{lm}(r') r'^{(l+2)} dr' \, Y_{lm}(\theta, \varphi). \tag{20.37}$$

Problem i Eliminate ρ_{lm} using the result of Problem a and show that the potential is finally given by:

$$\begin{aligned} V(r, \theta, \varphi) = &-\sum_{l=0}^\infty \sum_{m=-l}^l \frac{4\pi G}{(2l+1)} \frac{1}{r^{l+1}} \\ &\times \int_0^a \rho(r', \theta', \varphi') r'^l Y_{lm}^*(\theta', \varphi') \, dV' \, Y_{lm}(\theta, \varphi). \end{aligned} \tag{20.38}$$

Note that the integration in this expression is over the volume of the Earth rather than over the distance r' to the Earth's center.

Let us reflect on the relation between this result and the derivation of upward continuation in a Cartesian geometry of Section 20.2. Equation (20.38) can be compared with (20.15). In (20.15) the potential is written as an integration over wavenumber and the potential is expanded in basis functions e^{ikx}, whereas in (20.38) the potential is written as a summation over the degree l and order m and the potential is expanded in basis functions Y_{lm}. In both expressions the potential is written as a sum over basis functions with increasingly shorter wavelength as the summation index l or the integration variable k increases. The decay of the potential with height is in both geometries faster for a potential with rapid horizontal variations than for a potential with smooth horizontal variations. In both cases the potential decreases when the height z (or the radius r) increases. In a flat geometry the potential decreases as $e^{-|k|z}$, whereas in a spherical geometry the potential decreases as $r^{-(l+1)}$. This difference in the reduction of the potential with distance is due to the difference in the geometry in the two problems.

Expressions (20.9) and (20.38) both express the gravitational potential due to the same density distribution $\rho(\mathbf{r})$; therefore, these expressions must be identical.

Problem j Use the equivalence of these expressions to derive that for $r' < r$ the following identity holds:

$$\frac{1}{|\mathbf{r} - \mathbf{r}'|} = \sum_{l=0}^{\infty} \sum_{m=-l}^{l} \frac{4\pi}{(2l+1)} Y_{lm}(\theta, \varphi) Y_{lm}^*(\theta', \varphi') \frac{r'^l}{r^{l+1}}. \tag{20.39}$$

The derivation in this section could also have been made using (20.39) as a starting point because this expression can be derived by using the *generating function* of Legendre polynomials and by using the addition theorem to obtain the m-summation (Arfken and Weber, 2005; Jackson, 1998). However, these concepts are not needed in the treatment in this section, which is based only on the expansion of functions in spherical harmonics and on the use of Green's functions.

As a last exercise let us consider the special case of a spherically symmetric mass distribution: $\rho = \rho(r)$ in a sphere with radius R. In this case the density does not depend on the angles θ and φ. According to Table 19.2 the spherical harmonic Y_{00} is a constant; hence, only the terms $l = m = 0$ contribute to the sum (20.27).

Problem k Show that in this case

$$\rho_{00}(r) = \sqrt{4\pi} \rho(r). \tag{20.40}$$

Problem l Use this result in expressions (20.36) and (20.28) to show that outside of the mass ($r > R$)

$$V(r) = -\frac{4\pi G}{r} \int_0^R \rho(r') r'^2 dr'. \tag{20.41}$$

For a spherically symmetric mass distribution, $4\pi r'^2 dr'$ is the volume element dV' of a spherical shell; hence, $4\pi \int_0^R \rho(r') r'^2 dr' = \int \rho(r') dV' = M$, where M is the mass of the body. This means that the potential (20.41) is given by

$$V(\mathbf{r}) = -\frac{GM}{r}. \tag{20.42}$$

The gradient of this potential is indeed equal to the gravitational acceleration given in (8.6) for a spherically symmetric mass M.

20.5 Monopoles, dipoles, and quadrupoles

We have seen in Section 8.2 that a spherically symmetric mass leads to a gravitational field $\mathbf{g}(\mathbf{r}) = -GM\hat{\mathbf{r}}/r^2$, which corresponds to a gravitational potential $V(\mathbf{r}) = -GM/r$. Similarly, the electric potential due to a spherically symmetric charge distribution is given by $V(\mathbf{r}) = q/4\pi\epsilon_0 r$, where q is the total charge. In this section we investigate what happens if we place a positive charge and a negative charge close together. Since there is no negative mass, we treat for the moment the electric potential, but we will see in Section 20.6 that the results also have a bearing on the gravitational potential. The theory developed here is not only important in electrostatics; it also accounts for the measurable effect of the ellipsoidal shape of the Earth on its gravitational field. This application is treated in more detail in Section 20.7.

Consider a charge distribution that consists of a positive charge $+q$ placed at position $\mathbf{a}/2$ and a negative charge $-q$ placed at position $-\mathbf{a}/2$ as shown in Figure 20.4.

Problem a The total charge of this system is zero. What would you expect the electric potential to be at positions that are very far from the charges compared to the distance between the charges?

Problem b The potential follows by adding the potentials for the two point charges. Show that the electric potential generated by these two charges is given by

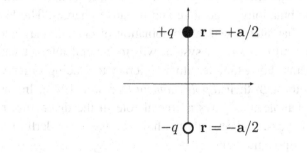

Figure 20.4 Two opposite charges that constitute an electric dipole.

$$4\pi\epsilon_0 V(\mathbf{r}) = \frac{q}{|\mathbf{r} - \mathbf{a}/2|} - \frac{q}{|\mathbf{r} + \mathbf{a}/2|}. \tag{20.43}$$

Problem c Ultimately we will place the charges very close to the origin by taking the limit $a \to 0$. We can therefore restrict our attention to the special case that $a \ll r$. Use a first-order Taylor expansion to show that up to order a:

$$\frac{1}{|\mathbf{r} - \mathbf{a}/2|} = \frac{1}{r} + \frac{1}{2r^3}(\mathbf{r} \cdot \mathbf{a}). \tag{20.44}$$

Hint: Use that $|\mathbf{r} - \mathbf{a}/2|^{-1} = [(\mathbf{r} - \mathbf{a}/2) \cdot (\mathbf{r} - \mathbf{a}/2)]^{-1/2}$.

Problem d Insert this into (20.43) and derive that the electric potential is given by:

$$4\pi\epsilon_0 V(\mathbf{r}) = \frac{q(\mathbf{r} \cdot \mathbf{a})}{r^3}. \tag{20.45}$$

Now suppose we bring the charges in Figure 20.4 closer and closer together, and suppose we let the charge q increase so that the product $\mathbf{p} = q\mathbf{a}$ is constant, then the electric potential is given by:

$$V(\mathbf{r}) = \frac{(\hat{\mathbf{r}} \cdot \mathbf{p})}{4\pi\epsilon_0 r^2}, \tag{20.46}$$

where we have used that $\mathbf{r} = r\hat{\mathbf{r}}$. The vector $\mathbf{p}$ is called the *dipole vector*.

We will see in the next section how the dipole vector can be defined for arbitrary charge or mass distributions. In Problem a you might have guessed that the electric potential would go to zero at great distance. Of course, the potential due to the combined charges goes to zero much faster than the potential due to a single charge only: the electric potential of a dipole vanishes as $1/r^2$ compared to the $1/r$ decay of the potential for a single charge. Many physical systems, such as neutral atoms, consist of neutral combinations of positive and negative charges. The lesson we learn from (20.45) is that such a neutral combination of charges may generate a nonzero electric field and that such a system will in general interact with other electromagnetic systems. For example, atoms interact to leading order with the radiation field (light) through their *dipole moment* (Sakurai, 1978). In chemistry, the dipole moment of molecules plays a crucial role in the distinction between polar and apolar substances. Water would not have its many wonderful properties if H_2O did not have a dipole moment.

Let us now consider the electric field generated by an electric dipole.

Problem e Take the gradient of (20.46) to show that this field is given by

$$\mathbf{E}(\mathbf{r}) = \frac{1}{4\pi\epsilon_0 r^3} \left[3\hat{\mathbf{r}} \left(\hat{\mathbf{r}} \cdot \mathbf{p} - \mathbf{p} \right) \right]. \tag{20.47}$$

Hint: Either use the expression of the gradient in spherical coordinates or take the gradient in Cartesian coordinates and use (6.12).

The electric field generated by an electric dipole has the same form as the magnetic field generated by a magnetic dipole as shown in (6.5). The mathematical reason for this is that the magnetic field satisfies (6.29), which states that $(\nabla \cdot \mathbf{B}) = 0$, while the electric field in free space satisfies according to (6.24) the field equation: $(\nabla \cdot \mathbf{E}) = 0$. However, there is an important difference. The electric field is generated by electric charges, and this field satisfies the equation $(\nabla \cdot \mathbf{E}) = \rho(\mathbf{r})/\epsilon_0$. In the example in this section we created a dipole field by taking two opposite charges and putting them closer and closer together. However, the magnetic field satisfies $(\nabla \cdot \mathbf{B}) = 0$ *everywhere*. The reason for this is that the magnetic equivalent of the electric charge, the *magnetic monopole*, has not been discovered in nature.

It seems puzzling that magnetic monopoles have not been observed in nature, because we have seen that the magnetic dipole field has the same form as the electric dipole field, which was constructed by putting two opposite electric charges close together. The reason for the analogy between an electric and a magnetic dipole field is not that the magnetic dipole can be seen as a combination of positive and negative magnetic "charges" placed close together. In the context of classical electromagnetism, the magnetic dipole field is generated by a current that runs in a small circular loop. On a microscopic scale a magnetic dipole is generated by the *spin* of particles. This means that the electric and magnetic dipole fields have fundamentally different origins.

The starting point of the derivation in this section is the electric field of a single point charge. Such a single point charge is called a monopole (Figure 20.5). The field of this charge decays as $1/r^2$. If we put two opposite charges close together, we can create a dipole (Figure 20.5). Its electric field is derived in (20.47); this field decays as $1/r^3$. We can also put two opposite dipoles together as shown in Figure 20.5. The resulting charge distribution is called a quadrupole. To leading order, the electric fields of the dipoles that constitute the quadrupole cancel, and we will see in the next section that the electric potential for a quadrupole decays as $1/r^3$ so that the electric field decays with distance as $1/r^4$.

You may wonder whether the concept of a dipole or quadrupole can also be used for the gravity field because for the electric field these concepts are based on the presence of both positive and negative charges, whereas we know that only positive mass occurs in nature. However, there is nothing to keep us from computing the

Monopole   Dipole   Quadrupole

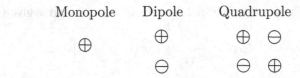

Figure 20.5   The definition of the monopole, dipole, and quadrupole in terms of electric charges.

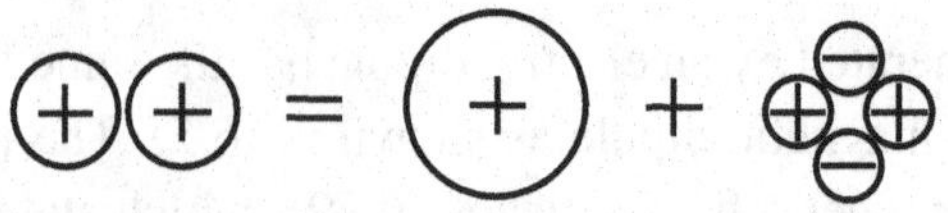

$$\left(+\right)\left(+\right) = \left(+\right) + \oplus\ominus\oplus$$

Figure 20.6   The decomposition of a binary star in a gravitational monopole and a gravitational quadrupole.

gravitational field for a negative mass, and this is actually quite useful. As an example, let us consider a binary star that consists of two heavy stars that rotate around their joint center of gravity. The first-order field is the monopole field that is generated by the joint mass of the stars. However, as shown in Figure 20.6, the mass of the two stars can be seen as approximately the sum of a monopole and a quadrupole consisting of two positive and two negative masses. Since the stars rotate, the gravitational quadrupole rotates as well, and this is the reason why rotating binary stars are seen as a prime source for the generation of *gravitational waves* (Ohanian and Ruffini, 1976). However, gravitational waves that spread in space with time cannot be described by the classic expression (20.1) for the gravitational potential.

Problem f   Can you explain why (20.1) cannot account for propagating gravitational waves?

A proper description of gravitational waves depends on the general theory of relativity (Ohanian and Ruffini, 1976). Huge detectors for gravitational waves are being developed (Barish and Weiss, 1999), because these waves can be used to investigate the theory of general relativity as well as the astronomical objects that generate gravitational waves.

20.6 Multipole expansion

Now that we have learned that the concepts of the monopole, dipole, and quadrupole are relevant for both the electric field and the gravity field, we continue the analysis with the gravitational field. In this section we derive the *multipole expansion* in which the total field is written as a superposition of a monopole field, a dipole field, a quadrupole field, an octupole field, etc.

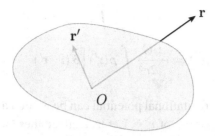

Figure 20.7 Definition of the integration variable $\mathbf{r}'$ within the mass and the observation point $\mathbf{r}$ outside the mass.

Consider the situation shown in Figure 20.7 in which a finite body has a mass-density $\rho(\mathbf{r}')$. The gravitational potential generated by this mass is given by:

$$V(\mathbf{r}) = -G \int \frac{\rho(\mathbf{r}')}{|\mathbf{r} - \mathbf{r}'|} dV'. \tag{20.9}$$

We consider the potential at a distance that is much larger than the size of the body. Since the integration variable $\mathbf{r}'$ is limited by the size of the body, a "large distance" means in this context that $r \gg r'$. We therefore make a Taylor expansion of the term $1/|\mathbf{r} - \mathbf{r}'|$ in the parameter (r'/r), which is much smaller than unity.

Problem a Show that

$$|\mathbf{r} - \mathbf{r}'| = \sqrt{r^2 - 2(\mathbf{r} \cdot \mathbf{r}') + r'^2}. \tag{20.48}$$

Problem b Use a Taylor expansion in the small parameter r'/r to derive that:

$$\frac{1}{|\mathbf{r} - \mathbf{r}'|} = \frac{1}{r} \left\{ 1 + \frac{1}{r}(\hat{\mathbf{r}} \cdot \mathbf{r}') + \frac{1}{2r^2}\left[3(\hat{\mathbf{r}} \cdot \mathbf{r}')^2 - r'^2\right] + \mathcal{O}\left(\frac{r'}{r}\right)^3 \right\}. \tag{20.49}$$

Be careful that you account for all the terms of order r' correctly. Also be aware of the distinction between the position vector $\mathbf{r}$ and the unit vector $\hat{\mathbf{r}}$.

From this point on we will ignore the terms of order $(r'/r)^3$.

Problem c Insert the expansion (20.49) into (20.9) and show that the gravitational potential can be written as a sum of different contributions:

$$V(\mathbf{r}) = V_{mon}(\mathbf{r}) + V_{dip}(\mathbf{r}) + V_{qua}(\mathbf{r}) + \cdots, \tag{20.50}$$

with

$$V_{mon}(\mathbf{r}) = -\frac{G}{r} \int \rho(\mathbf{r}') dV', \tag{20.51}$$

$$V_{dip}(\mathbf{r}) = -\frac{G}{r^2} \int \rho(\mathbf{r}') \left(\hat{\mathbf{r}} \cdot \mathbf{r}' \right) dV', \qquad (20.52)$$

$$V_{qua}(\mathbf{r}) = -\frac{G}{2r^3} \int \rho(\mathbf{r}') \left[3 \left(\hat{\mathbf{r}} \cdot \mathbf{r}' \right)^2 - r'^2 \right] dV'. \qquad (20.53)$$

It thus follows that the gravitational potential can be written as the sum of terms that decay with increasing powers of r^{-n}. Let us analyze these terms in turn. The term $V_{mon}(\mathbf{r})$ in (20.51) is the simplest, since the volume integral of the mass-density is simply the total mass of the body: $\int \rho(\mathbf{r}')dV' = M$. This means that this term is given by

$$V_{mon}(\mathbf{r}) = -\frac{GM}{r}. \qquad (20.54)$$

This is the potential generated by a point mass M. To leading order, the gravitational field is the same as if all the mass of the body were concentrated in the origin. The mass distribution within the body does not affect this part of the gravitational field at all. Because the resulting field is the same as for a point mass, this field is called the *monopole field*.

For the analysis of the term $V_{dip}(\mathbf{r})$ in (20.52), it is useful to define the center of gravity $\mathbf{r}_g$ of the body:

$$\mathbf{r}_g \equiv \frac{\int \rho(\mathbf{r}')\mathbf{r}'dV'}{\int \rho(\mathbf{r}')\,dV'}. \qquad (20.55)$$

This is simply a weighted average of the position vector with the mass-density as weight functions. Note that the word "weight" here has a double meaning!

Problem d Show that $V_{dip}(\mathbf{r})$ is given by:

$$V_{dip}(\mathbf{r}) = -\frac{GM}{r^2} \left(\hat{\mathbf{r}} \cdot \mathbf{r}_g \right). \qquad (20.56)$$

Note that this potential has exactly the same form as the potential (20.46) for an electric dipole.[1] For this reason $V_{dip}(\mathbf{r})$ is called the dipole field.

Problem e Compared to the monopole term, the dipole term decays as $1/r^2$ rather than $1/r$. The monopole term does not depend on $\hat{\mathbf{r}}$, the direction of observation. Show that the dipole term varies with the direction of observation as $\cos \theta$ and show how the angle θ must be defined.

---

[1] The sign difference between the electric dipole and this expression stems from the opposite sign in the source terms for gravity and electrostatics.

Problem f You may be puzzled by the fact that the gravitational potential contains a dipole term, despite the fact that there is no negative mass. Draw a figure similar to Figure 20.6 to show that a displaced mass can be written as an undisplaced mass plus a mass dipole.

Of course, one is free in the choice of the origin of the coordinate system. If one chooses the origin to be at the center of mass of the body, then $\mathbf{r}_g = 0$ and the dipole term vanishes.

We now analyze the term $V_{qua}(\mathbf{r})$ in (20.53). It can be seen from this expression that this term decays with distance as $1/r^3$. For this reason, this term is called the quadrupole field. The dependence of the quadrupole field on the direction is more complex than for the monopole field and the dipole field. In the determination of the directional dependence of the quadrupole term, it is useful to use the double contraction between two tensors. Tensors are treated in Chapter 26. For the moment you only need to know that the double contraction is defined as:

$$(\mathbf{A} : \mathbf{B}) \equiv \sum_{i,j} A_{ij} B_{ij}. \tag{20.57}$$

The double contraction generalizes the concept of the inner product of two vectors $(\mathbf{a} \cdot \mathbf{b}) = \sum_i a_i b_i$ to matrices or tensors of rank two. A double contraction occurs for example in the following identity

$$1 = (\hat{\mathbf{r}} \cdot \hat{\mathbf{r}}) = (\hat{\mathbf{r}} \cdot \mathbf{I}\hat{\mathbf{r}}) = (\hat{\mathbf{r}}\hat{\mathbf{r}} : \mathbf{I}), \tag{20.58}$$

where $\mathbf{I}$ is the identity operator. Note that the term $\hat{\mathbf{r}}\hat{\mathbf{r}}$ is a dyad. If you are unfamiliar with the concept of a dyad, you may want to look at Section 12.1 before continuing with this chapter.

Problem g Use these results to show that $V_{qua}(\mathbf{r})$ can be written as:

$$V_{qua}(\mathbf{r}) = -\frac{G}{2r^3} (\hat{\mathbf{r}}\hat{\mathbf{r}} : \mathbf{T}), \tag{20.59}$$

where $\mathbf{T}$ is the *inertia tensor* defined as

$$\mathbf{T} = \int \rho(\mathbf{r}) \left(3\mathbf{r}\mathbf{r} - \mathbf{I}r^2\right) dV. \tag{20.60}$$

Note that we have renamed the integration variable $\mathbf{r}'$ in the inertia tensor as $\mathbf{r}$.

Problem h   Show that in explicit matrix notation **T** is given by:

$$\mathbf{T} = \int \rho(\mathbf{r}) \begin{pmatrix} 2x^2 - y^2 - z^2 & 3xy & 3xz \\ 3xy & 2y^2 - x^2 - z^2 & 3yz \\ 3xz & 3yz & 2z^2 - x^2 - y^2 \end{pmatrix} dV.$$

(20.61)

Note the resemblance between (20.59) and (20.56). For the dipole field the directional dependence is described by the single contraction $(\hat{\mathbf{r}} \cdot \mathbf{r}_g)$, whereas for the quadrupole field directional dependence is now given by the double contraction $(\hat{\mathbf{r}}\hat{\mathbf{r}} : \mathbf{T})$. This double contraction leads to a greater angular dependence of the quadrupole term than for the monopole term and the dipole term.

To find the angular dependence, we use that the inertia tensor **T** is a real symmetric 3×3 matrix. This matrix therefore has three orthogonal eigenvectors $\hat{\mathbf{v}}^{(i)}$ with corresponding eigenvalues λ_i. Using expression (12.39) this implies that the inertia tensor can be written as:

$$\mathbf{T} = \sum_{i=1}^{3} \lambda_i \hat{\mathbf{v}}^{(i)} \hat{\mathbf{v}}^{(i)}.$$

(20.62)

Problem i   Use this result to show that the quadrupole field can be written as

$$V_{qua}(\mathbf{r}) = -\frac{G}{2r^3} \sum_{i=1}^{3} \lambda_i \cos^2 \Psi_i,$$

(20.63)

where the Ψ_i denote the angles between the eigenvectors $\hat{\mathbf{v}}^{(i)}$ and the observation direction $\hat{\mathbf{r}}$; see Figure 20.8 for the definition of these angles.

The directional dependence of (20.63) varies as $\cos^2 \Psi_i = (\cos 2\Psi_i + 1)/2$; this implies that the quadrupole field varies through two periods when Ψ_i increases from 0 to 2π. This contrasts with the monopole field, which does not depend

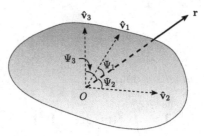

Figure 20.8   Definition of the angles Ψ_i between the observation point **r** and the eigenvectors **v** of the inertia tensor **T**.

on the direction at all, as well as with the dipole field that varies according to Problem e as $\cos\theta$. There is actually a close connection between the different terms in the multipole expansion and spherical harmonics. This can be seen by comparing the multipole terms (20.51)–(20.53) with (20.38) for the gravitational potential. In (20.38), the different terms decay with distance as $r^{-(l+1)}$ and have an angular dependence $Y_{lm}(\theta, \varphi)$. Similarly, the multipole terms decay as r^{-1}, r^{-2}, and r^{-3}, respectively, and depend on the direction as $\cos 0$, $\cos\theta$, and $\cos 2\Psi$, respectively.

20.7 Quadrupole field of the Earth

Let us now investigate what the multipole expansion implies for the gravity field of the Earth. The monopole term is by far the dominant term. It explains why an apple falls from a tree, why the Moon orbits the Earth, and most other manifestations of gravity that we observe in daily life. The dipole term has in this context no physical meaning whatsoever. This can be seen from (20.56), which states that the dipole term only depends on the distance from the Earth's center of gravity to the origin of the coordinate system. Since we are free in choosing the origin, the dipole term can be made to vanish by choosing the origin of the coordinate system as the Earth's center of gravity. It is through the quadrupole field that some of the subtleties of the Earth's gravity field become manifest.

Problem a The quadrupole field vanishes when the mass distribution in the Earth is spherically symmetric. Show this by computing the inertia tensor $\mathbf{T}$ when $\rho = \rho(r)$.

The dominant departure of the shape of the Earth from spherical is the flattening of the Earth due to its rotation. If that is the case, then by symmetry one eigenvector of $\mathbf{T}$ must be aligned with the Earth's axis of rotation, and the two other eigenvectors must be perpendicular to the axis of rotation. By symmetry these other eigenvectors must correspond to equal eigenvalues. When we choose a coordinate system with the z-axis along the Earth's axis of rotation, the eigenvectors are therefore given by the unit vectors $\hat{\mathbf{z}}$, $\hat{\mathbf{x}}$, and $\hat{\mathbf{y}}$ with eigenvalues λ_z, λ_x, and λ_y, respectively. The last two eigenvalues are identical because of the rotational symmetry around the Earth's axis of rotation; hence, $\lambda_y = \lambda_x$.

Let us first determine the eigenvalues. Once the eigenvalues are known, the quadrupole moment tensor follows from (20.63). The eigenvalues could be found in the standard way by solving $\det(\mathbf{T} - \lambda\mathbf{I}) = 0$, but this is unnecessarily difficult. Once we know the eigenvectors, the eigenvalues can easily be found from (20.62).

Problem b   Take twice the inner product of (20.62) with the eigenvector $\hat{\mathbf{v}}^{(j)}$ to show that

$$\lambda_j = \hat{\mathbf{v}}^{(j)} \cdot \mathbf{T} \cdot \hat{\mathbf{v}}^{(j)}. \tag{20.64}$$

Problem c   Use this with (20.61) to show that the eigenvalues are given by:

$$\left. \begin{aligned} \lambda_x &= \int \rho(\mathbf{r}) \left(2x^2 - y^2 - z^2\right) dV, \\ \lambda_y &= \int \rho(\mathbf{r}) \left(2y^2 - x^2 - z^2\right) dV, \\ \lambda_z &= \int \rho(\mathbf{r}) \left(2z^2 - x^2 - y^2\right) dV. \end{aligned} \right\} \tag{20.65}$$

It is useful to relate these eigenvalues to the Earth's *moments of inertia* (Stacey, 1992). Physically, the moment of inertia for a certain axis is a measure of the resistance to changes in the rotation rate around that axis. The moment of inertia of the Earth around the z-axis is defined as:

$$C \equiv \int \rho(\mathbf{r}) \left(x^2 + y^2\right) dV, \tag{20.66}$$

whereas the moment of inertia around the x-axis is defined as

$$A \equiv \int \rho(\mathbf{r}) \left(y^2 + z^2\right) dV. \tag{20.67}$$

By symmetry the moment of inertia around the y-axis is given by the same moment A. These moments describe the rotational inertia around the coordinate axes as shown in Figure 20.9. The eigenvalues in (20.65) can be related to these moments of inertia. Because of the assumed axisymmetric density distribution, the integral of y^2 in (20.65) is equal to the integral of x^2. The eigenvalue λ_x in (20.65) is therefore given by: $\lambda_x = \int \rho(\mathbf{r}) \left(x^2 - z^2\right) dV = \int \rho(\mathbf{r}) \left(x^2 + y^2 - y^2 - z^2\right) dV = C - A$.

Problem d   Apply a similar treatment to the other eigenvalues to show that:

$$\lambda_x = \lambda_y = C - A, \qquad \lambda_z = -2(C - A). \tag{20.68}$$

Problem e   Use these eigenvalues in (20.62) together with (20.59) and expression (4.7) for the unit vector $\hat{\mathbf{r}}$ to show that the quadrupole field is given by:

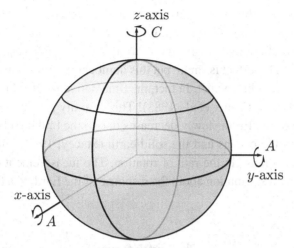

z-axis

x-axis

y-axis

Figure 20.9 Definition of the moments of inertia A and C for an Earth with cylindrical symmetry around the rotation (z-)axis.

$$V_{qua}(\mathbf{r}) = \frac{G}{2r^3}(C - A)\left(3\cos^2\theta - 1\right). \qquad (20.69)$$

The Legendre polynomial of order 2 is given by: $P_2^0(x) = \frac{1}{2}\left(3x^2 - 1\right)$. The quadrupole field can therefore be written as:

$$V_{qua}(\mathbf{r}) = \frac{G}{r^3}(C - A)\, P_2^0(\cos\theta). \qquad (20.70)$$

The term $(C - A)$ denotes the difference in the moments of inertia of the Earth around the rotation axis and around an axis through the equator, see Figure 20.9. If the Earth were a perfect sphere, these moments of inertia would be identical and the quadrupole field would vanish. However, the rotation of the Earth causes the Earth to bulge at the equator. This departure from spherical symmetry is responsible for the Earth's quadrupole field.

If the Earth were spherical, the motion of satellites orbiting the Earth would satisfy Kepler's laws (Goldstein, 1980). The quadrupole term in the potential effects a measurable deviation of the trajectories of satellites from the orbits predicted by Kepler's laws. For example, if the potential is spherically symmetric, a satellite orbits in a fixed plane. The quadrupole field causes the plane in which the satellite orbits to precess slightly. Observations of the orbits of satellites can therefore be used to deduce the departure of the Earth's shape from spherical symmetry (Lambeck, 1988). Using these techniques, it has been found that the difference $(C - A)$ in the moments of inertia has the numerical value (Stacey, 1992):

$$J_2 = \frac{(C - A)}{Ma^2} = 1.082626 \times 10^{-3}. \qquad (20.71)$$

In this expression a is the radius of the Earth, and the term Ma^2 is a measure of the average moment of inertia of the Earth. Expression (20.71) states therefore that the relative departure of the mass distribution of the Earth from spherical symmetry is of the order 10^{-3}. This effect is small, but this number carries important information about the dynamics of our planet. In fact, the time derivative $\dot{J}_2$ of this quantity has been measured as well (Yoder et al., 1983)! This quantity is of importance because the rotation rate of the Earth slowly decreases due to the braking effect of the tidal forces that act on the oceans and the solid earth (Stacey, 1992); the Earth adjusts its shape to this decrease in the rate of rotation. The measurement of $\dot{J}_2$ therefore provides important information about the response of the Earth to a time-dependent loading.

20.8 Fifth force

Gravity is the force in nature that was first understood by mankind through the discovery by Newton of the law of gravitational attraction. The reason that the gravitational force was understood first is that this force manifests itself in the macroscopic world in the motion of the Sun, Moon, and planets. Later the electromagnetic force, and the strong and weak interactions were discovered. This means that presently four forces are thought to be operative in nature. Of these four forces, the electromagnetic force and the weak nuclear force can now be described by a single unified theory.

In the 1980s, geophysical measurements of gravity suggested that the gravitational field behaves in a different way over geophysical length scales (between meters and kilometers) than over astronomical length scales ($>10^4$ km). This has led to the speculation that this discrepancy is due to a fifth force in nature. This speculation and the observations that fuelled this idea are clearly described by Fischbach and Talmadge (1992). The central idea is that in Newton's theory of gravity the gravitational potential generated by a point mass M is given by (20.42):

$$V_N(\mathbf{r}) = -\frac{GM}{r}. \tag{20.72}$$

The hypothesis of the fifth force presumes that a new potential should be added to this Newtonian potential: this fifth potential is given by

$$V_5(\mathbf{r}) = -\alpha \frac{GM}{r} e^{-r/\lambda}. \tag{20.73}$$

Note that this potential has almost the same form as the Newtonian potential $V_N(\mathbf{r})$, the main differences are that the fifth force decays exponentially with distance over a length λ and that it is weaker than Newtonian gravity by a factor α. This idea was prompted by measurements of gravity in mines, in the ice cap of Greenland, on a 600-m high telecommunication tower, and by a number of other experimental results that seemed to disagree with the gravitational force that follows from the Newtonian potential $V_N(\mathbf{r})$.

Problem a Effectively, the fifth force leads to a change of the gravitational constant G with distance. Compute the gravitational acceleration $\mathbf{g}(r)$ for the combined potential $V_N + V_5$ by taking the gradient and write the result as $-G(r)M\hat{\mathbf{r}}/r^2$ to show that the effective gravitational constant is given by:

$$G(r) = G\left[1 + \alpha\left(1 + \frac{r}{\lambda}\right)e^{-r/\lambda}\right]. \tag{20.74}$$

The fifth force thus effectively leads to a change of the gravitational constant over a characteristic distance λ. This effect is small: in 1991 the value of α was estimated to be less than 10^{-3} for all estimates of λ longer than 1 cm (Fischbach and Talmadge, 1992).

In doing geophysical measurements of gravity, one has to correct for perturbing effects such as the topography of the Earth's surface and density variations within the Earth's crust. It has now been shown that the uncertainties in these corrections are much larger than the observed discrepancy between the gravity measurements and Newtonian gravity (Parker and Zumberge, 1989). This means that the issue of the fifth force seems be put to rest for the moment, and that the physical world appears again to be governed by four fundamental forces.

21

Probability and statistics

Probability and statistics help us characterize the degree to which quantities may vary for reasons that are unknown to us. Perhaps it is a fear of the unknown that makes many shun the topic of statistics. Yet, dealing with uncertainty is an essential aspect of being a scientist; all measurements have errors, some processes are too complicated to characterize deterministically, and according to quantum mechanics, the concept of probability is woven into the basic properties of light, matter, and even vacuum (Merzbacher, 1961; Sakurai, 1978). In this chapter we first show that the concept of probability has different connotations across the scientific community. We introduce concepts such as statistical moments, specifically the mean and standard deviation, and show how to estimate these from measurements. In addition, we present several probability density functions that are of importance in the sciences. In the last section we introduce Bayes' theorem, which allows us to quantify how uncertainties change if we add new information.

21.1 Basic concepts in probability

Probability is a concept that appears to be straightforward, but that is not so trivial when we try to clarify what probability really means. In fact, the concept of probability has a different meaning depending on the context in which it is used, and on the person who uses it.

Let us consider the first meaning of probability using the game of roulette as an example (Figure 21.1). In the European version of this gambling game, the ball on a wheel lands in one of 37 positions that are numbered from 0 to 36.[1] When the ball lands on slot number 0, the bets made by players go to the bank (usually the casino). One out of the 37 positions in which the ball can end has the number 0, and the probability of the ball ending in this position is defined as the relative number of events with this outcome. The probability that the ball ends in position 0 thus

---

[1] The American roulette wheel has an additional "00" slot, while Russian roulette has no slots at all!

Figure 21.1 A European roulette wheel with alternating black and dark grey slots and a light grey slot with the number 0. This image is from the open clipart library in Inkscape.

is $1/37$. There are 36 out of 37 possibilities that the ball ends in a position with a number different than 0. Denoting the number that the ball ends on by n, these probabilities are given by

$$P(n = 0) = \frac{1}{37} \quad P(n \neq 0) = \frac{36}{37}. \tag{21.1}$$

In this interpretation, probability is defined as the relative occurrence of events. This is called a *frequentist* interpretation of probability, because the relative frequency of events defines the probability. This frequentist interpretation is most popular when introducing probability and leads to exercises involving the computation of the probability that one draws a pair of matching socks from a drawer with a given number of socks of different colors. We will spare the reader such mind-numbing exercises.

In roulette the ball either ends up at number 0, or it ends up at a number that is nonzero. These two options exhaust all possibilities, and therefore the probability that the ball lands either on the number 0, or at a nonzero number is equal to 1. Indeed, expression (21.1) gives $P(n = 0) + P(n \neq 0) = 1/37 + 36/37 = 1$. In general the probability satisfies

$$\sum_n P(n) = 1, \tag{21.2}$$

where the sum $\sum_n$ is over all possible outcomes.

The concept of probability is, however, sometimes used in a different way. Suppose the weatherman says that the probability of rain on March 20, 2012, is equal to 70%. This statement obviously cannot be interpreted in a frequentist way, because there is only one day with the date of March 20, 2012. In this case, the statement of the weatherman reflects her subjective idea that it is more likely that it will rain than that it remains dry, but she cannot make a hard promise of any rain.

Perhaps the statement of the weatherman might mean that on all days for which she forecasted a chance precipitation of 70%, there actually is precipitation 70% of the time. This interpretation would bridge the gap between probability as a subjective assessment of outcomes of events, and a frequentist interpretation, but few people would interpret the statement of the weatherman in this way. This frequentist interpretation is especially difficult when it concerns events that rarely occur, such as devastating earthquakes, for example. This makes it difficult to verify or refute statistical statements made in earthquake prediction about the occurrence of large earthquakes (Lomnitz, 1994).

In the game of roulette, the outcome is discrete in the sense that the number n on which the ball lands can only have a finite number of values that are integer values from 0 to 37. Sometimes the outcome of an experiment can be any real number. For example, when measuring the length of a table, the outcome can be any positive real number x. In that case, one cannot use the frequentist definition of probability defined earlier. After all, the probability of the length being given exactly by 98.32175336 cm is equal to zero. However, the probability that the length of the table lies between, say, 98.0 cm and 98.5 cm is finite and well defined. For this reason, one defines a *probability density function* $p(x)$ such that the probability that the variable lies between x and $x + dx$ is defined to be equal to $p(x)dx$. The probability that the value of x lies between two bounds a and b is given by

$$P(a < x < b) = \int_a^b p(x)dx. \tag{21.3}$$

The probability that x can take on any positive value is equal to 1; hence, the counterpart of expression (21.2) for a continuous probability density function is given by

$$\int_{-\infty}^{\infty} p(x)dx = 1. \tag{21.4}$$

Probabilities are important, but in general one is more interested in the expected outcome of an event than in a specific probability. For example, in the casino the bottom line for players is how much they have gained or lost. In order to quantify expected outcomes, we introduce the concept *expectation value* with the example of the game of roulette. Suppose one bets $100 in roulette that the ball ends on an even number. According to the rules of the game there are three possible outcomes: (1) if the ball ends on an even number, the player wins $100, (2) if the ball ends on an odd number, the player loses $100, and (3) if the ball ends on the "0," the player loses $100. There are 18 possibilities for the ball to end on an even number different from 0, there are 18 possibilities to end up on an odd number, and there is one possibility to end up on 0. The expected profit follows by summing over these three possibilities:

$$\langle profit \rangle = \frac{18}{37} \times \$100 + \frac{18}{37} \times (-\$100) + \frac{1}{37} \times (-\$100) = -\$2.70. \quad (21.5)$$

In this chapter the angled brackets $\langle \cdots \rangle$ denote the expectation value. It follows from expression (21.5) that for the even/odd bet in roulette the player is expected to lose 2.7% each time the ball is rolled.[2] In general, the expectation value of a quantity f_n for a probability distribution P_n is given by

$$\langle f \rangle = \sum_n f_n P_n, \quad (21.6)$$

while the expectation value of a function $f(x)$ for a continuous probability density function $p(x)$ is given by

$$\langle f \rangle = \int_{-\infty}^{\infty} f(x) p(x) dx. \quad (21.7)$$

Note that the arguments to introduce the expectation value were those of a frequentist. The expected gain or loss in a casino is only meaningful when viewed as the outcome of averaging over many events. When probability is interpreted as an estimation that an event will occur, one may not be able to average over many occurrences, as one does in a casino. This may affect the interpretation of the concept of expectation value.

21.2 Statistical moments

In this section we study the expectation value of powers of random variables for a continuous probability density function $p(x)$. By replacing $p(x)dx$ by p_n, these results are easily generalized to a variable that can only assume discrete values. The *moment* defined as a the expectation value of x^n:

$$\langle x^n \rangle = \int x^n p(x) dx. \quad (21.8)$$

In this expression, and the following, we assume the integration to be carried out over all possible values of the variable x.

The concept of moments is related to important statistical quantities. For example, the *mean* μ is the first moment:

$$\mu = \langle x \rangle = \int x p(x) dx. \quad (21.9)$$

The concepts of statistical moments is akin to ideas used in the multipole expansion in Section 20.6. For a mass distribution $\rho(\mathbf{r})$, the center of mass is according

---

[2] This money goes to the casino, which explains why running a casino is a good bet. It does raise the question, however, why anybody would spend any money on roulette.

expression (20.55) proportional to $\mathbf{r}_g \propto \int \mathbf{r}\rho(\mathbf{r})dV$. This equation is similar to expression (21.9) where the mass density plays the same role as the probability density. The mean of a probability density function thus plays the same role as the center of mass for an extended mass distribution. Similarly, according to expression (20.60), the xx-moment of inertia of a mass distribution depends on $\int x^2\rho(\mathbf{r})dV$, which corresponds to expression (21.8) for $n = 2$.

To familiarize ourselves with distributions and moments, consider the *uniform distribution*, which states that x lies in the interval $a < x < b$ with constant probability:

$$p(x) = \begin{cases} \dfrac{1}{b-a} & \text{for} \quad a < x < b, \\ 0 & \text{for} \quad x < a \quad \text{or} \quad x > b. \end{cases} \tag{21.10}$$

Problem a   Show that this distribution is normalized. In other words, it satisfies expression (21.4).

Problem b   Show that the mean of this distribution is given by

$$\mu = \frac{1}{2}(b + a). \tag{21.11}$$

The solid line in Figure 21.2 is an example of a uniform distribution, where the dashed vertical line is the mean value.

We next consider the Gauss (or *normal*) distribution:

$$p(x) = \sqrt{\frac{c}{\pi}}\, e^{-c(x-x_0)^2}. \tag{21.12}$$

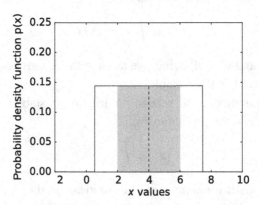

Figure 21.2   The probability density function for a uniform distribution with $a = 0.536$ and $b = 7.464$, which corresponds to mean $\mu = 4$ and standard deviation $\sigma = 2$. Values within one standard deviation from the mean are shaded.

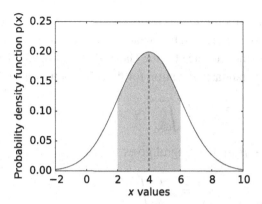

Figure 21.3 The probability density function for a normal distribution with mean $\mu = 4$ and standard deviation $\sigma = 2$. Values within one standard deviation from the mean are shaded.

Problem c Use equation (4.49) to show that this distribution is normalized.

Problem d Derive that the mean of this distribution is given by

$$\mu = x_0. \tag{21.13}$$

The solid line in Figure 21.3 is an example of a normal distribution, where the dashed vertical line is the mean value.

The spread of a distribution is quantified by the *standard deviation* σ that is defined by

$$\sigma^2 = \langle (x - \mu)^2 \rangle. \tag{21.14}$$

The square of the standard deviation (σ^2) is called the *variance*.

Problem e Use the definition $\mu = \langle x \rangle$ to show that

$$\sigma^2 = \langle x^2 \rangle - \langle x \rangle^2. \tag{21.15}$$

The variance thus can be expressed in the first and second moments of the distribution. Note that using expression (21.9), equation (21.15) can also be written as

$$\langle x^2 \rangle = \sigma^2 + \mu^2. \tag{21.16}$$

This identity is reminiscent of the quadratic addition used in Pythagoras rule.

As an example we return to the uniform distribution (21.10).

Problem f Show that the variance of this distribution is given by

$$\sigma^2 = \frac{1}{12}(b - a)^2. \tag{21.17}$$

The variance is thus directly related to the width $b - a$ of this distribution. The gray box in Figure 21.2 is centered on the mean value and two standard deviations wide.

We next compute the variance for the Gauss distribution (21.12). From equations (21.12)–(21.14) the standard deviation for this distribution is given by

$$\sigma^2 = \sqrt{\frac{c}{\pi}} \int_{-\infty}^{+\infty} (x - \mu)^2 e^{-c(x-\mu)^2} dx. \qquad (21.18)$$

Replacing $x \to x + \mu$ in the integral gives

$$\sigma^2 = \sqrt{\frac{c}{\pi}} \int_{-\infty}^{+\infty} x^2 e^{-cx^2} dx. \qquad (21.19)$$

To evaluate the integral, we introduce the following notation for the Gaussian integral

$$I(c) = \int_{-\infty}^{+\infty} e^{-cx^2} dx. \qquad (21.20)$$

Problem g   Use this definition and expression (21.19) to show that

$$\sigma^2 = -\sqrt{\frac{c}{\pi}} \frac{dI(c)}{dc}. \qquad (21.21)$$

Problem h   Use expression (4.49) to evaluate the derivative in the previous expression and show that $c = 1/2\sigma^2$.

The gray box in Figure 21.3 is centered on the mean value and two standard deviations wide. This last identity can be used to express the *Gauss distribution* (21.12) explicitly in its mean and variance:

$$p(x) = \frac{1}{\sqrt{2\pi\sigma^2}} e^{-(x-\mu)^2/2\sigma^2}. \qquad (21.22)$$

This distribution is also known as the *normal distribution*.

We next consider the probability that the variable x lies within a half standard deviation from the mean.

Problem i   Show that for the uniform distribution the probability is given by

$$P(\mu - \sigma/2 < x < \mu + \sigma/2) = \frac{1}{\sqrt{12}} \approx 29\%. \qquad (21.23)$$

As shown in Table 24.1, the corresponding probability for the Gauss distribution is equal to 35.2%, which differs from the probability computed in expression (21.23) for the uniform distribution. This means that the interpretation of the standard deviation as measure of error depends on the statistical distribution used.

Problem j According to Table 24.2 the probability $P(x > \mu + 3\sigma)$ that a variable that obeys the Gauss distribution lies more than 3σ beyond the mean is 0.135%. Show that this probability is exactly equal to zero for the uniform distribution and explain this answer by drawing the distribution and the point $\mu + 3\sigma$.

21.3 The moment-generating function

A useful tool for computing all moments is the *moment-generating function* that is for a probability density function $p(x)$ defined as

$$M(k) \equiv \langle e^{ikx} \rangle = \int p(x) e^{ikx} dx. \tag{21.24}$$

A comparison with equation (14.28) shows that $M(k)$ is the Fourier transform of $p(x)$. (The sign of the exponential and the factor of $1/(2\pi)$ are different than in expression (14.28), but as shown in Section 14.5 this is irrelevant.) We next explore how the moment-generating function can be used to compute the statistical moments

Problem a Use a Taylor expansion of the exponent in equation (21.24) to show that

$$M(k) = \sum_{n=0}^{\infty} \frac{(ik)^n}{n!} \langle x^n \rangle = 1 + ik\langle x \rangle - \frac{1}{2}k^2 \langle x^2 \rangle - \frac{i}{6}k^3 \langle x^3 \rangle + \cdots \tag{21.25}$$

This expression shows that all statistical moments can be found by computing the moment-generating function, and by making a Taylor expansion of $M(k)$ in the variable k.

Problem b Show that the Taylor expansion (21.25) is related to the mean and standard deviation in the following way:

$$M(k) = 1 + ik\mu - \frac{1}{2}k^2 \left(\sigma^2 + \mu^2 \right) + \mathcal{O}(k^3). \tag{21.26}$$

We next apply the moment-generating function to the exponential distribution that is defined as

$$p(x) = \begin{cases} 0 & \text{for } x < 0, \\ \dfrac{1}{a} e^{-x/a} & \text{for } x > 0. \end{cases} \tag{21.27}$$

Problem c Verify that this distribution is normalized.

Problem d   Derive that

$$M(k) = \frac{1}{1 - ika}.$$ (21.28)

Problem e   Use a Taylor expansion of this expression with equation (21.24) to show that

$$\langle x^n \rangle = n! a^n.$$ (21.29)

Problem f   Use the previous expression to derive that

$$\mu = a \quad \text{and} \quad \sigma = a.$$ (21.30)

The moment-generating function is an efficient tool for computing all the moments for this distribution. As an alternative, the moments could have been computed using the gamma function of equation (24.19). Note that according to equation (21.29) all the moments depend on the same parameter a. The reason is that this is the only parameter that occurs in the probability density function.

21.4 Estimating the mean and variance from measurements

In this section we consider the example where we have a number of repeated measurements $x_1, x_2, \cdots, x_N$ of the same variable x with an unknown probability density function. From the measurements, we seek to find the mean and standard deviation of this distribution. For example, measurement errors of the period of pendulum results in N different values. We assume that all measurements x_i are drawn from the same distribution with mean μ and standard deviation σ:

$$\langle x_i \rangle = \mu \quad \text{and} \quad \langle (x_i - \mu)^2 \rangle = \sigma^2.$$ (21.31)

We also assume that the measurements are independent, which means that the fluctuations of different measurements x_i and x_j around the mean are uncorrelated:

$$\langle (x_i - \mu)(x_j - \mu) \rangle = 0 \quad \text{for} \quad i \neq j.$$ (21.32)

Expressions (21.31) and (21.32) can be combined into a single form:

$$\langle (x_i - \mu)(x_j - \mu) \rangle = \sigma^2 \delta_{ij}.$$ (21.33)

The left-hand side of this expression is called the *covariance matrix*; it measures the degree to which statistical fluctuations in x_i and x_j are related.

We define the sample mean $\bar{x}$ as the number that is as close as possible to all measurements. We need to be more precise than "as close as possible," and will define it to be the number that minimizes the least-squares misfit between $\bar{x}$ and all measurements:

$$S = \frac{1}{2} \sum_{i=1}^{N} (x_i - \bar{x})^2. \tag{21.34}$$

Problem a The minimum of S as a function of $\bar{x}$ can be found by setting the derivative of S with respect to $\bar{x}$ equal to zero. Do this and show that S is minimized by

$$\bar{x} = \frac{1}{N} \sum_{i=1}^{N} x_i. \tag{21.35}$$

In the following we refer to $\bar{x}$ as the *sample mean*.

Problem b The variable $\bar{x}$ is a function of the random variables $x_1, x_2, \cdots, x_N$, and therefore is a statistical variable as well. Show that its expectation value satisfies

$$\langle \bar{x} \rangle = \mu. \tag{21.36}$$

This important result shows that the sample mean is, on average, equal to the true mean. In statistical jargon it states that the sample mean is an *unbiased estimator* of the true mean, in the sense that its expectation value is equal to the parameter it seeks to estimate.

Let us next express the variance $\sigma_{\bar{x}}^2$ of $\bar{x}$ into the variance of the individual measurements. From equations (21.15) and (21.36), we find that

$$\sigma_{\bar{x}}^2 = \langle \bar{x}^2 \rangle - \langle \bar{x} \rangle^2 = \langle \bar{x}^2 \rangle - \mu^2. \tag{21.37}$$

Problem c When we insert expression (21.35) in the right-hand side, we need the expectation value of $\langle x_i x_j \rangle$. Use equation (21.33) to show that this expectation value is given by

$$\langle x_i x_j \rangle = \sigma^2 \delta_{ij} + \mu^2. \tag{21.38}$$

Problem d Use this result and the definition (21.35) to show that

$$\langle \bar{x}^2 \rangle = \frac{1}{N^2} \sum_{i,j=1}^{N} \left(\sigma^2 \delta_{ij} + \mu^2 \right). \tag{21.39}$$

Problem e Evalue the double sum to derive that

$$\langle \bar{x}^2 \rangle = \frac{\sigma^2}{N} + \mu^2. \tag{21.40}$$

Problem f   Show that together with equation (21.37) this expression implies that the variance of the sample mean is given by

$$\sigma_{\bar{x}} = \frac{\sigma}{\sqrt{N}}. \tag{21.41}$$

This important result states that by averaging N repeated measurements, the standard deviation of the estimated mean is reduced by a factor $1/\sqrt{N}$. This is a formal justification of the well-known experimental procedure to average over repeated measurements. Note, however, that because of the behavior of the square root, this is a game of diminishing returns as the number of measurements increases; by taking 11 measurements, the standard deviation is reduced by a factor 0.3 – one needs 25 measurements to reduce the variance with a factor 0.2, while 100 measurements are needed to reduce the variance with a factor 0.1. Expression (21.41) explains why averaging over repeated measurements is useful. However, as long as we do not know the standard deviation σ of the individual measurements, we do not know the variance of the sample mean either.

One might think that because of equation (21.14), the standard deviation is given by the spread around the mean, and that an estimator of the variance of the individual measurements is given by

$$s^2 = \frac{1}{N} \sum_{i=1}^{N} \langle (x_i - \bar{x})^2 \rangle. \tag{21.42}$$

We show, however, in the following that this quantity underestimates the standard deviation.

Problem g   In expression (21.42) write $x_i - \bar{x} = (x_i - \mu) - (\bar{x} - \mu)$, and expand the square to show that

$$s^2 = \frac{1}{N} \sum_{i=1}^{N} \langle (x_i - \mu)^2 \rangle - \frac{2}{N} \sum_{i=1}^{N} \langle (x_i - \mu)(\bar{x} - \mu) \rangle + \frac{1}{N} \sum_{i=1}^{N} \langle (\bar{x} - \mu)^2 \rangle. \tag{21.43}$$

Problem h   The expectation value in the first term is equal to σ^2, while the expectation value in the last term is equal to $\sigma_{\bar{x}}^2$. Use this to show that

$$s^2 = \sigma^2 + \sigma_{\bar{x}}^2 - \frac{2}{N} \sum_{i=1}^{N} \langle (x_i - \mu)(\bar{x} - \mu) \rangle. \tag{21.44}$$

Problem i Use the definition (21.35) and expression (21.33) to derive that

$$\langle (x_i - \mu)(\bar{x} - \mu) \rangle = \frac{1}{N} \sum_{j=1}^{N} \langle (x_i - \mu)(x_j - \mu) \rangle = \frac{1}{N} \sum_{j=1}^{N} \sigma^2 \delta_{ij} = \frac{\sigma^2}{N}. \quad (21.45)$$

Problem j Insert this result into equation (21.44), use equation (21.41) to eliminate $\sigma_{\bar{x}}^2$, and show that

$$s^2 = \frac{N-1}{N} \sigma^2. \quad (21.46)$$

This equation implies that s^2 is not a good estimator of the variance σ^2, except when the sample size N is very large and $(N-1)/N \to 1$.

Problem k Use the definition (21.42) to derive that

$$\sigma^2 = \frac{1}{N-1} \sum_{i=1}^{N} \langle (x_i - \bar{x})^2 \rangle. \quad (21.47)$$

This expression shows that, perhaps counterintuitively, one needs to divide by $N-1$ instead of N to find the correct value for σ. This change accounts for the fact that the sample mean $\bar{x}$ is not necessarily equal to the true mean μ. As shown in Problem a, $\bar{x}$ is the number that minimizes the least-squares misfit $\sum_{i=1}^{N} (x_i - \bar{x})^2$. Since $\bar{x}$ minimizes the least-squares misfit, $\langle \sum_{i=1}^{N} (x_i - \bar{x})^2 \rangle \leq \langle \sum_{i=1}^{N} (x_i - \mu)^2 \rangle$. According to the right-hand side of expression (21.33), $\langle (x_i - \mu)^2 \rangle = \sigma^2$; hence, $\langle \sum_{i=1}^{N} (x_i - \bar{x})^2 \rangle \leq N\sigma^2$. The denominator $1/(N-1)$ instead of $1/N$ in equation (21.47) corrects for this inequality. In the limit $N \to \infty$, $\bar{x}$ approaches μ, and $N/(N-1) \to 1$, so that the distinction between s^2 and σ^2 in expression (21.46) disappears.

21.5 Error propagation for linear problems

In this section we study the problem that we have a number of random variables $x_1, x_2, \cdots, x_N$. The variables have a mean that is not necessarily the same for all variables:

$$\langle x_i \rangle = \mu_i. \quad (21.48)$$

The covariance between the different variables is denoted as

$$\langle (x_i - \mu_i)(x_j - \mu_j) \rangle = C_{ij}^{(x)}. \quad (21.49)$$

In contrast to the problem in the previous section, in particular expression (21.32), we don't assume that the variables are uncorrelated. The *covariance matrix*

$\mathbf{C}^{(x)}$ therefore is not necessarily diagonal. Let us assume that from the variables $x_1, x_2, \cdots, x_N$ we create new variables $y_1, y_2, \cdots, y_M$ by linear superposition through a matrix $\mathbf{A}$:

$$\mathbf{y} = \mathbf{A}\mathbf{x}. \tag{21.50}$$

Problem a We denote the mean of the new variables y_i by $v_i = \langle y_i \rangle$. Take the expectation value of equation (21.50) to derive that

$$v_i = \sum A_{ij} \mu_j. \tag{21.51}$$

By analogy with equation (21.49), we define the covariance matrix for the new variables by

$$\langle (y_i - v_i)(y_j - v_j) \rangle = C_{ij}^{(y)}, \tag{21.52}$$

and we seek to find the relation between $\mathbf{C}^{(y)}$ and $\mathbf{C}^{(x)}$.

Problem b Insert expression (21.50) into the previous expression to show that

$$C_{ij}^{(y)} = \sum_{k,l} A_{ik} A_{jl} C_{kl}^{(x)}. \tag{21.53}$$

Note that this is the same transformation rule as expression (26.45) for the transformation of a matrix under a unitary coordinate transformation.

Problem c Show that expression (21.53) can also be written as

$$\mathbf{C}^{(y)} = \mathbf{A} \mathbf{C}^{(x)} \mathbf{A}^T, \tag{21.54}$$

where the superscript T denotes the transpose.

Expression (21.54) describes how errors in the original parameters propagate into errors of the new parameters. This concept returns in Chapter 22, when we assess inverse solutions in the presence of errors in the data.

21.6 The Gauss distribution from a random walk

We introduced the *Gauss distribution* that is given by

$$p(x) = \frac{1}{\sqrt{2\pi\sigma^2}} e^{-(x-\mu)^2/2\sigma^2}, \tag{21.22}$$

where μ and σ are the mean and standard deviation, respectively, of this distribution. The Gauss distribution is ubiquitous in statistics; Jaynes (2003) gives several reasons why the Gauss distribution describes so many statistical processes. One of these reasons is the *central limit theorem*, which states that the sum of N independent random numbers approaches the Gauss distribution as $N \to \infty$, irrespective of the distribution of the individual random numbers. Random processes often consist of the superposition of numerous underlying statistical variations; hence, the central limit theorem is an explanation why so many statistical variables follow the Gauss distribution. In this section we derive the central limit theorem as a result of a random walk process. For simplicity we assume the mean of the each step in the random walk to vanish.

We consider a random walk process where a particle can be at discrete locations along the x-axis that are given by

$$x_i = i \Delta. \tag{21.55}$$

The particle jumps after each time interval δ to a new location. After n jumps, the particle has probability $p_n(x_i)$ to be at location x_i. At $t = 0$ the particle is at the origin; hence,

$$p_0(x_i) = \delta_{i0}, \tag{21.56}$$

where δ_{i0} is the Kronecker delta. When the particle jumps to the next position after each time δ, it does so with a transition probability T_{i-j}. This quantity gives the probability that the particle jumps from position x_j to position x_i. We assume that the transition probability depends only on the relative location of these positions; hence, this probability depends only on the difference $i - j$. The probability at step $n + 1$ follows from the probability at step n and the transition probability:

$$p_{n+1}(x_i) = \sum_j T_{i-j} p_n(x_j). \tag{21.57}$$

The transition probability can be anything, but it must, of course, ensure that probability is conserved. This means that the total probability to jump to location x_i from every possible location x_j must be equal to one; hence, $\sum_j T_{i-j} = 1$. Renaming the summation variable $i - j \to k$, this normalization condition can also be written as

$$\sum_k T_k = 1. \tag{21.58}$$

We next study the random walk as a function of the number of steps taken. Making the substitution $j \to i - k$ changes expression (21.57) into

$$p_{n+1}(x_i) = \sum_j T_k p_n(x_{i-k}). \tag{21.59}$$

Problem a Using equation (21.55), x_{i-k} can be rewritten as

$$x_{i-k} = x_i - k \Delta. \tag{21.60}$$

Insert this in equation (21.59) and carry out a second-order Taylor expansion in p_n to show that to second order in Δ

$$p_{n+1}(x_i) = \sum_k T_k p_n(x_i) - \Delta \sum_k k T_k p_n'(x_i) + \frac{1}{2}\Delta^2 \sum_k k^2 T_k p_n''(x_i), \quad (21.61)$$

where the prime denotes the derivative with respect to x.

Each of the terms in the right-hand side is the product of a sum $\sum_k k^m T_k$ times the probability $p_n(x_i)$ or its derivatives; hence, the k-summation can be factored out in each term. The first term in the right-hand side contains the sum $\sum_k T_k$, which by virtue of expression (21.58) is equal to one. The second term in the right-hand side contains a sum $\Delta \sum_k k T_k$. In this quantity, $k\Delta$ is the step taken in the random walk. When weighted with the transition probability T_k for taking this step, the sum gives the mean displacement take in one step the random walk:

$$\sum_k k \Delta T_k = \mu. \quad (21.62)$$

In the following we assume, for simplicity, that this mean step is equal to zero. Finally, the last term in the right-hand side of equation (21.61) contains the sum $\sum_k k^2 \Delta^2 T_k$. This is the expectation value of the square of the step length $\langle k^2 \Delta^2 \rangle$; hence,

$$\sum_k k^2 \Delta^2 T_k = \langle k^2 \Delta^2 \rangle = \sigma^2 + \mu^2, \quad (21.63)$$

where σ^2 is the variance of the distance covered in one step, and where we used expression (21.16) in the last identity. When the mean step length vanishes, as we assume here, $\mu = 0$, so that $\sum_k k^2 \Delta^2 T_k = \sigma^2$. Inserting these results in expression (21.61) then gives

$$p_{n+1}(x_i) = p_n(x_i) + \frac{1}{2}\sigma^2 p_n''(x_i). \quad (21.64)$$

We can next rename the location $x_i \to x$, rename the time of step n in the random walk as $n\delta \to t$, and make the replacement $p_n(x_i) \to p(x, t)$. In this notation step $n+1$ is reached at time $(n+1)\delta = t+\delta$; hence, $p_{n+1}(x_i)$ should in this new notation be replaced by $p(x, t + \delta)$. With these replacements, equation (21.64) changes into

$$p(x, t + \delta) = p(x, t) + \frac{1}{2}\sigma^2 \frac{\partial^2 p(x, t)}{\partial x^2}. \quad (21.65)$$

Problem b   Make a first-order Taylor expansion of the time dependence of the left-hand side of this expression to show that $p(x, t)$ satisfies

$$\frac{\partial p(x, t)}{\partial t} = \frac{\sigma^2}{2\delta} \frac{\partial^2 p(x, t)}{\partial x^2}. \quad (21.66)$$

This equation is simply the diffusion equation (25.31) with diffusion constant

$$\kappa = \sigma^2/(2\delta). \tag{21.67}$$

The initial condition (21.56) translates in this new notation into

$$p(x, t = 0) = \delta(x). \tag{21.68}$$

The solution to the diffusion equation with this initial condition is according to equation (18.15) equal to

$$p(x, t) = \frac{1}{\sqrt{4\pi\kappa t}} e^{-x^2/4\kappa t}. \tag{21.69}$$

Problem c This expression gives the probability in terms of continuous variables. We can revert to the discrete random walk process by using that $t = n\delta$ and by using the diffusion constant of equation (21.67). Make these substitutions to show that

$$p_n(x) = \frac{1}{\sqrt{2\pi\sigma_n^2}} e^{-x^2/2\sigma_n^2}, \tag{21.70}$$

with

$$\sigma_n^2 = n\sigma^2. \tag{21.71}$$

Equation (21.70) implies that the result of a random walk with zero mean is a Gaussian centered on the starting point. We can conclude that the ultimate probability density distribution is Gaussian *regardless of the details of the transition probability* T_k of the random walk. The variance of the position after n steps is given by $\sigma_n^2 = n\sigma^2$. The random walk consists of adding random numbers that give the movement in each step of the random walk. Expression (21.70) thus also gives the probability density function for the sum of n uncorrelated random numbers drawn from a probability density function with zero mean. When the mean of each step is equal to μ, the mean of the sum shifts on average over a distance $n\mu$, and expression (21.70) should be replaced by

$$p_n(x) = \frac{1}{\sqrt{2\pi\sigma_n^2}} e^{-(x-n\mu)^2/2\sigma_n^2}. \tag{21.72}$$

This is a Gauss distribution with mean $n\mu$ and standard deviation $\sigma_n = \sqrt{n}\,\sigma$.

21.7 Derivation of the Poisson distribution

The Poison distribution describes the probability of random discrete events. The events can be the number of radioactive particles that decay in a given time interval, the number of cars passing a road per minute, the number of a certain type of accidents occurring, or the number of defective lightbulbs in a box with 2,000 bulbs.

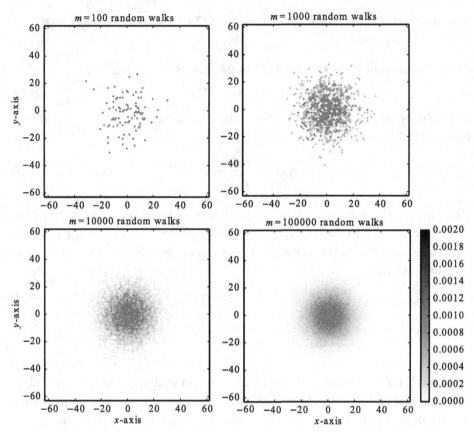

Figure 21.4   The probability density of m random walks with 250 steps from the origin. For large values of m, this approximates a two-dimensional Gaussian with a zero mean and a variance of 250, describing the 2D diffusion of these random walkers.

In the following we refer to these occurrences as *events*. The mean number of events per unit time is denoted by λ. A major assumption of the Poisson distribution is that the events are uncorrelated. This means that the probability of an event occurring does not depend on the presence or absence of earlier events.

In the following $P_n(t)$ denotes the probability on n events occurring between zero time and time t. At time zero no event has occurred; hence, the probability of zero events at that time is equal to 1, while the probability of 1 or more events vanishes:

$$P_0(0) = 1,$$
$$P_n(0) = 0 \text{ for } n \geq 1.$$

(21.73)

We next derive a differential equation for $P_n(t)$. The probability of n events at time $t + \Delta t$ is given by

$$P_n(t + \Delta t) = P_n(t) + \lambda \Delta t \, P_{n-1}(t) - \lambda \Delta t \, P_n(t).$$

(21.74)

The first term on the right-hand side gives the probability of having n events at time t. The second gives the probability that an event occurs between times t and $t + \Delta t$. This probability is equal to the probability $\lambda \Delta t$ that an event occurs times the probability that we started out with $n - 1$ events. In the same way, the last term gives the probability that we change from n to $n + 1$ events in the time interval. Since we next proceed by taking the limit $\Delta t \to 0$, we don't have to worry about two or more events occurring in the time interval.

Problem a Make a Taylor expansion of $P_n(t + \Delta t)$ in Δt, take the limit $\Delta t \to 0$, and show that equation (21.74) reduces to

$$\frac{d P_n(t)}{dt} = \lambda \left(P_{n-1}(t) - P_n(t) \right) \quad (n \geq 1). \tag{21.75}$$

Problem b This expression only holds for $n \geq 1$. Repeat this reasoning for the case $n = 0$ and show that $P_0(t)$ satisfies

$$\frac{d P_0(t)}{dt} = -\lambda P_0(t). \tag{21.76}$$

Problem c Show that this differential equation with the initial condition (21.73) has the solution

$$P_0(t) = e^{-\lambda t}. \tag{21.77}$$

Problem d We next solve equation (21.75). Make the substitution $P_n(t) = f_n(t) \exp(\lambda t)$ to show that $f_n(t)$ satisfies

$$\frac{df_n(t)}{dt} = \lambda f_{n-1}(t). \tag{21.78}$$

Problem e This differential equation suggests that $f_n(t)$ is a polynomial in t. For this reason substitute $f_n(t) = a_n t^n$ and show that the coefficients a_n satisfy the recursive relation $a_n = (\lambda/n)a_{n-1}$. Show that this implies that

$$a_n = \frac{\lambda^n}{n!} a_0. \tag{21.79}$$

Problem f It follows from the solution (21.77) that $a_0 = 1$. Use this to show that

$$P_n(t) = \frac{(\lambda t)^n}{n!} e^{-\lambda t}, \tag{21.80}$$

which is called the Poisson distribution.

We next determine the mean number of events predicted by the Poisson distribution. First, the probability that any number of events should occur an arbitrary time t should be equal to one.

Problem g Verify this by showing that

$$\sum_{n=0}^{\infty} P_n(t) = 1. \tag{21.81}$$

Hint: Use a Taylor expansion of $e^{\lambda t}$.

Problem h The probability that $n \neq 0$ events occur should peak as a function of time, because at $t = 0$ no events have occurred while for $t \rightarrow \infty$ infinitely many events have occurred. There must be an intermediary time where the probability to have n events is maximized. Show that this time is given by

$$t_{max}^{(n)} = \frac{n}{\lambda}. \tag{21.82}$$

This equation makes sense by writing it as $\lambda t_{max}^{(n)} = n$. The parameter λ gives the average number of events per unit time. By multiplying it with the time, we get the number of events.

Problem i We denote the mean number of events that occur in a time t by

$$\mu = \langle n \rangle = \sum_{n=0}^{\infty} n P_n(t). \tag{21.83}$$

Show that

$$\mu = \lambda t. \tag{21.84}$$

This equation simply states that the mean number of events in the interval t is given by the probability to have an event per unit time, multiplied with the time interval. Note that the Poisson distribution (21.80) depends only on the product λt, which is equal to μ. This means that the Poisson distribution can also be written as

$$P_n = \frac{\mu^n e^{-\mu}}{n!}. \tag{21.85}$$

We can now do away with the length of the time interval. Equation (21.85) simply gives the probability of n uncorrelated events occurring when the average number of events is equal to μ. The histogram outlines a Poisson distribution, while the dashed vertical line is the mean value.

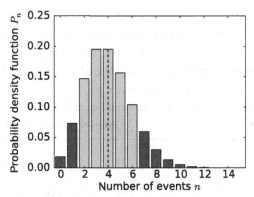

Figure 21.5   The probability density function for a Poisson distribution with mean $\mu = 4$ and standard deviation $\sigma = 2$. Values within one standard deviation from the mean are in lighter gray.

21.8 Properties of the Poisson distribution

In this section we investigate properties of the Poisson distribution (21.85). We first compute the mean and the variance.

Problem a   In computing the mean and the variance, the following identity is useful:

$$n\mu^n = \mu \frac{d\mu^n}{d\mu}. \tag{21.86}$$

Show that this identity holds.

Problem b   Use this identity to show that

$$\langle n \rangle = \sum_{n=0}^{\infty} n P_n = \mu, \tag{21.87}$$

and

$$\langle n^2 \rangle = \sum_{n=0}^{\infty} n^2 P_n = \mu^2 + \mu. \tag{21.88}$$

Problem c   Use these identities and equation (21.16) to show that the standard deviation of the Poisson distribution is given by

$$\sigma = \sqrt{\mu}. \tag{21.89}$$

Expressions (21.88) and (21.89) imply that the relative standard deviation is given by

Table 21.1   *The average number of earthquakes*
occurring per year as a function of magnitude as
given by the U.S. Geological Survey.
Data from `http://earthquake.usgs.gov/`
`earthquakes/search/`.

Magnitude	Annual number of events
> 8	1
> 7	16
> 6	150
> 5	1334

$$\frac{\sigma}{\mu} = \frac{1}{\sqrt{\mu}}. \tag{21.90}$$

This ratio decreases as μ increases, which means that relatively speaking the fluctuations become less important as the events occur more often.

As an example we consider the number of earthquakes per year with a magnitude larger than a certain threshold that occur anywhere in the world as shown in Table 21.1. Strong earthquakes occur less frequently than weak earthquakes. In fact, when the threshold magnitude goes up by one point, the number of earthquakes with a magnitude larger than that threshold becomes about 10 times as small. This dependence on the number of earthquakes is called the *Gutenberg-Richter law* (Gutenberg and Richter, 1956; Stein and Wysession, 2003).

Problem d   Show that according to Poisson statistics the number of events in a year with magnitude > 7 is equal to 16 ± 4, where the number after the $\pm$ is the standard deviation.

Problem e   Show that the number of events with magnitude > 7 per month is approximately equal to 1.33 ± 1.15.

Problem f   Use this result to explain that when two such earthquakes occur within the same month, this has no special significance. (As seismologists, we have often been asked by journalists whether two strong earthquakes in a row are related.)

The important consequence of Poisson statistics is that when an event is more rare (small μ), the relative fluctuations in the number of events (σ/μ) is larger. This is of particular importance for rare, but dramatic, events such as plane crashes or volcanic eruptions.

Table 21.2 *Left column: the number of random walks used to construct Figure 21.4. Second column: (n) the number of walks that visit a cell at $(x, y) = (5, 5)$ with width and height $dx = dy = 1$. Third column: the standard deviation in the number of visits $\sigma = \sqrt{n}$ predicted by equation (21.89). Right column: the relative error in the number of visits.*

random walks (*m*)	*n*	σ	σ/n
100	1	1.0	100%
1,000	2	1.4	71%
10,000	24	4.9	21%
100,000	220	14.0	6%

In the example of Figure 21.4 we needed 100,000 random walks to create a Gauss distribution that was reasonably clear. Why did we need so many random walks? Poisson statistics has the answer. Figure 21.4 was created by counting the number of random walks that visit cells with size $dx = dy = 1$. The number of walks that visit a cell centered on $(x, y) = (5, 5)$ is shown in the second column of Table 21.2. Because the cells are small, the number of random walks n that visit the cell is only about 0.2% of the total number of random walks (shown in the left column). The number of walks hitting a given cell is a Poisson process, and according to expression (21.89), the standard deviation in the number of visits by random walks is equal to $\sigma = \sqrt{n}$ as shown in the third column of Table 21.2. The relative error, defined as the ratio of the standard deviation to the average number of visits ($\sigma/n = 1/\sqrt{n}$), is shown the right column of the table. According to the right column of Table 21.2 for 100 random walks, the relative error in the number of visits is 100%. Indeed, the top left panel of Figure 21.4 shows a random cluster of dots that is vaguely centered on the origin. Even for 10,000 random walks, the relative error is equal to 21%. This is reflected in the speckle that is present in the bottom left panel of Figure 21.4. For 100,000 random walks the relative error is equal to only 6%, and the bottom right panel of Figure 21.4 looks like a Gauss distribution in the (x, y)-plane. One could have used Poisson statistics before doing the numerical simulation to estimate how many random walks would be needed with a given cell size.

21.9 Bayes' theorem

In many situations one knows something, albeit with a given uncertainty, and then does a measurement that provides additional information. How do we combine the information before we did the measurement with the new measurement to quantify the information that we have after the measurement? This is the topic of Bayes' theorem.

For example, the average person has a probability of 10% to carry a certain disease. The state of having the disease is denoted here by A; hence, in this example $P(A) = 10\%$. This probability is called the *prior probability*, because it characterizes our knowledge before we do a measurement. There is a test for the disease, and B is the state of the test having a positive outcome. We next introduce the *conditional probability* $P(A|B)$, which is the probability for getting A given that B is true. In the example, this is the probability of having the disease while testing positive. If the test were perfect, that probability would be 100%, but since the test is not necessarily perfect, this probability may be less. The goal is to estimate this probability. There is another conditional probability $P(B|A)$, which is the probability that the test is positive when the person has the disease. This probability can be determined by testing a sample of persons that carry the disease.

To derive Bayes' theorem, consider $P(A|B)P(B)$. This is the probability that B is true, multiplied by the probability that A is true given that B is true. This combination of probabilities gives the probability that both A and B are true, we denote this by $P(A \cap B)$; hence,

$$P(A \cap B) = P(A|B)P(B). \qquad (21.91)$$

The roles of A and B can be reversed, so that

$$P(B \cap A) = P(B|A)P(A). \qquad (21.92)$$

Problem a   The quantities $P(A \cap B)$ and $P(B \cap A)$ both give the probability that A and B are true; hence, they are equal. Use this with the previous expressions to derive that

$$P(A|B) = \frac{P(B|A)P(A)}{P(B)}. \qquad (21.93)$$

This is the simplest form of *Bayes' theorem*.

Let us next apply Bayes' theorem to our example and assume that $P(B) = 12\%$ and $P(B|A) = 75\%$. The fact that $P(B|A)$ is not equal to 100% means that the test does not always give a positive outcome for an ill person; hence, the test is not perfect. Moreover, the probability to testing positive $P(B)$ is 12%, while only 10% of the population carries the disease. This means that the test has a positive bias.

Problem b   Use Bayes' theorem (21.93) to show that in this example the probability of carrying the disease when testing positive is equal to $P(A|B) = 62\%$.

Table 21.3 *The probabilities of carrying a disease when testing either positive or negative.*

	Carry the disease (A)	Be healthy ($\overline{A}$)
Test positive (B)	62%	38%
Test negative ($\overline{B}$)	3%	97%

This means that the probability of actually having the disease while testing positive for the disease is far from 100%! In fact, this probability is fairly close to flipping a coin, which has a 50% probability for one of the two outcomes.

We can also ask for the probability of having the disease while testing negative. From the patient's point of view, this is an issue of concern, because the negative test might erroneously lead to a lack of treatment. The *complement* $\overline{B}$ of a state B is the situation where B is not true; in our example, it denotes the probability that the test is negative. Since either B or its complement is true, we must always have for any state B:

$$P(B) + P(\overline{B}) = 1. \tag{21.94}$$

Problem c   Use this expression to show that in our example the probability of testing negative is given by $\overline{B} = 88\%$.

Problem d   Bayes' theorem (21.93) can also be used to compute $P(A|\overline{B})$ by replacing B by $\overline{B}$. Use this to show that for our example $P(A|\overline{B})=3\%$.

The quantity $P(A|\overline{B})$ is the probability of carrying the disease while testing negative. This is fairly small at 3%, but keep in mind that the probability of having the disease is only 10%. The probability of carrying the disease when testing either positive or negative is shown in the left column of Table 21.3.

Regardless of whether the test is positive or negative, one either carries the disease or one is healthy. This means that $P(A|B) + P(\overline{A}|B) = 1$ and $P(A|\overline{B}) + P(\overline{A}|\overline{B}) = 1$.

Problem e   Use these relations to compute the probabilities in the right column of Table 21.3 that give the probability of being healthy when testing either positive or negative.

Table 21.3 carries important information. Ideally the test is positive only for persons that carry the disease, and is negative only for persons that don't carry the disease. In that case the diagonal elements in the table would be equal to 100%, and the off-diagonal elements would be equal to 0%. The number in the lower left gives the probability of carrying the disease while testing negative; this outcome is called a *false negative*. The number in the upper right is the probability of testing positive while being healthy; this is called a *false positive*. This probability is fairly large, 38%, which may be a problem when treatment or follow-up diagnostics is either expensive or may carry a significant health risk for a perfectly healthy person.

Bayes' theorem is a tool to quantify what is learned by adding information to a prior state of knowledge. The methodology used in the example of testing for a disease can be used in numerous situations where it is essential to characterize the reliability of additional information and the probability of false positives and false negatives. Examples include forecasting earthquakes and nondestructive testing of materials.

We next consider the situation where A is not either true or untrue, but can have several outcomes $A_1, A_2, \cdots, A_N$. We assume that these outcomes are mutually exhaustive, which means that one of them must be true. Bayes' theorem applies, of course, to each of these outcomes:

$$P(A_i|B) = \frac{P(B|A_i)P(A_i)}{P(B)}. \tag{21.95}$$

We also know that the probability of B being true is the sum of the probabilities of B being true while each of the A_i is true:

$$P(B) = \sum_{j=1}^{N} P(B|A_j)P(A_j). \tag{21.96}$$

Problem f   Use the previous expression to show that for mutually exhaustive outcomes A_i, Bayes' theorem takes the form

$$P(A_i|B) = \frac{P(B|A_i)P(A_i)}{\sum_{j=1}^{N} P(B|A_j)P(A_j)}. \tag{21.97}$$

In the case of a continuous variable A one can simply replace the A_i by the continuous variable A and replace the summation over the different outcomes in the denominator with an integration:

$$p(A|B) = \frac{p(B|A)p(A)}{\int p(B|A')p(A')dA'}, \tag{21.98}$$

where we use the lowercase p to denote a continuous distribution. Note that we integrate over the variable A' in the denominator while A is fixed in the left-hand side and in the numerator. For this reason we use a different symbol in the denominator than in the rest of the expression, just like we sum over A_j in expression (21.97) while A_i is fixed.

Problem g Integrate expression (21.98) over the variable A and show that

$$\int p(A|B)dA = 1. \qquad (21.99)$$

Note that the denominator of the continuous form of Bayes' theorem (21.98) does not depend on A, but only on the integration variable A'. Since we use Bayes' theorem to assess the probability for A, we can treat denominator as a constant, and expression (21.98) can be written as

$$p(A|B) = C \; p(B|A)p(A), \qquad (21.100)$$

where the constant C follows from the normalization condition (21.99). We show in Section 22.9 that Bayes' theorem can be used solve inverse problems.

22

Inverse problems

In many areas of the physical sciences, we do not directly measure the quantity of interest. Instead, the goal is to infer that quantity from some other measurement(s). For example, in medical imaging, we measure electromagnetic or elastic waves propagated through the body. Doctor and patient are after the internal structure of the body, but this structure must be estimated from the speed, scattering, or attenuation of these waves. In the *forward problem*, we know the internal properties of the body and calculate how the waves propagate. In the *inverse problem* we make observations of the waves, and then try to infer the internal body structure responsible for these observations. This may sound a bit abstract. That is why we are going to work through a small example in geophysics; a field where almost every problem is an inverse problem.

22.1 Finding a sinkhole: a geophysical inverse problem

Careful measurements of the gravitational acceleration at the surface of the Earth reveal small variations, indicative of underground heterogeneity. For example, a sinkhole filled with air or water has a mass density ρ that is lower than the surrounding material[1] (Figure 22.1). The contribution of the sinkhole to the gravitational acceleration at a location $\mathbf{r}$ follows from equations (17.15) and (17.16) and making the replacement $1/4\pi\epsilon_0 \rightarrow -G$ to adapt the electrostatics problem to gravity:

$$\mathbf{g}(\rho, \mathbf{r}) = -\int_V \frac{G\rho(\mathbf{r}')}{|\mathbf{r}' - \mathbf{r}|^2} \frac{\mathbf{r} - \mathbf{r}'}{|\mathbf{r} - \mathbf{r}'|} dV', \qquad (22.1)$$

where V' is the volume the sinkhole occupies, and G is the universal gravitational constant. If we assume that the density distribution in the sinkhole is spherically symmetric and centered on $\mathbf{r}' = 0$, then according to expression (8.6) the contribution of the sinkhole to the gravitational acceleration is

$$\mathbf{g}(M, \mathbf{r}) = -\frac{GM}{r^2}\hat{\mathbf{r}}, \qquad (22.2)$$

---

[1] Technically, we are talking about a "karst." The sinkhole is not a sinkhole until this underground feature has collapsed.

Figure 22.1   Schematic of a sinkhole with mass M centered at a depth of r below the surface.

where M is the mass and r the distance from the observer to the center of the sinkhole. Let us only consider the gravitational field measurement at the surface, straight above the sinkhole:

$$g = \mathbf{g} \cdot \hat{\mathbf{z}}. \tag{22.3}$$

When the positive z-direction is chosen downward, $(\hat{\mathbf{z}} \cdot \hat{\mathbf{r}}) = -1$, and

$$g = \frac{GM}{r^2}. \tag{22.4}$$

Of course, the total planet contributes a much greater gravitational acceleration. For a sinkhole, we seek a deficiency in the total gravitational acceleration. We aim to estimate the mass and depth of the sinkhole from very small deviations of the average gravitational acceleration. How small? Geophysicists looking for sinkholes and other small gravitational anomalies, express the value of the gravitational acceleration in units of Gal (1 cm/s^2), named after the Italian philosopher and scientist Galileo (February 15, 1564 – January 8, 1642). Modern gravimeters can easily detect gravity anomalies, associated with sinkholes or other features, on the order of a μGal (Kearey et al., 2002).

Problem a   Assume realistic parameters for a sinkhole and convince yourself that the imprint of a sinkhole on the total gravitational field is on the order of 1-billionth of the average value for the Earth.

We consider the measured anomalies in the gravitational acceleration to be our observations, or data. While the implications of tiny effects of the sinkhole on the overall gravitational field are of no direct concern to us, we aim to use these to gain information about the sinkhole. This defines our inverse problem, and is the typical situation in geophysics: we measure deviations from the average gravitational acceleration on Earth, and want to know the depth r and mass M of the sinkhole responsible for the gravity anomaly.

Based on expression (22.2), we know that the gravitational acceleration depends linearly on the mass M of the sinkhole, but not linearly on its depth r. Again, this is a common situation in geophysical inverse problems: the arrival time of a seismic wave depends on subsurface parameters in a nonlinear way, as well (Section 23.3). As we adjust the seismic velocity model to match our travel-time observation, the path of the seismic ray changes, according to Snell's law. Seismologists typically *linearize* the problem (see expression (23.41)), which is what we will do here too in search of the sinkhole.

Let us assume the measurement of gravity over sinkholes in the area has a long history. The depths and mass of sinkholes average r_0 and M_0, respectively, giving rise to an average measured gravitational acceleration g_0 over sinkholes. We use the subscript 0 to denote a background – or average – value of parameters. Next, we look for sinkholes with parameters close to this background value:

$$r = r_0 + \delta r \quad \text{and} \quad M = M_0 + \delta M, \tag{22.5}$$

and decompose our observations, similarly:

$$g = g_0 + \delta g. \tag{22.6}$$

Problem b   Insert expressions (22.5) and (22.6) into equation (22.2) to show that a first-order Taylor expansion in δr and δM gives

$$\delta g/g_0 = \delta M/M_0 - 2\delta r/r_0. \tag{22.7}$$

Problem c   Rewrite equation (22.7) in the form

$$\mathbf{d} = \mathbf{Am}, \tag{22.8}$$

with

$$\mathbf{d} = \frac{\delta g}{g_0}, \quad \mathbf{A} = \begin{pmatrix} 1 & -2 \end{pmatrix}, \quad \mathbf{m} = \begin{pmatrix} \delta M/M_0 \\ \delta r/r_0 \end{pmatrix}. \tag{22.9}$$

Note that the "data vector" $\mathbf{d}$ has only one component, $\mathbf{A}$ is a 1×2 matrix, and $\mathbf{m}$ is a 2×1 vector containing the model parameters.

The matrix $\mathbf{A}$ is called the forward operator and captures the physics of how the model parameters relate to the data. In this example, we have one datum and two model parameters, which means that the system we are trying to solve is under-determined. As a result, the model parameters are coupled in a way that cause us some grief: increasing the mass and depth of this sinkhole in the right proportions leaves the gravitational anomaly unaffected. Generally, geophysical problems are underdetermined; the Earth is heterogeneous on all length scales, and the number of observations is – by definition – finite.

At first, you might consider multiplying equation (22.8) by the inverse matrix:

$$\mathbf{A}^{-1}\mathbf{A}\mathbf{m} = \mathbf{m} = \mathbf{A}^{-1}\mathbf{d}. \tag{22.10}$$

However, only square matrices can have an inverse, so that $\mathbf{A}\mathbf{A}^{-1} = \mathbf{A}^{-1}\mathbf{A} = \mathbf{I}$. The matrix $\mathbf{A}$ in expression (22.9) is not square, so let us try a different approach.

22.2 Least-squares optimization

Ultimately, we want to try to find a model, or better yet: all the possible models that fit our data. We will later discuss what "fitting the data" really means, but for now we minimize the difference between observed $\mathbf{d}$ and the data $\mathbf{A}\mathbf{m}$ predicted by a model $\mathbf{m}$. In a least-squares sense, we can write this as

$$\min_{\mathbf{m}} S(\mathbf{m}) = \|\mathbf{A}\mathbf{m} - \mathbf{d}\|^2. \tag{22.11}$$

To find the minimum of $S(\mathbf{m})$ as a function of the model parameters, we set the derivative of this function with respect to each of the model parameters equal to zero: $\partial S(\mathbf{m})/\partial m_p = 0$, with m_p one of the model parameters. In order to evaluate this derivative, we first need to express $S(\mathbf{m})$ into the components of the model parameters.

Problem a Complete the square in equation (22.11) and show that the least-squares misfit can be written as

$$S(\mathbf{m}) = \sum_{i,j,k} A_{ij} A_{ik} m_j m_k - 2 \sum_{i,j} d_i A_{ij} m_j + \sum_i d_i^2. \tag{22.12}$$

The last term does not depend on the model parameters at all, so its derivative with respect to m_p vanishes: $\partial \left(\sum_i d_i^2 \right) / \partial m_p = 0$. To take the derivate of the next to last term in equation (22.12), we need the derivative $\partial m_j/\partial m_p$. When $j \neq p$, this derivative vanishes because the different model parameters are independent. When $j = p$, this derivative reduces to $\partial m_p/\partial m_p = 1$. These different situations can be combined into

$$\frac{\partial m_j}{\partial m_p} = \delta_{jp}, \tag{22.13}$$

where δ_{jp} is the Kronecker delta.

Problem b   Use this relation to show that

$$\frac{\partial}{\partial m_p}\left(\sum_{i,j} d_i A_{ij} m_j\right) = \sum_i d_i A_{ip}, \tag{22.14}$$

and

$$\frac{\partial}{\partial m_p}\left(\sum_{i,j,k} A_{ij} A_{ik} m_j m_k\right) = \sum_{i,k} A_{ip} A_{ik} m_k + \sum_{i,j} A_{ij} A_{ip} m_j. \tag{22.15}$$

By renaming the index $k \to j$ in the last term of this expression, the two terms can be seen to be equal, so that

$$\frac{\partial}{\partial m_p}\left(\sum_{i,j,k} A_{ij} A_{ik} m_j m_k\right) = 2\sum_{i,j} A_{ij} A_{ip} m_j. \tag{22.16}$$

Problem c   Use the property $A_{ip} = A^T_{pi}$ to show that expressions (22.14) and (22.16) can be written as

$$\frac{\partial}{\partial m_p}\left(\sum_{i,j} d_i A_{ij} m_j\right) = \sum_i A^T_{pi} d_i, \tag{22.17}$$

and

$$\frac{\partial}{\partial m_p}\left(\sum_{i,j,k} A_{ij} A_{ik} m_j m_k\right) = 2\sum_{ij} A^T_{pi} A_{ij} m_j. \tag{22.18}$$

Problem d   Insert these results in equation (22.12) and show that the minimization requirement $\partial S/\partial m_p = 0$ can be written in vector form as

$$\left(\mathbf{A}^T \mathbf{A}\right) \mathbf{m}^{ls} = \mathbf{A}^T \mathbf{d}, \tag{22.19}$$

where we added the superscript *ls* to indicate that the model vector thus obtained in the least-squares solution.

The equations (22.19) are called the *normal equations*. The accompanying least-squares solution is

$$\mathbf{m}^{ls} = (\mathbf{A}^T \mathbf{A})^{-1} \mathbf{A}^T \mathbf{d}. \tag{22.20}$$

A matrix $\mathbf{A}$ only has an analytic inverse if the matrix has as many rows as columns (i.e., $\mathbf{A}$ is "square"), and if its rows/columns are linearly independent. The amount of independent rows or columns of a matrix is called the *rank* of $\mathbf{A}$. Because of the nature of geophysical inverse problems, there never are enough data to match the infinitely dimensional Earth model parameters. Therefore, matrix $\mathbf{A}$ is typically not square, but by construction $\mathbf{A}^T\mathbf{A}$ is. Computations involving large matrices such as $\mathbf{A}^T\mathbf{A}$ can be computationally intensive, but methods exist to optimize these, particularly for sparse matrices (Pissanetzky, 1984). Luckily, our problem here is about as small as they come.

Problem e Show that for the example of the sinkhole that

$$\mathbf{A}^T\mathbf{A} = \begin{pmatrix} 1 & -2 \\ -2 & 4 \end{pmatrix}. \tag{22.21}$$

Problem f Even though this matrix is square, the inverse of $\mathbf{A}^T\mathbf{A}$ does not exist. The second requirement for invertable matrices is that *all* its rows and columns are independent. Is that the case here?

There is a cure to make our matrix invertable, but we will see in the next section that this comes at a price.

22.3 Damped least squares

Maybe you found that the previous two sections ended somewhat disappointing. So far, we have been unable to find a model that can predict our data. The underlying reason is that our problem as stated does not have a unique solution. Let us therefore add a constraint on the model:

$$\min_{\mathbf{m}} S_\lambda(\mathbf{m}) = \|\mathbf{Am} - \mathbf{d}\|^2 + \lambda\|\mathbf{Rm}\|^2. \tag{22.22}$$

This equation is an optimization process that balances a data misfit $\|\mathbf{Am} - \mathbf{d}\|^2$ with another term $\lambda\|\mathbf{Rm}\|^2$. If the matrix $\mathbf{R}$ is the identity matrix, the equation penalizes the norm of the model: $\|\mathbf{m}\|^2$. This means the optimization process favors models with a small norm, which results in damping of the model. The matrix $\mathbf{R}$ can also be a discrete version of a derivative; in that case expression (22.22) favors "smooth" models. In any case, the added constraint $\lambda\|\mathbf{Rm}\|^2$ balances the data misfit and a particular penalty on the model. The addition of the last term to expression (22.22) is called *regularization*. Note that λ must be positive to ensure that the misfit function is positive. How to find a reasonable balance is discussed by Hansen (1992).

Problem a Show that expression (22.22) is minimized by

$$\mathbf{m}_\lambda^{ls} = (\mathbf{A}^T\mathbf{A} + \lambda\mathbf{R}^T\mathbf{R})^{-1}\mathbf{A}^T\mathbf{d}. \tag{22.23}$$

Hint: You can write the damped least-squares system as a new system $\tilde{\mathbf{A}}\mathbf{m} = \tilde{\mathbf{d}}$, where $\tilde{\mathbf{A}} = \begin{pmatrix} \mathbf{A} \\ \sqrt{\lambda}\mathbf{R} \end{pmatrix}$ and $\tilde{\mathbf{d}} = \begin{pmatrix} \mathbf{d} \\ 0 \end{pmatrix}$.

For suitably chosen matrices $\mathbf{R}$, the addition of the term $\lambda\|\mathbf{Rm}\|^2$ to expression (22.22) results in a clearly defined minimum of $S_\lambda(\mathbf{m})$, providing a "solution" to the inverse problem. We use quotation marks, because the details of this "solution" are defined by the regularization term $\lambda\|\mathbf{Rm}\|^2$; the choice of this term often is fairly arbitrary.

Problem b   Consider the case of straight damping, for which $\mathbf{R} = \mathbf{I}$. Show that for the example of the sinkhole

$$(\mathbf{A}^T\mathbf{A} + \lambda\mathbf{R}^T\mathbf{R})^{-1} = \frac{1}{\lambda(\lambda+5)}\begin{pmatrix} \lambda+4 & 2 \\ 2 & \lambda+1 \end{pmatrix}$$

and that

$$\mathbf{m}_\lambda^{ls} = \frac{\mathbf{d}}{5+\lambda}\begin{pmatrix} 1 \\ -2 \end{pmatrix} = \frac{\delta g/g_0}{5+\lambda}\begin{pmatrix} 1 \\ -2 \end{pmatrix}. \tag{22.24}$$

The resulting model estimate explains why this is called "damped" least-squares optimization. The factor $(\lambda+5)$ in the denominator of expression (22.24) acts as a damping factor: the larger the value of λ, the smaller the norm of the model $\mathbf{m}$.

Problem c   Compute $\|\mathbf{m}_\lambda^{ls}\|^2$ for $\lambda = 0, 5$ and 10, respectively.

Increasing the damping parameter λ reduces the model vector, thus enforcing less "wild" solutions, but these solutions may not reduce the data misfit as much.

Problem d   Compute the data misfit $\|\mathbf{d} - \mathbf{A}\mathbf{m}_\lambda^{ls}\|^2$ for $\lambda = 0, 5$, and 10, respectively.

From the previous exercises, you can deduce that the amount of damping depends on a trade-off between two things. First, how far do you believe the model parameters can stray from the average sinkhole with parameters r_0 and M_0? Second, how accurately do you believe your answer should match your observations? You can clearly see the end-members of our model: we now have an answer if the datum is noise-free (λ vanishes), and we have one in case we think the datum is noisy. The larger the parameter λ, the less our resulting model parameters stray from the background model (r_0, M_0).

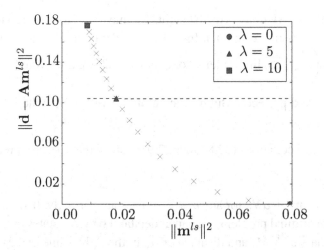

Figure 22.2 Data misfit versus model size, as a function of the damping parameter.

If the errors in the data **d** are uncorrelated and follow a normal distribution with a standard deviation $\sigma_{d,i}$, we can evaluate the statistical significance of these models by normalizing the data misfit with the standard deviation:

$$\chi^2 = \frac{1}{N} \left(\sum_{i=1}^{N} \frac{A_{ij}m_j - d_i}{\sigma_{d,i}} \right)^2. \tag{22.25}$$

This quantity, called *Chi-squared*, measures how large the data misfit is compared to the standard deviation of the data. While it may be hard to know the standard deviations $\sigma_{d,i}$, an estimate allows one to help identify models that "fit" the data in a statistical sense. For example, if the horizontally dashed line in Figure 22.2 represents $\chi^2 = 1$, this damped least-squares model with $\lambda = 5$ – and every model below this horizontal line – fits the data on average within one standard deviation σ_d. For smaller values of λ, the predicted data match the observations even better. However, it requires models more unlike the average sinkhole in the area to match the data better. And because data are noisy, this raises the question whether we are now also fitting noise in the data. We discuss the issue of noisy data later in this chapter.

22.4 The model null space

In the previous section, we found models that range from fitting the data perfectly to models that are in agreement with the average sinkhole parameters in the area. We discussed that the errors in your data help you decide which models are acceptable. So are we done? At first sight, we may want to declare victory, but let us revisit the least-squares model. In that model the damping vanishes ($\lambda = 0$), so that $\mathbf{m}^{ls} = \mathbf{d} \begin{pmatrix} 1 \\ -2 \end{pmatrix} /5$. In Problem d of Section 22.3, we discovered that this model predicts the data $\mathbf{d}$ perfectly. But how about $\mathbf{m}_0^{ls} = \mathbf{m}^{ls} + \mathbf{V}_0$, where $\mathbf{V}_0 = \begin{pmatrix} 2 \\ 1 \end{pmatrix}$?

Problem a   Use equations (22.8) and (22.9) to show that $\mathbf{m}_0^{ls}$ also predicts the data perfectly.

In fact, models $\mathbf{m}^{ls} + \alpha \mathbf{V}_0$ for any $\alpha \in \mathbb{R}$ fit the data perfectly. If we back up to the physics of the original problem, we can understand why. $\mathbf{V}_0$ states that if we change the relative mass $\delta M / M_0$ and distance $\delta r / r_0$ in the right – linearized – proportions of 2 to 1, respectively, the gravity anomaly over the sinkhole is unaffected. Components of the model that prove insensitive to data fit form the model *null space*.

The model null space is particularly intuitive in seismic tomography. As discussed in Section 23.3, seismic tomography involves a parameterization of the Earth in small slices to estimate the wave speed in each slice from the arrival times of seismic waves. (The Greek "tomos" means slice.) The observed travel times of seismic waves are matched by adjusting the wave speed in each slice. However, it is not uncommon that seismic rays miss certain slices, all together. This means that the wave speeds assigned to slices with no ray coverage have no effect on the data misfit. These slices are in the model null space. It is important to be aware of this problem, because you may find that subsets of your model parameters have no bearing on how well your model can predict the data. William of Ockham (c. 1287–1347), an English Franciscan friar, scholastic philosopher, and theologian, proposed that one should adopt the simplest model that explains the data. This principle we call now *Occam's Razor*. More than 700 years later, when you are presented a colorful image as the outcome of an inverse problem, you may want to ask the question: "Does a simpler – but less colorful – model fit the data equally well?"

22.5 The pseudo inverse from singular value decomposition

In the previous section, we argued on the basis of the physics of the problem that if we increase the depth of the sinkhole and its mass in the right proportions, it has no affect on the gravity anomaly over the sinkhole. If we use the matrix decomposition described in Section 12.6, we can let linear algebra lead us to the same conclusions. Using singular value decomposition, we decompose matrix $\mathbf{A}$ in a multiplication of three distinct matrices $\mathbf{A} = \mathbf{U}\Lambda\mathbf{V}^T$, where the columns of $\mathbf{U}$ are the eigenvectors of $\mathbf{A}\mathbf{A}^T$ with nonzero eigenvalues, the columns of $\mathbf{V}$ are the eigenvectors of $\mathbf{A}^T\mathbf{A}$ with nonzero eigenvalues. The elements on the main diagonal of the diagonal matrix Λ are the square root of the nonzero eigenvalues of $\mathbf{A}\mathbf{A}^T$ and $\mathbf{A}^T\mathbf{A}$.

We will see that the matrix decomposition allows us to construct a good substitute for the inverse of the matrix $\mathbf{A}$. Let us first decompose our matrix $\mathbf{A} = \begin{pmatrix} 1 & -2 \end{pmatrix}$:

Problem a   Show that $\mathbf{U}$ is a 1×1 matrix that is given by $\mathbf{U} = 1$.

$\mathbf{U}$ spans the data space. In other words, we can decompose the data $\mathbf{d}$ into the columns of $\mathbf{U}$. Because our data is a scalar in this example, so is $\mathbf{U}$. The model space is spanned by the eigenvectors of $\mathbf{A}^T\mathbf{A}$.

Problem b   Show that $\mathbf{A}^T\mathbf{A}$ has two eigenvalues, but only one of these is nonzero.

Problem c   Confirm that the nonzero eigenvalue corresponds to the normalized eigenvector $\mathbf{V} = \begin{pmatrix} 1 \\ -2 \end{pmatrix} / \sqrt{5}$. Because there is only one nonzero eigenvalue, the matrix $\mathbf{V}$ reduces to a vector.

Problem d   Show that $\Lambda = \sqrt{5}$, and that $\mathbf{A} = \mathbf{U}\Lambda\mathbf{V}^T$.

From this singular value decomposition, we can construct a matrix that is close to an inverse, even when there is no analytic inverse to the problem. We call this the *pseudo-inverse* solution (Strang, 2003):

$$\mathbf{A}^{-g} = \mathbf{V}\Lambda^{-1}\mathbf{U}^T. \tag{22.26}$$

Problem e   Show that the pseudo inverse $\mathbf{A}^{-g} = \mathbf{V}\Lambda^{-1}\mathbf{U}^T = \begin{pmatrix} 1 \\ -2 \end{pmatrix} / 5$.

The pseudo-inverse model is built from the model eigenvectors associated with nonzero eigenvalues. In this case, there is only one:

$$\mathbf{m}^{-g} = \mathbf{A}^{-g}\mathbf{d} = \begin{pmatrix} 1 \\ -2 \end{pmatrix} \mathbf{d}/5. \tag{22.27}$$

Note that the pseudo inverse equals the least-squares solution with the smallest norm, that is, without a component in the model null space $\mathbf{V}_0$.

Problem f   Show that our pseudo-inverse model predicts the data *perfectly*: $\mathbf{A}\mathbf{m}^{-g} = \mathbf{d}$.

This is no coincidence, but the direct result of the physics of our problem. If we call our predicted data

$$\mathbf{d}^{-g} = \mathbf{A}\mathbf{m}^{-g} = \mathbf{A}\mathbf{A}^{-g}\mathbf{d}. \tag{22.28}$$

This means $\mathbf{A}\mathbf{A}^{-g}$ is the filter acting upon the data. Using our singular value decomposition, we can write:

$$\mathbf{d}^{-g} = \mathbf{A}\mathbf{A}^{-g}\mathbf{d} = \mathbf{U}\mathbf{\Lambda}\mathbf{V}^T\mathbf{V}\mathbf{\Lambda}^{-1}\mathbf{U}^T\mathbf{d} = \mathbf{U}\mathbf{U}^T\mathbf{d}. \tag{22.29}$$

This filter $\mathbf{U}\mathbf{U}^T$ is called the data resolution matrix. In our case, $\mathbf{U} = 1$, which means the filter is a diagonal matrix (ok, of rank 1 in our one-dimensional example, but still!) and our data is predicted perfectly.

If the data vector is written as $\mathbf{d} = \mathbf{A}\mathbf{m}_T$, where $\mathbf{m}_T$ is the true model, then

$$\mathbf{m}^{-g} = \mathbf{A}^{-g}\mathbf{d} = \mathbf{A}^{-g}\mathbf{A}\mathbf{m}_T, \tag{22.30}$$

and the matrix $\mathbf{A}^{-g}\mathbf{A}$ defines how well the true model $\mathbf{m}_T$ is resolved by the estimated model. Again, using singular value decomposition:

$$\mathbf{m}^{-g} = \mathbf{A}^{-g}\mathbf{A}\mathbf{m}_T = \mathbf{V}\mathbf{\Lambda}^{-1}\mathbf{U}^T\mathbf{U}\mathbf{\Lambda}\mathbf{V}^T\mathbf{m}_T = \mathbf{V}\mathbf{V}^T\mathbf{m}_T, \tag{22.31}$$

and we reduced the model resolution matrix to $\mathbf{V}\mathbf{V}^T$.

Problem g   Show that in our example, $\mathbf{V}\mathbf{V}^T = \dfrac{1}{5}\begin{pmatrix} 1 & -2 \\ -2 & 4 \end{pmatrix}$.

The nonzero off-diagonal terms mean that the estimated model parameters are formed from a combination of the true model parameters. Ideally, our estimates of the mass M of the sinkhole is only dependent on the true mass M, and not on the depth r, but alas...

What we can conclude from the resolution matrices is that for an under-determined system, the pseudo inverse matrix $\mathbf{A}^{-g}$ is a right inverse of $\mathbf{A}$, but not a left inverse. In symbols, $\mathbf{A}^{-g}\mathbf{A} \neq \mathbf{I}$, but $\mathbf{A}\mathbf{A}^{-g} = \mathbf{I}$.

We can now define the model null space in this framework of linear algebra. It is spanned by the vector(s) $\mathbf{V}_0$ for which: $\mathbf{A}^T\mathbf{A}\mathbf{V}_0 = 0$. This is the eigenvector associated with the singular value that equals zero.

Problem h   Confirm that in our example $\mathbf{V}_0 = \begin{pmatrix} 2 \\ 1 \end{pmatrix}/\sqrt{5}$.

This confirms our earlier intuition: if we increase the depth and mass of the sink-hole – in the right proportions – our measured value of the gravitational acceleration remains the same.

22.6 Adding data

The treatment thus far of our sinkhole problem reveals the existence of models that fit the data equally well. We call the problem *non-unique*, or *ill-posed*. As discussed, this situation is the rule in imaging the subsurface, not the exception, because the subsurface is heterogeneous at all length scales and the amount of data is finite. However, in the example of the sinkhole we have reduced the inverse problem to only two unknowns: depth and mass of a sinkhole. So far, we made one measurement, but let us see what happens when we add a second measurement of the gravitational acceleration g_2. We assume this measurement is also taken directly over the sinkhole, but this time on a ladder of height r_L. For this measurement, the distance to the sinkhole is

$$r_2 = r_L + r_0 + \delta r = \alpha r_0 + \delta r, \tag{22.32}$$

where $\alpha = (r_L + r_0)/r_0$.

Problem a   Analogous to Problem b in Section 22.1, show that for the measurement on a ladder, we find

$$\alpha^2 g_2/g_0 - 1 = \delta M/M_0 - (2/\alpha)\delta r/r_0. \tag{22.33}$$

The linear(ized) system now contains two equations and two unknowns, still captured by

$$\mathbf{d} = \mathbf{Am}, \tag{22.34}$$

where the model vector remains

$$\mathbf{m} = \begin{pmatrix} \delta M/M_0 \\ \delta r/r_0 \end{pmatrix}, \tag{22.35}$$

but now the data vector is

$$\mathbf{d} = \begin{pmatrix} g_1/g_0 - 1 \\ \alpha^2 g_2/g_0 - 1 \end{pmatrix}, \tag{22.36}$$

and the forward operator

$$\mathbf{A} = \begin{pmatrix} 1 & -2 \\ 1 & -2/\alpha \end{pmatrix}. \tag{22.37}$$

This matrix is square and has independent rows/columns, as long as the ladder has a nonzero length ($\alpha \neq 1$). As a result, this matrix has an analytic inverse.

Let us assign some numbers to this example to bring it alive. We assume this particular sinkhole is at a true depth $r_T = 15$ m and $M_T = 5 \times 10^5$ kg, as opposed to the average sinkhole in the area at $r_0 = 10$ m and $M_0 = 6 \times 10^5$ kg. Our first measurement is at the surface over the sinkhole, and our second one is on a ladder of length $r_L = r_0 = 10$ m.

Problem b Show that for this particular sinkhole, equations (22.4) and (22.36) lead to

$$\mathbf{d} = \begin{pmatrix} -17/27 \\ -7/15 \end{pmatrix}. \qquad (22.38)$$

Problem c Now that $\mathbf{A} = \begin{pmatrix} 1 & -2 \\ 1 & -1 \end{pmatrix}$, confirm that $\mathbf{A}^{-1} = \begin{pmatrix} -1 & 2 \\ -1 & 1 \end{pmatrix}$.

There are many ways to obtain the inverse of this matrix, but one way would be to do a full singular value decomposition (Section 12.6). We now have a unique solution $\mathbf{m}_2 = \mathbf{A}^{-1}\mathbf{d}$, but is it the right one?

Problem d Calculate that our inverse solution $\mathbf{m}_2$ results in an estimate of $r \approx 12$ m, and the mass $M \approx 4.2 \times 10^5$ kg. This is closer to the correct answer, but not there (yet)!

This solution $\mathbf{m}_2$ predicts our data $\mathbf{d}$, but is far removed from the true model parameters $\mathbf{m}_T$. This is the solution to a linearized problem, but the gravity experiment is nonlinear in nature. In the next section we discuss how to minimize this linearization error.

22.7 Nonlinear inversion

The previous example with two data points provided a unique solution, yet it is not the one that represents the correct depth and mass of this particular sinkhole. Our data so far have been noise-free, and we had two equations and two unknowns, but to get to that point we linearized the physics underlying the experiment in Section 22.1. It is the error associated with the linearization that stands between us and an accurate solution. However, the model from the previous section was literally a step in the right direction. We can now use that model as the starting model (r_0, M_0) of a next iteration of the inversion process. Because this new starting model is closer to the true sinkhole model parameters than our first one, the error in the linearization process is reduced. Figure 22.3 illustrates this: after only a few iterations, the solution converges to the correct parameters of our sinkhole. In each new iteration, we take the previous estimate of M and r as the new reference values M_0 and r_0, and repeat the linear inversion.

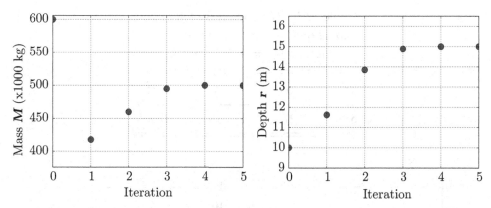

Figure 22.3   Iterative linear inversions to solve the nonlinear problem of a sink-hole of mass (left) and depth (right). The starting values were $M_0 = 6 \times 10^5$ kg and $r_0 = 10$ m, while the true sinkhole is at $M = 5 \times 10^5$ kg at $r = 15$ m depth.

Problem a   Write your own short computer program to reproduce Figure 22.3.

22.8 Real data are noisy

In reality, all measurements are noisy. Even if we *could* fit the data perfectly, we should not: we would perfectly fit the signal *and* data errors. Ideally, we fit only the component of our data that makes up the signal, and not the noise. This constitutes the difference between inversion and optimization. In the latter case, we aim to minimize the misfit between the observed and predicted data. In inverse theory, we seek the whole range of model parameters that "fit" the data. If we believe our data to be accurate, we need a better fit than if we had less confidence in our data.

We next add 5% noise to our observations from the previous section. This means we can represent our data by

$$\mathbf{d} = \begin{pmatrix} d_1 \pm \sigma_{d,1} \\ d_2 \pm \sigma_{d,2} \end{pmatrix}. \tag{22.39}$$

Following equation (21.33) we define the data covariance matrix as

$$C_{d,ij} = \langle (d_i - \langle d_i \rangle)(d_j - \langle d_j \rangle) \rangle. \tag{22.40}$$

If the noise is uncorrelated, we can write the data covariance matrix using expression (21.49):

$$\mathbf{C}_d = \begin{pmatrix} \sigma_{d,1} & 0 \\ 0 & \sigma_{d,2} \end{pmatrix}. \tag{22.41}$$

Because this was a numerical example, and we were the ones adding the noise, we know its characteristics. It is much trickier to separate signal from noise in

Table 22.1   *The goodness of fit of the model represented in terms of χ_i^2, for iterations $i = 0, 1, 2,$ and 3.*

χ_0^2	χ_1^2	χ_2^2	χ_3^2
558	27.9	1.18	0.006

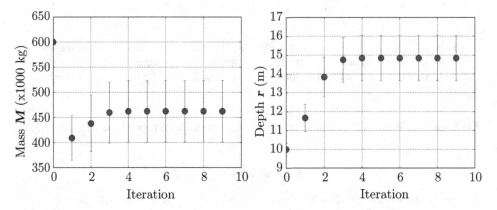

Figure 22.4   Iterative linear least-squares inversions to solve the nonlinear problem of a sinkhole of mass $M = 5 \times 10^5$ kg at $r = 15$ m depth, from noisy data.

real observations. Repeated measurements may help establish the magnitude of random variations in the data; see, for example, expression (21.41), and attempts at data-driven estimates of the data variance may be second-best (van Wijk et al., 2002). Let us explore inversion with noisy data. At each iteration, we evaluate our goodness of fit with the χ^2 criterion in Table 22.1, as well as the model parameters in Figure 22.4. The error bars in the model parameter estimates are computed using the propagation of errors in a linear system in expression (21.50). According to expression (21.54), the covariance of the estimated model parameters is given by

$$\mathbf{C}_m = \mathbf{A}^{-g}\mathbf{C}_d\mathbf{A}^{-g,T}. \tag{22.42}$$

As shown in Figure 22.4, we fit the data after three iterations on average within one standard deviation, and the model parameters plus its error bars at this iteration include the true value of the sinkhole parameters. Our results suggest a range of values for the mass and depth of the sinkhole that explain the data equally well. It will be up to other information – maybe there is a seismic survey, or Ground Penetrating Radar data – to further reduce the uncertainty in our estimate.

In truth, our error bars may be larger than we would like. In search of a real sinkhole, we would have readings of the gravitational acceleration at other locations than directly over the sinkhole. We would ideally collect a dense horizontal grid of measurements. This way, we would have spatial information on the shape of our anomaly, providing more information about the origin of the gravitational anomaly. A "sharp" anomaly (a large gradient of the anomalous value as we move around in the (x, y)-plane) indicates a shallow source (remember upward continuation in Section 20.2). However, even densely sampled data do not take anything away from the reality of geophysical inverse problems: geophysical models are non-unique, and we need to accept that answers to our probings of the Earth should be cast in ranges of model parameter values that all fit the data.

22.9 Bayes' theorem and inverse problems

In Section 21.9 we introduced Bayes' theorem. This theorem is based on the conditional probability $p(A|B)$ that describes the probability of getting an outcome A given B. The concept of conditional probability is so useful for inverse problems, because in such problems one aims to determine a model $\mathbf{m}$ given data $\mathbf{d}$. In a statistical approach this is described by the conditional probability $p(\mathbf{m}|\mathbf{d})$. The forward problem consists of estimating the data given the model. This relationship is described by the conditional probability $p(\mathbf{d}|\mathbf{m})$. This probability is determined by the physics that relates the data to the model. (We presume we know this physics, at least to a certain extent.) The quantity $p(\mathbf{d}|\mathbf{m})$ is statistical because (1) the data are contaminated with random errors, and (2) the physics of the forward problem may not be completely known or an approximation may be used to solve the forward problem. From a Bayesian point of view, inversion consists of determining $p(\mathbf{m}|\mathbf{d})$ from $p(\mathbf{d}|\mathbf{m})$. Connecting these conditional probabilities is exactly what Bayes' theorem does! To see this, we consider equation (21.100) replacing A by the model $\mathbf{m}$ and B by the data $\mathbf{d}$:

$$p(\mathbf{m}|\mathbf{d}) \propto p(\mathbf{d}|\mathbf{m})\,p(\mathbf{m}), \qquad (22.43)$$

where we replaced the proportionality constant C by the proportionality symbol $\propto$.

So how do we apply this in practice? The probability $p(\mathbf{m})$ is the probability of the model without us knowing of any data. In the case of the example of the sinkhole, we may might have an idea of the mass and depth of realistic sinkholes, and of the variations around these values for different sinkholes. In mathematical language we would say that $\mathbf{m}$ varies around an *a-priori* value $\mathbf{m}_0$ with a certain uncertainty. As an example we consider the simplest case where these fluctuations follow a Gauss distribution (21.22) with equal standard deviation σ_m for all model parameters:

$$p(\mathbf{m}) \propto \exp\left(-\frac{\|\mathbf{m} - \mathbf{m}_0\|^2}{2\sigma_m^2}\right). \tag{22.44}$$

This distribution is called the *a-priori probability* because it describes the model probability *before* we have used the data.

For the conditional probability $p(\mathbf{d}|\mathbf{m})$, we assume that in the absence of data errors, the data are explained by the forward problem. We denote the forward problem as $\mathbf{d} = \mathbf{F}(\mathbf{m})$. In the absence of errors, $\mathbf{d} - \mathbf{F}(\mathbf{m}) = 0$. Data errors cause the right-hand side to be nonzero, and the statistics of the data errors determines the right-hand side. For simplicity we consider the simplest case that the data errors are Gaussian as well and have equal standard deviation σ_d, so that

$$p(\mathbf{d}|\mathbf{m}) \propto \exp\left(-\frac{\|\mathbf{d} - \mathbf{F}(\mathbf{m})\|^2}{2\sigma_d^2}\right). \tag{22.45}$$

Problem a   Use Bayes' theorem (22.43) to show that for this example

$$p(\mathbf{m}|\mathbf{d}) \propto \exp\left(-\frac{\|\mathbf{d} - \mathbf{F}(\mathbf{m})\|^2}{2\sigma_d^2} - \frac{\|\mathbf{m} - \mathbf{m}_0\|^2}{2\sigma_m^2}\right). \tag{22.46}$$

This distribution is called the *a-posteriori probability* because it gives the probability for the model *after* we have used the information contained in the data.

One approach to move forward is to compute this probability for many values of the model vector $\mathbf{m}$. In practice, this is not very practical because the number of computations needed to sample this probability density function adequately can be prohibitive, especially when the model vector $\mathbf{m}$ has many components. As an alternative, approaches based on a random walk through model space can be used (Mosegaard and Tarantola, 1995; Sambridge and Mosegaard, 2002). Here we follow a simpler approach. We assume that the "best" model is the model that is most likely. This is called the *maximum likelihood estimator*.

Problem b   Show that for the probability (22.46) the maximum likelihood model follows by minimizing the following quantity

$$\min_{\mathbf{m}} S_{ml}(\mathbf{m}) = \frac{\|\mathbf{d} - \mathbf{F}(\mathbf{m})\|^2}{2\sigma_d^2} + \frac{\|\mathbf{m} - \mathbf{m}_0\|^2}{2\sigma_m^2}. \tag{22.47}$$

The minimization of the first term on the right-hand side corresponds to optimally fitting the data, while the second term acts as a regularization. The maximum likelihood solution to Bayesian inversion thus leads to damped least-squares minimization of the data misfit. For a linear forward problem, $\mathbf{F}(\mathbf{m}) = \mathbf{Am}$, with $\mathbf{A}$ a matrix. Using this in expression (22.47) and multiplying the result with the constant $2\sigma_d^2$ reduces the condition for the maximum likelihood solution to

$$\min_{\mathbf{m}} S_{ml}(\mathbf{m}) = \|\mathbf{d} - \mathbf{Am}\|^2 + \frac{\sigma_d^2}{\sigma_m^2}\|\mathbf{m} - \mathbf{m}_0\|^2. \tag{22.48}$$

Note the resemblance of this expression with equation (22.22) for the damped least-squares solution. The maximum likelihood solution for a Gaussian distribution thus is equivalent to damped least-squares. Because of the ubiquity of the Gauss distribution, see Section 21.6, least-squares inversion is widely used.

Bayes' theorem is a powerful tool for solving inverse problems, in essence because it expresses $p(\mathbf{m}|\mathbf{d})$ in $p(\mathbf{d}|\mathbf{m})$. The example in this section was based on the simplest case of Gaussian distributions for the model and data, with uncorrelated variables and equal standard deviations. These restrictions can be relaxed, and one can specify arbitrary covariance matrices for the data and model (Tarantola and Valette, 1982; Tarantola, 1987). A general framework for inverse problems was formulated and passionately promoted by Albert Tarantola in the 1980s, and his work has had a strong impact on the field of inverse problems (Mosegaard, 2011).

23

Perturbation theory

From this book and most other books on mathematical physics, you may have obtained the impression that most equations in the physical sciences can be solved. This is actually not true; most textbooks (including this book) give an unrepresentative state of affairs by only showing the problems that *can* be solved in closed form. It is an interesting paradox that as our theories of the physical world become more accurate, the resulting equations become more difficult to solve. In classical mechanics the problem of two particles that interact with a central force can be solved in closed form, but the three-body problem in which three particles interact has no analytical solution. In quantum mechanics, the one-body problem of a particle that moves in a potential can be solved for a limited number of situations only: for the free particle, the particle in a box, the harmonic oscillator, and the hydrogen atom. In this sense the one-body problem in quantum mechanics has no general solution. This shows that as a theory becomes more accurate, the resulting complexity of the equations makes it often more difficult to actually find solutions.

One way to proceed is to compute numerical solutions of the equations. Computers are a powerful tool and can be extremely useful in solving physical problems. Another approach is to find approximate solutions to the equations. In Chapter 11, scale analysis was used to drop from the equations terms that are of minor importance. In this chapter, a systematic method is introduced to account for terms in the equations that are small but that make the equations difficult to solve. The idea is that a complex problem is compared to a simpler problem that can be solved in closed form, and to consider these small terms as a perturbation to the original equation. The theory of this chapter then makes it possible to determine how the solution is perturbed by the perturbation in the original equation; this technique is called *perturbation theory*. A classic reference on perturbation theory has been written by Nayfeh (2011). Bender and Orszag (1999) give a useful and illustrative overview of a wide variety of perturbation methods.

The central idea of perturbation theory is introduced for an algebraic equation in Section 23.1. Sections 23.2, 23.3, and 23.5 contain important applications of perturbation theory to differential equations. As shown in Section 23.4, perturbation theory has a limited domain of applicability, and this may depend on the way the perturbation problem is formulated. Finally, it is shown in Section 23.7 that not every perturbation problem is well behaved; this leads to singular perturbation theory.

23.1 Regular perturbation theory

As an introduction to perturbation theory, let us consider the following equation

$$x^3 - 4x^2 + 4x = 0.01. \tag{23.1}$$

Let us for the moment assume that we do not know how to find the roots of a third-order polynomial, so we cannot solve this equation. The problem is the small term 0.01 on the right-hand side. If this term were equal to zero, the resulting equation can be solved; $x^3 - 4x^2 + 4x = 0$ is equivalent to $x(x^2 - 4x + 4) = x(x - 2)^2 = 0$, which has the solutions $x = 0$ and $x = 2$. In Figure 23.1 the polynomial of (23.1) is shown by the thick solid line; it is indeed equal to zero for $x = 0$ and $x = 2$.

The problem that we face is that the right-hand side of (23.1) is *not* equal to zero. In perturbation theory one studies the perturbation of the solution under a perturbation of the original equation. In order to do this, we replace the original equation (23.1) by the more general equation

$$x^3 - 4x^2 + 4x = \epsilon. \tag{23.2}$$

When $\epsilon = 0.01$ this equation is identical to the original problem, while for $\epsilon = 0$, it reduces to the unperturbed problem that we can solve in closed form. It may appear

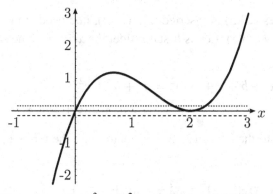

Figure 23.1   The polynomial $x^3 - 4x^2 + 4x$ (thick solid line) and the lines $\epsilon = 0.15$ (dotted line) and $\epsilon = -0.15$ (dashed line).

that we have made the problem more complex because we still need to solve the same equation as our original equation, but it now contains a new variable ϵ as well! However, this is also the strength of this approach.

The solution of (23.2) is a function of ϵ so that

$$x = x(\epsilon). \tag{23.3}$$

In Section 3.1 the Taylor series was used to approximate a function $f(x)$ by a power series in the variable x:

$$f(x) = \sum_{n=0}^{\infty} \frac{x^n}{n!} \frac{d^n f}{dx^n}(x=0) = f(0) + x\frac{df}{dx}(x=0) + \frac{x^2}{2}\frac{d^2 f}{dx^2}(x=0) + \cdots. \tag{3.11}$$

When the solution x of (23.2) depends in a regular way on ϵ, this solution can also be written as a similar power series by making the substitutions $x \to \epsilon$ and $f \to x$ in (3.11):

$$x(\epsilon) = x(0) + \epsilon\frac{dx}{d\epsilon}(\epsilon=0) + \frac{\epsilon^2}{2!}\frac{d^2 x}{d\epsilon^2}(\epsilon=0) + \cdots. \tag{23.4}$$

This expression is not very useful because we need the derivative $dx/d\epsilon$ and higher derivatives $d^n x/d\epsilon^n$ as well; in order to compute these derivatives, we need to find the solution $x(\epsilon)$ first, but this is just what we are trying to do. There is, however, another way to determine the series (23.4). Let us write the solution $x(\epsilon)$ as a power series in ϵ

$$x(\epsilon) = x_0 + \epsilon x_1 + \epsilon^2 x_2 + \cdots. \tag{23.5}$$

The coefficients x_n are not known at this point, but once we know them, the solution x can be found by inserting the numerical value $\epsilon = 0.01$. In practice one truncates the series (23.5); it is this truncation that makes perturbation theory an approximation.

When the series (23.5) is inserted into (23.2), one needs to compute x^2 and x^3 when x is given by (23.5). Let us first consider the x^2-term. The square of a sum of terms is given by

$$(a + b + c + \cdots)^2 = a^2 + b^2 + c^2 + \cdots \\ + 2ab + 2ac + 2bc + \cdots. \tag{23.6}$$

Let us apply this to the series (23.5) and retain only the terms up to order ϵ^2; this gives

$$\left(x_0 + \epsilon x_1 + \epsilon^2 x_2 + \cdots\right)^2 = x_0^2 + \epsilon^2 x_1^2 + \epsilon^4 x_2^2 + \cdots \\ + 2\epsilon x_0 x_1 + 2\epsilon^2 x_0 x_2 + 2\epsilon^3 x_1 x_2 + \cdots. \tag{23.7}$$

If we are only interested in retaining the terms up to order ϵ^2, the terms $\epsilon^4 x_2^2$ and $2\epsilon^3 x_1 x_2$ in this expression can be ignored. Collecting terms of equal powers of ϵ then gives

$$\left(x_0 + \epsilon x_1 + \epsilon^2 x_2 + \cdots\right)^2 = x_0^2 + 2\epsilon x_0 x_1 + \epsilon^2 \left(x_1^2 + 2x_0 x_2\right) + \mathcal{O}(\epsilon^3). \quad (23.8)$$

A similar expansion in powers of ϵ can be used for the term x^3. This expansion is based on the identity

$$(a + b + c + \cdots)^3 = a^3 + b^3 + c^3 + \cdots$$
$$+ 3a^2 b + 3ab^2 + 3a^2 c + 3ac^2 + 3b^2 c + 3bc^2 + \cdots. \quad (23.9)$$

Problem a Apply this identity to the series (23.5), collect together all the terms with equal powers of ϵ and show that up to order ϵ^2 the result is given by

$$\left(x_0 + \epsilon x_1 + \epsilon^2 x_2 + \cdots\right)^3 = x_0^3 + 3\epsilon x_0^2 x_1 + 3\epsilon^2 \left(x_0 x_1^2 + x_0^2 x_2\right) + \mathcal{O}(\epsilon^3). \quad (23.10)$$

Problem b At this point we can express all the terms in (23.2) in a power series of ϵ. Insert (23.5), (23.8), and (23.10) into the original equation (23.2) and collect together terms of equal powers of ϵ to derive that

$$x_0^3 - 4x_0^2 + 4x_0$$
$$+ \epsilon \left(3x_0^2 x_1 - 8x_0 x_1 + 4x_1 - 1\right)$$
$$+ \epsilon^2 \left(3x_0 x_1^2 + 3x_0^2 x_2 - 4x_1^2 - 8x_0 x_2 + 4x_2\right) + \cdots = 0. \quad (23.11)$$

In this and subsequent expressions the dots denote terms of order $\mathcal{O}(\epsilon^3)$. The term -1 in the term that multiplies ϵ comes from the right-hand side of (23.2).

At this point we use that ϵ does not have a fixed value, but that it can take any value within certain bounds. Expression (23.11) thus must be satisfied for a range of values of ϵ. This can only be the case when the coefficients that multiply the different powers ϵ^n are equal to zero. This means that (23.11) is equivalent to the following system of equations, which consists of the terms that multiply the terms ϵ^0, ϵ^1, and ϵ^2, respectively:

$$\left.\begin{array}{ll} \mathcal{O}(1)\text{-terms:} & x_0^3 - 4x_0^2 + 4x_0 = 0, \\ \mathcal{O}(\epsilon)\text{-terms:} & 3x_0^2 x_1 - 8x_0 x_1 + 4x_1 - 1 = 0, \\ \mathcal{O}(\epsilon^2)\text{-terms:} & 3x_0 x_1^2 + 3x_0^2 x_2 - 4x_1^2 - 8x_0 x_2 + 4x_2 = 0. \end{array}\right\} \quad (23.12)$$

You may wonder whether we have not made the problem more complicated. We started with a single equation for a single variable x, and now we have a system of coupled equations for many variables. However, we could not solve (23.2) for the single variable x, while it is not difficult to solve (23.12).

Problem c   Show that (23.12) can be rewritten in the following form:

$$\left.\begin{array}{l} x_0^3 - 4x_0^2 + 4x_0 = 0, \\ \left(3x_0^2 - 8x_0 + 4\right) x_1 = 1, \\ \left(3x_0^2 - 8x_0 + 4\right) x_2 = (4 - 3x_0)\, x_1^2. \end{array}\right\} \qquad (23.13)$$

The first equation is simply the unperturbed problem, this has the solutions $x_0 = 0$ and $x_0 = 2$. For reasons that will become clear in Section 23.7 we focus here on the solution $x_0 = 0$ only. Given x_0, the parameter x_1 follows from the second equation because this is a linear equation in x_1. The last equation is a linear equation in the unknown x_2, which can be solved once x_0 and x_1 are known.

Problem d   Solve (23.13) in this way to show that the solution near $x = 0$ is given by

$$x_0 = 0, \qquad x_1 = \frac{1}{4}, \qquad x_2 = \frac{1}{16}. \qquad (23.14)$$

Now we are close to the final solution of our problem. The coefficients of the previous expression can be inserted into the perturbation series (23.5) so that the solution as a function of ϵ is given by

$$x = 0 + \frac{1}{4}\epsilon + \frac{1}{16}\epsilon^2 + \mathcal{O}(\epsilon^3). \qquad (23.15)$$

At this point we can revert to the original equation (23.1) by inserting the numerical value $\epsilon = 0.01$, which gives:

$$x = \frac{1}{4} \times 10^{-2} + \frac{1}{16} \times 10^{-4} + \mathcal{O}(10^{-6}) = 0.002506. \qquad (23.16)$$

It should be noted that this is an approximate solution because the terms of order ϵ^3 and higher have been ignored. This is indicated by the term $\mathcal{O}(10^{-6})$ in (23.16). Assuming that the error made by truncating the perturbation series is of the same order as the first term that is truncated, the error in the solution (23.16) is of the order 10^{-6}. For this reason the number on the right-hand side of (23.16) is given to six decimals; the last decimal is of the same order as the truncation error.

If this result is not sufficiently accurate for the application that one has in mind, then one can extend the analysis to higher powers ϵ^n in order to reduce the truncation error of the truncated perturbation series. Although the algebra resulting from doing this can be tedious, there is no reason why this analysis cannot be extended to higher orders.

A truly formal analysis of perturbation problems can be difficult. For example, the perturbation series (23.5) converges only for sufficiently small values of ϵ. It is often not clear whether the employed value of ϵ (in this case, $\epsilon = 0.01$) is sufficiently small to ensure convergence. Even when a perturbation series does not converge for a given value of ϵ, one can often obtain a useful approximation to the solution by truncating the perturbation series at a suitably chosen order (Bender and Orszag, 1999). In this case one speaks of an *asymptotic series*.

When one has obtained an approximate solution of a perturbation problem, one can sometimes substitute it back into the original equation to verify whether this solution indeed satisfies the equation with an acceptable accuracy. For example, inserting the numerical value $x = 0.002506$ in (23.1) gives

$$x^3 - 4x^2 + 4x = 0.0099989 = 0.01 - 0.0000011. \tag{23.17}$$

This means that the approximate solution satisfies (23.1) with a *relative* error that is given by $0.0000011/0.01 = 10^{-4}$. This is a very accurate result given the fact that only three terms were retained in the perturbation analysis of this section.

23.2 Born approximation

In many scattering problems one wants to account for the scattering of waves by the heterogeneities in the medium. Usually these problems are so complex that they cannot be solved in closed form. Suppose one has a background medium in which scatterers are embedded. When the background medium is sufficiently simple, one can solve the wave propagation problem for this background medium. For example, in Section 18.3 we computed the Green's function for the Helmholtz equation in a homogeneous medium.

The wave equation (6.41) for waves with a time dependence $e^{-i\omega t}$ reduces to the Helmholtz equation. In this section we consider the Helmholtz equation with a *variable* velocity $c(\mathbf{r})$ as an example of the application of perturbation theory to scattering problems. This means we consider the wave field $p(\mathbf{r}, \omega)$ in the frequency domain that satisfies the following equation:

$$\nabla^2 p(\mathbf{r}, \omega) + \frac{\omega^2}{c^2(\mathbf{r})} p(\mathbf{r}, \omega) = S(\mathbf{r}, \omega). \tag{23.18}$$

In this expression $S(\mathbf{r}, \omega)$ denotes the source that generates the wave field. In order to facilitate a systematic perturbation analysis, we decompose $1/c^2(\mathbf{r})$ into a term $1/c_0^2$ that accounts for a homogeneous reference model and a perturbation:

$$\frac{1}{c^2(\mathbf{r})} = \frac{1}{c_0^2} [1 + \epsilon n(\mathbf{r})]. \tag{23.19}$$

In this expression ϵ is a small parameter that measures the strength of the heterogeneity. The function $n(\mathbf{r})$ gives the spatial distribution of the heterogeneity. Combining the previous expressions it follows that the wave field satisfies the following expression:

$$\nabla^2 p(\mathbf{r}, \omega) + \frac{\omega^2}{c_0^2} \left[1 + \epsilon n(\mathbf{r})\right] \, p(\mathbf{r}, \omega) = S(\mathbf{r}, \omega). \qquad (23.20)$$

The solution $p(\mathbf{r}, \omega)$ of this expression is a function of the scattering strength ϵ; for sufficiently small values of ϵ, it can be written as a power series in ϵ:

$$p = p_0 + \epsilon p_1 + \epsilon^2 p_2 + \cdots. \qquad (23.21)$$

Problem a   Insert the perturbation series (23.21) into (23.20), collect together the terms that multiply equal powers of ϵ and show that the terms that multiply the different powers of ϵ give the following equations:

$$\left.\begin{aligned}
\mathcal{O}(1): \quad & \nabla^2 p_0(\mathbf{r}, \omega) + \frac{\omega^2}{c_0^2} p_0(\mathbf{r}, \omega) = S(\mathbf{r}, \omega), \\[6pt]
\mathcal{O}(\epsilon): \quad & \nabla^2 p_1(\mathbf{r}, \omega) + \frac{\omega^2}{c_0^2} p_1(\mathbf{r}, \omega) = -\frac{\omega^2}{c_0^2} n(\mathbf{r}) p_0(\mathbf{r}, \omega), \\[6pt]
\mathcal{O}(\epsilon^2): \quad & \nabla^2 p_2(\mathbf{r}, \omega) + \frac{\omega^2}{c_0^2} p_2(\mathbf{r}, \omega) = -\frac{\omega^2}{c_0^2} n(\mathbf{r}) p_1(\mathbf{r}, \omega), \\
& \qquad \vdots
\end{aligned}\right\} \qquad (23.22)$$

The left-hand side of each of these expressions gives the Helmholtz equation for a homogeneous medium, and the right-hand sides contain source terms that are different for each order. The source of the unperturbed wave p_0 is given by $S(\mathbf{r}, \omega)$, which is also the source of the perturbed problem. The source of the first-order perturbation p_1 is given by the right-hand side of the second equation; hence, the source of p_1 is given by $-\left(\omega^2/c_0^2\right) n(\mathbf{r}) p_0(\mathbf{r}, \omega)$. This means that the source of p_1 is proportional to the inhomogeneity $n(\mathbf{r})$ of the medium. Physically this corresponds to the fact that the heterogeneity is the source of the scattered waves. The source of p_1 is also proportional to the unperturbed wave field p_0. The reason for this is that the generation of the scattered waves depends on the local perturbation of the medium as well as on the strength of the wave field at the location of the scatterers.

Each of the equations in (23.22) is of the form $\nabla^2 p(\mathbf{r}, \omega) + (\omega^2/c_0^2) p(\mathbf{r}, \omega) = F(\mathbf{r}, \omega)$. According to the theory of Section 17.4, the solution to this expression is given by

$$p(\mathbf{r}, \omega) = \int G_0(\mathbf{r}, \mathbf{r}'; \omega) F(\mathbf{r}', \omega) \, dV', \qquad (23.23)$$

where the unperturbed Green's function $G_0(\mathbf{r}, \mathbf{r}'; \omega)$ is the response in a homogeneous medium at location $\mathbf{r}$ due to a point source at location $\mathbf{r}'$:

$$\nabla^2 G_0(\mathbf{r}, \mathbf{r}'; \omega) + \frac{\omega^2}{c_0^2} G_0(\mathbf{r}, \mathbf{r}'; \omega) = \delta(\mathbf{r} - \mathbf{r}'). \tag{23.24}$$

The specific form of the unperturbed Green's function in one, two, and three dimensions is given in (18.46). From this point on, it is not shown explicitly that the solution and the Green's function depend on the angular frequency ω, but it should be kept in mind that all the results in this section depend on frequency.

Problem b   Use these results to show that the solution of (23.22) is given by

$$\left.\begin{aligned}
p_0(\mathbf{r}) &= \int G_0(\mathbf{r}, \mathbf{r}')S(\mathbf{r}')\,dV', \\[4pt]
p_1(\mathbf{r}) &= -\frac{\omega^2}{c_0^2}\int G_0(\mathbf{r}, \mathbf{r}')n(\mathbf{r}')p_0(\mathbf{r}')\,dV', \\[4pt]
p_2(\mathbf{r}) &= -\frac{\omega^2}{c_0^2}\int G_0(\mathbf{r}, \mathbf{r}')n(\mathbf{r}')p_1(\mathbf{r}')\,dV', \\
&\;\;\vdots
\end{aligned}\right\} \tag{23.25}$$

Problem c   Insert the expression for the unperturbed wave p_0 into the second equation (23.25) to show that the first-order perturbation is given by

$$p_1(\mathbf{r}) = -\frac{\omega^2}{c_0^2}\iint G_0(\mathbf{r}, \mathbf{r}_1)n(\mathbf{r}_1)G_0(\mathbf{r}_1, \mathbf{r}_0)S(\mathbf{r}_0)\,dV_1dV_0. \tag{23.26}$$

Note that the integration variable $\mathbf{r}'$ has been relabeled as $\mathbf{r}_0$ and $\mathbf{r}_1$, respectively.

Problem d   Insert this result in the last line of (23.25) to derive that the second-order perturbation is given by

$$p_2(\mathbf{r}) = \frac{\omega^4}{c_0^4}\iiint G_0(\mathbf{r}, \mathbf{r}_2)n(\mathbf{r}_2)G_0(\mathbf{r}_2, \mathbf{r}_1)n(\mathbf{r}_1)G_0(\mathbf{r}_1, \mathbf{r}_0)S(\mathbf{r}_0)\,dV_2dV_1dV_0. \tag{23.27}$$

Inserting this result and (23.25) in the perturbation series (23.21) finally gives the following perturbation series for the scattered waves:

$$\begin{aligned}
p(\mathbf{r}, \omega) = &\int G_0(\mathbf{r}, \mathbf{r}_0)S(\mathbf{r}_0)\,dV_0 \\[4pt]
&- \frac{\omega^2}{c_0^2}\iint G_0(\mathbf{r}, \mathbf{r}_1)n(\mathbf{r}_1)G_0(\mathbf{r}_1, \mathbf{r}_0)S(\mathbf{r}_0)\,dV_1dV_0 \\[4pt]
&+ \frac{\omega^4}{c_0^4}\iiint G_0(\mathbf{r}, \mathbf{r}_2)n(\mathbf{r}_2)G_0(\mathbf{r}_2, \mathbf{r}_1)n(\mathbf{r}_1)G_0(\mathbf{r}_1, \mathbf{r}_0)S(\mathbf{r}_0)\,dV_2dV_1dV_0 \\[4pt]
&+ \cdots.
\end{aligned} \tag{23.28}$$

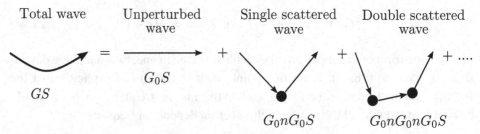

Figure 23.2   Decomposition of the total wave field (thick solid line) in the unperturbed wave, the single scattered wave, the double scattered wave, and higher-order scattering events. The total Green's function G is shown by a thick line, the unperturbed Green's function G_0 by thin lines, and each scattering event by the heterogeneity n is indicated by a black circle.

This expansion is shown graphically in Figure 23.2. Reading each of the lines in expression (23.28) from right to left, one can follow the "life-history" of the waves that are scattered in the medium. The top line of the right-hand side gives the unperturbed wave. This wave is excited by the source at location $\mathbf{r}_0$; this is described by the source term $S(\mathbf{r}_0)$. The wave then propagates through the unperturbed medium to the point $\mathbf{r}$; this is accounted for by the term $G_0(\mathbf{r}, \mathbf{r}_0)$. Graphically this is shown by the first diagram after the equality sign in Figure 23.2. In this figure the thin arrows denote the unperturbed Green's function G_0. The second line in (23.28) physically describes a wave that is generated at the source $S(\mathbf{r}_0)$; this wave then propagates through the unperturbed medium with the Green's function $G_0(\mathbf{r}_1, \mathbf{r}_0)$ to a scatterer at location $\mathbf{r}_1$. The wave is then scattered; this is accounted for by the terms $-\left(\omega^2/c_0^2\right) n(\mathbf{r}_1)$. In Figure 23.2 this scattering interaction is indicated by a solid dot. The wave then travels through the unperturbed medium to the point $\mathbf{r}$; this is accounted for by the term $G_0(\mathbf{r}, \mathbf{r}_1)$.

Problem e   Describe in a similar way the "life-history" of the double scattered wave that is given by the last line of (23.28) and convince yourself that this corresponds to the right-most diagram in Figure 23.2.

The analysis in this section can be continued to any order. The resulting series is called the *Neumann series*. It gives a decomposition of the total wave field into single scattered waves, double scattered waves, triple scattered waves, etc. In practice it is often difficult to compute the waves that are scattered more than once. In the *Born approximation* one simply truncates the perturbation series after the second term. Using the notation of (23.25), this means that in the Born approximation the wave field is given by

$$p_B(\mathbf{r}) = p_0(\mathbf{r}) - \frac{\omega^2}{c_0^2} \int G_0(\mathbf{r}, \mathbf{r}') n(\mathbf{r}') p_0(\mathbf{r}') \, dV'. \tag{23.29}$$

This expression is extremely useful for a large variety of applications. For sufficiently weak perturbations $\epsilon n(\mathbf{r})$, it allows the analytical computation of the (single) scattered waves. In the Born approximation the scattered waves are given by the last term in (23.29). This last term gives a linear(ized) relation between the scattered waves and the perturbation of the medium. In many applications, one measures the scattered waves and one wants to retrieve the perturbation of the medium. The Born approximation provides a linear relation between the scattered waves and the perturbation of the medium. Methods from linear algebra can then be used to infer the perturbation $n(\mathbf{r})$ of the medium from measurements of the scattered waves. This is an example of a linear inverse problem, such problems are treated in Chapter 22. The Born approximation provides the basis for most of the techniques used in reflection seismology for the detection of hydrocarbons in the Earth (Claerbout, 1985; Yilmaz, 2001). The Born approximation also forms the basis of the imaging techniques used with radar (Ishimaru, 1978) and a variety of other applications. In fact, it has been argued that imaging with multiple-scattered waves is not feasible in practice (Claerbout, 1985); a discussion of this controversial issue can be found in Scales and Snieder (1997).

In this section the Born approximation for the Helmholtz equation was derived. However, this derivation can readily be generalized to other scattering problems. The only required ingredient is that, when one divides the medium into an unperturbed medium and a perturbation, one can compute the Green's function for the unperturbed medium. The Born approximation is used in quantum mechanics (Merzbacher, 1961), electromagnetic wave scattering (Jackson, 1998), scattering of elastic body waves (Wu and Aki, 1985), and elastic surface waves (Snieder, 1986).

There is a famous application of the Born approximation. According to (23.29) the scattered waves are multiplied by ω^2 compared to the unperturbed waves. This is also the case for the scattering of electromagnetic waves (Jackson, 1998). This term ω^2 explains why the sky is blue. The scattered waves are proportional to ω^2 compared to the unperturbed waves. This means that light with a high frequency is scattered more strongly than light with a lower frequency. In other words, blue light is scattered more strongly than red light. The blue light that comes from the Sun is scattered more effectively out of the light beam from the Sun to an observer than the red light. When this blue light is scattered again by small particles in the atmosphere, it travels to an observer as blue light that comes from the sky from a different location than the Sun. We perceive this as "the blue sky." For the same reason, the Sun is red as it rises or sets. When light from the Sun propagates through the atmosphere at an oblique angle, the path length of the light through the atmosphere is long. The blue light is preferentially scattered, and as a result, the transmitted light that gives the color of the Sun is enriched in red colors.

23.3 Linear travel time tomography

An important tool for determining the interior structure of the Earth and other bodies is travel time tomography. In this technique one measures the travel time of waves between a large number of sources and receivers. When the coverage with rays is sufficiently dense, one can determine the velocity of seismic waves in the Earth from the recorded travel times. Detailed descriptions of seismic tomography can be found in textbooks (Iyer, 1993; Nolet, 1987). The travel time along a ray is given by the integral

$$\tau = \int \frac{1}{c(\mathbf{r})} \, ds. \tag{23.30}$$

Since the integral is proportional to $1/c(\mathbf{r})$, it is convenient to use the *slowness* $u(\mathbf{r}) = 1/c(\mathbf{r})$ rather than the velocity. Using this quantity the travel time is given by

$$\tau = \int_{\mathbf{r}[u]} u(\mathbf{r}) \, ds. \tag{23.31}$$

The last expression suggests a linear relation between the measured travel time τ and the unknown slowness $u(\mathbf{r})$. Such a linear relation is ideal for the inverse problem of estimating the subsurface slowness from the observed travel times. However, integral (23.31) is taken over the ray that joins the source and the receiver. These rays are curves of stationary travel time; as a result, the rays depend on the slowness as well. This dependence effectively makes the relation between the slowness and the travel time nonlinear. Inverse problems in general are discussed in Chapter 22, and specifically the issue of nonlinearity is addressed by perturbing the depth to the source of a gravity anomaly. In this section we perturb both the slowness and the travel time to derive a linearized relation between the travel time *perturbation* and the slowness *perturbation*. The travel time along rays follows from geometric ray theory as shown in Section 11.4. It is shown in that section that the travel time $\tau(\mathbf{r})$ from a given source to location $\mathbf{r}$ is given by the eikonal equation (11.25), which can be written as

$$|\nabla \tau(\mathbf{r})|^2 = u^2(\mathbf{r}). \tag{23.32}$$

Let us assume that we have a reasonable guess $u_0(\mathbf{r})$ for the slowness, and that we seek a small perturbation of the slowness; this perturbation is denoted as $\epsilon u_1(\mathbf{r})$. The slowness can then be written as

$$u(\mathbf{r}) = u_0(\mathbf{r}) + \epsilon u_1(\mathbf{r}). \tag{23.33}$$

Again, the parameter ϵ only serves to systematically set up the perturbation treatment. When the slowness is perturbed, the travel time changes as well and it can be expanded in a perturbation series of the parameter ϵ:

$$\tau = \tau_0 + \epsilon \tau_1 + \epsilon^2 \tau_2 + \cdots . \tag{23.34}$$

In this section we seek the relation between the first-order travel time perturbation τ_1 and the slowness perturbation u_1.

Problem a Insert (23.33) and (23.34) into the eikonal equation (23.32), use that $|\nabla \tau|^2 = (\nabla \tau \cdot \nabla \tau)$ and collect the terms in $\mathcal{O}(1)$ and $\mathcal{O}(\epsilon)$ to show that the first and zeroth-order travel time perturbation are given by

$$|\nabla \tau_0(\mathbf{r})|^2 = u_0^2(\mathbf{r}), \tag{23.35}$$

$$(\nabla \tau_0 \cdot \nabla \tau_1) = u_0 u_1. \tag{23.36}$$

The first equation is nothing but the eikonal equation for the unperturbed problem. This expression states that the length of the vector $\nabla \tau_0$ is equal to u_0. This means that $\nabla \tau_0$ can be written as

$$\nabla \tau_0 = u_0 \hat{\mathbf{n}}_0. \tag{23.37}$$

In this expression the unit vector $\hat{\mathbf{n}}_0$ gives the direction of $\nabla \tau_0$. We showed in Section 5.1 that the gradient $\nabla \tau_0$ is perpendicular to the surfaces of constant value of the function. In Figure 23.3 the solid lines are curves of constant travel time τ. The shape of these curves is determined by the underlying heterogeneity in the velocity structure, and the rays are the curves that are everywhere perpendicular to the surfaces of constant travel time.

We parameterize the location along a ray by the length of the ray between the source and the point under consideration. In the following we use the parameter s_0 to denote the length along the reference ray. We show in the following derivation that to first order in ϵ we don't need the location of the perturbed ray, so there is no reason to distinguish between the length s_0 along the reference ray and the length s along the perturbed ray. In higher-order perturbation theory, the distinction between s and s_0 is essential and nontrivial (Snieder and Sambridge, 1992, 1993), but we can ignore this subtlety to first order in ϵ.

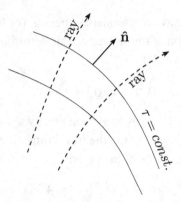

Figure 23.3   Wavefronts as the surfaces of constant travel time τ (solid lines), and the rays that are the curves perpendicular to the travel time surfaces (dashed lines). The unit vector $\hat{\mathbf{n}}$ is perpendicular to the wavefronts.

Problem b   Take the dot product of expression (23.37) with $\hat{\mathbf{n}}_0$ and use (5.22) to show that the directional derivative of τ_0 along the ray in the reference medium is given by

$$\frac{d\tau_0}{ds_0} = u_0, \tag{23.38}$$

where s_0 denotes the arclength along the ray in the reference medium.

This last expression can be integrated to give

$$\tau_0 = \int_{\mathbf{r}_0[u_0]} u_0(\mathbf{r}) \, ds_0. \tag{23.39}$$

This expression is identical to (23.31) with the exception that all quantities are for the reference medium u_0 and its associated rays $\mathbf{r}_0[u_0]$.

Problem c   In order to derive the first-order travel time perturbation, insert (23.37) into (23.36) to derive that

$$\left(\hat{\mathbf{n}}_0 \cdot \nabla \tau_1\right) = u_1. \tag{23.40}$$

Note that the unit vector $\hat{\mathbf{n}}_0$ is directed along the rays in the reference medium u_0 and that it is therefore independent of the slowness perturbation u_1.

Problem d   Use expression (5.22) for the directional derivative d/ds_0 and integrate the result to give

$$\tau_1 = \int_{\mathbf{r}_0[u_0]} u_1(\mathbf{r}) \, ds_0. \tag{23.41}$$

In this expression the integration is along the rays $\mathbf{r}_0[u_0]$ in the reference medium. Since these rays are assumed to be known, (23.41) constitutes a linearized relation between the travel time perturbation τ_1 and the slowness perturbation u_1. We can now write (23.41) in the form (22.8), after which techniques described in Chapter 22 can be used to determine the unknown slowness perturbation from the measured travel time perturbations τ_1. In many textbooks, for example, Nolet (1987), this result is derived from Fermat's theorem. However, the treatment in this section, as proposed by Aldridge (1994), is conceptually much simpler. In fact, the treatment in this section can easily be extended to compute the travel time perturbation to *any* order (Snieder and Aldridge, 1995).

23.4 Limits on perturbation theory

Perturbation theory is a powerful tool; in principle it provides a systematic way to derive the perturbation to any desired order. In this section we will discover that for a given order of truncation of the perturbation series the accuracy of the obtained result may depend strongly on the value of certain parameters of the problem that one is considering. This is illustrated with a simple problem we can also solve analytically. We consider the differential equation

$$\ddot{x} + \omega_0^2 (1 + \epsilon) x = 0, \tag{23.42}$$

with the initial conditions

$$x(0) = 1, \qquad \dot{x}(0) = 0. \tag{23.43}$$

This equation describes a harmonic oscillator in which the frequency is perturbed.

Problem a Show that the exact solution of this problem is given by

$$x(t) = \cos\left(\omega_0\sqrt{1+\epsilon}\, t\right). \tag{23.44}$$

Problem b The solution $x(t)$ is a function of the perturbation parameter ϵ; it can therefore be written as a perturbation series in this parameter:

$$x(t) = x_0(t) + \epsilon x_1(t) + \epsilon^2 x_2(t) + \cdots. \tag{23.45}$$

Insert this series into (23.42) and collect together the terms of equal powers in ϵ to show that the terms $x_n(t)$ satisfy the following equations:

$$\left.\begin{aligned}
\ddot{x}_0 + \omega_0^2 x_0 &= 0, \\
\ddot{x}_1 + \omega_0^2 x_1 &= -\omega_0^2 x_0, \\
\ddot{x}_2 + \omega_0^2 x_2 &= -\omega_0^2 x_1, \\
&\vdots
\end{aligned}\right\} \tag{23.46}$$

Problem c In order to solve these equations one must also consider the initial conditions of $x_n(t)$. Obtain these conditions by inserting the perturbation series (23.45) into the initial conditions (23.43) and derive that

$$\left.\begin{aligned}
x_0(0) &= 1, & \dot{x}_0(0) &= 0, \\
x_n(0) &= 0, & \dot{x}_n(0) &= 0 \quad \text{for } n \geq 1.
\end{aligned}\right\} \tag{23.47}$$

Problem d Solve the differential equation for x_0 for the boundary condition of (23.47) and show that the solution is given by

$$x_0(t) = \cos(\omega_0 t). \tag{23.48}$$

Note that this expression is identical to the exact solution (23.44) when one switches off the perturbation by setting $\epsilon = 0$. Inserting this solution into the second line of (23.46), one finds that the first-order perturbation satisfies the following differential equation

$$\ddot{x}_1 + \omega_0^2 x_1 = -\omega_0^2 \cos(\omega_0 t). \tag{23.49}$$

This equation describes a harmonic oscillator with eigenfrequency ω_0, which is driven by a force on the right-hand side. This driving force also oscillates with frequency ω_0. This means that the oscillator x_1 is driven at its resonance frequency, which in general leads to a motion that grows with time.

Problem e In order to solve (23.49), write the perturbation $x_1(t)$ as

$$x_1(t) = f(t)\cos(\omega_0 t) + g(t)\sin(\omega_0 t). \tag{23.50}$$

Insert this expression into (23.49) and collect the terms that multiply $\cos(\omega_0 t)$ and $\sin(\omega_0 t)$ to show that the unknown functions $f(t)$ and $g(t)$ obey the following differential equations:

$$\left.\begin{aligned}
\ddot{f} + 2\omega_0 \dot{g} &= -\omega_0^2, \\
\ddot{g} - 2\omega_0 \dot{f} &= 0.
\end{aligned}\right\} \tag{23.51}$$

Problem f Show that these equations are satisfied by the following solution:

$$\dot{f} = 0, \qquad \dot{g} = -\frac{\omega_0}{2}, \tag{23.52}$$

and integrate these equations to derive a particular solution that is given by

$$f = A, \qquad g(t) = -\frac{1}{2}\omega_0 t + B, \tag{23.53}$$

with A and B integration constants.

These expressions for f and g lead to the general solution

$$x_1(t) = -\frac{1}{2}\omega_0 t \, \sin(\omega_0 t) + A \cos(\omega_0 t) + B \sin(\omega_0 t). \tag{23.54}$$

It follows from (23.47) that the initial conditions for $x_1(t)$ are given by $x_1(0) = 0$, $\dot{x}_1(0) = 0$.

Problem g Derive from these initial conditions that the integration constants are given by $A = B = 0$, so that the solution is given by

$$x_1(t) = -\frac{1}{2}\omega_0 t \, \sin(\omega_0 t). \tag{23.55}$$

It was noted earlier that (23.49) describes an oscillator that is driven at its resonance frequency. Solution (23.55) grows linearly with time. This growth with time is called the *secular growth*. It is an artifact of the perturbation technique employed because the original problem (23.42) does not contain a resonant driving force at all. Inserting the first-order perturbation (23.55) and the unperturbed solution (23.48) into the perturbation series (23.45) finally gives

$$x(t) = \cos(\omega_0 t) - \frac{\epsilon}{2}\omega_0 t \, \sin(\omega_0 t) + \mathcal{O}(\epsilon^2). \tag{23.56}$$

Let us first verify that this expression is indeed the first-order expansion of the exact solution (23.44). Using the series (3.12), (3.13), and (3.16), the exact solution (23.44) is to first order in ϵ given by

$$\begin{aligned}
x(t) &= \cos\left(\omega_0\sqrt{1+\epsilon}\,t\right) \\
&= \cos\left\{\omega_0\left[1+\frac{\epsilon}{2}+\mathcal{O}(\epsilon^2)\right]t\right\} \\
&= \cos(\omega_0 t)\cos\left(\frac{\epsilon}{2}\omega_0 t\right) - \sin(\omega_0 t)\sin\left(\frac{\epsilon}{2}\omega_0 t\right) + \mathcal{O}(\epsilon^2) \\
&= \cos(\omega_0 t) - \frac{\epsilon}{2}\omega_0 t \sin(\omega_0 t) + \mathcal{O}(\epsilon^2). \tag{23.57}
\end{aligned}$$

This result is indeed identical to the first-order expansion (23.56) that was obtained from perturbation theory. The fact that this result is correct, however, does not imply that this result is also useful. Perturbation theory is based on the premise that the truncated perturbation series gives a good approximation to the true solution. The truncation of this series only makes sense when the subsequent terms in the perturbation expansion rapidly become smaller. However, the first-order term $(\epsilon\omega_0 t/2)\sin(\omega_0 t)$ in (23.46) is as large as the zeroth-order term $\cos(\omega_0 t)$ when $\epsilon\omega_0 t/2 \sim 1$. This means that the first-order perturbation series (23.56) ceases to be a good approximation to the true solution when

$$t \sim \frac{1}{\epsilon \omega_0}. \qquad (23.58)$$

The upshot of this example is that even though a truncated perturbation series may be correct, it may only be useful for a restricted range of parameters. In this example, the first-order perturbation series is a good approximation only when $t \ll 1/\epsilon\omega_0$. For the problem in this section, it would be more appropriate to make a perturbation series of the phase and the amplitude of the oscillator. Systematic techniques such as *multiple-scale analysis* (Bender and Orszag, 1999) have been developed for this purpose. In the following section we carry out a change of variables to derive the perturbations of the amplitude and phase of a wave that propagates through an inhomogeneous medium.

23.5 WKB approximation

In this section we analyze the propagation of a wave through a one-dimensional acoustic medium, where according to (6.41) the pressure satisfies in the frequency domain and in one space dimension the following differential equation:

$$\rho \frac{d}{dx} \left(\frac{1}{\rho} \frac{dp}{dx} \right) + \frac{\omega^2}{c^2} p = 0, \qquad (23.59)$$

where we used that in one space dimension $\rho \nabla (\rho^{-1} \nabla p) = \rho \partial_x (\rho^{-1} \partial_x p)$. In expression (23.59) both the density ρ and the velocity c vary with the position x. This equation can only be solved in closed form for a special form of the functions $\rho(x)$ and $c(x)$. It should be noted that the treatment in this section is also applicable to the Schrödinger equation

$$\frac{d^2 \psi}{dx^2} + \frac{2m}{\hbar^2} [E - V(x)] \psi = 0, \qquad (23.60)$$

by making the substitutions

$$\rho(x) \to 1, \qquad p \to \psi, \qquad 1/c(x) \to \sqrt{2m (E - V(x))}/\hbar\omega. \qquad (23.61)$$

As of yet, there is no small perturbation parameter. We restrict our attention to media in which the length scale of the variation in both ρ and c is much larger than the wavelength λ of the wave. Physically this type of medium does not contain strong inhomogeneities on the scale of a wavelength, so that the waves are not reflected strongly by the heterogeneity. Let the length scale of the heterogeneity be denoted by L. When this length scale is much larger than the wavelength $\lambda = 2\pi c/\omega$, the following parameter is small:

$$\epsilon = \frac{c}{\omega L} \ll 1. \tag{23.62}$$

As noted in the previous section, when the perturbation affects mostly the phase of a wave, it is advantageous to perturb the phase (and amplitude) of the wave rather than the solution p itself. This can be achieved by making the transformation

$$p(x) = p_0 \, e^{S(x)}, \tag{23.63}$$

where p_0 is constant of dimension pressure to make this equation dimensionally correct. When $S(x)$ is complex, this transformation is without any loss of generality.

Problem a Insert this relation into (23.59) and derive that $S(x)$ satisfies the following differential equation:

$$\frac{d^2 S}{dx^2} - \frac{1}{\rho} \frac{d\rho}{dx} \frac{dS}{dx} + \left(\frac{dS}{dx}\right)^2 + \frac{\omega^2}{c^2} = 0. \tag{23.64}$$

At this point we have made the problem more complicated because this equation is nonlinear in the unknown function $S(x)$, whereas the original equation (23.59) is linear in the pressure $p(x)$. However, we have not yet applied the perturbation technique. Before we do this, let us first reflect on the transformation (23.63). If the medium were homogeneous, that is, if both ρ and c were constant, the density terms in the left-hand side of (23.59) cancel, and the solution to (23.59) would be given by $p = p_0 e^{ikx}$, with the wavenumber given by $k = \omega/c$. This special solution corresponds to $S = \ln(p_0) + ikx$, so that $dS/dx = ik = i\omega/c$. For an inhomogeneous medium, one may expect the derivative of the phase to be close to this value; therefore, we make the following substitution:

$$\frac{dS}{dx} = \frac{i\omega}{c(x)} F(x). \tag{23.65}$$

Problem b Show that this transformation transforms (23.64) into the following differential equation for F:

$$\frac{dF}{dx} - \frac{1}{\rho c} \frac{d(\rho c)}{dx} F + \frac{i\omega}{c} F^2 = \frac{i\omega}{c}. \tag{23.66}$$

It is not clear yet how to apply perturbation analysis to this expression. This can be achieved by transforming the distance x to a dimensionless distance ξ defined by

$$\xi \equiv x/L, \tag{23.67}$$

where L is the characteristic length scale of the velocity and density variations.

Problem c Under this transformation, the derivative d/dx changes to $d/dx = d/d(\xi L) = (1/L)\,d/d\xi$. Use this to show that $F(\xi)$ satisfies the following differential equation:

$$\frac{c}{\omega L}\frac{dF}{d\xi} - \frac{c}{\omega L}\frac{1}{\rho c}\frac{d\,(\rho c)}{d\xi}F + iF^2 = i. \qquad (23.68)$$

This equation contains the small dimensionless parameter $c/\omega L$ defined in (23.62), so that this equation is equivalent to

$$\epsilon\frac{dF}{d\xi} - \epsilon\frac{1}{\rho c}\frac{d\,(\rho c)}{d\xi}F + iF^2 = i. \qquad (23.69)$$

Now we have an equation that looks similar to those in the perturbation problems we have seen in this chapter. We solve this equation by inserting the following perturbation series for $F(\xi)$:

$$F(\xi) = F_0(\xi) + \epsilon F_1(\xi) + \cdots. \qquad (23.70)$$

Problem d Insert this perturbation series into the differential equation (23.69) to derive that $F_0(\xi)$ and $F_1(\xi)$ satisfy the following equations:

$$\left.\begin{aligned} F_0^2 &= 1, \\ 2i F_0 F_1 &= -\frac{dF_0}{d\xi} + \frac{1}{\rho c}\frac{d\,(\rho c)}{d\xi}F_0. \end{aligned}\right\} \qquad (23.71)$$

The first of these equations has the solutions $F_0 = \pm 1$. According to (23.65) this corresponds to the phase derivative $dS/dx = \pm i\omega/c(x)$. The plus sign denotes a right-going wave, and the minus sign a left-going wave. We focus here on a right-going wave so that $F_0 = +1$.

Problem e Insert this solution into the second line of (23.71) and show that $F_1(\xi)$ is given by

$$F_1(\xi) = -\frac{i}{2}\frac{1}{\rho c}\frac{d\,(\rho c)}{d\xi}. \qquad (23.72)$$

This means that the first-order perturbation series for $F(\xi)$ is given by

$$F(\xi) = 1 - \frac{i\epsilon}{2}\frac{1}{\rho c}\frac{d\,(\rho c)}{d\xi} + O(\epsilon^2). \qquad (23.73)$$

Problem f Now that we have obtained this solution as a function of the transformed distance ξ, transform back to the original distance x by using (23.67). Use (23.62) to show that the solution (23.73) is equivalent to

$$F(x) = 1 - \frac{i}{2} \frac{1}{\rho\omega} \frac{d\,(\rho c)}{dx} + \cdots . \qquad (23.74)$$

Problem g Use (23.65) to convert this into an equation for dS/dx. Integrate this equation to show that the solution $S(x)$ is given by

$$S(x) = i \int_{-\infty}^{x} \frac{\omega}{c(x')} dx' + \frac{1}{2} \ln \left[\rho(x)c(x) \right] + B. \qquad (23.75)$$

Problem h Use the transformation (23.63) to show that this solution corresponds to the following pressure field

$$p(x) = A\sqrt{\rho(x)c(x)} \, \exp \left(i \int_{-\infty}^{x} \frac{\omega}{c(x')} dx' \right), \qquad (23.76)$$

with A a new constant.

This solution states that the wave propagates to the right with an amplitude that is proportional to $\sqrt{\rho(x)c(x)}$. The local wavenumber $k(x)$ of the wave is given by the derivative of the phase of the wave (Whitham, 2011); it is therefore given by

$$k(x) = \frac{d}{dx} \int_{-\infty}^{x} \frac{\omega}{c(x')} dx' = \frac{\omega}{c(x)}. \qquad (23.77)$$

This means that the local wavenumber at position x is given by the wavenumber $\omega/c(x)$. This is as if the medium were homogeneous with the properties of the medium at that location x. The solution (23.76) is known as the WKB solution (named after Wentzel, Kramers, and Brillouin). Seismologists prefer to call this solution the WKBJ approximation because of the contribution of Lord Jeffreys in Jeffreys (1925).

In most textbooks (e.g., Merzbacher, 1961), this solution is derived for the Schrödinger equation rather than the acoustic wave equation; with the transformation (23.61), the derivations are equivalent.

Problem i Use the correspondence (23.61) to show that in quantum mechanics the WKB solution is given by

$$\psi(x) = \frac{A}{[E - V(x)]^{1/4}} \exp \left\{ i \int_{-\infty}^{x} \frac{\sqrt{2m\left[E - V(x')\right]}}{\hbar} dx' \right\}. \qquad (23.78)$$

Problem j Show that this approximation is infinite at the *turning points* of the motion. These are the points where the total energy of the particle is equal to the potential energy. This means that the WKB solution breaks down at the turning points.

A clear account of the WKB approximation with a large number of applications is given by Bender and Orszag (1999).

23.6 The need for consistency

In perturbation theory one derives an approximate solution to a problem. In many applications this approximation is then used in subsequent calculations. In doing so, one must keep in mind that the solution obtained from perturbation theory is not the true solution, and that it is pointless to carry out the subsequent calculations with an accuracy that is higher than the accuracy of the solution obtained from perturbation theory.

As an example we consider in this section the WKB solution (23.76) for the pressure field $p(x)$ and compute the particle velocity $v(x)$ that is associated with this pressure field. The particle velocity follows from Newton's law (6.34), which for one dimension and in the absence of an external force reduces to $\rho \partial v/\partial t = -\partial p/\partial x$.

Problem a   Show that with the Fourier convention (14.42) the corresponding equation is in the frequency domain given by

$$i\omega\rho v = \partial p/\partial x. \qquad (23.79)$$

Problem b   Apply this result to the WKB solution (23.76) and show that the velocity is given by

$$i\omega\rho v = \frac{1}{2}\frac{1}{\rho c}\frac{d\,(\rho c)}{dx}p + \frac{i\omega}{c}p. \qquad (23.80)$$

Problem c   Use the estimate of the derivative in Section 11.2 to show that the first term on the right-hand side is of the order p/L, where L is the characteristic length scale over which the density and the velocity vary.

The second term on the right-hand side of (23.80) is of the order $\omega p/c$. This means that the ratio of the first term to the second term is given by $(p/L)/(\omega p/c) = c/(\omega L) = \epsilon$, where the parameter ϵ is defined in (23.62). In the previous section we assumed that the medium varies so smoothly that $\epsilon \ll 1$. This means that under the assumptions that underlie the WKB approximation the first term on the right-hand side of (23.80) can be ignored with respect to the second term.

Problem d   Show that in this approximation the velocity is given by

$$v = \frac{p}{\rho c}. \qquad (23.81)$$

Note that ignoring the first term is consistent with the terms that we have ignored in the WKB approximation.

This last expression has an interesting interpretation. The quantity ρc is called the *acoustic impedance*. This term is reminiscent of the theory of electromagnetism. For a resistor, Ohm's law $I = V/R$ relates the current I that is generated by a voltage V. For a general linear electric system, the current and the voltage are related by

$$I = \frac{V}{Z},$$
(23.82)

where Z is a generalization of the resistance that is called the *impedance*. The impedance gives the strength of the current for a given potential. Similarly, the acoustic impedance ρc in (23.81) gives the particle velocity for a given pressure.

Combining (23.76) with (23.81) shows that the velocity, and hence the particle motion, is proportional to $1/\sqrt{\rho c}$. This means that the particle motion increases when the acoustic impedance decreases. This has important implications for earthquake hazards. For soft soils, both the density ρ and the wave velocity c are small. This means that the acoustic impedance is much smaller in soft soils than in hard rock. This in turn means that the ground motion during earthquakes is much more severe in soft soils than in hard rock. The motion in the Earth is governed by the elastic wave equation rather than the acoustic wave equation. However, one can show (Aki and Richards, 2002) that for elastic waves also the displacement is inversely proportional to $1/\sqrt{\rho c}$, where c is the propagation velocity of the elastic wave under consideration.

The fact that the ground motion is inversely proportional to the square root of the impedance is one of the factors that made the 1985 earthquake along the west coast of Mexico cause so much damage in Mexico City. This city is constructed on soft sediments that have filled the swamp onto which the city is built. The small value of the associated elastic impedance was one of the causes of the extensive damage in Mexico City after this earthquake.

23.7 Singular perturbation theory

In Section 23.1 we analyzed the behavior of the root of the equation $x^3 - 4x^2 + 4x = \epsilon$ that was located near $x = 0$. As shown in that section, the unperturbed problem also has a root $x = 2$. The roots $x = 0$ and $x = 2$ can be seen graphically in Figure 23.1 because for these values of x the polynomial shown by the thick solid line is equal to zero. In Figure 23.1 the value $\epsilon = +0.15$ is shown by a dotted line, while the value $\epsilon = -0.15$ is indicated by the dashed line. There is a profound difference between the two roots when the parameter ϵ is nonzero. The root near $x = 0$ depends in a continuous way on ϵ, and (23.2) has for the root near $x = 0$ a solution regardless of whether ϵ is positive or negative. This situation is completely different for the root near $x = 2$. When ϵ is positive (the dotted line), the polynomial has *two* intersections with the dotted line, whereas when ϵ is negative, the polynomial does *not* intersect the dashed line at all. This means that depending on whether ϵ is positive or negative, the solution has two or zero solutions, respectively. This behavior cannot be described by a regular perturbation series of the form (23.5) because this expansion assigns *one* solution to each value of the perturbation parameter ϵ.

Let us first diagnose where the treatment of Section 23.1 breaks down when we apply it to the root near $x = 2$.

Problem a   Insert the unperturbed solution $x_0 = 2$ into the second line of (23.13) and show that the resulting equation for x_1 is

$$0 \cdot x_1 = 1. \tag{23.83}$$

This equation obviously has no finite solution. This is related to the fact that the tangent of the polynomial at $x = 2$ is horizontal. First-order perturbation theory effectively replaces the polynomial by the straight line that is tangent to the polynomial. When this tangent line is horizontal, it can never have a value that is nonzero.

This means that the regular perturbation series (23.5) is not the appropriate way to study the behavior of the root near $x = 2$. In order to find out how this root behaves, let us set

$$x = 2 + y. \tag{23.84}$$

Problem b   Show that under the substitution (23.84) the original problem (23.2) transforms to

$$y^3 + 2y^2 = \epsilon. \tag{23.85}$$

We will not yet carry out a systematic perturbation analysis, but we will first determine the dependence of the solution y on the parameter ϵ. For small values of ϵ, the parameter y is also small. This means that the term y^3 can be ignored with respect to the term y^2. Under this assumption (23.85) is approximately equal to $2y^2 \approx \epsilon$ so that $y \approx \sqrt{\epsilon/2}$. This means that the solution does not depend on integer powers of ϵ as in the perturbation series (23.5), but that it does depend on the square root of ϵ. The square root of ϵ is shown in Figure 23.4. Note that for $\epsilon = 0$ the tangent of this curve is vertical and that for $\epsilon < 0$ the real function $\sqrt{\epsilon}$ is

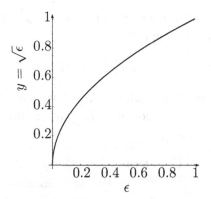

Figure 23.4   The function $y = \sqrt{\epsilon}$ approaches a vertical line near the origin.

not defined for negative values of ϵ.[1] This reflects the fact that the roots near $x = 2$ depend in a very different way on ϵ than the root near $x = 0$.

We know now that a regular perturbation series (23.5) is not the correct tool to use to analyze the root near $x = 2$. However, we do not yet know what type of perturbation series we should use for the root near $x = 2$; we only know that the perturbation depends to leading order on $\sqrt{\epsilon}$. This prompts us to make the following substitution:

$$x = 2 + \sqrt{\epsilon}\, z. \tag{23.86}$$

Problem c   Insert this solution into (23.2) and show that z satisfies the following equation:

$$\sqrt{\epsilon}\, z^3 + 2z^2 = 1. \tag{23.87}$$

Now we have a new perturbation problem with a small parameter. However, this small parameter is not the original perturbation parameter ϵ, but it is the square root $\sqrt{\epsilon}$. The perturbation problem in this section is a *singular perturbation problem*. In a singular perturbation problem, the solution is not a well-behaved function of the perturbation parameter. This has the result that the corresponding perturbation series cannot be expressed in powers ϵ^n, where n is a positive real integer. Instead, negative or fractional powers of ϵ are present in the perturbation series of a singular perturbation problem.

---

[1] When one allows a complex solution $x(\epsilon)$ of the equation, there are always two roots near $x = 2$. However, these complex solutions also display a fundamental change in their behavior when $\epsilon = 0$, which is characterized by a bifurcation.

Problem d   Since the small parameter in (23.87) is $\sqrt{\epsilon}$, it makes sense to seek an expansion of z in this parameter:

$$z = z_0 + \epsilon^{1/2} z_1 + \epsilon z_2 + \cdots . \tag{23.88}$$

Collect together the coefficients of equal powers of ϵ when this series is inserted into (23.87) and show that this leads to the following equations for the coefficients z_0 and z_1:

$$\left.\begin{array}{ll} \mathcal{O}(1)\text{-terms:} & 2z_0^2 - 1 = 0, \\ \mathcal{O}(\epsilon^{1/2})\text{-terms:} & z_0^3 + 4z_0 z_1 = 0. \end{array}\right\} \tag{23.89}$$

Problem e   The first equation of (23.89) has the solution $z_0 = \pm 1/\sqrt{2}$. Show that for both the plus and the minus signs $z_1 = -1/8$. Use these results to derive that the roots near $x = 2$ are given by

$$x = 2 \pm \frac{1}{\sqrt{2}}\sqrt{\epsilon} - \frac{1}{8}\epsilon + \mathcal{O}(\epsilon^{3/2}). \tag{23.90}$$

It is illustrative to compute the numerical values of these roots for the original problem (23.1), where $\epsilon = 0.01$; this gives for the two roots:

$$x = 1.928 \quad \text{and} \quad x = 2.069. \tag{23.91}$$

In these numbers only four decimals are shown. The reason is that the error in the truncated perturbation series is of the order of the first truncated term; hence, the error is of the order $(0.01)^{3/2} = 0.001$. When these solutions are compared with the perturbation solution (23.16) for the root near $x = 0$, it is striking that the singular perturbation series for the root near $x = 2$ converges much less rapidly than the regular perturbation series (23.16) for the root near $x = 0$. This is a consequence of the fact that the solution near $x = 2$ is a perturbation series in $\sqrt{\epsilon}(= 0.1)$ rather than $\epsilon(= 0.01)$. When the roots (23.91) are inserted into the polynomial (23.1) the following solutions are obtained for the two roots:

$$\left.\begin{array}{ll} x = 1.924: & x^3 - 4x^2 + 4x = 0.01 - 5 \times 10^{-6}, \\ x = 2.065: & x^3 - 4x^2 + 4x = 0.01 - 1 \times 10^{-4}. \end{array}\right\} \tag{23.92}$$

Note how accurate these approximations are.

The singular behavior of the roots of the polynomial (23.1) near $x = 2$ corresponds to the fact that the solution changes in a discontinuous way when the perturbation parameter ϵ goes to zero. Figure 23.1 shows that for the perturbation problem in this section the problem has one root near $x = 2$ when $\epsilon = 0$, it has no roots when $\epsilon < 0$, and there are two roots when $\epsilon > 0$. Such a discontinuous change in the character of the solution also occurs in fluid mechanics in which the equation of motion is given by

$$\frac{\partial(\rho\mathbf{v})}{\partial t} + \nabla \cdot (\rho\mathbf{v}\mathbf{v}) = \mu\nabla^2\mathbf{v} + \mathbf{F}. \tag{25.48}$$

In this expression the viscosity of the fluid gives a contribution $\mu\nabla^2\mathbf{v}$, where μ is the viscosity. This viscous term contains the highest spatial derivatives of the velocity that are present in the equation. When the viscosity μ goes to zero, the equation for fluid flow becomes a first-order differential equation rather than a second-order differential equation. This changes the number of boundary conditions that are needed for the solution, and hence it drastically affects the mathematical structure of the solution. This is relevant in boundary layer problems where fluid flows along a boundary and where a strong velocity gradient occurs in a thin boundary layer. Such problems are, in general, singular perturbation problems (Van Dyke, 1964).

When waves propagate through an inhomogeneous medium, they may be focused onto focal points or focal surfaces (Berry and Upstill, 1980). These regions in space where the wave amplitude is large are called *caustics*. The formation of caustics depends on $\epsilon^{2/3}$, where ϵ is a measure of the variations in the wave velocity (Kulkarny and White, 1982; Spetzler and Snieder, 2001). The non-integer power of ϵ indicates that the formation of caustics constitutes a singular perturbation problem.

24

Asymptotic evaluation of integrals

In mathematical physics, the result of a computation often is expressed as an integral. Frequently, an analytical solution to these integrals is not known or it is so complex that it gives little insight into the physical nature of the solution. As an alternative, one can often find an approximate solution to the integral that in many cases works surprisingly well. The approximations that are treated here exploit that there is a parameter in the problem that is either large or small. The corresponding approximation is called an asymptotic solution to the integral, because the approximation holds for asymptotically large or small values of that parameter. Excellent treatments of the asymptotic evaluation of integrals are given by Bender and Orszag (1999) and by Bleistein and Handelsman (1975).

24.1 Simplest tricks

In general there is no simple recipe for integrating a function. For this reason, there is no simple trick for approximating integrals that always works. The asymptotic evaluation of integrals is more like a bag of tricks. In this section we treat the simplest tricks: the Taylor series and integration by parts. As an example, let us consider the following integral

$$F(x) \equiv \int_0^x e^{-u^2} du. \tag{24.1}$$

For small values of x, u is restricted to small values as well, and the integrand can be approximated well by a Taylor series. Each term in this Taylor series can be integrated separately.

Problem a Expand the integrand in a Taylor series around $u = 0$, using (3.14) and integrate this series term by term to derive that

$$\int_0^x e^{-u^2} du = x - \frac{1}{3}x^3 + \frac{1}{10}x^5 - \frac{1}{42}x^7 + \cdots \tag{24.2}$$

442

Problem b Show that this result can also be written as

$$\int_0^x e^{-u^2} du = \sum_{n=0}^{\infty} \frac{(-1)^n}{(2n+1)\,n!} x^{2n+1}. \tag{24.3}$$

In general, such an infinite series is just as difficult to evaluate as the original integral. The approximation to this integral consists in truncating the integral at a certain point. For example, retaining the first two terms gives the approximation

$$\int_0^x e^{-u^2} du \approx x - \frac{1}{3} x^3 \qquad \text{for } x \ll 1. \tag{24.4}$$

The approximation sign $\approx$ is mathematically not very precise; for this reason the following notation is preferred by many

$$\int_0^x e^{-u^2} du = x - \frac{1}{3} x^3 + \mathcal{O}(x^5). \tag{24.5}$$

This notation has the advantage that is shows that the error made in (24.4) goes to zero as x^5 (or faster) as $x \to 0$. This type of mathematical rigor may appear to be attractive. However, the notation (24.5) still does not tell us how good the approximation (24.4) is for a given *finite* value of x. Since the mathematical rigor of (24.5) is not very informative for this problem, we will often use the more sloppy, but equally uninformative, notation (24.4). We return to this issue at the end of this section when we apply this result to the computation of probabilities for the Gaussian distribution.

In the following example we consider the integral

$$I(x) \equiv \int_x^{\infty} e^{-u^2} du, \tag{24.6}$$

for $x \gg 1$.

Problem c Show that we can write the integrand as

$$e^{-u^2} = \frac{-1}{2u} \frac{d e^{-u^2}}{du}. \tag{24.7}$$

The approximation that we will derive is based on this identity.

Problem d Insert (24.7) into (24.6) and carry out an integration by parts to show that

$$I(x) = \frac{1}{2x} e^{-x^2} - \frac{1}{2} \int_x^{\infty} \frac{1}{u^2} e^{-u^2} du. \tag{24.8}$$

The last integral satisfies

$$\int_x^\infty \frac{1}{u^2} e^{-u^2} du \le \frac{1}{x^2} \int_x^\infty e^{-u^2} du = \frac{1}{x^2} I(x) \ll I(x), \qquad (24.9)$$

where $u \ge x$ is used in the first inequality and $x \gg 1$ is used in the last inequality. This means that the integral in (24.8) is much smaller than the original integral (24.6). The first term in the right-hand side of (24.8) therefore is a good approximation to the integral:

$$\int_x^\infty e^{-u^2} du \approx \frac{1}{2x} e^{-x^2}. \qquad (24.10)$$

This approximation can, however, be refined further.

Problem e   Insert the identity (24.7) into the integral in (24.8) and carry out another integration by parts to show that

$$I(x) = \frac{1}{2x} e^{-x^2} - \frac{1}{4x^3} e^{-x^2} + \frac{3}{4} \int_x^\infty \frac{1}{u^4} e^{-u^2} du. \qquad (24.11)$$

Since $u \gg 1$, the integral in (24.11) is much smaller than the integral in (24.8). In fact:

$$\int_x^\infty \frac{1}{u^4} e^{-u^2} du \le \frac{1}{x^4} \int_x^\infty e^{-u^2} du \le \frac{1}{x^4} \int_0^\infty e^{-u^2} du = \frac{\sqrt{\pi}}{2x^4}, \qquad (24.12)$$

where the equality in this expression is derived in (4.48). This leads to the two-term approximation

$$\int_x^\infty e^{-u^2} du = \frac{1}{2x} e^{-x^2} \left(1 - \frac{1}{2x^2} + \mathcal{O}\left(\frac{1}{x^4}\right) \right). \qquad (24.13)$$

The tricks shown in this section can be applied to many other integrals. Here we discuss an application of the approximations (24.5) and (24.13). The function e^{-x^2} is an essential part of the Gauss distribution (21.22), and according to equation (21.3), integrals of a statistical distribution over an interval give the probability that a statistical variable lies within that interval. Integral (24.1) is related to the *error function*

$$\mathrm{erf}(x) \equiv \frac{2}{\sqrt{\pi}} \int_0^x e^{-u^2} du, \qquad (24.14)$$

while the integral (24.6) is related to the *complementary error function*

$$\mathrm{erfc}(x) \equiv \frac{2}{\sqrt{\pi}} \int_x^\infty e^{-u^2} du. \qquad (24.15)$$

Table 24.1   *The probability $P(|y| < \sigma/2)$ and various approximations to this probability.*

$P(|y| < \sigma/2)$		Relative error
True value	35.2%	
Leading-order asymptotic expansion	39.8%	0.13
Second-order asymptotic expansion	38.2%	0.086

Suppose that a random variable y follows a Gauss distribution with zero mean and standard deviation σ. Then according to (21.22) the corresponding probability density function is given by

$$p(y) = \frac{1}{\sqrt{2\pi}\sigma} \exp\left(-\frac{y^2}{2\sigma^2}\right). \tag{24.16}$$

The probability that y lies between a and b is given by

$$P(a < y < b) = \int_a^b p(y)\, dy. \tag{21.3}$$

The integral of (24.16) is not known in closed form. Tables of the integral of the Gaussian distribution exist. Robinson and Bevington (1992) give a table for the probability $P(|y| < a\sigma)$. For example, the probability that $|y| < \sigma/2$ is equal to 35.2%. We can estimate this probability with the small-x expansion (24.4) by using that

$$P(|y| < a\sigma) = 2P(0 < y < a\sigma) = (2/\sqrt{2\pi}\sigma) \int_0^{a\sigma} \exp(-y^2/2\sigma^2)\, dy. \tag{24.17}$$

Problem f   Use this result to compute the probability $P(|y| < \sigma/2)$ with the one-term Taylor expansion and the two-term Taylor expansion (24.4) and show that these approximations give the estimated probabilities shown in Table 24.1.

Note that the Taylor approximation gives a reasonably accurate estimate for this probability. Taking more terms of the Taylor series into account leads to more accurate estimates of this probability.

The large-x expansion (24.13) can be used to estimate the probability that y exceeds a given value $a\sigma$: $P(y > a\sigma)$. Here we estimate the probability that y is larger than three standard deviations from the mean by using that $P(y > 3\sigma) = (1/\sqrt{2\pi}\sigma) \int_{3\sigma}^\infty \exp(-y^2/2\sigma^2)\, dy$. According to Robinson and Bevington (1992) this probability is equal to 0.135%.

Table 24.2   *The probability $P(y > 3\sigma)$ and two approximations to this probability.*

$P(y > 3\sigma)$		Relative error
True value	0.135%	
Leading-order asymptotic expansion	0.148%	0.09
Second-order asymptotic expansion	0.132%	0.025

Problem g   Use the large-x expansion (24.13) to show that the leading-order and second-order asymptotic expansion give the estimates of this probability shown in Table 24.2.

Note that the estimates from the asymptotic expansion (24.13) are quite accurate, despite the fact that the employed value in its argument was $x = 3/\sqrt{2} = 2.12$. It is difficult to maintain that this value of x is much larger than one!

24.2 What does $n!$ have to do with e and $\sqrt{\pi}$?

The factorial function is defined as

$$n! \equiv 1 \cdot 2 \cdot 3 \cdots (n - 1) \cdot n. \qquad (24.18)$$

In this section we investigate how $n!$ behaves for large values of n. At first you might be puzzled that we seek an approximation to this function, because the recipe (24.18) is simple. However, the statistical mechanics of a many-particle system depends on the number of ways in which particles can be permuted (Lavenda, 1991). Since a macroscopic system typically contains a mole of particles, one needs to evaluate $n!$ when n is of the order 10^{23}. For these types of problems, the asymptotic behavior of $n!$ is very useful.

In order to derive this asymptotic behavior we first express $n!$ in an integral by using the gamma function that is defined as

$$\Gamma(x) \equiv \int_0^\infty u^{x-1} e^{-u} du. \qquad (24.19)$$

With an integration by parts, this function can be rewritten as

$$\Gamma(x + 1) = -\int_0^\infty u^x \frac{de^{-u}}{du} du = -\left[u^x e^{-u} \right]_{u=0}^{u=\infty} + \int_0^\infty \frac{du^x}{du} e^{-u} du. \qquad (24.20)$$

Problem a Evaluate the boundary term in the right-hand side and use the definition (24.19) to show that

$$\Gamma(x+1) = x\Gamma(x). \tag{24.21}$$

Problem b Show by direct integration that $\Gamma(1) = 1$. Use this result and (24.21) to show that when n is a positive integer

$$\Gamma(n+1) = n! \tag{24.22}$$

The combination of (24.19) and (24.22) implies that $n!$ can be expressed in the following integral

$$n! = \int_0^\infty F_n(u)\,du, \tag{24.23}$$

with

$$F_n(u) = u^n e^{-u}. \tag{24.24}$$

It is instructive to consider the behavior of this function for large values of n. The term u^n is a rapidly growing function of u, while the term e^{-u} is a decreasing function of u. For small values of u, the first term dominates; while for large values of u, the last term dominates. This means that $F_n(u)$ must have a maximum for some intermediate value of u.

Problem c Show that $F_n(u)$ has a maximum for $u_{\max} = n$.

The function $F_n(u)$ is shown in Figure 24.1 for $n = 2$, 10, and 20. The maximum of the function occurs for $u = n$. Note that as n increases, $F_n(u)$ becomes more and more a symmetrically peaked function around its maximum.

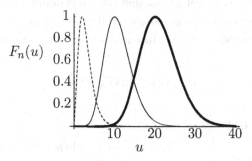

Figure 24.1 The function $F_n(u)$ normalized by its maximum value for $n = 2$ (dashed line), $n = 10$ (thin solid line), and $n = 20$ (thick solid line).

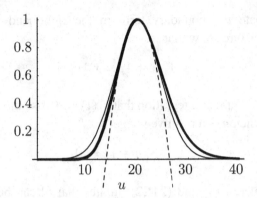

Figure 24.2   The function $F_{20}(u)$ normalized by its maximum value (thick solid line), the parabolic approximation (dashed line), and the Gaussian approximation (thin solid line).

Since for large values of n the integrand $F_n(u)$ is a peaked function, the value of the integral will mostly be determined by the behavior of the integrand near the peak. Let us first consider a Taylor approximation of the function near its maximum.

Problem d   Show that the second-order Taylor approximation around its maximum $u = n$ is given by

$$F_n(u) = n^n e^{-n} - \frac{1}{2} n^{n-1} e^{-n} (u - n)^2. \tag{24.25}$$

What we have done in this expression is approximate the function $F_n(u)$ by a parabola in the integration variable u. As shown by the dashed line in Figure 24.2, for $n = 20$, this is good approximation of the function near its maximum, but this is not a good approximation of the tails of the function as it approaches zero. The second-order Taylor approximation (24.25) does not account for the fact that $F_n(u)$ roughly has the shape of a Gaussian (which is defined as $\exp(-a(u - u_0)^2)$). In order to describe the Gaussian shape of the integrand, we carry out a Taylor expansion of the exponent of the function rather than the original function by defining

$$h_n(u) = \ln(F_n(u)). \tag{24.26}$$

Problem e   Show that

$$h_n(u) = -u + n \ln u, \tag{24.27}$$

and that

$$n! = \int_0^\infty e^{h_n(u)} du. \tag{24.28}$$

Problem f   The maximum of $F_n(u)$ at $u = n$ is also the maximum of $h_n(u)$. Use a second-order Taylor expansion of $h_n(u)$ around its maximum to derive that

$$n! \approx n^n e^{-n} \int_0^\infty e^{-(u-n)^2/2n^2} du. \tag{24.29}$$

The approximation sign means that terms $\mathcal{O}((u-n)^3)$ in the exponent have been ignored.

Problem g   The integrand of this integral is a Gaussian function with its peak at $u = n$. Show that the width of the peak (the "standard deviation") is given by $\sqrt{n}$.

Problem h   The last result implies that the width of the Gaussian ($\sqrt{n}$) for large values of n is small compared to the location of the maximum (n). Compute the value of the Gaussian function at the lower end of the integration interval ($u = 0$) for $n = 5, 10$, and 100, respectively.

The integral (24.29) runs from $u = 0$ to $u = \infty$. If the integration ran from $u = -\infty$ to $u = \infty$, then we could solve the integral with

$$\int_{-\infty}^{+\infty} e^{-bx^2} dx = \sqrt{\frac{\pi}{b}}. \tag{4.49}$$

However, Figure 24.2 shows that the integrand in (24.29) for large values of n is small for negative values of u. Also, as shown in the last problem, even for moderately large values of n, the integrand is extremely small at the integration boundary $u = 0$. For this reason, we make a small error by extending the integration limit in (24.29) to $-\infty$:

$$n! \approx n^n e^{-n} \int_{-\infty}^\infty e^{-(u-n)^2/2n^2} du. \tag{24.30}$$

Problem i   Use the Gaussian integral (4.49) to show that

$$n! \approx n^n e^{-n} \sqrt{2\pi n}. \tag{24.31}$$

This approximation is called *Stirling's formula*. In this expression the meaning of the $\approx$ sign is not precise. In the derivation of (24.31) we have made two approximations; the second-order Taylor approximation of Problem f and the extension of the lower limit of the integral (24.29) from 0 to $-\infty$. The first approximation can be improved by using a higher-order Taylor approximation, while the second approximation can be improved by using an asymptotic expansion of the error function

(24.14). A more detailed analysis of this problem shows that $n!$ can be written as (Bender and Orszag, 1999):

$$n! \approx n^n e^{-n} \sqrt{2\pi n} \left(1 + \frac{c_1}{n} + \frac{c_2}{n^2} + \cdots \right). \tag{24.32}$$

This series behaves differently to the series we have encountered so far since it has the following properties (Bender and Orszag, 1999):

- For small n, the c_n decrease with increasing values of n. However, for larger values of n, the c_n increase so rapidly with n that the series diverges. Trying to sum the series therefore is pointless.
- When the series is truncated at a certain point, the truncated series becomes more and more accurate as n increases.
- For a given value of n, there is an optimal truncation point.

With these remarks, you may have some reservation about using the approximation (24.31). Table 24.3 shows this approximation to $n!$ for several values of n. For $n \geq 10$, the error made by (24.31) is less than 1%. The approximation (24.31) is made for the case $n \gg 1$. However, for $n = 1$, the approximation is accurate within 8%, whereas one certainly cannot maintain that in that case $n \gg 1$. Try to imagine how accurate the approximation is when n is equal to Avogadro's number (6×10^{23}); this is the number typically used in application of (24.31) in statistical mechanics. Lavenda (1991) gives a wonderful historical account of how Max Planck discovered "Planck's law." A crucial step in his analysis was that Planck asked himself in how many ways a given amount of energy can be divided among a number of oscillators that each carry a given amount of energy. This question leads to the introduction of the binomial coefficients

Table 24.3   *The values of $n!$ and its approximation by Stirling's formula for different values of n.*

n	$n!$	$n^n e^{-n} \sqrt{2\pi n}$	Relative error (%)
1	1	0.922	7.8
2	2	1.919	4.1
3	6	5.836	2.7
4	24	23.51	2.1
5	120	118.0	1.6
10	3.629×10^6	3.598×10^6	0.83
15	1.307×10^{12}	1.300×10^{12}	0.55
20	2.432×10^{19}	2.423×10^{19}	0.42

$$\binom{n}{m} \equiv \frac{n!}{m!(n-m)!}. \qquad (24.33)$$

Analyzing the factorials with Stirling's equation gave an expression for the entropy that is crucial for explaining the radiation of black bodies.

Problem j Estimate $n!$ when n is equal to Avogadro's number. Hint: Take the logarithm of (24.31).

24.3 Method of steepest descent

In this section we treat the method of steepest descent. This method for the asymptotic evaluation of integrals is based on the idea that many functions in the complex plane have a stationary point. There is one direction at that stationary point in which the function decreases rapidly, and there is an orthogonal direction in which the function increases rapidly. By deforming the integration path so that it goes through the stationary point in the direction in which the function decreases, one can evaluate the integral asymptotically. As an example we apply this idea to the following integral

$$I = \int_{-\infty}^{\infty} e^{iax^2} dx, \qquad \text{with } a > 0 \text{ a real number.} \qquad (24.34)$$

Let us first investigate the behavior of the integrand $\exp\left(iz^2\right)$ in the complex plane with $z = x + iy$ for the special case where $a = 1$. The real part of the integrand $(\cos x^2)$ along the real axis is shown in Figure 24.3. Along this axis, the integrand is a rapidly oscillating function, except near the point $x = 0$ where the function is stationary. The real part of $\exp\left(iz^2\right)$ in the complex plane is shown in Figure 24.4. It can be seen from this figure that along the real axis and along

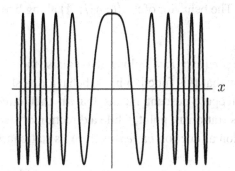

Figure 24.3 The function $\Re e\left(\exp\left(ix^2\right)\right) = \cos(x^2)$ for $-6 < x < 6$. The vertical axis runs from -1 to 1.

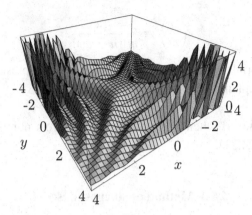

Figure 24.4   The function $\Re e\left(\exp\left(iz^2\right)\right)$ in the complex plane. For clarity the exponential growth has been diminished by a factor 10, by plotting the function $e^{-0.1\times2xy}\cos(x^2-y^2)$.

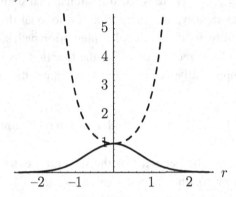

Figure 24.5   The function $\Re e\left(\exp\left(iz^2\right)\right)$ in the complex plane along the lines $x=y$ (solid line) and $x=-y$ (dashed line).

the imaginary axis, the function is oscillatory. Along the line $y=x$, the function has a *maximum* at the origin, while along the line $y=-x$ the function has a *minimum* at the origin. The behavior of $\Re e\left(\exp\left(iz^2\right)\right)$ along lines $y=\pm x$ is shown in Figure 24.5.

Problem a   Compute $\Re e\left(\exp\left(iz^2\right)\right)$ along the x-axis, the y-axis, and the lines $y=\pm x$ and show that this function has a behavior as shown in Figure 24.6. A direction of steepest descent means that in that direction the function *decreases* from its stationary point, while a direction of steepest ascent means that in that direction a function *increases* from its stationary point.

There is a good reason why there are steepest descent and steepest ascent directions that are perpendicular. As shown in Section 15.1, the real and imaginary parts of an analytic function satisfy Laplace's equation: $\nabla^2 f = 0$. Let us momentarily

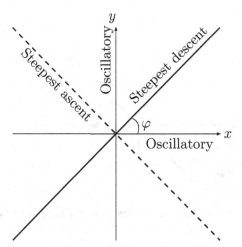

Figure 24.6 The behavior of the function $\exp(iz^2)$ in the complex plane.

assume that the x-axis at the stationary point is aligned with the steepest descent direction. This means that the stationary point is a maximum, and according to Section 10.1 this means that $\partial^2 f/\partial x^2 < 0$. Since f satisfies Laplace's equation $\partial^2 f/\partial x^2 + \partial^2 f/\partial y^2 = 0$, this means that at the stationary point $\partial^2 f/\partial y^2 > 0$; hence, the stationary point is a minimum in the y-direction. The stationary point is a *saddle point*; the function decreases in one direction while it increases in another direction. This can clearly be seen in Figure 24.4. This saddle point behavior forms the basis of Earnshaw's theorem that was treated in Section 10.5.

We now use the steepest descent path $y = x$ to evaluate the integral (24.34). To do this, we deform the integration path C_{real} into the steepest descent path $C_{descent}$ as shown in Figure 24.7. In doing so we need to close the contour with two circle segments C_R with a radius R that goes to infinity. To be general, we consider arbitrary real values of $a > 0$.

Problem b Use the fact that e^{iaz^2} is analytic within the closed contour of Figure 24.7 to show that

$$\int_{-\infty}^{\infty} e^{iax^2}\,dx + \int_{C_R} e^{iaz^2}\,dz - \int_{C_{descent}} e^{iaz^2}\,dz = 0. \tag{24.35}$$

Problem c At the contour C_R, $z = R\,e^{i\varphi}$ with $0 < \varphi < \pi/4$ or $\pi < \varphi < 5\pi/4$. Show that along the contour C_R:

$$\left|e^{iaz^2}\right| = e^{-aR^2 \sin 2\varphi}, \tag{24.36}$$

and that for the employed values of φ this function decays exponentially as $R \to \infty$.

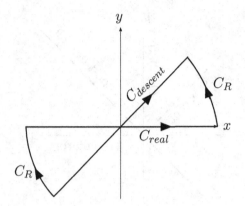

Figure 24.7   Definition of the integrations paths for the steepest descent integration.

This result implies that the contribution of the contours C_R vanishes as $R \to \infty$, so that according to (24.35):

$$\int_{-\infty}^{\infty} e^{iax^2} dx = \int_{C_{descent}} e^{iaz^2} dz. \qquad (24.37)$$

Problem d   Evaluate the integral on the right-hand side by substituting $z = ue^{i\pi/4}$ and use (4.49) to show that

$$\int_{-\infty}^{\infty} e^{iax^2} dx = \sqrt{\frac{\pi}{a}}\, e^{i\pi/4} \qquad (a > 0). \qquad (24.38)$$

Problem e   Show that this result can also be obtained by simply substituting $b = -ia$ into the integral (4.49). This substitution does not address any convergence issues of the resulting integral, but provides a simple trick to memorize (24.38).

Problem f   Repeat the analysis that leads to Problem d for $a < 0$. Show that in this case the steepest descent direction is along the line $y = -x$ and that

$$\int_{-\infty}^{\infty} e^{iax^2} dx = \sqrt{\frac{\pi}{|a|}}\, e^{-i\pi/4} \qquad (a < 0). \qquad (24.39)$$

Note that the expressions (24.38) and (24.39) can be summarized by the single expression

$$\int_{-\infty}^{\infty} e^{iax^2} dx = \sqrt{\frac{\pi}{|a|}} e^{i\pi/4\, \text{sgn}(a)}, \qquad (24.40)$$

with

$$sgn(a) \equiv \begin{cases} +1 \text{ when } a > 0, \\ -1 \text{ when } a < 0. \end{cases} \qquad (24.41)$$

The integration in this section was over a single variable. Bleistein and Handelsman (1975) show how this analysis can be extended to multidimensional integrals:

$$\int \exp(i\mathbf{x} \cdot \mathbf{A} \cdot \mathbf{x}) d^N \mathbf{x} = \sqrt{\frac{\pi^N}{|\det \mathbf{A}|}} \, e^{i\pi/4 \, sgn(\mathbf{A})}, \qquad (24.42)$$

where $sgn(\mathbf{A})$ = *number of positive eigenvalues of* $\mathbf{A}-$ *number of negative eigenvalues of* $\mathbf{A}$.

The advantage of using expression (24.40) is that we do not need to know the steepest descent path once we know this integral. The integral (24.40) forms the basis of the method of stationary phase that will be treated in the next section. Since the method of stationary phase uses the original integration path along the real axis, one can say that *the method of stationary phase is the steepest descent method for those that are too lazy to deform contours in the complex plane* (including the authors).

24.4 Group velocity and the method of stationary phase

Before treating the method of stationary phase, let us briefly examine a common explanation of the concept of group velocity. Suppose a wave consists of two propagating waves of equal amplitudes with frequencies ω_1 and ω_2 that are close:

$$p(x, t) = \cos(k(\omega_1)x - \omega_1 t) + \cos(k(\omega_2)x - \omega_2 t). \qquad (24.43)$$

Note that the wave number is a function of frequency: $k = k(\omega)$. In the following we use

$$\omega_1 = \omega_0 - \Delta \qquad \text{and} \qquad \omega_2 = \omega_0 + \Delta, \qquad (24.44)$$

so that ω_0 is the center frequency and 2Δ the frequency separation.

Problem a Use a first-order Taylor expansion of $k(\omega)$ around the point ω_0 to show that

$$\cos(k(\omega_1)x - \omega_1 t) = \cos(k_0 x - \omega_0 t) \cos\left(\Delta \left(\frac{\partial k}{\partial \omega}x - t\right)\right)$$

$$\qquad (24.45)$$

$$+ \sin(k_0 x - \omega_0 t) \sin\left(\Delta \left(\frac{\partial k}{\partial \omega}x - t\right)\right),$$

where $k_0 = k(\omega_0)$ and $\partial k/\partial \omega$ is evaluated at ω_0.

Problem b   Apply the same analysis to the last term of (24.43) and derive that

$$p(x, t) = 2\cos\left(\omega_0\left(t - \frac{x}{c}\right)\right)\cos\left(\Delta\left(t - \frac{x}{U}\right)\right),$$
(24.46)

with

$$c = \frac{\omega}{k} \quad \text{and} \quad U = \frac{\partial\omega}{\partial k}.$$
(24.47)

Since $\Delta \ll \omega_0$, the total wave field (24.46) consists of a wave $\cos(\omega_0(t - (x/c)))$ that propagates with a velocity c that is modulated by a slowly varying amplitude variation $\cos(\Delta(t - x/U))$ that propagates with a velocity U. The velocity U of the amplitude variation is called the *group velocity*, while the velocity c with which the carrier wave of frequency ω_0 propagates is called the *phase velocity*.

Even though this analysis is relatively simple, it is not very realistic. A propagating wave rarely consists of two frequency components that have equal amplitudes. In general, a wave consists of the superposition of all frequency components within a certain frequency band, and the amplitude spectrum $A(\omega)$ in general is not a constant. In the frequency domain such a wave is given by $A(\omega)\exp(ik(\omega)x)$, and after a Fourier transform (14.42) the wave in the time domain is given by

$$p(x, t) = \frac{1}{2\pi}\int_{-\infty}^{\infty} A(\omega)e^{i(k(\omega)x - \omega t)}d\omega.$$
(24.48)

In the following we assume that the amplitude $A(\omega)$ varies slowly with frequency compared to the phase

$$\psi(\omega) = k(\omega)x - \omega t.$$
(24.49)

In general, the phase is a rapidly oscillating function of frequency. We argued in the previous section that the dominant contribution to the integral $\int_{-\infty}^{\infty} e^{iax^2}dx$ came from the saddle point $x = 0$. In Figure 24.3 the saddle point $x = 0$ corresponds to the point where the function ceases to oscillate. This corresponds to the point where the phase is stationary; that is, the point where the phase is either a maximum or a minimum. For the function e^{iax^2} the phase is given by ax^2, the phase is stationary when $d\left(ax^2\right)/dx = 0$; this implies that $x = 0$, which is indeed the point of stationary phase in Figure 24.3.

Problem c   Show that the phase (24.49) is stationary for variations in ω when

$$\frac{\partial k}{\partial\omega}x - t = 0.$$
(24.50)

With the definition (24.47) for the group velocity, this result can also be written as

$$U(\omega_0) = \frac{x}{t}.$$
(24.51)

Before we proceed, let us briefly reflect on this result. The Fourier integral (24.48) states that all frequencies contribute to the wave field. However, for given values of x and t, the dominant contribution to this frequency integral comes from the frequency ω_0 for which the phase of the integrand is stationary. This frequency is defined by expression (24.51). When the group velocity is known as a function of frequency, this expression implicitly defines ω_0. Note that this frequency depends on the location x and the time t for which we want to compute the wave field. The right-hand side x/t is the distance x covered in a time t; the principle of stationary phase simply states that this distance is covered with a velocity given by the group velocity.

We next approximate the integral (24.48) here with the method of stationary phase. In this approximation we assume that:

- The amplitude $A(\omega)$ varies slowly with ω, so that we can replace the amplitude by its value at the stationary point:

$$A(\omega) \approx A(\omega_0). \tag{24.52}$$

- For the phase we use a second-order Taylor expansion around the stationary point:

$$\psi(\omega) \approx \psi(\omega_0) + \frac{1}{2} \frac{\partial^2 \psi}{\partial \omega^2} (\omega - \omega_0)^2, \tag{24.53}$$

where the second derivative is evaluated at the stationary point ω_0. Note that the first derivative does not appear in this Taylor expansion because at the stationary point we have by definition that $\partial \psi / \partial \omega = 0$.

We then insert these approximations into (24.48) and evaluate the remaining integral with expression (24.40).

Problem d The first step in this approach is trivial. In the second step we need $\partial^2 \psi / \partial \omega^2$ at the stationary point. Compute this derivative and use (24.47) to show that

$$\frac{\partial^2 \psi}{\partial \omega^2} = \frac{\partial^2 k}{\partial \omega^2} x = -\frac{1}{U^2} \frac{\partial U}{\partial \omega} x, \tag{24.54}$$

where all quantities are evaluated at frequency ω_0.

Problem e Insert (24.52)–(24.54) into the integral (24.48) and solve the remaining integral with (24.40) to show that

$$p(x, t) \approx A(\omega_0) e^{i(k_0 x - \omega_0 t)} \sqrt{\frac{2\pi}{|\partial U/\partial \omega| x}} \, U(\omega_0) e^{-i\pi/4 \, sgn(\partial U/\partial \omega)}. \tag{24.55}$$

This expression generalizes (24.46) for a wave with an arbitrary amplitude spectrum $A(\omega)$. The wave (24.55) consists of a carrier wave $\exp(i(k_0 x - \omega_0 t))$ with amplitude $A(\omega_0)$. The frequency ω_0 is determined by the condition that the group velocity $U(\omega_0)$ is the velocity needed to cover the distance x in time t. The amplitude of the wave is also determined by the term $1/\sqrt{|\partial U/\partial \omega| x}$. Let us first consider the x-dependence. The wave we consider here has a group velocity that depends on frequency. This means that the different frequency components travel with a different velocity. Because of this, the amplitude of the wave is reduced as the wave propagates, as different frequency components are spread out over space during the propagation. This phenomenon is called *dispersion*. The decay of the amplitude with propagation distance is given by the term $1/\sqrt{x}$. Suppose that the group velocity depends weakly on frequency. In that case, $\partial U/\partial \omega$ is small. In that situation, many different frequency components interfere constructively in the Fourier integral. This leads to a large amplitude that is described by the term $1/\sqrt{|\partial U/\partial \omega|}$. An extensive treatment of the concept of group velocity is given by Brillouin and Sommerfeld (1960).

You have seen wave dispersion in Figure 19.12, although you may not have realized it! If we zoom in on the Rayleigh wave arrival from $t = 15.05$ UTC and $t = 15.11$ UTC displayed in Figure 24.8, the frequencies around $t = 15.09$ are lower than those arriving around $t = 15.10$. This is because the seismic velocity in the Earth increases, in general, with depth. Lower-frequency surface waves penetrate deeper in the Earth than the high-frequency components, sampling parts of the Earth with higher velocities. As a result these low-frequency surface waves propagate faster than the high-frequency surface waves.

There may be applications where the group velocity does not depend on frequency. According to (24.54), this means that $\partial^2 \psi/\partial \omega^2 = 0$ and the stationary phase integral (24.55) is infinite. In that case the Taylor expansion (24.53) must be replaced by the third-order Taylor expansion

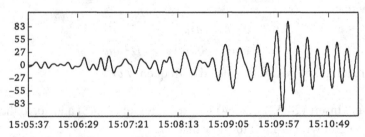

Figure 24.8   Part of the wave field from an earthquake in Costa Rica recorded at Hillside Junior High School (Figure 19.12), highlighting the dispersive Rayleigh-wave arrival.

$$\psi(\omega) \approx \psi(\omega_0) + \frac{1}{6}\frac{\partial^3 \psi}{\partial \omega^3}(\omega - \omega_0)^3. \qquad (24.56)$$

The resulting integral can be integrated to give an Airy function (Bender and Orszag, 1999). The corresponding wave arrivals are called *Airy phases* (Vallina, 1999); these waves have, in general, a strong amplitude because many different frequency components interfere constructively.

24.5 Asymptotic behavior of the Bessel function $J_0(x)$

In Section 19.5 we showed that for $x \gg n$ the Bessel function $J_n(x)$ behaves as a decaying cosine:

$$J_n(x) \approx \frac{A}{\sqrt{x}}\cos(x + \varphi). \qquad (24.57)$$

This relation followed from an analysis of the differential equation of the Bessel function for large values of x. The analysis of Section 19.5 did not tell us what the constants A and φ are. Here we find these constants for $J_0(x)$ by applying a stationary phase analysis to the following integral representation of the Bessel function (Arfken and Weber, 2005):

$$J_0(x) = \frac{1}{\pi}\int_0^{\pi} \cos(x \sin \varphi)\, d\varphi. \qquad (24.58)$$

Before we solve this integral in the stationary phase approximation, let us verify that this is a valid representation of the Bessel function. In order to do this, we need to check if the representation (24.58) satisfies the Bessel equation (19.16), with the following initial conditions:

$$J_0(x = 0) = 1, \qquad \frac{dJ_0}{dx}(x = 0) = 0. \qquad (24.59)$$

Problem a Derive from (24.58) that

$$\frac{dJ_0}{dx} = \frac{-1}{\pi}\int_0^{\pi} \sin(x \sin \varphi) \sin \varphi\, d\varphi, \qquad (24.60)$$

and

$$\frac{d^2 J_0}{dx^2} = \frac{-1}{\pi}\int_0^{\pi} \cos(x \sin \varphi) \sin^2 \varphi\, d\varphi. \qquad (24.61)$$

Using the identity $\sin^2 \varphi = 1 - \cos^2 \varphi$, the last integral can be written as $-\pi^{-1}\int_0^{\pi} \cos(x \sin \varphi)\, d\varphi + \pi^{-1}\int_0^{\pi} \cos(x \sin \varphi)\cos^2 \varphi\, d\varphi$. In the last term we can use that

$$\cos(x \sin \varphi) = \frac{1}{x \cos \varphi}\frac{d \sin(x \sin \varphi)}{d\varphi}. \qquad (24.62)$$

Problem b   Insert these results into (24.61), and use an integration by parts of the last term to derive that

$$\frac{d^2 J_0}{dx^2} = \frac{-1}{\pi} \int_0^{\pi} \cos{(x \sin \varphi)} \, d\varphi + \frac{1}{\pi x} \int_0^{\pi} \sin{(x \sin \varphi)} \sin \varphi \, d\varphi. \quad (24.63)$$

Problem c   Use this result with (24.58) and (24.60) to show that the representation (24.58) satisfies the Bessel equation (19.16).

Problem d   We next check if the integral representation (24.58) for $J_0(x)$ indeed satisfies the initial conditions (24.59). Apply the Taylor approximation technique of Section 24.1 to the integral (24.58) and show that for small values of x:

$$J_0(x) = 1 + 0 \cdot x + \mathcal{O}(x^2). \quad (24.64)$$

This result implies that the initial conditions (24.59) are satisfied, so that (24.58) indeed is a valid representation of the Bessel function $J_0(x)$.

In order to apply a stationary phase analysis for $x \gg 1$ to (24.58), we first rewrite it in the following form:

$$J_0(x) = \frac{1}{\pi} \Re \int_0^{\pi} e^{ix \psi(\varphi)} d\varphi, \quad (24.65)$$

with

$$\psi(\varphi) = \sin \varphi. \quad (24.66)$$

Problem e   Show that the phase ψ is stationary for $\varphi = \pi/2$ and that at that point $d^2 \psi / d\varphi^2 = -1$.

Problem f   Use a second-order Taylor expansion around the stationary point to rewrite (24.65) as

$$J_0(x) \approx \frac{1}{\pi} \Re e^{ix} \int_0^{\pi} \exp{\left(-\frac{ix}{2} \left(\varphi - \frac{\pi}{2} \right)^2 \right)} d\varphi. \quad (24.67)$$

The integrand has the behavior shown in Figure 24.3; away from the stationary point the integrand fluctuates rapidly for $x \gg 1$ and the positive contributions are canceled by the negative contributions so that the dominant contribution to the integral comes from a region near the stationary point.

Problem g   We can estimate the region of stationary phase by finding the values of φ for which the real part of the integrand of (24.67) vanishes. These points correspond to the zero crossing in Figure 24.3 closest to the stationary point. Show that the corresponding values of φ are given by

$$\varphi = \frac{\pi}{2} \pm \sqrt{\frac{\pi}{x}}. \tag{24.68}$$

For $x \gg 1$, these values of φ are close to the stationary point. Since the dominant contribution to the stationary phase integral (24.67) comes from the region close to the stationary point, we can extend the integration interval to infinity:

$$J_0(x) \approx \frac{1}{\pi} \Re e^{ix} \int_{-\infty}^{\infty} \exp \left(-\frac{ix}{2} \left(\varphi - \frac{\pi}{2} \right)^2 \right) d\varphi. \tag{24.69}$$

Problem h Use (24.40) to show that

$$J_0(x) \approx \frac{1}{\pi} \Re e \sqrt{\frac{2\pi}{x}} e^{i(x-\pi/4)}, \tag{24.70}$$

and that

$$J_0(x) \approx \sqrt{\frac{2}{\pi x}} \cos \left(x - \frac{\pi}{4} \right). \tag{24.71}$$

Bender and Orszag (1999) derive the same result with a steepest descent analysis of the integral (24.58). Their analysis is much more complicated than the stationary phase analysis of this section.

24.6 Image source

In this section we consider waves in three dimensions that are reflected from a plane $z = 0$. The geometry of this problem is shown in Figure 24.9. A source at $\mathbf{r}_s$ emits waves that propagate to a reflection point $\mathbf{r}$ in the reflection plane. The waves are reflected at that point with reflection coefficient $R(x, y)$, and then propagate to the receiver $\mathbf{r}_r$. We assume here that the reflection coefficient $R(x, y)$ varies slowly with the position on the reflecting plane. The total reflected wave follows by an integration over the reflection surface. In this section we use a coordinate system as shown in Figure 24.9. The reflection plane is defined by $z = 0$, and the y-axis of the coordinate system is aligned with the source–receiver positions in such a way that:

$$\mathbf{r}_s = \begin{pmatrix} x_s \\ 0 \\ z_s \end{pmatrix}, \quad \mathbf{r} = \begin{pmatrix} x \\ y \\ 0 \end{pmatrix}, \quad \mathbf{r}_r = \begin{pmatrix} x_r \\ 0 \\ z_r \end{pmatrix}. \tag{24.72}$$

For scalar waves in the frequency domain in a 3D homogeneous medium, the propagation of the waves from the source to the reflection point, and from the reflection

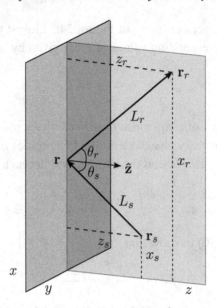

Figure 24.9   Definition of the geometric variables for a wave from source point $\mathbf{r}_s$ that reflects at point $\mathbf{r}$ in a plane to receiver point $\mathbf{r}_r$.

point to the receiver, is given by the Green's function (18.40). The total wave field is given by the integral over the reflecting plane, so that

$$p(\mathbf{r}_r) = \iint_{-\infty}^{\infty} G(|\mathbf{r}_r - \mathbf{r}|)R(x, y)G(|\mathbf{r} - \mathbf{r}_s|) \, dxdy, \qquad (24.73)$$

with G given by expression (18.40).

Problem a   Show that this integral can be written as

$$p(\mathbf{r}_r) = \iint_{-\infty}^{\infty} A(x, y)R(x, y)e^{ik\psi(x,y)} \, dxdy, \qquad (24.74)$$

with

$$A(x, y) = \frac{1}{(4\pi)^2} \left((x_r - x)^2 + y^2 + z_r^2 \right)^{-1/2} \left((x_s - x)^2 + y^2 + z_s^2 \right)^{-1/2}, \qquad (24.75)$$

and

$$\psi(x, y) = \sqrt{(x_r - x)^2 + y^2 + z_r^2} + \sqrt{(x_s - x)^2 + y^2 + z_s^2}. \qquad (24.76)$$

We will solve the integral (24.74) in the stationary phase approximation, as it is unclear how to solve this equation analytically. In order to do this, we evaluate both the x- and the y-integral with a stationary phase approximation. Before we can do this, we first need to find the stationary point in the (x, y)-plane.

Problem b To find this point, show first that

$$\frac{\partial \psi}{\partial x} = \frac{x - x_r}{\sqrt{(x_r - x)^2 + y^2 + z_r^2}} + \frac{x - x_s}{\sqrt{(x_s - x)^2 + y^2 + z_s^2}}, \tag{24.77}$$

and

$$\frac{\partial \psi}{\partial y} = \frac{y}{\sqrt{(x_r - x)^2 + y^2 + z_r^2}} + \frac{y}{\sqrt{(x_s - x)^2 + y^2 + z_s^2}}. \tag{24.78}$$

Problem c At the stationary point, both derivatives vanish; show that the stationary point is given by

$$\frac{x - x_r}{L_r} + \frac{x - x_s}{L_s} = 0 \qquad \text{and} \qquad y = 0, \tag{24.79}$$

with L_r and L_s defined in Figure 24.9.

Let us first analyze this stationary phase condition. The condition $y = 0$ states that the line from the source to the reflection point and the line from the reflection point to the receiver lie in the same plane. The phase is thus stationary for *in-plane* reflection. The first equality in (24.79) states for which point in that plane the phase is stationary.

Problem d Use the angles θ_s and θ_r defined in Figure 24.9 to show that the point of stationary phase as defined by the first equality in (24.79) satisfies

$$\theta_s = \theta_r. \tag{24.80}$$

This identity implies that at the point of stationary phase the angles of the incoming and the outgoing waves are equal. If the waves could be described by rays, the reflection point would be defined by (24.80) and the condition $y = 0$. The condition of stationary phase thus gives the same reflection point as would be given by ray theory. Expression (24.80) can be viewed as Snell's law (11.32) for reflected waves. A similar condition holds when the incoming and outgoing waves travel at a different velocity. Snieder (1986) shows that for the reflection of surface waves where the incoming surface wave mode travels with velocity c_{in} and the outgoing surface wave mode with velocity c_{out} the point of stationary phase satisfies

$$\frac{\sin \theta_{in}}{c_{in}} = \frac{\sin \theta_{out}}{c_{out}}. \tag{24.81}$$

This is nothing but Snell's law (11.32) for mode-converted waves.

The condition of stationarity thus gives us the location of the ray-geometric reflection point. Let is now compute the reflected wavefield. In order to evaluate (24.74) in the stationary phase approximation, we need the second derivatives of the phase.

Problem e   Show for an arbitrary point (x, y) that

$$\frac{\partial^2 \psi}{\partial x^2} = \frac{y^2 + z_r^2}{\left((x_r - x)^2 + y^2 + z_r^2\right)^{3/2}} + \frac{y^2 + z_s^2}{\left((x_s - x)^2 + y^2 + z_s^2\right)^{3/2}}, \qquad (24.82)$$

and

$$\frac{\partial^2 \psi}{\partial y^2} = \frac{(x_r - x)^2 + z_r^2}{\left((x_r - x)^2 + y^2 + z_r^2\right)^{3/2}} + \frac{(x_s - x)^2 + z_s^2}{\left((x_s - x)^2 + y^2 + z_s^2\right)^{3/2}}. \qquad (24.83)$$

Problem f   Show that at the stationary point:

$$\frac{\partial^2 \psi}{\partial x^2} = \cos^2 \theta \left(\frac{1}{L_r} + \frac{1}{L_s} \right), \qquad (24.84)$$

and

$$\frac{\partial^2 \psi}{\partial y^2} = \left(\frac{1}{L_r} + \frac{1}{L_s} \right), \qquad (24.85)$$

where $\theta = \theta_r = \theta_s$ at the reflection point, and L_s and L_r are the distance from the reflection point to the source and receiver, respectively.

Problem g   Show that near the reflection point the phase is approximately given by

$$\psi(x, y) \approx L_r + L_s + \frac{1}{2} \cos^2 \theta \left(\frac{1}{L_r} + \frac{1}{L_s} \right) (x - x_{refl})^2$$

$$+ \frac{1}{2} \left(\frac{1}{L_r} + \frac{1}{L_s} \right) (y - y_{refl})^2, \qquad (24.86)$$

with (x_{refl}, y_{refl}) the coordinates of the reflection point.

Problem h   Show that the amplitude A near the reflection point is given by

$$A(x, y) = \frac{1}{(4\pi)^2 \, L_r L_s}. \qquad (24.87)$$

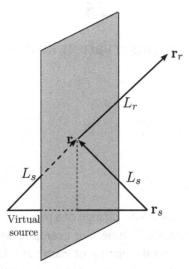

Figure 24.10   The position of an image source that is the mirror image of the real source at position $\mathbf{r}_s$.

Problem i   Evaluate the x and y integral in (24.74) each with the stationary phase approximation, and show that in this approximation

$$p(\mathbf{r}_r) = r\frac{\exp\left(ik(L_r + L_s)\right)}{4\pi(L_r + L_s)},\tag{24.88}$$

with

$$r = \frac{iR}{2k\cos\theta}.\tag{24.89}$$

In this expression R is the reflection coefficient at the reflection point at an angle of incidence θ.

The term $\exp\left(ik(L_r + L_s)\right)/4\pi(L_r + L_s) = -G(L_r + L_s)$ is, up to a minus sign, the Green's function for a distance $L_r + L_s$. The wave field (24.88) can therefore be interpreted as the wave field generated by an image source that is the mirror image of the original source in the reflector plane (Figure 24.10). This corresponding Green's function $G(L_r + L_s)$ is multiplied in (24.88) with an effective reflection coefficient r that accounts for the net reflectivity of the reflection plane. The stationary phase approximation thus not only accounts for the ray-geometric reflection point, it also describes how the original source can be replaced by an image source at the other side of the reflection point. We basically have here the method of images (Jackson, 1998) to find the Green function for waves propagating in a medium with a reflector.

25

Conservation laws

In physics one frequently handles the change of a property with time by considering properties that do *not* change with time. For example, when two particles collide elastically, the momentum and the energy of each particle may change. However, this change can be found from the consideration that the total momentum and energy of the system are conserved. Often in physics, such conservation laws are the main ingredients for describing a system. In this chapter we deal with conservation laws for continuous systems. These are systems in which the physical properties are a continuous function of the space coordinates. Examples are the motion in a fluid or solid, and the temperature distribution in a body. The introduced conservation laws are not only of great importance in physics, they also provide worthwhile exercises in the use of vector calculus introduced in previous chapters.

25.1 General form of conservation laws

In this section a general derivation of conservation laws is given. Suppose we consider a physical quantity Q. This quantity could denote the mass density of a fluid, the heat content within a solid or any other type of physical variable. In fact, there is no reason why Q should be a scalar, it could also be a vector (such as the momentum density) or a higher-order tensor. Let us consider a volume V in space that does not change with time. This volume is bounded by a surface ∂V. The total amount of Q within this volume is given by the integral $\int_V Q dV$. The rate of change of this quantity with time is given by $\frac{\partial}{\partial t} \int_V Q dV$.

In general, there are two reasons for the quantity $\int_V Q dV$ to change with time. First, the field Q may have sources or sinks within the volume V. The net source of the field Q per unit volume is denoted by the symbol S: the source density. The total source of Q within the volume is simply the volume integral $\int_V S dV$ of the source density. Second, it may be that a quantity Q is transported in – or out of – the volume V. With this transport process is associated a current $\mathbf{J}$ of the quantity Q. As an example one can think of Q as being the mass-density of a fluid. In that case $\int_V QdV$ is the total mass of the fluid in the volume. This total mass can change because there is a source or sink of fluid within the volume (i.e., a tap or a bathroom sink), or because there is flow through the boundary of the volume.

The rate of change of $\int_V QdV$ by the current is given by the *inward* flux of the current $\mathbf{J}$ through the surface ∂V. (The flux corresponds to the net flow through a surface; this quantity is defined in Section 6.1.) If we retain the convention that the surface element $d\mathbf{S}$ points out of the volume, the inward flux is given by $-\oint_{\partial V} \mathbf{J} \cdot d\mathbf{S}$. Together with the rate of change due to the source density S within the volume, the rate of change of the total amount of Q within the volume satisfies:

$$\frac{\partial}{\partial t} \int_V QdV = -\oint_{\partial V} \mathbf{J} \cdot d\mathbf{S} + \int_V SdV. \tag{25.1}$$

Using Gauss' theorem (8.1), the surface integral can be written as $-\int_V (\nabla \cdot \mathbf{J})dV$, so that the previous expression becomes

$$\frac{\partial}{\partial t} \int_V QdV + \int_V (\nabla \cdot \mathbf{J})dV = \int_V SdV. \tag{25.2}$$

Because the volume V is assumed to be fixed with time, the time derivative of the volume integral is the volume integral of the time derivative. Furthermore, expression (25.2) holds for *any* volume V. If the volume is an infinitesimal volume, the volume integrals in (25.2) can be replaced by the integrand multiplied by the infinitesimal volume. This means that (25.2) is equivalent to:

$$\frac{\partial Q}{\partial t} + (\nabla \cdot \mathbf{J}) = S. \tag{25.3}$$

This is the general form of a conservation law in physics; it simply states that the rate of change of a quantity is due to the sources (or sinks) of that quantity and due to the divergence of the current of that quantity. Of course, the general conservation law (25.3) is not very meaningful as long as we do not provide expressions for the current $\mathbf{J}$ and the source S. In this section we will see some examples in which the current and the source follow from physical theory, but we will also encounter examples in which they follow from an "educated" guess.

Equation (25.3) will not be completely new to you. In Section 8.4 the probability-density current for a quantum-mechanical system was derived.

Problem a   Use the derivation in this section to show that expression (8.16) can be written as

$$\frac{\partial}{\partial t} |\psi|^2 + (\nabla \cdot \mathbf{J}) = 0, \tag{25.4}$$

with $\mathbf{J}$ given by (8.18).

This equation constitutes a conservation law for the probability density of a quantum particle. Note that (8.16) could be derived rigorously from the Schrödinger equation (8.14), so that the conservation law (25.4) and the expression for the current **J** follow from the basic equation of the system.

25.2 Continuity equation

Let us now consider the conservation of mass in a continuous medium, such as a gas, fluid, or solid. In that case, the quantity Q is the mass-density ρ. If we assume that mass is neither created nor destroyed, the source term vanishes: $S = 0$. The vector **J** is the mass current and denotes the flow of mass per unit volume. To see that this is the case, consider a small volume δV. The mass within this volume is equal to $\rho \delta V$. If the velocity of the medium is denoted by **v**, the mass flow is given by $\rho \delta V \mathbf{v}$. Dividing this by the volume δV one obtains the mass flow per unit volume; this quantity is called the mass-density current:

$$\mathbf{J} = \rho \mathbf{v}. \tag{25.5}$$

From expression (25.3) with $Q = \rho$ and (25.5), the principle of the conservation of mass can be expressed as

$$\frac{\partial \rho}{\partial t} + \nabla \cdot (\rho \mathbf{v}) = 0. \tag{25.6}$$

This expression plays an important role in continuum mechanics, and is called the *continuity equation*.

Up to this point, the reasoning has been based on a volume V that did not change with time. This means that our treatment was strictly Eulerian: we considered the change of physical properties at a fixed location. Alternatively, a Lagrangian description of the same process can be given. In such an approach one specifies how physical properties change as they are moved along by the flow. In this case, one seeks an expression for the total time derivative d/dt of physical properties, rather than expressions for the partial derivative $\partial/\partial t$. It follows from (5.41) that these two derivatives are related in the following way:

$$\frac{d}{dt} = \frac{\partial}{\partial t} + (\mathbf{v} \cdot \nabla). \tag{25.7}$$

This distinction between the total time derivative and the partial time derivative, and between Eulerian and Lagrangian descriptions, is treated in detail in Section 5.5.

Problem a   Show that the total derivative of the mass-density obeys

$$\frac{d\rho}{dt} + \rho(\nabla \cdot \mathbf{v}) = 0. \tag{25.8}$$

Hint: Use equation (7.22) for $\nabla \cdot (\rho \mathbf{v})$.

This expression gives the change in the density when one follows the flow. Let us consider an infinitesimal volume δV that is carried around with the flow. The mass of this volume is given by $\delta m = \rho \delta V$. The volume moves with the flow; therefore, there is no mass transport across the boundary of the volume.

Problem b This means that the mass within that volume is conserved: $\delta \dot{m} = 0$. The dot denotes the total time derivative, *not* the partial time derivative. Use this expression and (25.8) to show that $(\nabla \cdot \mathbf{v})$ is the rate of change of the volume normalized by the size of the volume:

$$\frac{\delta \dot{V}}{\delta V} = (\nabla \cdot \mathbf{v}). \qquad (25.9)$$

We have learned a new meaning of the divergence of the velocity field: it equals the relative change in volume per unit time.

25.3 Conservation of momentum and energy

We showed in Section 5.4 for a point mass in classical mechanics that Newton's law is equivalent to the conservation of energy. The same is true for a continuous medium such as a fluid or a solid. In order to formulate Newton's law for a continuous medium, we start with a Lagrangian point of view and consider a volume δV *that moves with the flow*. The mass of this volume is given by $\delta m = \rho \delta V$. This mass is constant because the volume is defined to move with the flow; hence, mass cannot flow into or out of the volume. Let the force per unit volume be denoted by $\mathbf{F}$, so that the total force acting on the volume is $\mathbf{F} \delta V$. The force $\mathbf{F}$ contains both forces generated by external agents (such as gravity) and internal agents such as the pressure force $-\nabla p$ or the effect of internal stresses $(\nabla \cdot \boldsymbol{\sigma})$. The stress tensor $\boldsymbol{\sigma}$ is treated in Section 26.10. Newton's law applied to the volume δV takes the form:

$$\frac{d}{dt}(\rho \delta V \mathbf{v}) = \mathbf{F} \delta V. \qquad (25.10)$$

Since the mass $\delta m = \rho \delta V$ is constant with time, it can be taken outside the derivative in (25.10). Dividing the resulting expression by δV leads to the Lagrangian form of the equation of motion:

$$\rho \frac{d\mathbf{v}}{dt} = \mathbf{F}. \qquad (25.11)$$

Note that the density appears *outside* the total time derivative, whereas in expression (25.10) the density is inside the derivative. Using the prescription (25.7), one obtains the Eulerian form of Newton's law for a continuous medium:

$$\rho \frac{\partial \mathbf{v}}{\partial t} + \rho \mathbf{v} \cdot \nabla \mathbf{v} = \mathbf{F}. \tag{25.12}$$

This equation is not yet in the general form (25.3) of conservation laws, because in the first term on the left-hand side we have the density *times* a time derivative, and because the second term on the left-hand side is not the divergence of some current.

Problem a   Use (25.12) and the continuity equation (25.6) to show that:

$$\frac{\partial (\rho \mathbf{v})}{\partial t} + \nabla \cdot (\rho \mathbf{v} \mathbf{v}) = \mathbf{F}. \tag{25.13}$$

This expression does take the form of a conservation law. In classical mechanics, the product of the mass and velocity is called the *momentum*: $\mathbf{p} = m\mathbf{v}$ (Goldstein, 1980). The momentum per unit volume is called the *momentum density* $\rho \mathbf{v}$. For brevity we will often drop the affix "density" in the description of the different quantities, but remember that all quantities are given per unit volume. According to expression (25.13), the momentum density satisfies a conservation law. The source of momentum is given by the force density $\mathbf{F}$, and this reflects that forces are the cause of changes in momentum. In addition there is a momentum current $\mathbf{J} = \rho \mathbf{v} \mathbf{v}$ that describes the transport of momentum by the flow. This momentum current is not a simple vector; it is called a dyad and represented by a 3×3 matrix. The concept of the dyad is introduced in Section 12.1, but for now it suffices to know that in index notation $J_{ij} = \rho v_i v_j$. The fact that the momentum current is a 3×3 matrix is due to the fact the momentum is a vector with three components and that each component can be transported in three spatial directions.

It can be a bit arbitrary what we call the current and what we call the source. As an example let us consider (25.13) again. According to (5.13), the pressure force is given by $\mathbf{F} = -\nabla p$. If there are no other forces acting on the fluid, the right-hand side of (25.13) is given by $-\nabla p$. This term does not have the form of the divergence, but we can rewrite it by using

$$\frac{\partial p}{\partial x_i} = \sum_j \frac{\partial}{\partial x_j} (p \delta_{ij}), \tag{25.14}$$

where δ_{ij} is the Kronecker delta. This quantity is defined as follows:

$$\delta_{ij} = \begin{cases} 1 & \text{when} \quad i = j, \\ 0 & \text{when} \quad i \neq j. \end{cases} \tag{25.15}$$

The matrix elements of the identity matrix $\mathbf{I}$ are given by the Kronecker delta: $I_{ij} = \delta_{ij}$.

Problem b Show that expression (25.14) may be written in vector form as $\nabla p = \nabla \cdot (p\mathbf{I})$. Show that the law of conservation of momentum is given by

$$\frac{\partial(\rho\mathbf{v})}{\partial t} + \nabla \cdot \mathbf{S} = 0, \tag{25.16}$$

where $\mathbf{S}$ is defined by

$$\mathbf{S} = \rho\mathbf{v}\mathbf{v} + p\mathbf{I}. \tag{25.17}$$

This quantity $\mathbf{S}$ describes the *radiation stress* (Beissner, 1998), which accounts for the internal stress in an acoustic medium that is generated by the waves that propagate through the medium. Of course, when other external forces are present, they will lead to a right-hand side of (25.16) that is nonzero. This example shows that it can be arbitrary whether a certain physical effect is accounted for as a source term or as a current. This arbitrariness is caused by the fact that it is not clear what is internal to the system and what is external. In (25.13) the pressure force is treated as an external force, whereas in (25.17) the pressure contributes to a current within the system. There is no objective reason to prefer one or the other of the two formulations.

You may find the inner products of vectors and the ∇-operator in expressions such as (25.12) confusing, and indeed a notation such as $\rho\mathbf{v} \cdot \nabla\mathbf{v}$ can be a source of error and confusion. When working with quantities like this, it is clearer to explicitly write out the components of all vectors or tensors. In component form an equation such as (25.12) is written as:

$$\rho\frac{\partial v_i}{\partial t} + \sum_j \rho v_j \partial_j v_i = F_i. \tag{25.18}$$

Problem c Redo the derivation of Problem a with all equations in component form to arrive at the conservation law of momentum (25.6) in component form:

$$\frac{\partial(\rho v_i)}{\partial t} + \sum_j \partial_j(\rho v_j v_i) = F_i. \tag{25.19}$$

To derive the law of energy conservation, we start by deriving the conservation law for the kinetic energy (density).

Problem d Express the partial time derivative $\partial(\rho v^2)/\partial t$ in the time derivatives $\partial(\rho v_i)/\partial t$ and $\partial v_i/\partial t$, use (25.18) and (25.19) to eliminate these time derivatives and write the final results as:

$$\sum_i \frac{\partial(\frac{1}{2}\rho v_i v_i)}{\partial t} = -\sum_{i,j} \partial_j \left(\frac{1}{2}\rho v_i v_i v_j\right) + \sum_j v_j F_j. \qquad (25.20)$$

The kinetic energy is

$$E_K = \frac{1}{2}\rho v^2 = \sum_i \frac{1}{2}\rho v_i v_i. \qquad (25.21)$$

Problem e   Use expressions (25.20) and (25.21) to write the conservation law of kinetic energy:

$$\frac{\partial E_K}{\partial t} + \nabla \cdot (\mathbf{v} E_K) = (\mathbf{v} \cdot \mathbf{F}). \qquad (25.22)$$

This equation states that the kinetic energy current is given by $\mathbf{J} = \mathbf{v} E_K$, this term describes how kinetic energy is transported by the flow. The term $(\mathbf{v} \cdot \mathbf{F})$ on the right-hand side denotes the source of kinetic energy. It was shown in Section 5.4 that $(\mathbf{v} \cdot \mathbf{F})$ is the power delivered by the force $\mathbf{F}$. This means that (25.22) states that the power produced by the force $\mathbf{F}$ is the source of kinetic energy.

To invoke the potential energy, as well, we assume for the moment that the force $\mathbf{F}$ is the gravitational force. Suppose there is a gravitational potential $V(\mathbf{r})$, then the gravitational force is given by

$$\mathbf{F} = -\rho \nabla V, \qquad (25.23)$$

and the potential energy E_P is given by

$$E_P = \rho V, \qquad (25.24)$$

as discussed in Jackson (1998) and Section 13.6.

Problem f   Take the (partial) time derivative of (25.24), use the continuity equation (25.6) to eliminate $\partial \rho / \partial t$, use that the potential $V(\mathbf{r})$ does not depend explicitly on time, to derive the conservation law of potential energy:

$$\frac{\partial E_P}{\partial t} + \nabla \cdot (\mathbf{v} E_P) = -(\mathbf{v} \cdot \mathbf{F}). \qquad (25.25)$$

You may want to consult Section 7.7 for the divergence and gradient of products.

This conservation law is similar to the conservation law (25.22) for kinetic energy. The meaning of the second term on the left-hand side will be clear to you by now; it denotes the divergence of the current $\mathbf{v} E_P$ of potential energy. Note that the right-hand side of (25.24) has the opposite sign to the right-hand side of (25.22). This reflects the fact that when the force **F** acts as a source of kinetic energy, it acts as a sink of potential energy; the opposite signs imply that kinetic and potential energy are converted into each other. However, the total energy $E = E_K + E_P$ should have no source or sink.

Problem g Show that the conservation law for the total energy is source-free:

$$\frac{\partial E}{\partial t} + \nabla \cdot (\mathbf{v} E) = 0. \tag{25.26}$$

We showed in this section that mass, momentum, and energy in a continuous medium are conserved. The expression that state the conservation of these quantities are all of the general form (25.3).

25.4 Heat equation

In the previous section we saw that the momentum and energy current could be derived from Newton's law. Such a rigorous derivation is not always possible. In this section the transport of heat is treated, and we will see that the law for heat transport cannot be derived rigorously. Consider the general conservation equation (25.3), where T is the temperature. Strictly speaking we should derive the heat equation using a conservation law for the thermal energy rather than the temperature. The thermal energy per unit volume Q is given by $\rho C T$, with C the specific heat capacity and ρ the mass density. When the specific heat and the density are constant, the distinction between thermal energy and temperature implies multiplication by a constant, but for simplicity this multiplication is left out here.

The source term in the conservation equation is simply the amount of heat (normalized by the heat capacity) supplied to the medium. An example of such a source is the decay of radioactive isotopes that forms a major source in the heat budget of the Earth. The transport of heat is affected by the heat current **J**. In the Earth, heat can be transported by two mechanisms: heat advection and heat conduction

$$\mathbf{J} = \mathbf{J}^{conduction} + \mathbf{J}^{advection}. \tag{25.27}$$

The first process accounts for the heat that is transported by the flow field **v** in the medium:

$$\mathbf{J}^{advection} = \mathbf{v} T. \tag{25.28}$$

In the Earth, the advective heat transport is caused by the convection in the Earth's mantle (Section 11.5) and outer core.

This viewpoint of the process of heat transport is in fact too simplistic in many situations. Fletcher (1996) describes how the human body loses heat during outdoor activities through four processes: conduction, advection, evaporation, and radiation. He describes in detail the conditions under which each of these processes dominate, and how the associated heat loss can be reduced. In the physics of the atmosphere, energy transport by radiation and by evaporation (or condensation) also plays a crucial role.

The other transport mechanism for heat is heat conduction, which is similar to diffusion; it accounts for the fact that heat flows from warm regions to colder regions. The vector ∇T points from cold regions to warmer regions. It is therefore logical that the heat conduction points in the opposite direction from the temperature gradient:

$$\mathbf{J}^{conduction} = -\kappa \nabla T, \tag{25.29}$$

as can be seen in the left-hand panel of Figure 25.1. This expression is identical to Fick's law, which accounts for diffusion processes. The constant κ is the heat conductivity. For a given value of ∇T, the heat conduction increases when κ increases; hence, it indeed measures the conductivity. However, the simple law (25.29) does not hold for every medium. Consider a medium consisting of alternating layers of a good heat conductor (such as copper) and a poor heat conductor (such as styrofoam). In such a medium the heat will be preferentially transported along the planes of the good heat conductor and the conductive heat flow $\mathbf{J}^{conduction}$ and the temperature gradient are not antiparallel, as depicted in the right-hand panel in Figure 25.1. In that case there is a matrix operator that relates $\mathbf{J}^{conduction}$ and ∇T: $J_i^{conduction} = -\sum_j \kappa_{ij} \partial_j T$, with κ_{ij} the heat conductivity tensor. In this section we restrict ourselves to the simple conduction law (25.29). Combining this law with the expressions (25.27), (25.28), and the conservation law (25.3) for heat gives:

$$\frac{\partial T}{\partial t} + \nabla \cdot (\mathbf{v}T - \kappa \nabla T) = S, \tag{25.30}$$

where S is now a source.

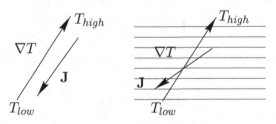

Figure 25.1   Heat flow and temperature gradient in an isotropic medium (left-hand panel) and in a medium consisting of alternating layers of copper and styrofoam (right-hand panel).

As a first example we consider a solid in which there is no flow ($\mathbf{v} = 0$) and thus only conduction. For a constant heat conductivity κ, (25.30) reduces to:

$$\frac{\partial T}{\partial t} = \kappa \nabla^2 T + S. \tag{25.31}$$

The expression is called the *heat equation*, despite the fact that it holds only under these special conditions of heat transport. This is not surprising since heat is transported by a diffusive process in the absence of advection. For this reason, equation (25.31) is also called the *diffusion equation*. We analyze in Sections 17.5 and 18.1 solutions to the diffusion is one and three dimensions, respectively.

We now consider heat transport in a one-dimensional medium (such as an infinitely long bar) when there is no source of heat. In that case the heat equation reduces to

$$\frac{\partial T}{\partial t} = \kappa \frac{\partial^2 T}{\partial x^2}. \tag{25.32}$$

If we know the temperature throughout the medium at some initial time (i.e., $T(x, t = 0)$ is known), then (25.32) can be used to compute the temperature at later times. As a special case we consider a Gaussian-shaped temperature distribution at $t = 0$:

$$T(x, t = 0) = T_0 \exp\left(-\frac{x^2}{L^2}\right). \tag{25.33}$$

Problem a Sketch this temperature distribution and indicate its dependence on the constants T_0 and L.

We assume that the temperature profile maintains a Gaussian shape at later times but that the peak value and the width may change; that is, we consider a solution of the following form:

$$T(x, t) = F(t)e^{-H(t)x^2}. \tag{25.34}$$

At this point the functions $F(t)$ and $H(t)$ are not yet known.

Problem b Show that these functions satisfy the initial conditions:

$$F(0) = T_0, \qquad H(0) = 1/L^2. \tag{25.35}$$

Problem c Show that for the special solution (25.34) the heat equation reduces to:

$$\frac{\partial F}{\partial t} - x^2 F \frac{\partial H}{\partial t} = \kappa \left(4FH^2x^2 - 2FH\right). \tag{25.36}$$

It is possible to derive equations for the time evolution of $F(t)$ and $H(t)$ by recognizing that (25.36) can only be satisfied for *all values of x* when all terms proportional to x^2 balance and when the terms independent of x balance.

Problem d   Use this to show that $F(t)$ and $H(t)$ satisfy the following differential equations:

$$\frac{\partial F}{\partial t} = -2\kappa FH, \tag{25.37}$$

$$\frac{\partial H}{\partial t} = -4\kappa H^2. \tag{25.38}$$

It is easiest to solve the last equation first because it contains only $H(t)$ whereas (25.37) contains both $F(t)$ and $H(t)$.

Problem e   Solve (25.38) with the initial condition (25.35) and show that:

$$H(t) = \frac{1}{4\kappa t + L^2}. \tag{25.39}$$

Hint: Solve expression (25.38) by introducing a new variable $J = 1/H$.

Problem f   Solve (25.37) with the initial condition (25.35) and show that:

$$F(t) = T_0 \frac{L}{\sqrt{4\kappa t + L^2}}. \tag{25.40}$$

Hint: Divide equation (25.37) by F, and use that $(1/F)\partial F/\partial t = \partial \ln F/\partial t$. Insert equation (25.39) for H and integrate the result.

Inserting these solutions into (25.34) gives the temperature field at all times $t \geq 0$:

$$T(x, t) = T_0 \frac{L}{\sqrt{4\kappa t + L^2}} \exp\left(-\frac{x^2}{4\kappa t + L^2}\right). \tag{25.41}$$

Problem g   Sketch the temperature for several later times and describe with solution (25.41), how the temperature profile changes as time progresses.

The temperature in expression (25.41) has the form of a Gauss distribution. We treated the Gaussian dependence of diffusive processes, such as heat conduction, in more detail in Section 18.1. Note that the solution (25.41) reduces to the one-dimensional Green's function (18.15) for the heat equation by taking $L \to 0$ and $T_0 L \to 1$. The condition $L \to 0$ states that at time $t = 0$ the heat distribution has zero width, as it should for the temperature caused by a delta function heating.

The total heat at time t is $Q(t)$, the heat per unit volume, integrated over that volume. In our one-dimensional bar, this is $Q^{total}(t) = \rho C \int_{-\infty}^{\infty} T(x, t)dx$.

Problem h   Use equation (4.49) to show that the total heat does not change with time for the solution (25.41).

Problem i   Show that in fact for *any* solution of the heat equation (25.32), where the heat flux vanishes at the endpoints ($\kappa \partial_x T (x = \pm\infty, t) = 0$), the total heat $Q^{total}(t)$ is constant in time.

Problem j   The peak value of the temperature field (25.41) decays as $1/\sqrt{4\kappa t + L^2}$ with time. Do you expect that in more dimensions this decay will be more rapid or slower with time? Do not do any calculations but use only common sense!

Up to this point, we have considered the conduction of heat in a medium without flow ($\mathbf{v} = 0$). In many applications the flow in the medium plays a crucial role in redistributing heat. This is particularly the case when heat is the source of convective motion, as for example in the Earth's mantle, the atmosphere, and the central heating systems in buildings. As an example of the role of advection we consider the cooling model of the oceanic lithosphere proposed by Parsons and Sclater (1977).

At the mid-oceanic ridges, lithospheric material with thickness H is produced. At a ridge the temperature of this material is essentially the temperature T_m of mantle material. As shown in Figure 25.2, this implies that at $x = 0$ and at depth $z = H$ the temperature is given by the mantle temperature: $T(x = 0, z) = T(x, z = H) = T_m$. We assume that the velocity with which the plate moves away from the ridge is constant:

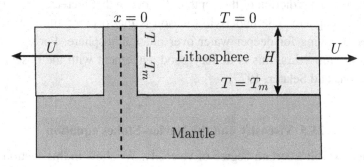

Figure 25.2   Cooling model of the oceanic lithosphere of thickness H, spreading horizontally with speed U. At the mid-oceanic ridge ($x = 0$) and at the bottom of the lithosphere the temperature is that of the mantle $T = T_m$. At the surface, the temperature $T = 0$.

$$\mathbf{v} = U\hat{\mathbf{x}}. \tag{25.42}$$

We consider the situation in which the temperature is stationary. This does not imply that the flow vanishes; it means that the partial time derivatives vanish: $\partial T/\partial t = 0$, $\partial \mathbf{v}/\partial t = 0$.

Problem k   Show that in the absence of heat sources ($S = 0$) the conservation equation (25.30) reduces to:

$$U\frac{\partial T}{\partial x} = \kappa \left(\frac{\partial^2 T}{\partial x^2} + \frac{\partial^2 T}{\partial z^2} \right). \tag{25.43}$$

In general the thickness of the oceanic lithosphere is less than 100 km, whereas the width of ocean basins is several thousand kilometers.

Problem l   Use this to explain that the following expression is a reasonable approximation to (25.43):

$$U\frac{\partial T}{\partial x} = \kappa \frac{\partial^2 T}{\partial z^2}. \tag{25.44}$$

Problem m   Show that with the replacement $\tau = x/U$ this expression is identical to the heat equation (25.32).

Note that τ is the time it has taken the oceanic plate to move from its point of creation ($x = 0$) to the point under consideration (x); hence, the time τ is simply the *age* of the oceanic lithosphere. This implies that solutions of the one-dimensional heat equation can be used to describe the cooling of oceanic lithosphere with the age of the lithosphere taken as the time variable. Accounting for cooling with such a model leads to a prediction of the depth of the ocean that increases as $\sqrt{t}$ with the age of the lithosphere. The underlying reason is that oceanic lithosphere shrinks as it cools, allowing for deeper water over older lithosphere. For ages less than about 100 million years, this is in very good agreement with the observed ocean depth (Parsons and Sclater, 1977).

25.5 Viscosity and the Navier–Stokes equation

Many fluids exhibit a certain degree of viscosity. We show in this section that viscosity can be seen as an ad-hoc description of the momentum current in a fluid due to small-scale movements in the fluid. The starting point of the analysis is the equation of momentum conservation in a fluid:

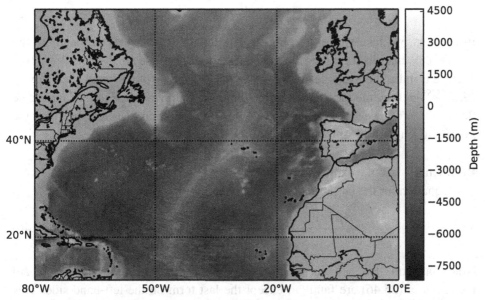

Figure 25.3 Bathymetry of the Atlantic Ocean: water depth at the mid-oceanic ridge running approximately north-south is smaller than away from the ridge.

$$\frac{\partial(\rho \mathbf{v})}{\partial t} + \nabla \cdot (\rho \mathbf{v}\mathbf{v}) = \mathbf{F}. \tag{25.13}$$

In a real fluid, motion takes place at a large range of length scales from microscopic eddies to organized motions with a size comparable to the size of the fluid body. Whenever we describe a fluid, it is impossible to account for the motion at the very small length scales. This is not only so in analytical descriptions, but it is in particular the case in numerical simulations of fluid flow. For example, in current weather prediction schemes, the motion of the air is computed on a grid with a distance of about 100 km between the grid points. When you look at the weather, it is obvious that there is considerable motion at smaller length scales (e.g., cumulus clouds indicating convection, fronts). In general one cannot simply ignore the motion at these short-length scales because these small-scale fluid motions transport significant amounts of momentum, heat, and other quantities such as moisture (Peixoto and Oort, 1992).

One way to account for the effect of the small-scale motion is to express the small-scale motion in the large-scale motion. It is not obvious that this is consistent with reality, but it appears to be the only way to avoid a complete description of the small-scale motion of the fluid (which would be impossible).

In order to do this, we assume that there is some length scale that separates the small-scale flow from the large-scale flow, and we decompose the velocity into a long-wavelength component $\mathbf{v}^L$ and a short-wavelength component $\mathbf{v}^S$:

$$\mathbf{v} = \mathbf{v}^L + \mathbf{v}^S. \tag{25.45}$$

In addition, we take spatial averages over a length scale that corresponds to the length scale that distinguishes the large-scale flow from the small-scale flow. This average is indicated by angle brackets: $\langle \cdots \rangle$. The average of the small-scale flow is zero, $\langle \mathbf{v}^S \rangle = 0$, while the average of the large-scale flow is equal to the large-scale flow, $\langle \mathbf{v}^L \rangle = \mathbf{v}^L$, because the large-scale flow by definition does not vary over the averaging length. For simplicity we assume that the density does not vary.

Problem a   Use the expressions (25.13) and (25.45) to show that the momentum equation for the large-scale flow is given by:

$$\frac{\partial(\rho \mathbf{v}^L)}{\partial t} + \nabla \cdot (\rho \mathbf{v}^L \mathbf{v}^L) + \nabla \cdot (\langle \rho \mathbf{v}^S \mathbf{v}^S \rangle) = \mathbf{F}. \tag{25.46}$$

Note that terms that are linear in $\mathbf{v}^S$ do not contribute because $\langle \mathbf{v}^S \rangle = 0$. All the terms in (25.46) are familiar, except the last term on the left-hand side. This term exemplifies the effect of the small-scale flow on the large-scale flow since it accounts for the transport of momentum by the small-scale flow. It seems that at this point further progress is impossible without knowing the small-scale flow $\mathbf{v}^S$. One way to make further progress is to express the small-scale momentum current $\langle \rho \mathbf{v}^S \mathbf{v}^S \rangle$ in the large-scale flow.

Consider the large-scale flow shown in Figure 25.4. Whatever the small-scale motions are, in general they will have the character of mixing. In the example in the figure, the momentum is large at the top of the figure and smaller at the bottom. As a first approximation one may assume that the small-scale motions transport momentum in the direction opposite to the momentum gradient of the large-scale flow. By analogy with (25.29) we can approximate the momentum transport by the small-scale flow by:

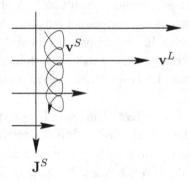

Figure 25.4   The direction of momentum transport within a large-scale flow by small-scale motions.

$$\mathbf{J}^S \equiv \langle \rho \mathbf{v}^S \mathbf{v}^S \rangle \approx -\mu \nabla \mathbf{v}^L. \tag{25.47}$$

This term account for the diffusion of velocity by the small-scale flow. Such a diffusive process is called *viscosity*, and for this reason the parameter μ is also called the viscosity.

Problem b Insert this relation into (25.46), and drop the superscript L of $\mathbf{v}^L$ to show that large-scale flow satisfies:

$$\frac{\partial(\rho \mathbf{v})}{\partial t} + \nabla \cdot (\rho \mathbf{v} \mathbf{v}) = \mu \nabla^2 \mathbf{v} + \mathbf{F}. \tag{25.48}$$

This equation is called the Navier–Stokes equation. The first term on the right-hand side accounts for the momentum transport by small-scale motions. Effectively this leads to the viscosity of the fluid.

Problem c Viscosity tends to damp motion at smaller length scales more than motion at larger length scales. Show that the term $\mu \nabla^2 \mathbf{v}$ indeed affects shorter length scales more than larger length scales.

Problem d Do you think this treatment of the momentum flux due to small-scale motion is realistic?

Despite reservations that you may (or may not) have against the treatment of viscosity in this section, you should realize that the Navier–Stokes equation (25.48) is widely used in fluid mechanics.

25.6 Quantum mechanics and hydrodynamics

As we saw in Section 8.4, the behavior of atomic particles is described by Schrödinger's equation:

$$i\hbar \frac{\partial \psi(\mathbf{r}, t)}{\partial t} = -\frac{\hbar^2}{2m} \nabla^2 \psi(\mathbf{r}, t) + V(\mathbf{r}) \psi(\mathbf{r}, t), \tag{8.14}$$

rather than Newton's law. In this section we reformulate the linear wave equation (8.14) as the laws of conservation of mass and momentum for a normal fluid. To do this, we write the wavefunction ψ as

$$\psi = \sqrt{\rho} \exp(i\varphi/\hbar). \tag{25.49}$$

This equation is simply the decomposition of a complex function into its absolute value and its phase; hence, ρ and φ are real functions. It follows from expression

(25.49) that $\rho = |\psi|^2$, and we have seen in Section 8.4 that $|\psi|^2$ this is probability density of the particle. The factor $\hbar$ in equation (25.49) is a constant added here for notational convenience. This constant is characteristic of quantum mechanics and is called Planck's constant.

Problem a   Insert the decomposition (25.49) into Schrödinger's equation (8.14), divide by $\sqrt{\rho} \exp(i\varphi/\hbar)$ and separate the result into real and imaginary parts to show that ρ and φ satisfy the following differential equations:

$$\frac{\partial \rho}{\partial t} + \nabla \cdot \left(\rho \frac{1}{m} \nabla \varphi \right) = 0, \qquad (25.50)$$

$$\frac{\partial \varphi}{\partial t} + \frac{1}{2m} |\nabla \varphi|^2 + \frac{\hbar^2}{8m} \left(\frac{1}{\rho^2} |\nabla \rho|^2 - \frac{2}{\rho} \nabla^2 \rho \right) = -V. \qquad (25.51)$$

The problem is that at this point we do not yet have a velocity. Let us define the following velocity vector:

$$\mathbf{v} \equiv \frac{1}{m} \nabla \varphi. \qquad (25.52)$$

Problem b   Show that this definition of the velocity is identical to the velocity obtained in (8.21) for a plane wave.

Problem c   Show that with this definition of the velocity, (25.50) is identical to the continuity equation:

$$\frac{\partial \rho}{\partial t} + \nabla \cdot (\rho \mathbf{v}) = 0. \qquad (25.6)$$

Problem d   To reformulate (25.51) as an equation of conservation of momentum, differentiate (25.51) with respect to x_i. Then use the definition (25.52) and the relation between force and potential ($\mathbf{F} = -\nabla V$) to write the result as:

$$\frac{\partial v_i}{\partial t} + \frac{1}{2} \sum_j \partial_i (v_j v_j) + \frac{\hbar^2}{8m^2} \left[\partial_i \left(\frac{1}{\rho^2} |\nabla \rho|^2 \right) - 2\partial_i \left(\frac{1}{\rho} \nabla^2 \rho \right) \right] = \frac{1}{m} F_i.$$

$$(25.53)$$

The second term on the left-hand side does not look very much like the term $\sum_j \partial_j (\rho v_j v_i)$ in the left-hand side of (25.13). To make progress, we need to rewrite the term $\sum_j \partial_i (v_j v_j)$ as a term of the form $\sum_j \partial_j (v_j v_i)$. In general these terms are different.

Problem e Show that for the special case that the velocity is the gradient of a scalar function (as in expression (25.52)):

$$\sum_j \frac{1}{2} \partial_i (v_j v_j) = \sum_j \partial_j (v_j v_i) - \sum_j (\partial_j v_j) v_i \,. \qquad (25.54)$$

With this step we can rewrite the second term on the left-hand side of (25.53). Part of the third term in (25.53) we will designate as Q_i:

$$Q_i \equiv -\frac{1}{8m} \left[\partial_i \left(\frac{1}{\rho^2} |\nabla \rho|^2 \right) - 2\partial_i \left(\frac{1}{\rho} \nabla^2 \rho \right) \right]. \qquad (25.55)$$

Problem f Use (25.6) and (25.53)–(25.55) to derive that:

$$\frac{\partial (\rho \mathbf{v})}{\partial t} + \nabla \cdot (\rho \mathbf{v} \mathbf{v}) = \frac{\rho}{m} \left(\mathbf{F} + \hbar^2 \mathbf{Q} \right). \qquad (25.56)$$

Note that this equation is identical to the momentum equation (25.13). This implies that the Schrödinger equation is equivalent to the continuity equation (25.6) and to the momentum equation (25.13) for a classical fluid. In Section 8.4 we saw that atomic particles behave as waves rather than as point-like particles. In this section we have discovered that these particles also behave like a fluid! This has led to hydrodynamic formulations of quantum mechanics (Halbwachs, 1960; Madelung, 1927). In general, quantum-mechanical phenomena depend critically on Planck's constant. Quantum mechanics reduces to classical mechanics in the limit $\hbar \to 0$.[1] The only place where Planck's constant occurs in (25.56) is in the additional force: $\mathbf{Q}$ multiplied by the square of Planck's constant. This implies that the action of the force term $\mathbf{Q}$ is fundamentally quantum mechanical; it has no analogue in classical mechanics.

Problem g Suppose we consider a particle in one dimension that is represented by the following wave function:

$$\psi(x, t) = e^{-x^2/L^2} e^{i(kx - \omega t)}. \qquad (25.57)$$

Sketch the corresponding probability-density $\rho = |\psi|^2$ and use expression (25.55) to deduce that the quantum force Q acts to broaden the wavefunction with time.

This example shows that (at least for this case) the quantum force $\mathbf{Q}$ makes the wavefunction "spread out" with time. This reflects the fact that if a particle propagates with time, its position becomes more and more uncertain.

---

[1] Of course, a fundamental constant such as $\hbar$ cannot be made to vanish, but the limit $\hbar \to 0$ denotes the situation where $\hbar \mathbf{Q}$ is negligible compared to $\mathbf{F}$.

The acoustic wave equation (6.41) cannot be transformed into the continuity equation (25.6) and the momentum equation (25.13), despite the fact that these equations describe the conservation of mass and momentum in an acoustic medium. The reason for this paradox is that in the derivation of the acoustic wave equation from the continuity equation and the momentum equation, the advective terms $\nabla \cdot (\rho \mathbf{v})$ and $\nabla \cdot (\rho \mathbf{v}\mathbf{v})$ have been ignored. Once these terms have been ignored, there is no transformation of variables that can bring them back. This contrasts the acoustic wave equation from the Schrödinger equation, which implicitly retains the physics of the advection of mass and momentum.

26

Cartesian tensors

In physics and mathematics, coordinate transformations play an important role because many problems are much simpler when a suitable coordinate system is used. Furthermore, the requirement that physical laws do not change under certain transformations imposes constraints on the physical laws. An example of this is presented in Section 26.11 where we show that the fact that the pressure in a fluid is isotropic follows from the requirement that some physical laws may not change under a rotation of the coordinate system. In this chapter it is shown how the change of vectors and matrices under coordinate transformations is derived. The derived transformation properties can be generalized to other mathematical objects called tensors. In this chapter, only transformations of rectangular coordinate systems are considered. Since these coordinate systems are called Cartesian coordinate systems, the associated tensors are called Cartesian tensors. The transformation properties of tensors in Cartesian and curvilinear coordinate systems are described in detail by Butkov (1968) and Riley et al. (2006).

26.1 Coordinate transforms

In this section we consider the transformation of a coordinate system in two dimensions. Figure 26.1 contains an old coordinate system with coordinates x^{old} and y^{old}. The unit vectors along the old coordinate axis are denoted by $\hat{e}^{x,old}$ and $\hat{e}^{y,old}$. In a coordinate transformation, these old unit vectors are transformed to new unit vectors $\hat{e}^{x,new}$ and $\hat{e}^{y,new}$, respectively. The matrix that maps the old unit vectors onto the new unit vectors is denoted by $\mathbf{C}^{-1}$:

$$\mathbf{C}^{-1}\hat{e}^{x,old} = \hat{e}^{x,new}, \qquad \mathbf{C}^{-1}\hat{e}^{y,old} = \hat{e}^{y,new}. \tag{26.1}$$

The fact that we use the inverse of the matrix has no significance; the only reason is that in the following sections we mostly use the inverse of $\mathbf{C}^{-1}$, which is equal to $\mathbf{C}$. The elements of the matrix $\mathbf{C}^{-1}$ are labeled as follows

$$\mathbf{C}^{-1} = \begin{pmatrix} C_{11}^{-1} & C_{12}^{-1} \\ C_{21}^{-1} & C_{22}^{-1} \end{pmatrix}, \tag{26.2}$$

485

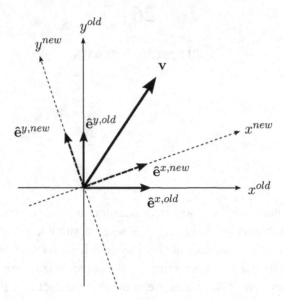

Figure 26.1 Definition of a vector **v** in the old (solid) and new (dashed) coordinate system.

where it is understood that C_{ij}^{-1} is not the inverse of the element C_{ij}, but is the i, j-element of the matrix $\mathbf{C}^{-1}$.

There is a close relation between the new basis vectors $\hat{\mathbf{e}}^{x,new}$ and $\hat{\mathbf{e}}^{y,new}$ and the matrix $\mathbf{C}^{-1}$. To see this we first use that in the old coordinate system the unit vector $\hat{\mathbf{e}}^{x,old}$ is the unit vector in the x-direction and that $\hat{\mathbf{e}}^{y,old}$ is the unit vector in the y-direction:

$$\hat{\mathbf{e}}^{x,old} = \begin{pmatrix} 1 \\ 0 \end{pmatrix}, \qquad \hat{\mathbf{e}}^{y,old} = \begin{pmatrix} 0 \\ 1 \end{pmatrix}. \tag{26.3}$$

Problem a   Let the matrix $\mathbf{C}^{-1}$ act on these expressions and use expression (26.1) to show that the first column of $\mathbf{C}^{-1}$ is given by $\hat{\mathbf{e}}^{x,new}$, while the second column is equal to $\hat{\mathbf{e}}^{y,new}$:

$$\hat{\mathbf{e}}^{x,new} = \begin{pmatrix} C_{11}^{-1} \\ C_{21}^{-1} \end{pmatrix}, \qquad \hat{\mathbf{e}}^{y,new} = \begin{pmatrix} C_{12}^{-1} \\ C_{22}^{-1} \end{pmatrix}. \tag{26.4}$$

Problem b   Use (26.2) to show that this implies that the matrix $\mathbf{C}^{-1}$ can also be written as

$$\mathbf{C}^{-1} = \begin{pmatrix} \hat{\mathbf{e}}^{x,new} & \hat{\mathbf{e}}^{y,new} \end{pmatrix}. \tag{26.5}$$

In other words, the transformed basis vectors are the columns of the matrix that describes the coordinate transform.

Let us now consider a vector **v** as shown in Figure 26.1. This vector is a physical quantity that denotes, for example, the position of an object relative to some specified origin. This vector is independent of any coordinate system since it is a physical quantity. However, when we describe this vector, we use a system of basis vectors to give the components of this vector in directions defined by the basis vectors. For example, using the old coordinate vectors $\hat{\mathbf{e}}^{x,old}$ and $\hat{\mathbf{e}}^{y,old}$ the vector **v** can be decomposed into its components v_x^{old} and v_y^{old}:

$$\mathbf{v} = v_x^{old}\hat{\mathbf{e}}^{x,old} + v_y^{old}\hat{\mathbf{e}}^{y,old}. \tag{26.6}$$

This same vector can also be decomposed into its components along the new coordinate system

$$\mathbf{v} = v_x^{new}\hat{\mathbf{e}}^{x,new} + v_y^{new}\hat{\mathbf{e}}^{y,new}. \tag{26.7}$$

The vector **v** is the same in each of these expressions. The goal of this section is to derive the relation between the old components v^{old} and the new components v^{new}.

Problem c Use that (26.6) and (26.7) are expressions for the same vector and insert (26.3) and (26.4) for the unit vectors to show that the old and the new components are related by

$$v_x^{old}\begin{pmatrix} 1 \\ 0 \end{pmatrix} + v_y^{old}\begin{pmatrix} 0 \\ 1 \end{pmatrix} = v_x^{new}\begin{pmatrix} C_{11}^{-1} \\ C_{21}^{-1} \end{pmatrix} + v_y^{new}\begin{pmatrix} C_{12}^{-1} \\ C_{22}^{-1} \end{pmatrix}. \tag{26.8}$$

Problem d Rewrite this result to derive that

$$\left. \begin{array}{l} C_{11}^{-1}v_x^{new} + C_{12}^{-1}v_y^{new} = v_x^{old} \\ C_{21}^{-1}v_x^{new} + C_{22}^{-1}v_y^{new} = v_y^{old} \end{array} \right\}. \tag{26.9}$$

At this point it is useful to introduce the vectors $\mathbf{v}^{old}$ and $\mathbf{v}^{new}$ with the components of the vector in the old and new coordinate system respectively:

$$\mathbf{v}^{old} \equiv \begin{pmatrix} v_x^{old} \\ v_y^{old} \end{pmatrix}, \qquad \mathbf{v}^{new} \equiv \begin{pmatrix} v_x^{new} \\ v_y^{new} \end{pmatrix}. \tag{26.10}$$

This harmless-looking definition belies a subtle complication. We now have three notations for the same vector. The notation **v** denotes the physical vector that is defined independently of any coordinate system, while the vectors $\mathbf{v}^{old}$ and $\mathbf{v}^{new}$ give the *representation* of this vector with respect to the old and the new coordinate systems, respectively. It is crucial to separate these three different vectors.

Problem e   Use the rules of matrix–vector multiplication to show that (26.9) implies that

$$\mathbf{C}^{-1}\mathbf{v}^{new} = \mathbf{v}^{old}, \qquad (26.11)$$

and derive from this that

$$\mathbf{v}^{new} = \mathbf{C}\mathbf{v}^{old}. \qquad (26.12)$$

This is the desired result because this expression prescribes how the old vector $\mathbf{v}^{old}$ is mapped in the coordinate transform onto the new vector $\mathbf{v}^{new}$. It is interesting to compare this expression with (26.1), which states that the old unit vectors are mapped onto the new basis vectors by the *inverse* matrix $\mathbf{C}^{-1}$. There is a good reason why the basis vectors transform with the inverse of the transformation matrix of the components. According to (26.6) the physical vector $\mathbf{v}$ is the sum of the product of the components of the vector and the basis vectors: $\mathbf{v} = v_x\hat{\mathbf{e}}^x + v_y\hat{\mathbf{e}}^y$. Since the physical vector $\mathbf{v}$ by definition is not changed by the coordinate transform, the sum $v_x\hat{\mathbf{e}}^x + v_y\hat{\mathbf{e}}^y$ of the *product* of the components and the basis vectors is invariant. This can only be the case when the transformation rule for the components is the inverse of the transformation rule of the basis vectors.

Problem f   As an example, consider the case where the coordinate vectors are rotated in the counterclockwise direction over an angle on 90 degrees. In that case

$$\mathbf{C}^{-1} = \begin{pmatrix} 0 & -1 \\ 1 & 0 \end{pmatrix} \quad \text{and} \quad \mathbf{C} = \begin{pmatrix} 0 & 1 \\ -1 & 0 \end{pmatrix}. \qquad (26.13)$$

Verify that the product of these matrices is the identity matrix: $\mathbf{C}^{-1}\mathbf{C} = \mathbf{C}\mathbf{C}^{-1} = \mathbf{I}$.

Problem g   Use expressions (26.4) and (26.12) to show that in this case:

$$\hat{\mathbf{e}}^{x,new} = \begin{pmatrix} 0 \\ 1 \end{pmatrix}, \quad \hat{\mathbf{e}}^{y,new} = \begin{pmatrix} -1 \\ 0 \end{pmatrix}, \quad \begin{pmatrix} v_x^{new} \\ v_y^{new} \end{pmatrix} = \begin{pmatrix} v_y^{old} \\ -v_x^{old} \end{pmatrix}. \qquad (26.14)$$

Problem h   Use the expressions above to show that

$$v_x^{new}\hat{\mathbf{e}}^{x,new} + v_y^{new}\hat{\mathbf{e}}^{y,new} = v_y^{old}\begin{pmatrix} 0 \\ 1 \end{pmatrix} + (-v_x^{old})\begin{pmatrix} -1 \\ 0 \end{pmatrix} = \begin{pmatrix} v_x^{old} \\ v_y^{old} \end{pmatrix}. \qquad (26.15)$$

This example shows that the new representation of the vector is equal to the old vector. Note the cancellation of the two minus signs in the last identity, this cancellation is due to the fact that the components and the basis vectors transform with $\mathbf{C}$ and $\mathbf{C}^{-1}$, respectively.

The treatment in this section was for a two-dimensional coordinate transform, but each step of the argument can be generalized to coordinate transforms in higher dimensions.

26.2 Unitary matrices

In the previous section, it was tacitly assumed that the matrix $\mathbf{C}$ was defined in such a way that the new basis vectors $\hat{\mathbf{e}}^{x,new}$ and $\hat{\mathbf{e}}^{y,new}$ were of unit length. For a general coordinate transform $\mathbf{C}$, this is not true. We are, however, interested in orthonormal coordinate systems only: these are coordinate systems in which the basis vectors have unit length and are mutually orthogonal.

Problem a Let the unit vectors in a coordinate system be denoted by $\hat{\mathbf{e}}^i$. Show that for an orthonormal coordinate system

$$\left(\hat{\mathbf{e}}^i \cdot \hat{\mathbf{e}}^j\right) = \delta_{ij}, \tag{26.16}$$

where δ_{ij} is the Kronecker delta is

$$\delta_{ij} = \begin{cases} 1 & \text{when } i = j, \\ 0 & \text{when } i \neq j. \end{cases} \tag{25.15}$$

When the old coordinate system and the new coordinate system are orthonormal, both coordinate systems must satisfy (26.16). The condition is certainly satisfied when the inner product of two vectors does not change under the coordinate transformation. Such coordinate transformations are called *unitary transformations*. To be more specific, let $\mathbf{u}$ and $\mathbf{v}$ be two vectors in the old coordinate system that are mapped by the coordinate transform $\mathbf{C}$ to new vectors $\mathbf{u}' = \mathbf{Cu}$ and $\mathbf{v}' = \mathbf{Cv}$, respectively. The coordinate transform $\mathbf{C}$ is called unitary when $(\mathbf{u}' \cdot \mathbf{v}') = (\mathbf{u} \cdot \mathbf{v})$, or equivalently

$$(\mathbf{Cu} \cdot \mathbf{Cv}) = (\mathbf{u} \cdot \mathbf{v}). \tag{26.17}$$

Problem b Show that the fact that the length of a vector and the angle between vectors is preserved implies that the inner product of two vectors is preserved as well. Examples of unitary coordinate transformations are rotations, reflections in a plane, or combinations of these. Draw a number of figures to convince yourself that these coordinate transformations do not change the length of the vector and the angle between two vectors so that these transformations are indeed unitary.

Problem c The requirement (26.17) can be used to impose a constraint on $\mathbf{C}$. Write the matrix products and the inner product in (26.17) in terms of the components of the matrix $\mathbf{C}$ and the vectors $\mathbf{u}$ and $\mathbf{v}$ to derive that it is equivalent to

$$\sum_{i,j,k} C_{ij} u_j C_{ik} v_k = \sum_i u_i v_i. \qquad (26.18)$$

Writing out a matrix expression in its components has an important advantage. In general one cannot interchange the order of two matrices in a multiplication, because the matrix product $\mathbf{AB}$ may differ from the matrix product $\mathbf{BA}$.

Problem d   Verify this statement by taking

$$\mathbf{A} = \begin{pmatrix} 0 & 1 & 0 \\ -1 & 0 & 0 \\ 0 & 0 & 1 \end{pmatrix} \quad \text{and} \quad \mathbf{B} = \begin{pmatrix} 1 & 0 & 0 \\ 0 & 0 & 1 \\ 0 & -1 & 0 \end{pmatrix} \qquad (26.19)$$

and showing that $\mathbf{AB} \neq \mathbf{BA}$.

Problem e   Geometrically the matrix $\mathbf{A}$ represents a rotation around the z-axis through 90 degrees, while $\mathbf{B}$ represents a rotation around the x-axis through 90 degrees. Take this book and carry out these rotations in the two different orders to see that the final orientation of the book is indeed different in each case.

This means that in general one cannot interchange the order of matrices in a product. However, when a term $A_{ij} B_{jk}$ occurs in a summation, the terms A_{ij} and B_{jk} refer to individual matrix elements. These matrix elements are regular numbers that can be interchanged in multiplication: $A_{ij} B_{jk} = B_{jk} A_{ij}$. This means that the decomposition of a matrix equation into its components allows us to move the individual components around in any way that we like.

In the following we will use the transpose $\mathbf{C}^T$ of a matrix $\mathbf{C}$ extensively. This is the matrix in which the rows and the columns are interchanged, so that

$$C_{ij}^T \equiv C_{ji}. \qquad (26.20)$$

Problem f   Use the results in this section to show that (26.18) can be written as

$$\sum_{i,j,k} C_{ji}^T C_{ik} u_j v_k = \sum_i u_i v_i. \qquad (26.21)$$

At this point it is important to realize that it is immaterial how we label the summation index. The right-hand side of (26.21) can just as well be written as $\sum_j u_j v_j$ because this sum and $\sum_i u_i v_i$ both mean: "take the corresponding elements of $\mathbf{u}$ and $\mathbf{v}$, multiply them and sum over all components."

Problem g Show that (26.21) is equivalent to

$$\sum_{j,k} \left(\sum_i C_{ji}^T C_{ik} \right) u_j v_k = \sum_j u_j v_j. \tag{26.22}$$

Problem h Use the definition (25.15) of the Kronecker delta to show that $\sum_j u_j v_j = \sum_{j,k} \delta_{jk} u_j v_k$.

Problem i Use this result to show that **C** is unitary when it satisfies the following condition

$$\left(\sum_i C_{ji}^T C_{ik} \right) = \delta_{jk}. \tag{26.23}$$

At this point we use that δ_{jk} is just the (j, k)-element of the identity matrix **I** because the elements of this matrix are equal to 1 on the diagonal ($j = k$) and equal to zero off the diagonal ($j \neq k$). In matrix notation, (26.23) for a unitary matrix therefore implies that

$$\mathbf{C}^T \mathbf{C} = \mathbf{I}. \tag{26.24}$$

Problem j Show that this implies that the inverse of a unitary matrix is equal to its transpose:

$$\mathbf{C}^{-1} = \mathbf{C}^T. \tag{26.25}$$

This result can save you a lot of work. Suppose you need to invert a matrix, and suppose that you suspect that the matrix is unitary, then according to (26.25) the inverse matrix is equal to the transpose. You can verify whether this is indeed the case by multiplying this "guess" by the original matrix **C** to check whether the result is equal to the identity matrix; this amounts to applying (26.24).

Problem k The matrix corresponding to a rotation over an angle φ is given by

$$\mathbf{R} = \begin{pmatrix} \cos \varphi & \sin \varphi \\ -\sin \varphi & \cos \varphi \end{pmatrix}. \tag{26.26}$$

Compute the inverse of this matrix with a minimal amount of work. (This is also called the principle of maximal laziness.)

26.3 Shear or dilatation/compression?

As an example of the importance of coordinate transforms we consider a geologist who shears a rock sample as shown in Figure 26.2. The height and the width of the

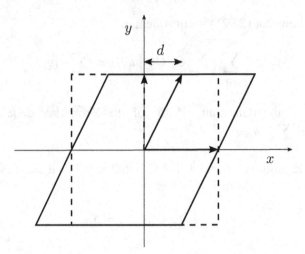

Figure 26.2   The shape of a rock before (dashed lines) and after (solid lines) shear deformation.

rock are initially of unit length. In the shear motion the upper side of the rock is displaced rightward over a distance d, while the lower side of the rock is displaced leftward over a distance $-d$.

Problem a   Use Figure 26.2 to show that the original unit vectors change during the deformation as

$$\begin{pmatrix} 1 \\ 0 \end{pmatrix} \to \begin{pmatrix} 1 \\ 0 \end{pmatrix} \quad \text{and} \quad \begin{pmatrix} 0 \\ 1 \end{pmatrix} \to \begin{pmatrix} d \\ 1 \end{pmatrix}. \tag{26.27}$$

Problem b   Let the deformation be described by a 2×2 matrix **D**. Show that this matrix is given by

$$\mathbf{D} = \begin{pmatrix} 1 & d \\ 0 & 1 \end{pmatrix}. \tag{26.28}$$

Hint: Check that this produces the deformation described by (26.27).

The geologist is interested in the internal deformation of the rock. However, the transformation shown in Figure 26.2 entails both a deformation within the rock sample, as well as a rotation of the sample. This can be seen from the unit vector in Figure 26.2 that is initially along the y-axis. In the transformation, this vector is rotated, but also extended from a unit length to $\sqrt{1 + d^2}$. We first need to unravel these two contributions. In the following we only considered very small shear ($d \ll 1$), so that the associated rotation is also small.

Problem c   Use a Taylor expansion of (26.26) to show that the rotation matrix over a small angle φ is to first order in φ given by

$$\mathbf{R} = \mathbf{I} + \begin{pmatrix} 0 & \varphi \\ -\varphi & 0 \end{pmatrix}. \tag{26.29}$$

The shear described by (26.28) can be decomposed into this rotation plus an internal deformation. To make this explicit, we decompose the matrix $\mathbf{D}$ into the matrix on the right-hand side of (26.29), which accounts for the rotation and a new matrix $\mathbf{E}$:

$$\mathbf{D} = \begin{pmatrix} 0 & \varphi \\ -\varphi & 0 \end{pmatrix} + \mathbf{E}. \tag{26.30}$$

The matrix $\mathbf{E}$ includes the identity matrix $\mathbf{I}$ from expression (26.29). At this point we need to find the parameter φ that characterizes the rotational component. It follows from (26.29) that the difference between the 12-component and the 21-component of the matrix is equal to 2φ. According to (26.28) this difference is given by $D_{12} - D_{21} = d$; since this is equal to 2φ, the rotation angle is given by $\varphi = d/2$.

Problem d   Use the expressions (26.28), (26.30), and the relation $\varphi = d/2$ to show that the deformation component of the transformation in Figure 26.2 is given by

$$\mathbf{E} = \begin{pmatrix} 1 & d/2 \\ d/2 & 1 \end{pmatrix}. \tag{26.31}$$

Note that this matrix is symmetric $(E_{ij} = E_{ji})$, whereas the matrix $\mathbf{R}$ in (26.29) is antisymmetric $(R_{ij} = -R_{ji})$. The matrix $\mathbf{D}$ has thus been decomposed into a symmetric part and an antisymmetric part. This corresponds to a decomposition of the transformation into a rotational component plus a deformation. We have seen this before in Chapter 7, where it was shown that the curl of a vector field in general has two contributions: rotation and shear. We return to this issue in Section 26.9.

Problem e   We now focus on the internal deformation of the sample. Show that Figure 26.3 displays the deformation $\mathbf{E}$ of the rock. Hint: Determine first what happens to the unit vectors in the x- and y-directions when they are multiplied with the matrix $\mathbf{E}$.

Let us now consider a second geologist, who is studying the same deforming rock with a coordinate system that is rotated through 45 degrees with respect to the coordinate system used by the geologist who views the world as

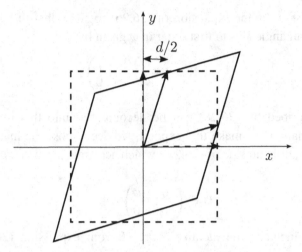

Figure 26.3   The same deformation as in the previous figure, but the rotational component is removed. The sample prior to deformation is shown by the dashed lines, the deformed sample by the solid lines.

seen in Figure 26.2. We ignore the effect of the rotation altogether because it does not lead to internal deformation of the rock. The x'-axis of the second geologist is aligned along the line $x = y$, whereas the y'-axis is aligned along the line $x = -y$. Hence, the basis vectors of the second geologist are given by

$$\hat{\mathbf{e}}^{x'} = \frac{1}{\sqrt{2}} \begin{pmatrix} 1 \\ 1 \end{pmatrix}, \qquad \hat{\mathbf{e}}^{y'} = \frac{1}{\sqrt{2}} \begin{pmatrix} 1 \\ -1 \end{pmatrix}. \tag{26.32}$$

Problem f   The simplest way to determine how the second geologist views the deformation is to determine what happens when $\mathbf{E}$ acts on the basis in the vectors of the second geologist. Show that this gives

$$\mathbf{E}\hat{\mathbf{e}}^{x'} = (1 + d/2)\,\hat{\mathbf{e}}^{x'}, \qquad \mathbf{E}\hat{\mathbf{e}}^{y'} = (1 - d/2)\,\hat{\mathbf{e}}^{y'}. \tag{26.33}$$

Problem g   Use this result to show that the second geologist sees the deformation process as depicted in Figure 26.4.

Note that the unit vectors of the second geologist are the eigenvectors of the matrix $\mathbf{E}$. This means that the second geologist would describe the deformation process seen in Figure 26.4 as follows. The sample is extended in the x'-direction because the unit vector $\hat{\mathbf{e}}^{x'}$ is mapped onto itself with an amplification factor $(1 + d/2)$ that is greater than 1; in the y'-direction the sample is compressed because the unit vector $\hat{\mathbf{e}}^{y'}$ is mapped onto itself with an amplification factor $(1 - d/2)$ that is smaller than 1. This means that (apart from a rotation) the first

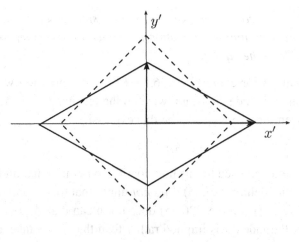

Figure 26.4   The deformation of the previous figure seen by a geologist who uses a coordinate system that is rotated through 45 degrees.

geologist describes the transformation as pure shear, whereas the second geologist describes the transformation as a combination of dilatation and compression in the rock. These two points of view must be equivalent. However, they can be reconciled only when we have techniques for switching from one coordinate system to another. This is the goal of tensor calculus. Before we derive the transformation rules for matrices and more general tensors, we first introduce a notational convention that turns out to be very convenient.

26.4 Summation convention

In this section the summation convention is introduced; this convention is sometimes also referred to as the "Einstein summation convention." Let us consider the multiplication of the matrix **A** by a vector **x** that in a vector notation gives a new vector **y** = **Ax**.

Problem a   Show that in component form this expression should be written as

$$\sum_j A_{ij} x_j = y_i. \tag{26.34}$$

You have seen in Section 26.2 that matrix and vector calculations involve a lot of summations. It is also shown in that section that working with the expressions in component form has advantages over working with the more abstract vector/matrix notation. We could save a lot of work by leaving out the summation signs in the equations. This is exactly what the summation convention does:

> *In the summation convention we leave out the summation signs $\left(\sum\right)$ with the understanding that a summation is implied over any repeated index on either side of the equation.*

What does this mean for expression (26.34)? In this expression we sum over the index j. Note that this index appears twice on the left-hand side. According to the summation convention, we would write this expression as

$$A_{ij}x_j = y_i. \qquad (26.35)$$

It takes some time to get used to this notation. Keep in mind that under the summation convention the identity (26.35) does *not* imply that this expression holds for a single value of j. For example, (26.35) does not mean that $A_{i2}x_2 = y_i$ because a summation over the index j is implied rather than that j can take any fixed value $j = 1$ or $j = 2$ or $j = 3$.

As another example of the summation convention, consider the inner product of two vectors. Normally one would write this as $(\mathbf{u} \cdot \mathbf{v}) = \sum_i u_i v_i$; using the summation convention one writes it as $(\mathbf{u} \cdot \mathbf{v}) = u_i v_i$. Note that the name of the summation index is irrelevant. This expression could therefore also be written as $(\mathbf{u} \cdot \mathbf{v}) = u_j v_j$. The matrix equation $\mathbf{C}^T\mathbf{C} = \mathbf{I}$ is written in component form as $\sum_k C_{ik}^T C_{kj} = \sum_k C_{ki} C_{kj} = \delta_{ij}$; in the summation convention this is written as $C_{ik}^T C_{kj} = C_{ki} C_{kj} = \delta_{ij}$.

Problem b   Write out the matrix product $\mathbf{AI}$ (with $\mathbf{I}$ as the identity matrix) in component form, and show that the result can be written as: $(\mathbf{AI})_{ij} = A_{ik}\delta_{kj}$.

In the last term a summation over the index k is implied. The Kronecker delta is always zero, except when $k = j$. This means that the only nonzero contribution from the k-summation comes from the term $k = j$, so that $(\mathbf{AI})_{ij} = A_{ik}\delta_{kj} = A_{ij}$. This is simply the identity $\mathbf{AI} = \mathbf{A}$ in component form.

Problem c   Show that $v^2 = v_i v_i$ where v is the length of the vector $\mathbf{v}$.

Problem d   The Laplacian is the divergence of the gradient; use this to write the Laplacian in the following form

$$\nabla^2 f = \partial_i \partial_i f, \qquad (26.36)$$

with $\partial_i \equiv \partial/\partial x_i$.

Problem e   Use the summation convention to rewrite expression (25.48) as

$$\frac{\partial(\rho v_i)}{\partial t} + \partial_j(\rho v_j v_i) = \mu \partial_k \partial_k v_i + F_i. \qquad (26.37)$$

Problem f In some applications, summations occur over several indices. As an example consider the double sum $\sum_{i,j} A_{ij}\delta_{ji}$. Carry out the summation and show that the result can be written as

$$A_{ij}\delta_{ji} = A_{ii}. \tag{26.38}$$

Note that on the left-hand side a summation over i and j is implied, whereas on the right-hand side a sum over i is implied. The term on the right-hand side is the sum of the diagonal elements of the matrix $\mathbf{A}$. This quantity plays an important role in a variety of applications and is called the *trace* of $\mathbf{A}$:

$$\text{tr } \mathbf{A} \equiv \sum_{i} A_{ii} = A_{ii}. \tag{26.39}$$

One caveat should be made with the summation convention. According to the summation convention $M_{ii} = 0$ means that the trace of a matrix $\mathbf{M}$ is equal to zero. However, it could also mean that all the diagonal elements of $\mathbf{M}$ are equal to zero: $M_{11} = M_{22} = M_{33} = 0$. This is an example where an expression can be ambiguous. It is important to state explicitly the use of the summation convention. Even when one does, a deviation from this convention in an equation can be stated. For example, to express that all the diagonal elements of $\mathbf{M}$ are equal to zero, you could write this in the summation convention as:

$$M_{ii} = 0 \qquad \text{(no summation over } i\text{)}. \tag{26.40}$$

26.5 Matrices and coordinate transforms

In this section we determine the transformation properties of matrices under a coordinate transform $\mathbf{C}$ as introduced in Section 26.1. In the remainder of this section we then restrict ourselves to unitary coordinate transformations. Let a matrix $\mathbf{D}$ map a vector $\mathbf{x}$ to a vector $\mathbf{y}$

$$\mathbf{y} = \mathbf{Dx}. \tag{26.41}$$

Consider this same operation in a new coordinate system and let us use a prime to denote the corresponding quantities in the new coordinate system. According to (26.12) the vector $\mathbf{x}$ in the new coordinate system is given by $\mathbf{x}' = \mathbf{Cx}$, and a similar expression holds for $\mathbf{y}$. Expression (26.41) is given in the new coordinate system by

$$\mathbf{y}' = \mathbf{D}'\mathbf{x}'. \tag{26.42}$$

In this section we determine the relation between the matrix $\mathbf{D}'$ in the new coordinate system and the old matrix $\mathbf{D}$.

Problem a   Use (26.12) to show that (26.42) can be written as

$$\mathbf{Cy} = \mathbf{D'Cx}. \tag{26.43}$$

Problem b   Multiply this expression on the left by $\mathbf{C}^{-1}$, compare the result with (26.41) to derive that $\mathbf{D} = \mathbf{C}^{-1}\mathbf{D'C}$. Multiply this last expression on the left and the right by suitable matrices to derive that

$$\mathbf{D'} = \mathbf{CDC}^{-1}. \tag{26.44}$$

This is the general transformation rule for matrices under a coordinate transform. It looks different from the transformation rule (26.12) for vectors, because in (26.44) the inverse $\mathbf{C}^{-1}$ of the coordinate transformation appears as well. However, for unitary coordinate transforms one can eliminate this inverse and write (26.44) in a way that is similar to (26.12).

Problem c   Use that $\mathbf{C}$ is unitary and rewrite (26.44) in component form as

$$D'_{ij} = \sum_{k,l} C_{ik} C_{jl} D_{kl}. \tag{26.45}$$

Remember that you cannot interchange the order of matrices in a multiplication, but that you can change the order of matrix *elements* that are multiplied.

Problem d   Redo the calculation in the last problem using the summation convention at every step and rewrite the preceding result as

$$D'_{ij} = C_{ik} C_{jl} D_{kl}. \tag{26.46}$$

Problem e   To compare this result with the transformation rule (26.12) for vectors, write $\mathbf{v}^{old} = \mathbf{v}$, $\mathbf{v}^{new} = \mathbf{v'}$, and use the summation convention to rewrite (26.12) using the summation convention as

$$v'_i = C_{ij} v_j. \tag{26.47}$$

Applying the matrix transformation (26.46) may seem complicated. The easiest way to apply it in practice is to write it as

$$\mathbf{D'} = \mathbf{CDC}^T. \tag{26.48}$$

This expression is identical to (26.44) with the only exception that it uses that $\mathbf{C}$ is unitary: $\mathbf{C}^{-1} = \mathbf{C}^T$.

According to (26.26) a rotation through 45 degrees of the coordinate system is described by the coordinate transform

$$\mathbf{C} = \frac{1}{\sqrt{2}} \begin{pmatrix} 1 & 1 \\ -1 & 1 \end{pmatrix}. \tag{26.49}$$

Problem f Check that this coordinate transformation transforms the matrix $\mathbf{E}$ in (26.31) to the matrix

$$\mathbf{E}' = \begin{pmatrix} 1+d/2 & 0 \\ 0 & 1-d/2 \end{pmatrix}. \tag{26.50}$$

Problem g Show that this matrix describes the deformation shown in Figure 26.4.

26.6 Definition of a tensor

Expressions (26.46) and (26.47) allow us to see a resemblance between the transformation rule (26.46) for a matrix and the rule (26.47) for a vector. In both equations the old vector or matrix is multiplied by the matrix $\mathbf{C}$; for the vector this happens once and for the matrix it happens twice. The *first* index of the quantity on the left-hand side (i in both cases) is the first index of the *first* matrix $\mathbf{C}$ on the right-hand side as indicated by the arrows in the following expressions:

$$v_i' = C_{ij} v_j, \qquad D_{ij}' = C_{ik} C_{jl} D_{kl}.$$

The *second* index of $\mathbf{D}'$ in (26.46) corresponds to the first index of the *second* term $\mathbf{C}$ as indicated by the arrows:

$$D_{ij}' = C_{ik} C_{jl} D_{kl}.$$

In both (26.46) and (26.47) the *second* index of each term $\mathbf{C}$ also occurs in the term $\mathbf{D}$ or $\mathbf{v}$ on which this matrix acts as shown by the arrows in the following expressions:

$$v_i' = C_{ij} v_j, \qquad D_{ij}' = C_{ik} C_{jl} D_{kl}, \qquad D_{ij}' = C_{ik} C_{jl} D_{kl}.$$

Note that according to the summation convention, a summation over these repeated indices is implied.

Some quantities that we use are not labeled by any index. Examples are temperature and pressure. Such a quantity is a pure number and it is called a *scalar*. In addition to a size, a vector has a direction and is labeled with one index. A matrix has two indices. In fact, we can define objects characterized by an arbitrary number of indices. In general a physical quantity can have any number of subscripts, and a quantity with n subscripts can be denoted as $T_{i_1 i_2 \cdots i_n}$. We call such an object a tensor when it follows a transformation rule similar to (26.46) and (26.47) for a matrix and a vector, respectively. Combining these, we get the following definition.

Definition   The quantity $T_{i_1 i_2 \cdots i_n}$ is called a tensor of rank n when it transforms under a unitary coordinate transform $\mathbf{C}$ with the following transformation rule:

$$T'_{i_1 i_2 \cdots i_n} = C_{i_1 j_1} C_{i_2 j_2} \cdots C_{i_n j_n} T_{j_1 j_2 \cdots j_n}. \qquad (26.51)$$

Note that the summation convention is used, so a summation over the indices $j_1, j_2, \ldots, j_n$ is implied.

Problem a   Convince yourself that the transformation rules (26.46) and (26.47) for a matrix and a vector, respectively, are special cases of this definition for the values $n = 2$ and $n = 1$, respectively.

This means that a vector that transforms according to (26.47) is a tensor of rank one and a matrix that transforms according to (26.46) is a tensor of rank two. Note that a scalar does not change under coordinate transformations. It has zero subscripts and one can say that in executing the coordinate transformation, (26.51) – the coordinate transform $\mathbf{C}$ – is applied zero times. For this reason a scalar is called a tensor of rank zero.

Let us consider some examples of tensors. The position vector $\mathbf{r}$ is a tensor of rank one, as follows from the transformation rule (26.12) that you derived in Section 26.1. From this it follows that the velocity vector $\mathbf{v}$ is a tensor of rank one provided that the coordinate transformation $\mathbf{C}$ does not depend on time.

Problem b   Differentiate the transformation law $x'_i = C_{ij} x_j$ for the position vector with respect to time and use the result to show that the velocity indeed transforms as a tensor of rank one.

Problem c   Show that acceleration is a tensor of rank one.

Now we can use Newton's law $\mathbf{F} = m\mathbf{a}$ and that the mass m is a scalar. Since the acceleration is a tensor of rank one, the force must also be a tensor of rank one.

It is interesting to see what happens when one works in a rotating coordinate system. In that case the transformation $\mathbf{C}$ that transforms the fixed system to the rotating system depends on time.

Problem d Go through the steps of Problems b and c and show that in that case the acceleration transforms as

$$a_i' = C_{ij}a_j + 2\dot{C}_{ij}v_j + \ddot{C}_{ij}x_j, \qquad (26.52)$$

where the dot denotes a time derivative.

This means that for such a time-dependent coordinate transform the acceleration does *not* transform as a vector of rank one. Note that additional terms appear that are proportional to the velocity and the position vector. These correspond to the Coriolis force and the centrifugal force as treated in Section 12.2.

Another example of a tensor of rank one is the gradient vector $\nabla_i = \partial/\partial x_i$. The chain rule of differentiation states that

$$\nabla_i' = \frac{\partial}{\partial x_i'} = \frac{\partial x_j}{\partial x_i'}\frac{\partial}{\partial x_j} = \frac{\partial x_j}{\partial x_i'}\nabla_j. \qquad (26.53)$$

To conform with the summation convention, a summation over j is implied in (26.53).

Problem e Use the transformation law of the position vector to derive that $\partial x_j/\partial x_i' = C_{ji}^{-1}$, and use the property that $\mathbf{C}$ is a unitary matrix to derive that

$$\nabla_i' = C_{ij}\nabla_j. \qquad (26.54)$$

In other words, the gradient vector is a tensor of rank one.

As an example of a tensor of rank two, we consider the identity matrix whose elements are the Kronecker delta δ_{ij}. At first sight you might believe that since the elements of the Kronecker delta are simply equal to zero or one, it is a scalar. However, the identity matrix really transforms like a tensor of rank two.

Problem f To see this, assume that the identity matrix follows the transformation rule (26.46) and use the property of the Kronecker delta to derive that

$$I_{ij}' = C_{ik}C_{jl}\delta_{kl} = C_{ik}C_{jk}. \qquad (26.55)$$

Problem g Write the last term as C_{kj}^T and use the property that $\mathbf{C}$ is unitary to derive that

$$I_{ij}' = \delta_{ij}. \qquad (26.56)$$

In other words, the identity matrix transforms as a tensor of rank two and yet it takes the same form in every coordinate system!

26.7 Not every vector is a tensor

At this point you may think that any object vector with n-components is a tensor of rank one. However, keep in mind that not every vector transforms according to the transformation rule (26.51) for a tensor. Let us first consider an example of the distribution of shoe sizes of students. In such a study, the shoe size of each student in the group and can be represented by a vector of the data:

$$\mathbf{d} = \begin{pmatrix} \text{shoe size of Marie} \\ \text{shoe size of Peter} \\ \text{shoe size of Klaas} \\ \vdots \end{pmatrix}. \tag{26.57}$$

This vector is not a tensor, for two reasons. The coordinate transformation matrix $\mathbf{C}$ is by definition a square matrix with a dimension equal to the dimension of the coordinate system (usually 3 and for some applications 4, see Section 26.12). The vector $\mathbf{d}$ can have any dimension and the action of $\mathbf{C}$ in this vector is not defined. The second reason why $\mathbf{d}$ is not a tensor of rank one is that the shoe size of students does not depend on the coordinate system. In other words, all the elements of $\mathbf{d}$ are scalars; hence, $\mathbf{d}$ is certainly not a tensor of rank one.

Another example of a vector that is not a tensor is the stress–displacement vector that is used in the description of elastic wave propagation in elastic media (Aki and Richards, 2002):

$$\mathbf{w} = \begin{pmatrix} u_y \\ \sigma_{yz} \end{pmatrix}, \tag{26.58}$$

where u_y and σ_{yz} are suitably chosen components of the displacement and the stress. This vector is not a tensor of rank one. It has a different dimension than $\mathbf{C}$; hence, the action of $\mathbf{C}$ on this vector is not defined. Furthermore, the components of $\mathbf{w}$ are of different dimension; hence, they can never be mixed in the transformation rule (26.51).

Let us now consider a more subtle example. At this point you might think that the magnetic field vector is a tensor of rank one. Consider the circular current in the (x, y)-plane shown in Figure 26.5 that generates a magnetic field. Given the direction of this current the magnetic field on the z-axis is oriented in the direction of the positive z-axis. This holds for the points P and Q in Figure 26.5.

Problem a   Let us perform the coordinate transformation in which the system is reflected in the (x, y)-plane. Make a drawing of the transformed current and make sure you understand that the current has not changed in the coordinate transform.

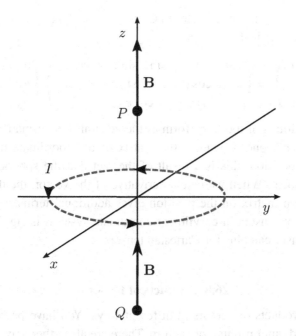

Figure 26.5   A circular current in the horizontal plane induces a vertical magnetic field **B**.

Problem b   Now that you know that the current has not changed, use Figure 26.5 to draw the associated magnetic field vector at the points P and Q.

Problem c   Now add to your figure the image of the vectors **B** of Figure 26.5 after they have been subjected to a reflection in the (x, y)-plane.

You probably found that Problems b and c predict transformed magnetic field vectors that differ by a minus sign. One of the problems must have given the wrong answer. Problem b gives the right answer, because the current defines the magnetic field and the current does not change under a reflection in the (x, y)-plane. Therefore, Problem c gives a wrong answer. This leads to the conclusion that the magnetic field does not behave like a tensor of rank one under coordinate transforms as described by the transformation rule (26.47). In fact, the magnetic field behaves in a different way. It is an example of a *pseudo-tensor* indicating that it shares some of the properties of the tensors that we have explored, but that its transformation rules differ from the transformation rule (26.47). More details about pseudo-tensors can be found in Riley et al. (2006), but the main thing to remember at this point is that not every vector is a tensor!

The final example in this section shows what happens when a coordinate transform is not a constant. In Section 4.2 we derived the transformation that maps the components u_x, u_y, and u_z of a vector in Cartesian coordinates into its components u_r, u_θ, and u_φ in spherical coordinates:

$$\begin{pmatrix} u_r \\ u_\theta \\ u_\varphi \end{pmatrix} = \begin{pmatrix} \sin\theta\cos\varphi & \sin\theta\sin\varphi & \cos\theta \\ \cos\theta\cos\varphi & \cos\theta\sin\varphi & -\sin\theta \\ -\sin\varphi & \cos\varphi & 0 \end{pmatrix} \begin{pmatrix} u_x \\ u_y \\ u_z \end{pmatrix}. \qquad (4.19)$$

This transformation is of the same form as the general transformation rule (26.47) for a tensor of rank one. However, the matrix of the coordinate transform now depends on the position; this is a result of the fact that the spherical coordinate system is curvilinear. When we take a derivative of the vector, the dependence of the transformation matrix on the position gives additional terms. For this reason tensor calculus for tensors in curvilinear coordinate systems is significantly more complex than tensor calculus for Cartesian tensors.

26.8 Products of tensors

One can form products of vectors in different ways. You have probably seen the inner product and outer product of vectors. There are also other ways in which one can make products of tensors. The first product of tensors that we consider is the *contraction*. You have already seen examples of this, such as the inner product of two vectors:

$$(\mathbf{u} \cdot \mathbf{v}) = u_i v_i. \qquad (26.59)$$

Another example is matrix–vector multiplication:

$$(\mathbf{M} \cdot \mathbf{v})_i = M_{ij} v_j. \qquad (26.60)$$

Note that according to the summation convention a summation over i is implied in (26.59) and a summation over j in (26.60).

These operations can be extended. Let $U_{i_1 i_2 \cdots i_n}$ be a tensor of rank n and $V_{j_1, j_2 \cdots j_m}$ a tensor of rank m. The contraction of $\mathbf{U}$ and $\mathbf{V}$ is defined by setting the last index of $\mathbf{U}$ equal to the first index of $\mathbf{V}$ and summing over this index:

$$(\mathbf{U} \cdot \mathbf{V})_{i_1 i_2 \cdots i_{n-1} k_2 \cdots k_m} = U_{i_1 \cdots i_{n-1} r} V_{r k_2 \cdots k_m}. \qquad (26.61)$$

Note that we sum over the index r.

Problem a   Show that the rank of this tensor is $n + m - 2$.

Problem b   Show that the inner product of two vectors and matrix–vector multiplication are special cases of the contraction (26.61). Make sure you define n and m in each example. Verify that the rank of the result is indeed $n + m - 2$.

One can also apply a multiple contraction by summing over two or more indices of the tensors that one contracts. For example, the double contraction is given by

$$(\mathbf{U} : \mathbf{V})_{i_1 i_2 \cdots i_{n-2} k_3 \cdots k_m} = U_{i_1 \cdots i_{n-2} rs} V_{srk_3 \cdots k_m}. \tag{26.62}$$

Problem c   Show that the resulting tensor is of rank $n + m - 4$.

A very important property is that *the contraction of two tensors is also a tensor*. The proof of this property is not difficult, but mostly a tedious bookkeeping exercise, the details of which can be found in Butkov (1968) or Riley et al. (2006).

Problem d   Show that the double contraction of a matrix $\mathbf{A}$ with the identity matrix is equal to the trace of $\mathbf{A}$ as defined in (26.39):

$$(\mathbf{A} : \mathbf{I}) = A_{ii} = \operatorname{tr} \mathbf{A}. \tag{26.63}$$

The fact that the contraction of a tensor is also a tensor implies that the trace of a matrix is a tensor of rank zero: the trace of a matrix is a scalar. Therefore, the trace of a matrix is invariant to unitary coordinate transformations.

Problem e   There is one coordinate transformation in which the new basis vectors are aligned with the eigenvectors of $\mathbf{A}$. Show that in that particular coordinate system the trace is equal to the sum of the eigenvectors λ_i. Use this result to derive that in any coordinate system the trace of a matrix is equal to the sum of the eigenvalues:

$$\operatorname{tr} \mathbf{A} = \sum_i \lambda_i. \tag{26.64}$$

A matrix that is often used is the *Hessian* of a function f, which is defined by the second partial derivatives of that function:

$$H_{ij} \equiv \frac{\partial^2 f}{\partial x_i \partial x_j}. \tag{26.65}$$

Problem f   Show that $\mathbf{H}$ is a tensor of rank two by generalizing the derivation of Problem e of Section 26.6 to the transformation properties of the second derivative.

Problem g   Show that the trace of the Hessian is the Laplacian of f

$$\operatorname{tr} H = \nabla^2 f. \tag{26.66}$$

The trace is invariant for unitary coordinate transformations and the Laplacian is the trace of a tensor; therefore, the Laplacian is invariant for unitary coordinate transformations as well.

The second type of product of tensors is the direct product. The direct product of two tensors is formed by multiplying the different elements of the two tensors without carrying out a summation over repeated indices. An example of this is the dyad of two vectors that you encountered for example in Section 12.1 where the projection operator is written as the direct product of the unit vector $\hat{\mathbf{n}}$ with itself:

$$\mathbf{P} = \hat{\mathbf{n}}\hat{\mathbf{n}}^T. \tag{12.5}$$

In component form this expression is given by

$$P_{ij} = n_i n_j. \tag{26.67}$$

This idea can be generalized to form the direct product of a tensor $\mathbf{U}$ of rank n and a tensor $\mathbf{V}$ of rank m. This direct product is usually denoted with the symbol $\otimes$:

$$(\mathbf{U} \otimes \mathbf{V})_{i_1 \cdots i_n j_1 \cdots j_m} = U_{i_1 \cdots i_n} V_{j_1 \cdots j_m}. \tag{26.68}$$

Note that there are no repeated indices over which one sums.

Problem h   Show that the rank of this direct product is $n + m$.

Problem i   Show that the dyad (12.5) is a special example of the direct product (26.68).

Problem j   The direct product of two tensors is also a tensor. Show this property by applying the transformation rule (26.51) to the tensors on the right-hand side of (26.68).

Problem k   The gradient of a function is described in Chapter 5. The gradient $\mathbf{G} = \nabla \otimes \mathbf{v}$ of a vector $\mathbf{v}$ is defined as

$$G_{ij} \equiv \frac{\partial v_j}{\partial x_i}. \tag{26.69}$$

Show that this is a tensor of rank two.

26.9 Deformation and rotation again

As an important example of the direct product we consider in this section the strain tensor, which measures the state of deformation in a medium. In a medium that is subject to some deformation, let us focus on two nearby points $\mathbf{r}$ and $\mathbf{r} + \delta\mathbf{r}$ as

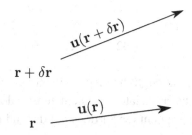

Figure 26.6   The deformation **u** of two nearby points **r** and **r** + δ**r**.

shown in Figure 26.6. During the deformation each point **r** is displaced by a vector **u**(**r**), which means that after the deformation the point **r** is located at **r** + **u**(**r**) and the point **r** + δ**r** is located at **r** + δ**r** + **u**(**r** + δ**r**). When the displacement of these neighboring points is the same (**u**(**r** + δ**r**) = **u**(**r**)), the relative positions of these points are not changed, and no deformation has taken place. Therefore, the deformation is associated with the *variation* of the displacement vector **u**(**r**) with position **r**. To describe the deformation, we use the gradient tensor **D** = ∇ ⊗ **u** of the displacement vector.

$$D_{ij} \equiv \frac{\partial u_j}{\partial x_i}. \tag{26.70}$$

Problem a   In a uniform translation (**u**(**r**) = *const.*) the medium is not deformed. Show that for this deformation the tensor **D** is indeed equal to zero.

The previous problem illustrates that when the medium is not deformed, the gradient tensor **D** is equal to zero. Unfortunately the reverse is not true; a displacement that entails no deformation can have a nonzero gradient tensor. We have seen in Section 26.3 an example of the decomposition of the displacement into a rotation and a deformation. Let us assume for the moment that the displacement is due to a rigid rotation around a point **r**₀ with rotation vector **Ω**. According to (12.18) the associated displacement is given by

$$\mathbf{u}^{rot}(\mathbf{r}) = \mathbf{\Omega} \times (\mathbf{r} - \mathbf{r}_0). \tag{26.71}$$

Problem b   Show that in component form this displacement is given by

$$\mathbf{u}^{rot}(\mathbf{r}) = \begin{pmatrix} \Omega_y(z - z_0) - \Omega_z(y - y_0) \\ \Omega_z(x - x_0) - \Omega_x(z - z_0) \\ \Omega_x(y - y_0) - \Omega_y(x - x_0) \end{pmatrix}. \tag{26.72}$$

Problem c   Compute the partial derivatives of this vector to show that the associated gradient tensor is given by

$$\mathbf{D}^{rot} = \begin{pmatrix} 0 & \Omega_z & -\Omega_y \\ -\Omega_z & 0 & \Omega_x \\ \Omega_y & -\Omega_x & 0 \end{pmatrix}. \tag{26.73}$$

This is a remarkable result; although the displacement associated with the rotation depends on the position, the associated gradient tensor does not depend on the position. Note also that the gradient tensor does not depend on the point $\mathbf{r}_0$ around which the rotation takes place.

The gradient tensor $\mathbf{D}^{rot}$ is antisymmetric: each element is equal to the element on the other side of the main diagonal with an opposite sign: $D^{rot}_{ij} = -D^{rot}_{ji}$. Stated differently, the sum of each element and the element on the other side of the diagonal is equal to zero:

$$D^{rot}_{ij} + D^{rot}_{ji} = 0. \tag{26.74}$$

In general, any tensor of rank two can be written as the sum of a symmetric tensor and an antisymmetric tensor by using the following identity:

$$D_{ij} = \underbrace{\frac{1}{2}\left(D_{ij} + D_{ji}\right)}_{\text{deformation}} + \underbrace{\frac{1}{2}\left(D_{ij} - D_{ji}\right)}_{\text{rigid rotation}}. \tag{26.75}$$

Problem d   Verify that this identity holds for any matrix **D**. Show that the first term is symmetric and that the second term is antisymmetric.

According to (26.74) the rotational component of the displacement does not contribute to the first term of (26.75). For this reason, the first term of this expression is used to characterize the deformation of the medium, while the second term characterizes the rotational component of the displacement. The strain tensor that characterizes the deformation of the medium ϵ is defined by the first term of (26.75).

Problem e   Show that according to this definition

$$\epsilon_{ij} = \frac{1}{2}\left(\frac{\partial u_j}{\partial x_i} + \frac{\partial u_i}{\partial x_j}\right). \tag{26.76}$$

Problem f   We know that **D** is a tensor because it is the direct product of the gradient vector and the displacement vector, but we have to show that ϵ is a tensor. Do this by showing that when **D** is a tensor of rank two, then $\mathbf{D}^T$ also transforms as a tensor of rank 2. Then use that $\epsilon = (1/2)\left(\mathbf{D} + \mathbf{D}^T\right)$ to show that ϵ is a tensor of rank two as well.

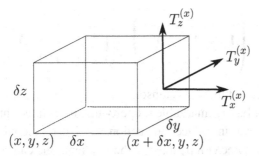

Figure 26.7   The traction acting on a surface perpendicular to the x-direction.

The strain tensor plays a crucial role in continuum mechanics because it is a measure of the degree of deformation in a medium. As shown in the tutorial of Lister and Williams (1983), the partitioning between rotation and shear plays a vital role in structural geology, since it is crucial in the generation of faults and shear zones.

26.10 The stress tensor

In general, when a medium is deformed, reaction forces are operative that tend to counteract the deformation. However, what do we mean by *reaction forces*? A force acts on something, but in a continuous medium there appears to be nothing to act on except a "point" in the medium. Since a point has zero mass, this would lead to an infinite acceleration. This paradox can be resolved by considering a hypothetical cube in the medium as shown in Figure 26.7. The cube has six sides and we consider the force exerted by the rest of the medium on these six sides. Let us focus on the side on the right that is perpendicular to the x-direction. In other words, the normal vector to this side is orientated in the x-direction.

The force on this surface depends on the size of the surface. A meaningful way to describe this is to use the *traction*, which is defined as the force per unit surface area. This traction $\mathbf{T}^{(x)}$ obviously has three components, which are shown in Figure 26.7; these are denoted by $T_x^{(x)}$, $T_y^{(x)}$, and $T_z^{(x)}$. The subscript refers to the component of the traction, the superscript refers to the fact that this is the traction on a surface perpendicular to the x-direction. The component $T_x^{(x)}$ is normal to the surface and corresponds to a normal force, while the components $T_y^{(x)}$ and $T_z^{(x)}$ are parallel to the surface. The latter components are called the shear tractions, because they cause shear motion of the medium.

On the surface perpendicular to the y-direction one also finds a traction operating on that surface with components $T_x^{(y)}$, $T_y^{(y)}$, and $T_z^{(y)}$, and it is not clear at this point whether the traction $\mathbf{T}^{(y)}$ has any relation to the traction $\mathbf{T}^{(x)}$ that acts on a surface perpendicular to the x-direction. The tractions acting on the three surfaces can be grouped in a 3×3 matrix:

$$\sigma \equiv \begin{pmatrix} \vdots & \vdots & \vdots \\ \mathbf{T}^{(x)} & \mathbf{T}^{(y)} & \mathbf{T}^{(z)} \\ \vdots & \vdots & \vdots \end{pmatrix} = \begin{pmatrix} T_x^{(x)} & T_x^{(y)} & T_x^{(z)} \\ T_y^{(x)} & T_y^{(y)} & T_y^{(z)} \\ T_z^{(x)} & T_z^{(y)} & T_z^{(z)} \end{pmatrix}. \tag{26.77}$$

The quantity σ is called the stress tensor.

The stress tensor has a remarkable property that is reminiscent of the property of the gradient, which was introduced in Section 5.1. The gradient is so useful because once one knows the three components of the gradient, one can compute the change of a function in *any* direction. The stress tensor gives the components of the traction to surfaces that are perpendicular to the coordinates' axes. One can show that the traction on a surface that is perpendicular to an arbitrary unit vector $\hat{\mathbf{n}}$ is given by

$$\mathbf{T} = \sigma \cdot \hat{\mathbf{n}}. \tag{26.78}$$

Problem a   The proof of this identity is actually very simple. Consult Section 16.3 of Butkov (1968) or your favorite book on continuum mechanics or mathematical physics (that is, aside from this book) for a proof.

We have called σ the stress tensor, but we have not shown that σ is a tensor. However, one can show that when a quantity is contracted with a tensor and the result is also a tensor, then this quantity must be a tensor as well. In (26.78) we know that $\mathbf{T}$ is a tensor because it is a force (normalized with surface area) and we have shown in Section 26.6 that the force is a tensor of rank one. The normal vector $\hat{\mathbf{n}}$ is also a tensor of rank one because it transforms in the same way as the position vector. Hence, we need to show from (26.78) and the fact that both $\mathbf{T}$ and $\hat{\mathbf{n}}$ are tensors of rank one that σ is a tensor as well.

In components (26.78) is written as

$$T_i = \sigma_{ij} n_j, \tag{26.79}$$

while the same expression in a transformed coordinate system is given by

$$T_i' = \sigma_{ij}' n_j'. \tag{26.80}$$

Before we determine the transformation property of the stress tensor σ, we first express the original vector $\mathbf{v}$ in terms of the transformed vector $\mathbf{v}'$ using the transformation rule

$$v_i' = C_{ij} v_j. \tag{26.47}$$

Problem b   Multiply this expression on the left by C_{ki}^{-1} and sum over i to obtain $C_{ki}^{-1} v_i' = C_{ki}^{-1} C_{ij} v_j$. Use the fact that $\mathbf{C}$ is unitary on the left-hand side and carry out the matrix multiplication on the right-hand side to show that $C_{ik}v'_i = v_k$. Finally rename $i \to j$ and $k \to i$ to obtain

$$v_i = C_{ji}v'_j. \tag{26.81}$$

It is interesting to compare this expression for the original vector $\mathbf{v}$ in terms of the transformed vector $\mathbf{v}'$ with (26.47), which gives the transformed vector given the original vector. The only difference is the order of the subscripts in the coordinate transform $\mathbf{C}$. This is due to the fact that this matrix is unitary so that inversion amounts to interchanging the indices: $C_{ij}^{-1} = C_{ij}^T = C_{ji}$.

Problem c Insert (26.81) into (26.79) in order to express the unprimed vectors $\mathbf{T}$ and $\hat{\mathbf{n}}$ in terms of their transformed vectors to obtain

$$C_{ki}T'_k = \sigma_{ij}C_{lj}n'_l. \tag{26.82}$$

Problem d Write $C_{ki} = C_{ik}^T = C_{ik}^{-1}$ on the left-hand side and multiply by C_{mi} to obtain

$$T'_m = C_{mi}C_{lj}\sigma_{ij}n'_l. \tag{26.83}$$

Problem e We want to compare this expression with (26.80) in order to find σ'. However, the indices are different. As we noted earlier, the names of indices are irrelevant. Rename the indices in (26.83) so that this expression can be compared directly with (26.80) and use this to show that

$$\sigma'_{ij} = C_{ir}C_{js}\sigma_{rs}. \tag{26.84}$$

This is just the transformation rule (26.51) for a tensor of rank two. We have thus shown that the stress tensor is indeed a tensor of rank two. This is due to the fact that when an object is contracted with a tensor to give another tensor, this object must be a tensor as well. Using this property we can use a bootstrap procedure to find higher-order tensors. An important example of this is the elasticity tensor c_{ijkl}, which relates stress to strain:

$$\sigma_{ij} = c_{ijkl}\epsilon_{kl}. \tag{26.85}$$

This expression generalizes the elastic force $F = -kx$ in a spring to continuous media; it is known as Hooke's law. The quantity c_{ijkl} must be a tensor because we know that both the stress σ and the strain ϵ are tensors. This means that c_{ijkl} is a tensor of rank four.

26.11 Why pressure in a fluid is isotropic

Finally we have reached the point where we can use tensors to learn about physics. As a first example we consider the pressure in a fluid. The pressure is the normal force per unit surface area. It is observed that in a fluid this force does not depend on the *orientation* of this surface. This is why the weather forecaster speaks about a pressure of 1020 mbar rather than saying that the pressure is 1015 mbar in the vertical direction and 1025 mbar in the horizontal direction. In this section we discover why pressure is independent of direction.

In order to understand this, we return to the stress tensor (26.77) and we consider the traction acting on a surface perpendicular to the x-direction. As shown in Figure 26.7, $T_x^{(x)}$ gives the traction normal to this surface, while $T_y^{(x)}$ and $T_z^{(x)}$ give the shear traction that acts on this surface. This reasoning can be used for all the surfaces; the diagonal elements $T_x^{(x)}$, $T_y^{(y)}$, and $T_z^{(z)}$ of the stress tensor give the normal tractions, while all the other elements give the shear tractions. In a fluid, there are no shear tractions because a fluid has zero shear strength. This means that in a fluid the stress tensor is diagonal:

$$\sigma = - \begin{pmatrix} p_x & 0 & 0 \\ 0 & p_y & 0 \\ 0 & 0 & p_z \end{pmatrix}. \tag{26.86}$$

The diagonal elements p_i denote the pressure in the three directions. The minus sign reflects the fact that a positive pressure corresponds to a force that is directed *inwards*. In this section we show that the pressure is isotropic. In other words, the diagonal elements of (26.86) are identical.

Let us see what happens to the stress tensor (26.86) when we rotate the coordinate system through 45 degrees around the z-axis. A rotation in two dimensions is given by (26.26). Setting the rotation angle φ to 45 degrees and extending the result to three dimensions gives the following matrix representation of this coordinate transformation:

$$\mathbf{C} = \begin{pmatrix} 1/\sqrt{2} & -1/\sqrt{2} & 0 \\ 1/\sqrt{2} & 1/\sqrt{2} & 0 \\ 0 & 0 & 1 \end{pmatrix}. \tag{26.87}$$

Problem a   Verify that this coordinate transformation is unitary.

Problem b   Use the transformation property of a tensor of rank two (26.84) that the stress tensor (26.86) in the rotated coordinate system is given by

$$\sigma' = -\begin{pmatrix} (p_x + p_y)/2 & (p_x - p_y)/2 & 0 \\ (p_x - p_y)/2 & (p_x + p_y)/2 & 0 \\ 0 & 0 & p_z \end{pmatrix}. \qquad (26.88)$$

In a fluid, the stress tensor is diagonal in *any* coordinate system, since the shear tractions vanish in any coordinate system. This means that the off-diagonal elements of σ' must be equal to zero; hence, $p_x = p_y$.

Problem c Find a suitable coordinate transform to show that $p_x = p_z$.

This means that all the diagonal elements are identical; this quantity is referred to as the pressure: $p_x = p_y = p_z = p$. In an acoustic medium such as a fluid or gas the stress tensor is therefore given by

$$\sigma = -p\,\mathbf{I}, \qquad (26.89)$$

so that pressure is indeed independent of direction.

Note that we have done something truly remarkable; we have derived a physical law ("the pressure is isotropic") from the invariance of a property ("the shear stress in a fluid is zero") under a coordinate transformation.

26.12 Special relativity

One of the most spectacular applications of tensor calculus is the theory of relativity, which describes the physics of objects and fields at very high speeds. The theory of general relativity accounts for the fact that mass in the universe leads to a non-Cartesian structure of space-time (Ohanian and Ruffini, 1976); by definition this cannot be treated with the Cartesian tensors used in this chapter. The theory of special relativity describes how different observers who both use Cartesian coordinate systems that move at great speeds with respect to each other describe the same physical phenomena. A clear physical description of the theory of special relativity is given by Wheeler and Taylor (1966). In this section we use the notation of Muirhead (1973) who uses a complex time variable.

Central to the theory of special relativity is the notion that space and time are intricately linked. The three position variables x, y, and z as well as time t are placed in a four-dimensional vector, called a four-vector:

$$\mathbf{x} = \begin{pmatrix} x \\ y \\ z \\ ict \end{pmatrix}. \qquad (26.90)$$

In this expression c is the speed of light and $i = \sqrt{-1}$. The fact that the last component is complex leads to the surprising result that the length of the four-vector can be negative because

$$|\mathbf{x}|^2 = (\mathbf{x} \cdot \mathbf{x}) = x^2 + y^2 + z^2 - c^2t^2 \qquad (26.91)$$

and there is no reason why the last term cannot dominate the other terms.

Suppose we have one observer who uses unprimed variables, and suppose that another observer moves in the x-direction with a relative velocity v with respect to the first observer. The two coordinate systems of the observers are then related by a *Lorentz transformation* (Muirhead, 1973):

$$\mathbf{L} = \begin{pmatrix} 1/\sqrt{1-v^2/c^2} & 0 & 0 & i\dfrac{v}{c}/\sqrt{1-v^2/c^2} \\ 0 & 1 & 0 & 0 \\ 0 & 0 & 1 & 0 \\ -i\dfrac{v}{c}/\sqrt{1-v^2/c^2} & 0 & 0 & 1/\sqrt{1-v^2/c^2} \end{pmatrix}. \qquad (26.92)$$

Problem a   Show that the Lorentz transform is unitary by showing that $\mathbf{L}^T \mathbf{L} = \mathbf{I}$.

Note that the transpose $\mathbf{L}^T$ is used here and *not* the Hermitian conjugate $\mathbf{L}^\dagger$ that is defined as the transpose and the complex conjugate: $L^\dagger_{ij} \equiv L^*_{ji}$.

Let us first consider the transformation of the space-time vector (26.90). The theory of special relativity states that the four-vector $\mathbf{x}$ is a tensor of rank one. This implies that $|\mathbf{x}|^2$ is a scalar, which means that both observers will assign the same value to this property, so that

$$x'^2 + y'^2 + z'^2 - c^2t'^2 = x^2 + y^2 + z^2 - c^2t^2. \qquad (26.93)$$

Problem b   Use the fact that $\mathbf{x}$ is a tensor of rank one to show that after a Lorentz transformation the x'- and t'-coordinates are given by

$$\left. \begin{aligned} x' &= (x - vt) / \sqrt{1 - v^2/c^2}, \\ t' &= \left(t - \frac{vx}{c^2}\right) / \sqrt{1 - v^2/c^2}. \end{aligned} \right\} \qquad (26.94)$$

Problem c   Verify that this transformation (together with $y = y'$ and $z = z'$) indeed satisfies the identity (26.93).

The transformation (26.94) means that space and time are mixed in a Lorentz transformation. The distinction between space and time largely disappears in the theory of relativity! The terms $\sqrt{1 - v^2/c^2}$ on the right-hand side leads to the clocks of the two observers running at different speeds, the fact that different observers assign different lengths to the same object, and other phenomena that are counterintuitive but have been confirmed experimentally (Wheeler and Taylor, 1966). Note that the contraction terms $\sqrt{1 - v^2/c^2}$ are crucial in Problem c to establish that (26.93) is satisfied.

The theory of special relativity has many surprises. We have already seen in Section 26.7 that a magnetic field does not behave as a tensor of rank one under coordinate transformations in three dimensions. In fact, when we also consider coordinate transformations between moving coordinate systems, the electric field and the magnetic fields are mixed in the sense that what one observer calls an electric field may be seen as a magnetic field by another observer. According to the theory of special relativity (Muirhead, 1973), the magnetic field **B** and the electric field **E** transform as the following tensor of rank two:

$$
\mathbf{F} = \begin{pmatrix} 0 & B_z & -B_y & -iE_x/c \\ -B_z & 0 & B_x & -iE_y/c \\ B_y & -B_x & 0 & -iE_z/c \\ iE_x/c & iE_y/c & iE_z/c & 0 \end{pmatrix}. \tag{26.95}
$$

The momentum of a particle is also given by a four-vector; this momentum-energy vector (Muirhead, 1973) is given by

$$
\mathbf{p} = \begin{pmatrix} p_x \\ p_y \\ p_z \\ iE/c \end{pmatrix}, \tag{26.96}
$$

where E is the energy of the particle. This four-vector transforms as a tensor of rank one under a Lorentz transformation (26.92) (see Muirhead, 1973).

Problem d Use the property that **p** is a tensor of rank one to show that $p^2 - E^2/c^2$ is invariant, where $p^2 = p_x^2 + p_y^2 + p_z^2$.

Problem e In general the energy of a body is given by the energy of that body at rest plus a contribution due to the motion of the body. This rest energy is denoted by E_0. Show that for the particle at rest the norm of the four-vector **p** is given by $|\mathbf{p}|^2 = -E_0^2/c^2$ and use the result of Problem d to show that

$$
E = \sqrt{E_0^2 + p^2 c^2}. \tag{26.97}
$$

This expression holds for any value of the momentum. For the moment we consider a particle that moves much slower than the speed of light ($p = mv \ll mc$). We will show that for such a particle the rest energy E_0 is much larger than the energy pc that is due to the motion.

Problem f   Make a Taylor series expansion of (26.97) in the parameter pc/E_0 to show that for slow speeds

$$E = E_0 + \frac{c^2 p^2}{2E_0} - \frac{c^4 p^4}{8E_0^3} + \cdots . \tag{26.98}$$

The first term on the right-hand side corresponds to the rest energy, which we do not yet know. The second term is quadratic in the momentum. For small velocities, the laws of classical mechanics hold and the kinetic energy is given by $\frac{1}{2}m_0 v^2$. Using the classical relation $p = m_0 v$, the kinetic energy is in classical mechanics given by $p^2/2m_0$. In classical mechanics the terms of order $(cp/E_0)^4$ are ignored. The kinetic energy $p^2/2m_0$ in classical mechanics is therefore equal to the second term on the right-hand side of (26.98).

Problem g   Derive from this statement that

$$E_0 = m_0 c^2 . \tag{26.99}$$

You have just derived the famous relation that relates the rest energy of a particle to its rest mass. Note that the only ingredients that you have used were the fact that the four-vector **p** transforms like a tensor of rank one under coordinate transforms and some elementary results from classical mechanics! The implications of (26.99) are profound. It reflects the fact that matter can be seen as a condensed form of energy. In nuclear reactions this is used for the benefit (or demise) of mankind because the mass of the end-result of some nuclear reactions is smaller than the mass of the ingredients for that reaction. This mass difference is released in the form of radiation and the kinetic energy (heat) of the resulting particles.

Problem h   In the derivation of (26.99) we assumed that the rest energy in the classical limit is much larger than the kinetic energy. Show that the ratio of the second term on the right-hand side of (26.98) to the rest mass is given by

$$\frac{c^2 p^2}{2E_0}/E_0 = \frac{1}{2}(v/c)^2 . \tag{26.100}$$

Problem i   The velocity with which a rocket can overcome the attraction of the Earth is called the *escape velocity* (Kermode et al., 1996); it has the numerical value of $11,84$ km/s. Compute the ratio v/c for a rocket that leaves the Earth at the escape velocity and compute the ratio of the third term to the second term on the right-hand side of (26.98).

The third term on the right-hand side of (26.98) is a relativistic correction term that gives the leading order correction to the classical kinetic energy. At this point you might think that special relativity only has implications for high-velocity bodies in cosmological problems. This is, however, not true; the electrons in atoms move so fast that relativistic effects leave an imprint on microscopic bodies as well. For example, the relativistic correction term $c^4 p^4 / 8 E_0^3$ in (26.98) leads in quantum theory to a measurable shift in the frequency of light emitted by excited hydrogen atoms that is called *fine-structure splitting* (Sakurai, 1978). The Global Positioning System (GPS) is a navigation system that utilizes the radio signals emitted from a number of satellites that move around the Earth at a velocity that is much smaller than the speed of light. According to expression (26.94) relativistic effects cause the clocks of these satellites to run at a slightly different rate $\sqrt{1 - v^2/c^2}$ than earth-bound clocks. This is a tiny effect, but the extreme time-accuracy that is required for the precision of navigation that is required makes it necessary to correct for the change in the clock rate due to relativistic effects (Ashby, 2002).

27

Variational calculus

Variational calculus is concerned with finding functions that minimize or maximize a prescribed property. As an example we showed in Section 10.2 that the shortest curve in a plane that connects two points is a straight line. This is, of course, a problem with an obvious answer, but often the answer is not obvious. Maximizing or minimizing properties of functions is not only of interest in mathematical problems such as, for example, finding the shortest curve on a sphere between two points as shown in Section 27.3. Variational calculus also sheds a new light on physics: we show in Section 27.5 that Newton's law can be derived from minimizing the time-average of the difference between kinetic and potential energy, and use variational calculus in Section 27.6 to show that rays are curves that render the travel time stationary. In the remainder of the section we treat optimization problems that are subject to a constraint. An example of this is shown in Section 27.9 where we derive the shape of the Gateway Arch in St. Louis. This last example shows that variational calculus is not just a mathematical topic; it has numerous applications in engineering.

27.1 Designing a can

In many applications the ratio between the volume and surface area of an object is an important parameter. Chemical reactivity is aided by relatively large surface areas, for example. Manufacturers of soda-drink cans want to minimize the amount of aluminum to use. In theory, both applications ask for spherical objects, as these have the largest volume to surface area ratio. Clearly, for drinks, a spherical can is not practical: how would you set your opened drink on the table, for example? We therefore restrict our can design for the moment to cans with a cylindrical shape. A squat can with a small height and a large radius has a small volume, but a tall can with a large height but a small radius cannot contain much either. Clearly there is a specific shape of the can that maximizes the volume for a given surface area. Consider the can shown in Figure 27.1; the top and bottom each have a surface

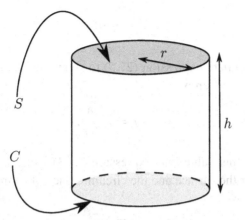

Figure 27.1 Definition of the surface area S, height h, circumference C, and radius r of a can.

area S, a height denoted by h, and a circumpherence denoted by C. The total surface area A is twice the surface area of the bottom, plus the surface area hC of the side:

$$A = 2S + hC. \tag{27.1}$$

The volume V of the can is the product of the surface area of the bottom times the height:

$$V = hS. \tag{27.2}$$

Problem a   We want to eliminate the height h from the problem. Use expression (27.1) to show that

$$h = \frac{(A - 2S)}{C}, \tag{27.3}$$

and use this result to show that

$$V = \frac{(A - 2S)\,S}{C}. \tag{27.4}$$

Problem b   For the moment we assume that the can has the shape of cylinder with radius r. Express the surface area S and the circumference C in this radius, and show that the volume is given by

$$V(r) = \frac{r}{2}(A - 2\pi r^2). \tag{27.5}$$

The volume of the can is now given as a function of r. The value of the radius that maximizes the volume can be found by requiring that the derivative of the volume with respect to the radius vanishes:

$$\frac{dV}{dr} = 0. \tag{27.6}$$

Problem c   Differentiate expression (27.5) to show that the radius that gives the
largest volume is given by

$$r = \sqrt{\frac{A}{6\pi}}. \tag{27.7}$$

Problem d   Insert this value into expression (27.3) while using the appropriate
expressions for the surface and the circumference of a circle to show that

$$h = \sqrt{\frac{2A}{3\pi}}, \tag{27.8}$$

and that the optimal ratio of the height and the radius is given by

$$\frac{h}{r} = 2. \tag{27.9}$$

This expression defines the aspect ratio of the can; a can whose height is equal
to its diameter $2r$ holds the largest volume for a given surface area. The treatment of this section is called an *optimization problem*. In such a problem some
quantity is being optimized. Often a straightforward differentiation such as in
expression (27.6) gives the optimum value of a parameter. Optimization problems
arise naturally in many design problems, but also in economic problems and other
situations where a strategy needs to be formulated. In Chapter 22 we will encounter
optimization problems, subject to constraints on model and data (mis)fit.

27.2 Why are cans round?

The title to this section may sound a bit silly, but is it obvious that a can with
a circular top (and bottom) contains the largest volume? Why should the cross
section of a can not be ellipsoidal or square? In the previous section we optimized
the shape of a cylindrical can by the simple differentiation (27.6) with respect to
the single parameter r. However, for an arbitrary shape of the can, the radius r is
a function of the angle φ as shown in Figure 27.2. In that situation the radius is
not a single parameter, and the problem is to find the *function* $r(\varphi)$ that maximizes
the volume of the can. This is a different problem that cannot be solved by simple
differentiation, because the radius $r(\varphi)$ is a function rather than a constant.

It follows from expression (27.4) that for a given value of the surface S, the
volume is largest when the circumference C is smallest. The problem of finding the
function $r(\varphi)$ that maximizes the volume thus is equivalent to finding the function

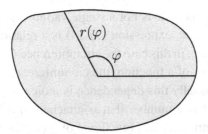

Figure 27.2   Top view of a can with its radius a function of the angle φ.

$r(\varphi)$ that minimizes the circumference C for a given surface area S. Let us first express the circumference C in the function $r(\varphi)$.

Problem a   The radius vector $\mathbf{r}$ is given by

$$\mathbf{r} = \begin{pmatrix} r(\varphi)\cos\varphi \\ r(\varphi)\sin\varphi \end{pmatrix}. \tag{27.10}$$

Show that

$$\frac{d\mathbf{r}}{d\varphi} = \begin{pmatrix} \dot{r}(\varphi)\cos\varphi - r(\varphi)\sin\varphi \\ \dot{r}(\varphi)\sin\varphi + r(\varphi)\cos\varphi \end{pmatrix}, \tag{27.11}$$

with

$$\dot{r} \equiv \frac{dr}{d\varphi}. \tag{27.12}$$

Previously we have used the dot exclusively for a time derivative, but in this section we use it to designate a derivative with respect to the independent parameter. In this section the independent parameter is the longitude φ.

Problem b   An increment ds along the circumference that corresponds with an increment $d\varphi$ satisfies $(ds/d\varphi)^2 = (d\mathbf{r}/d\varphi \cdot d\mathbf{r}/d\varphi)$. Use this relation and expression (27.11) to derive that

$$ds = \sqrt{\dot{r}^2 + r^2}\, d\varphi. \tag{27.13}$$

Integrating this expression over a full circle gives the circumference:

$$C = \int_0^{2\pi} \sqrt{\dot{r}^2 + r^2}\, d\varphi. \tag{27.14}$$

A function constitutes a recipe that maps one variable onto another variable. An example is the function $V(r)$ in expression (27.5) that gives the volume of a cylindrical can for a given value of the radius. The integral (27.14) is different. The quantity r in the right-hand side is not a single variable, since r now is a function itself of the angle φ. In fact, expression (27.14) is a relation that maps a function $r(\varphi)$ onto a single variable; in this case the circumference C. In order to contrast this from the normal behavior of a function, the circumference C is called a *functional* of the function $r(\varphi)$. Usually this dependence is indicated with the notation $C[r]$. In general, a functional is a number that is attached to a function. For example, the mean $\bar{f}$ of a function $f(x)$ is a functional of $f(x)$. In general, many different functionals can be attached to a function. In our problem the functional of interest is the circumference given in expression (27.14).

Here we want to find the function $r(\varphi)$ that minimizes the circumference for a given surface area. Let us find the solution by examining expression (27.14). The quantity $\dot{r}^2$ is always positive. This means that the circumference C is smallest when $\dot{r} = 0$, because this is the value that minimizes $\dot{r}^2$. The condition $\dot{r} = 0$ implies that $r = constant$, which in turn states that the base of the can is a circle; hence, the can that holds the largest volume for a given surface area is indeed cylindrical.

In this particular example we could see the solution rather easily. However, the reasoning used is not very rigorous, and in more complex problems the solution cannot be seen in a simple way. In this chapter we develop systematic tools to find the maximum or minimum of a functional. The mathematics that is involved is called *variational calculus*. We return to the problem of minimizing the circumference in Section 27.8. In this chapter we cover the basic principles of variational calculus. Smith (1974) gives a comprehensive and clear overview of this topic.

By the way, why don't cans have a hexagonal shape? In this way they can be stacked without having voids between the cans, and they are probably stronger. This shows that in design there often are different design criteria that must be balanced.

27.3 The great-circle

We actually introduced variational calculus in Section 10.2 where we showed that the shortest distance between points in a plane is a straight line. We encourage the reader to read Section 10.2 before proceeding. In this section we discuss the more challenging problem of finding the shortest distance between two points on a sphere.

When you live on a sphere, the shortest distance between two points is not a straight line. On an intercontinental flight between Europe and the United States, the flight path usually is curved toward the north pole. This is illustrated by the flight path from Amsterdam to San Francisco in Figure 27.3. Amsterdam is located at about 52 degrees north, the latitude of San Francisco is about 37 degrees north, while the flight flies over Greenland at about 60 degrees north. The reason for this

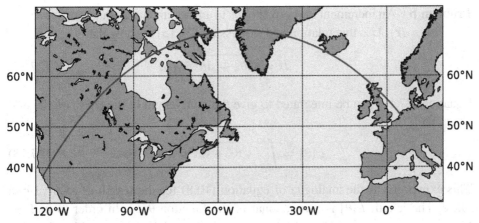

Figure 27.3   The shortest path of a flight from Amsterdam to San Francisco on a map with a Mercator projection.

apparent detour to the north is that the line with the shortest distance between two points on the sphere is not a straight line. This curve lies on a *great-circle*. The equator is an example of a great-circle, but on a sphere there are many other great-circles. In this section we determine the shortest curve that joins two points (φ_1, θ_1) and (φ_2, θ_2) on a sphere, where the angles φ and θ are the longitude and colatitude on the sphere. These angles are the ones used in spherical coordinates as defined in Section 4.1.

In this section we treat the longitude φ as the independent parameter and consider the colatitude $\theta(\varphi)$ as a function of the longitude. In order to solve this problem we first need to compute the length of the curve that joins the two endpoints for a given function $\theta(\varphi)$. For simplicity we consider the length of the curve on the unit sphere. This means that the radius r is equal to 1 and that the distance between the two points on the sphere is measured in radians.

Problem a   The position vector on the unit sphere is given by the vector $\hat{\mathbf{r}}$ in expression (4.7). Show that an increment in this vector due to a change $d\varphi$ in the independent parameter is given by

$$d\hat{\mathbf{r}} = \begin{pmatrix} \cos\varphi\cos\theta\,\dot{\theta} - \sin\varphi\sin\theta \\ \sin\varphi\cos\theta\,\dot{\theta} + \cos\varphi\sin\theta \\ -\sin\theta\,\dot{\theta} \end{pmatrix} d\varphi, \qquad (27.15)$$

with

$$\dot{\theta} \equiv \frac{d\theta}{d\varphi}. \qquad (27.16)$$

Problem b   An increment $d\varphi$ corresponds to an increments ds that satisfies $ds^2 = (d\hat{\mathbf{r}} \cdot d\hat{\mathbf{r}})$. Use this relation and expression (27.15) to show that

$$ds = \sqrt{\dot{\theta}^2 + \sin^2\theta}\, d\varphi. \qquad (27.17)$$

Equation (27.17) can be integrated to give the total length of the curve on the unit sphere:

$$L[\theta] = \int_{\varphi_1}^{\varphi_2} \sqrt{\dot{\theta}^2 + \sin^2\theta}\, d\varphi. \qquad (27.18)$$

This expression is the analogue of equation (10.8) for the length of a curve in a plane. The length $L[\theta]$ is a functional of the function $\theta(\varphi)$. In order to find the function $\theta(\varphi)$ that minimizes the length, we perturb this function with a function $\epsilon(\varphi)$ and require that the first-order variation of the length L with the perturbation $\epsilon(\varphi)$ vanishes. The perturbed curve is shown by the dashed line in Figure 27.4. Since the perturbed curve must go through the endpoints (φ_1, θ_1) and (φ_2, θ_2), the perturbation must satisfy the following boundary conditions

$$\epsilon(\varphi_1) = \epsilon(\varphi_2) = 0. \qquad (27.19)$$

Problem c   Show that under the perturbation $\theta \to \theta + \epsilon$ the following perturbations hold to first order in ϵ:

$$\dot{\theta}^2 \to \dot{\theta}^2 + 2\dot{\theta}\dot{\epsilon}, \qquad (27.20)$$

$$\sin\theta \to \sin(\theta + \epsilon) = \sin\theta + \epsilon\cos\theta, \qquad (27.21)$$

$$\sin^2\theta \to \sin^2(\theta + \epsilon) = \sin^2\theta + 2\epsilon\sin\theta\cos\theta. \qquad (27.22)$$

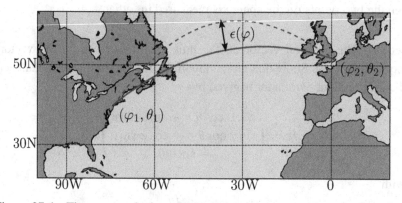

Figure 27.4   The unperturbed curve $\theta(\varphi)$ that joins two points on the sphere (solid line), the perturbation $\epsilon(\varphi)$, and the perturbed curve (dashed line).

Use these results to show that to first order in the perturbation ϵ the integrand of expression (27.18) is perturbed in the following way

$$\sqrt{\dot{\theta}^2 + \sin^2\theta} \rightarrow \sqrt{\dot{\theta}^2 + \sin^2\theta} + \frac{2\dot{\theta}\dot{\epsilon} + \epsilon\sin 2\theta}{2\sqrt{\dot{\theta}^2 + \sin^2\theta}}. \tag{27.23}$$

Integrating this expression between the endpoints φ_1 and φ_2 gives the following first-order perturbation of the length due to the perturbation $\epsilon(\varphi)$:

$$\delta L = \frac{1}{2} \int_{\varphi_1}^{\varphi_2} \frac{2\dot{\theta}\dot{\epsilon} + \epsilon\sin 2\theta}{\sqrt{\dot{\theta}^2 + \sin^2\theta}}\, d\varphi. \tag{27.24}$$

Problem d This perturbation depends on both ϵ and $\dot{\epsilon}$. The derivative $\dot{\epsilon}$ can be eliminated by carrying out an integration by parts of the first term of the numerator using the boundary conditions (27.19). Use this to show that

$$\delta L = \frac{1}{2} \int_{\varphi_1}^{\varphi_2} \left\{ -\frac{d}{d\varphi}\left(\frac{2\dot{\theta}}{\sqrt{\dot{\theta}^2 + \sin^2\theta}} \right) + \frac{\sin 2\theta}{\sqrt{\dot{\theta}^2 + \sin^2\theta}} \right\} \epsilon(\varphi)\, d\varphi. \tag{27.25}$$

For the curve that minimizes the distance, the length L is stationary and its first-order variation δL with a perturbation $\epsilon(\varphi)$ vanishes. This can only hold for every perturbation $\epsilon(\varphi)$ when

$$-\frac{d}{d\varphi}\left(\frac{2\dot{\theta}}{\sqrt{\dot{\theta}^2 + \sin^2\theta}} \right) + \frac{\sin 2\theta}{\sqrt{\dot{\theta}^2 + \sin^2\theta}} = 0. \tag{27.26}$$

This is a differential equation for the function $\theta(\varphi)$ that gives the shortest distance between two points on a sphere. This nonlinear differential equation does not have an obvious solution. In variational calculus the resulting differential equations often are complicated, so that solving the differential equation can be the hardest part of the problem.

Problem e Carry out the differentiation in the first term of expression (27.26) to show that this expression can be rewritten as

$$\ddot{\theta} - 2\cot\theta\,\dot{\theta}^2 - \frac{1}{2}\sin 2\theta = 0. \tag{27.27}$$

Problem f Despite the complicated appearance of the differential equation (27.26), or the equivalent form (27.27), this equation has the following simple solution

$$\theta(\varphi) = \arctan\left(\frac{A}{\cos(\varphi + B)} \right). \tag{27.28}$$

The integration constants A and B follow from the requirement that the curve goes through the fixed endpoints; hence, these constants follow from the boundary conditions $\theta(\varphi_1) = \theta_1$ and $\theta(\varphi_2) = \theta_2$. Show that the solution (27.28) indeed satisfies the differential equation (27.27). This is a lengthy calculation that requires extensive use of trigonometric identities.

The solution (27.28) gives the curve that minimizes the distance between two points on the sphere. Once the constants A and B are known, this solution can be used to compute the great-circle that goes through two fixed points. However, this function gives little insight into the shape of a great-circle. In order to understand the solution better, we rewrite the solution (27.28) as

$$\tan \theta = \frac{A}{\cos(\varphi + B)}. \tag{27.29}$$

Let us analyze this solution in Cartesian coordinates (x, y, z).

Problem g   Use Figure 4.1 to show that

$$\tan \theta = \frac{\sqrt{x^2 + y^2}}{z}. \tag{27.30}$$

Problem h   Show that

$$\cos(\varphi + B) = \cos \varphi \, (\cos B - \tan \varphi \sin B). \tag{27.31}$$

Use Figure 4.1 to show that $\cos \varphi = x/\sqrt{x^2 + y^2}$ and $\tan \varphi = y/x$, and use these results to derive that

$$\cos(\varphi + B) = \frac{x \cos B - y \sin B}{\sqrt{x^2 + y^2}}. \tag{27.32}$$

Problem i   Insert these relations into expression (27.29) and show that the solution for the great-circle in Cartesian coordinates is given by

$$x \cos B - y \sin B = Az. \tag{27.33}$$

Since A and B are constant, this is the equation for a plane that goes through the origin. The points (x, y, z) on a great-circle are confined to that plane. Since the great-circle is also confined to the surface of the unit sphere, we can conclude that the great-circle is given by the intersection of a plane that goes through the center of the sphere and the endpoints of the curve, with the unit sphere. This solution is shown in Figure 27.5.

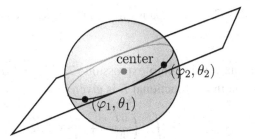

Figure 27.5   The great-circle as the intersection of the surface of the sphere with a plane that is spanned by the center of the sphere, and the two points that define the great-circle (indicated by black circles).

27.4 Euler–Lagrange equation

The approach taken in the Sections 10.2 and 27.3 was consistent. The problems in both sections were reformulated into equations (10.12) and (27.25) that both give a first-order perturbation of the form $\delta L = \int (\cdots) \epsilon(x)dx$. The requirement that this first-order perturbation vanishes for every perturbation $\epsilon(x)$ led to the requirement that the terms indicated by dots must be equal to zero. The similarity of the treatments in these sections suggests a systematic approach.

Let us consider a functional L of a function $y(\xi)$:

$$L[y] = \int_{\xi_1}^{\xi_2} F(y, \dot{y}, \xi)\, d\xi. \qquad (27.34)$$

In the functional (10.8) for the length of a line in a plane, the x-coordinate is the independent parameter, whereas in the functional (27.18) for the length of a curve on the sphere, φ is the independent parameter. Whatever the independent parameter is, we indicate it in this section with the variable ξ. The functional $L[y]$ depends on the function $y(\xi)$. The derivative of this function with respect to ξ is denoted by $\dot{y}$. The integrand F is assumed to depend on y, $\dot{y}$, and the independent variable ξ.

In order to find the function $y(\xi)$ that renders the functional $L[y]$ stationary, we perturb the function with a perturbation $\epsilon(\xi)$. Under this perturbation $y \to y + \epsilon$, and $\dot{y} \to \dot{y} + \dot{\epsilon}$. Since $y(\xi)$ is assumed to have fixed values at the endpoints of the interval, the perturbation must vanish at these endpoints:

$$\epsilon(\xi_1) = \epsilon(\xi_2) = 0. \qquad (27.35)$$

Problem a   Use a first-order Taylor expansion to show that under this perturbation the integrand F is to first order perturbed in the following way:

$$F(y, \dot{y}, \xi) \to F(y + \epsilon, \dot{y} + \dot{\epsilon}, \xi) = F(y, \dot{y}, \xi) + \frac{\partial F}{\partial y}\epsilon + \frac{\partial F}{\partial \dot{y}}\dot{\epsilon}. \qquad (27.36)$$

Note that in this Taylor expansion y and $\dot{y}$ are treated as independent parameters.

Problem b   Insert this relation into expression (27.34) and derive that the first-order perturbation in the functional L is given by

$$\delta L = \int_{\xi_1}^{\xi_2} \left(\frac{\partial F}{\partial y} \epsilon + \frac{\partial F}{\partial \dot{y}} \dot{\epsilon} \right) d\xi. \tag{27.37}$$

Problem c   Carry out an integration by parts of the second term, and show that with the boundary conditions (27.35) the first-order perturbation is given by

$$\delta L = \int_{\xi_1}^{\xi_2} \left(\frac{\partial F}{\partial y} - \frac{d}{d\xi} \left(\frac{\partial F}{\partial \dot{y}} \right) \right) \epsilon(\xi)\, d\xi. \tag{27.38}$$

This first-order perturbation vanishes for any perturbation $\epsilon(\xi)$ when

$$\frac{\partial F}{\partial y} - \frac{d}{d\xi} \left(\frac{\partial F}{\partial \dot{y}} \right) = 0. \tag{27.39}$$

This equation is called the *Euler–Lagrange equation*; it constitutes a differential equation for the function y that renders the functional L stationary.

The derivation of the Euler–Lagrange equation is not particularly difficult. However, when using this equation it is important to keep track of the different types of derivatives. The partial derivatives $\partial/\partial y$ and $\partial/\partial \dot{y}$ treat y and $\dot{y}$ as independent parameters. This may be confusing because $\dot{y}$ is the derivative of y with respect to ξ. The notation with partial derivatives is used to denote the dependence of F on y and $\dot{y}$, respectively. The derivative $d/d\xi$ is a total derivative. This means that the derivative acts on all quantities and the chain law must be used to compute this derivative.

Let us see how this works for the particular example of the minimization of the length of the curve in Section 10.2. According to expression (10.8) we have in that case

$$F(y, \dot{y}, x) = \sqrt{1 + \dot{y}^2}, \tag{27.40}$$

where we now denote h by y. Since F does not depend explicitly on y, the partial derivative of F with respect to y vanishes: $\partial F/\partial y = 0$. Furthermore, $\partial F/\partial \dot{y} = \dot{y}/\sqrt{1 + \dot{y}^2}$. Inserting these results in the Euler–Lagrange equation and using that x plays the role of the independent variable ξ gives

$$0 - \frac{d}{dx} \left(\frac{\dot{y}}{\sqrt{1 + \dot{y}^2}} \right) = 0. \tag{27.41}$$

This equation has the solution $\dot{y} = constant$, which corresponds to a straight line. The Euler–Lagrange equation thus gives the same result as the treatment of Section 10.2.

As a second example let us consider the determination of the curve with the shortest distance between two points on a sphere. According to expression (27.18), F is in that case given by

$$F(\theta, \dot{\theta}, \varphi) = \sqrt{\dot{\theta}^2 + \sin^2 \theta}. \tag{27.42}$$

In this problem, φ plays the role of the independent variable ξ, and the function y is denoted by θ. When encountering a problem with a different notation, it is often useful to reformulate the Euler–Lagrange equation for the variables that are used. In this particular problem the Euler–Lagrange equation (27.39) is given by

$$\frac{\partial F}{\partial \theta} - \frac{d}{d\varphi}\left(\frac{\partial F}{\partial \dot{\theta}}\right) = 0. \tag{27.43}$$

The partial derivatives with respect to θ and $\dot{\theta}$ are given by

$$\frac{\partial F}{\partial \theta} = \frac{\sin \theta \cos \theta}{\sqrt{\dot{\theta}^2 + \sin^2 \theta}},$$

$$\frac{\partial F}{\partial \dot{\theta}} = \frac{\dot{\theta}}{\sqrt{\dot{\theta}^2 + \sin^2 \theta}}. \tag{27.44}$$

Inserting these values in the Euler–Lagrange equation (27.43) gives

$$\frac{\sin \theta \cos \theta}{\sqrt{\dot{\theta}^2 + \sin^2 \theta}} - \frac{d}{d\varphi}\left(\frac{\dot{\theta}}{\sqrt{\dot{\theta}^2 + \sin^2 \theta}}\right) = 0. \tag{27.45}$$

Problem d   Verify that this expression is identical to equation (27.26).

The examples shown here imply that the Euler–Lagrange equation gives the same results as we derived in the previous sections. We have covered the simplest case where $y(\xi)$ depends on one variable only and where the endpoints of the function y are fixed. The treatment can be generalized to include functions y that depend on more than one independent variable, functions whose endpoints are not necessarily fixed, and functionals that also contain the second derivative $\ddot{y}$. Details can be found in Smith (1974).

When one wants to minimize a functional that depends on the trajectory $\mathbf{r}(\xi)$ that depends on three spatial coordinates. In that case, the functional depends on three spatial coordinates captured in $\mathbf{r}$:

$$F = F[\mathbf{r}], \tag{27.46}$$

where $\mathbf{r}$ is a function of the independent parameter ξ. The curve that renders this functional stationary can be found by changing the trajectory $\mathbf{r}(\xi)$ with a vector perturbation $\boldsymbol{\epsilon}(\xi)$, and by requiring that to first order in $\boldsymbol{\epsilon}(\xi)$ the functional does not change. This perturbation in general has three components as well. However, nothing keeps us from perturbing one of the components of $\boldsymbol{\epsilon}$ only, let us denote this component with ϵ_i. Let the components of the vector $\mathbf{r}$ be denoted by y_i. The derivation in expressions (27.35)–(27.39) can be repeated for that case, with the only change that the variable ϵ must be replaced by ϵ_i. This gives the Euler–Lagrange equation for each of the components y_i:

$$\frac{\partial F}{\partial y_i} - \frac{d}{d\xi}\left(\frac{\partial F}{\partial \dot{y}_i}\right) = 0. \tag{27.47}$$

27.5 Lagrangian formulation of classical mechanics

In this section we consider a particle with mass m that moves in one dimension. The mass m is not necessarily constant. In order to relate this problem to the notation of the previous section, we denote the location of the particle with the coordinate y. The particle starts at a location y_1 at time t_1 and moves to the endpoint y_2 where it arrives at time t_2. The kinetic energy of the particle is given by $E_K = m\dot{y}^2/2$, where the dot denotes the time derivative: $\dot{y} \equiv dy/dt$. The time t now plays the role of the independent variable ξ. The potential energy of the particle is given by $E_P = V(y)$, where V describes the potential in which the particle moves. Let us consider the difference of the kinetic and potential energy integrated over time:

$$A[y] \equiv \int_{t_1}^{t_2} \left(\frac{1}{2}m\dot{y}^2 - V(y)\right) dt, \tag{27.48}$$

this quantity is called the *action* (Goldstein, 1980). At this moment there is no specific reason yet to compute the action, but we will see that the trajectory $y(t)$ that renders the action stationary has a special meaning.

In the notation of equation (27.34) the integrand of the action is given by

$$L(y, \dot{y}, t) = \frac{1}{2}m\dot{y}^2 - V(y), \tag{27.49}$$

this quantity is called the *Lagrangian* (Goldstein, 1980).

Problem a   Verify that the partial derivatives of the Lagrangian with respect to y and $\dot{y}$ are given by

$$\frac{\partial L}{\partial y} = -\frac{\partial V}{\partial y},$$

$$\frac{\partial L}{\partial \dot{y}} = m\dot{y}. \tag{27.50}$$

Problem b Insert these results in the Euler–Lagrange equation to show that the trajectory $y(t)$ that renders the action stationary is given by

$$\frac{d}{dt}(m\dot{y}) = -\frac{\partial V}{\partial y}. \tag{27.51}$$

The derivative $-\partial V/\partial y$ is simply the force F that acts on the particle. Since $\dot{y}$ is the velocity, $m\dot{y}$ is the momentum, and equation (27.51) is equivalent to Newton's law: $d(mv)/dt = F$. This result gives us a new way to interpret the motion of a particle in classical mechanics. The Lagrangian (27.49) is the difference of the kinetic and potential energy. This means that in classical mechanics a particle follows a trajectory such that the time-averaged difference between the kinetic and potential energy is minimized. This is one example of the variational formulation of classical mechanics. More details and many examples can be found in textbooks (Goldstein, 1980; Lanczos, 1970).

Let us redo this problem in three dimensions. In that case the action is given by

$$A[\mathbf{y}] \equiv \int_{t_1}^{t_2} \left(\frac{1}{2}m\dot{\mathbf{y}}^2 - V(\mathbf{y}) \right) dt, \tag{27.52}$$

where $\mathbf{y}(t)$ describes the trajectory of the particle as a function of time.

Problem c Use the Euler–Lagrange equation (27.47) to show that this functional is stationary when

$$\frac{d}{dt}(m\dot{y}_i) = -\frac{\partial V}{\partial y_i}. \tag{27.53}$$

In vector-form this expression can be written as

$$\frac{d}{dt}(m\dot{\mathbf{y}}) = -\nabla V. \tag{27.54}$$

This is Newton's law in three dimensions for a force that is related to the potential by $\mathbf{F} = -\nabla V$.

27.6 Rays are curves of stationary travel time

In Section 11.4 we discussed geometric ray theory as a high-frequency approximation to solutions of the wave equation. In this section we develop an alternative view on ray theory. When a wave travels along a segment of a ray with length ds, and if the wave speed is denoted by c, then the time needed to cover this distance is given by $dt = c^{-1}ds$. In this section we use the symbol u for slowness, the reciprocal of speed:

$$u \equiv 1/c. \qquad (27.55)$$

Using this definition, the total travel time along the ray is given by

$$T = \int u(\mathbf{r})\, ds. \qquad (27.56)$$

In this section we consider the trajectories that render the travel time stationary.

The integral (27.56) suggests use of the arc-length s as independent parameter. This is, however, not a good idea since it is not known how long the ray that connects a source to a receiver will be. For this reason the endpoint condition (27.35) cannot be applied because ξ_2 is unknown. The travel time cannot be used as independent parameter either, because it is the travel time that we seek to optimize. Instead we will use the relative arc-length as independent parameter. Let the total length of the ray be denoted by S. The relative arc-length is defined as

$$\xi \equiv s/S. \qquad (27.57)$$

At the starting point of the ray $\xi = 0$, while at the endpoint $\xi = 1$.

At this moment we do not know S yet, but this quantity can be related to the curve $\mathbf{r}(\xi)$ by using that

$$ds = \left|\frac{d\mathbf{r}}{d\xi}\right| d\xi = |\dot{\mathbf{r}}|\, d\xi = \sqrt{\dot{x}^2 + \dot{y}^2 + \dot{z}^2}\, d\xi. \qquad (27.58)$$

Problem a   Use these results to show that the travel time can be written as

$$T[\mathbf{r}] = \int_0^1 F(\mathbf{r}, \dot{\mathbf{r}})\, d\xi, \qquad (27.59)$$

with

$$F(\mathbf{r}, \dot{\mathbf{r}}) = u(\mathbf{r})\sqrt{\dot{x}^2 + \dot{y}^2 + \dot{z}^2}. \qquad (27.60)$$

Problem b   The curve that renders the travel time stationary follows from the Euler–Lagrange equation (27.47). Show that

$$\frac{\partial F}{\partial x} = \frac{\partial u}{\partial x}\sqrt{\dot{x}^2 + \dot{y}^2 + \dot{z}^2}, \qquad (27.61)$$

and

$$\frac{\partial F}{\partial \dot{x}} = u\frac{\dot{x}}{\sqrt{\dot{x}^2 + \dot{y}^2 + \dot{z}^2}}. \qquad (27.62)$$

Problem c Use these results to show that the Euler–Lagrange equation (27.47) for the variable x is given by

$$\frac{d}{d\xi}\left(u\frac{\dot{x}}{\sqrt{\dot{x}^2 + \dot{y}^2 + \dot{z}^2}}\right) = \frac{\partial u}{\partial x}\sqrt{\dot{x}^2 + \dot{y}^2 + \dot{z}^2}. \tag{27.63}$$

Similar equations hold for the variables y and z. In order to interpret this equation, we introduce the unit vector $\hat{\mathbf{n}}$, whose x-component is defined by

$$n_x \equiv \frac{\dot{x}}{\sqrt{\dot{x}^2 + \dot{y}^2 + \dot{z}^2}}, \tag{27.64}$$

with a similar definition for the y- and z-components.

Problem d Show that this vector is of unit length: $|\hat{\mathbf{n}}| = 1$.

Problem e Expression (27.58) can be used to convert the ξ-derivative in expression (27.63) into a derivative along the curve, by using

$$\frac{d}{d\xi} = \frac{ds}{d\xi}\frac{d}{ds} = \sqrt{\dot{x}^2 + \dot{y}^2 + \dot{z}^2}\,\frac{d}{ds}. \tag{27.65}$$

Use these results to show that the Euler–Lagrange equation (27.63) can be written as

$$\frac{d}{ds}(un_x) = \frac{\partial u}{\partial x}. \tag{27.66}$$

Similar expressions hold for the y- and z-components of the Euler–Lagrange equation. In vector-form the Euler–Lagrange equation can be written as

$$\frac{d}{ds}(u\hat{\mathbf{n}}) = \nabla u. \tag{27.67}$$

With definition (27.64), the unit vector $\mathbf{n}$ can be rewritten in the following way

$$\hat{\mathbf{n}} = \frac{d\mathbf{r}/d\xi}{\sqrt{\dot{x}^2 + \dot{y}^2 + \dot{z}^2}} = \frac{d\mathbf{r}}{d\xi}\frac{d\xi}{ds} = \frac{d\mathbf{r}}{ds}, \tag{27.68}$$

where expression (27.65) is used in the second identity. Using this result in equation (27.67), and using that $u = c^{-1}$ gives

$$\frac{d}{ds}\left(\frac{1}{c}\frac{d\mathbf{r}}{ds}\right) = \nabla\left(\frac{1}{c}\right). \tag{27.69}$$

This equation is called the *equation of kinematic ray tracing*.

The equation of kinematic ray tracing can be derived from the eikonal equation (11.25) that was derived from the high-frequency approximation to the wave equation (Aki and Richards, 2002). The eikonal equation (11.25) describes the propagation of wavefronts, while the equation of kinematic ray tracing (27.69) describes the propagation of rays. These equations are equivalent and provide a complementary view of geometric ray theory. In this section we derived the equation of kinematic ray tracing from the requirement that the travel time is stationary. This means that rays are curves that render the travel time stationary. Most often, rays are curves that minimize the travel time. However, seismic waves that reflect once off the Earth's surface are so-called *minimax arrivals*. The travel time increases when the ray is perturbed in one direction and it decreases when the ray is perturbed in another direction. In that case the travel time has a saddle point as a function of the reflection point at the surface. Whether a ray is a minimum time arrival or a so-called minimax arrival can be detected in the phase of the arriving waves. Observations of this phenomenon are shown by Choy and Richards (1975).

It is interesting to convert the derivative d/ds in expression (27.69) into a time derivative by using that

$$\frac{d}{ds} = \frac{dt}{ds}\frac{d}{dt} = \frac{1}{c}\frac{d}{dt}. \tag{27.70}$$

This changes the equation of kinematic ray tracing into

$$\frac{d}{dt}\left(\frac{1}{c^2}\frac{d\mathbf{r}}{dt}\right) = c\nabla\left(\frac{1}{c}\right). \tag{27.71}$$

This expression can be compared with Newton's law (27.54) for a particle with a mass m that may vary with time:

$$\frac{d}{dt}\left(m\frac{d\mathbf{r}}{dt}\right) = \mathbf{F}. \tag{27.72}$$

These equations are identical when $1/c^2$ in (27.71) is equated to the mass m in (27.72), and when $c\nabla(1/c)$ takes the role of the force $\mathbf{F}$ in Newton's law. In classical mechanics it is the force that makes a trajectory curve. Therefore, in geometric ray theory it is the velocity gradient that makes a ray curve. The close analogy between geometric ray theory and classical mechanics has led to a wide body of theory that utilizes the same mathematical tools for classical mechanics and geometric ray theory (Bennett, 1973; Farra and Madariaga, 1987).

27.7 Lagrange multipliers

Often minimization involves a constraint that the solution should satisfy. In Section 27.9 we will consider the problem of a wire that is suspended between two points. The shape of the wire is dictated by the requirement that its potential energy in the gravitational field is minimized. Obviously, the potential energy is smallest when the wire curves down as much as possible. The constraint that the wire has a given length restricts the downward displacement of the wire. This amounts to a *constrained optimization problem* where the potential energy is minimized under the constraints that the wire has a fixed length. Problems like this can be solved with a technique called *Lagrange multipliers*. In this section we introduce this technique first with a finite-dimensional problem, and then generalize it to problems with infinitely many degrees of freedom.

To fix our mind we consider the following question: what is the point on the surface of a ball that has the lowest potential energy in a gravitational field? This is a trivial problem, and the solution obviously is given by the point at the bottom of the ball. Here we solve this problem while introducing the technique of Lagrange multipliers.

Let us first cast the condition that the point lies on the surface of the ball in a mathematical form. When the ball has a radius R and is centered on the origin, this constraint is given by

$$C(x, y, z) = x^2 + y^2 + z^2 = R^2. \tag{27.73}$$

The constraint thus implies that $C(x, y, z) = constant$. Suppose that we move a point $\mathbf{r} = (x, y, z)$ along the surface of the ball with an infinitesimal displacement $\delta \mathbf{r}$, as shown in Figure 27.6. Since the displaced point $\mathbf{r} + \delta \mathbf{r}$ must lie on the surface of the ball as well, it must satisfy

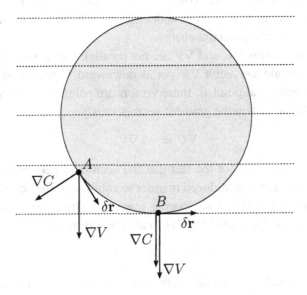

Figure 27.6 The direction of the vectors ∇C and ∇V at two points on a ball. The dashed lines indicate the equipotential surfaces where $V = const$.

$$C(\mathbf{r} + \delta\mathbf{r}) = C(\mathbf{r}).\tag{27.74}$$

Problem a   Use equation (5.9) to show that to first order in $\delta\mathbf{r}$ this implies that

$$\nabla C \cdot \delta\mathbf{r} = 0.\tag{27.75}$$

Geometrically this condition states that $\delta\mathbf{r}$ is perpendicular to the gradient ∇C. For our particular problem, ∇C points in the radial direction, so the condition (27.74) simply states that the point is perturbed along the surface of the sphere.

We want to minimize the potential energy $V(\mathbf{r})$ associated with the gravitational field as a function of the location $\mathbf{r}$. According to (5.9) the potential energy is to first order in the perturbation $\mathbf{r}$ changed by an amount

$$\delta V = (\nabla V \cdot \delta\mathbf{r}).\tag{27.76}$$

The potential energy is minimized when this quantity vanishes; this is the case when the displacement $\delta\mathbf{r}$ is perpendicular to ∇V.

How can we see that the point A in the sphere in Figure 27.6 does not minimize the potential energy, but that the point B does minimize the potential energy? Consider the displacement $\delta\mathbf{r}$ at point A along the surface of the sphere. At point A the displacement $\delta\mathbf{r}$ is not perpendicular to the gradient ∇V. According to expression (27.76) this implies that the potential energy changes to first order under this perturbation; hence, point A does not minimize the potential energy. At point B, the displacement $\delta\mathbf{r}$ is perpendicular to the vector ∇V, so that at that point the potential energy does not change to first order under this perturbation. In other words, point B is an extremum of the potential energy.

At point A the vectors ∇C and ∇V are not parallel; the vector ∇C points in the radial direction, while the vector ∇V points downward. However, at the minimum of the potential energy at point B, these vectors are parallel, see Figure 27.6. This means that at the minimum the gradient vectors satisfy

$$\nabla C = -\lambda \nabla V.\tag{27.77}$$

The constant $-\lambda$ accounts for the fact that the vectors in general have a different length. The minus sign is introduced in order to conform to the notation most often found in the literature; it has no specific meaning. The constant λ is called the *Lagrange multiplier*. Equation (27.77) can also be written as

$$\nabla(V + \lambda C) = 0.\tag{27.78}$$

For the unconstrained minimization of the potential energy, we just needed to solve the equation $\nabla V = 0$. The constraint that the point lies on the surface of the ball leads to the minimization problem (27.78). Note that this problem contains the unknown parameter λ; hence, the constrained minimization problem depends on the four variables x, y, z, and λ, whereas the original problem was dependent only on the three variables x, y, and z. However, the constrained minimization problem has an extra equation that should be satisfied: the constraint (27.73). This means that we now have four equations with four unknowns.

Yet another way to see that point B minimizes the potential energy is the following. The constraint states that the point lies on the surface of the sphere. Stated differently, the point must lie on the surface $C(x, y, z) = constant$. At the extremum of the potential energy, this surface must touch the surface $V(x, y, z) = constant$; if it would intersect that surface at an angle, the point could be moved along the surface $C(x, y, z) = constant$ while raising or lowering the potential energy. This contradicts the fact that the point maximizes or minimizes the potential energy. This means the extremum of the potential energy is attained at the points where the surfaces $C(x, y, z) = constant$ and $V(x, y, z) = constant$ are tangent. (When a ball lies on the ground, the surface of the ball and the ground touch each other.) A different way of stating this is that the vectors ∇C and ∇V are parallel, which is equivalent to condition (27.77).

Let us see how the machinery of the Lagrange multipliers works for finding the point on a ball that has the lowest potential energy. In this case the potential energy is given by

$$V(\mathbf{r}) = z, \tag{27.79}$$

and the constraint that the point lies on a sphere with radius R is given by (27.73). For this example, $V + \lambda C = z + \lambda(x^2 + y^2 + z^2)$.

Problem b Show that for this problem equation (27.78) leads to

$$\begin{pmatrix} 2\lambda x \\ 2\lambda y \\ 1 + 2\lambda z \end{pmatrix} = 0. \tag{27.80}$$

This expression constitutes three equations for the four unknowns x, y, z, and λ. The first two lines dictate that $x = y = 0$. The last line states that $z = -1/2\lambda$.

Problem c Use the constraint (27.73) to show that $\lambda = \pm 1/2R$.

This means that the solution is given by

$$\mathbf{r} = \begin{pmatrix} 0 \\ 0 \\ \pm R \end{pmatrix}. \tag{27.81}$$

The potential energy has extrema for the top and the bottom of the sphere. The top of the sphere maximizes the potential energy, while at the bottom of the sphere the potential energy has a minimum.

In the treatment of this section, it is not essential that the constraint $C(\mathbf{r})$ describes the surface of the sphere, and the $V(\mathbf{r})$ is the potential energy. The employed arguments hold for the minimization of any function $V(\mathbf{r})$ under a constraint $C(\mathbf{r})$. As long as the gradient of these functions is well defined, the constrained optimization problem can be solved with expression (27.78).

The extension of Lagrange multipliers to the constrained optimization of functionals is not trivial. Here we sketch the main idea. Suppose we want to minimize the functional

$$L[y] = \int F(y, \dot{y}, \xi) \, d\xi, \tag{27.82}$$

subject to the constraint

$$\int C(y, \dot{y}, \xi) d\xi = constant. \tag{27.83}$$

Suppose that $y(\xi)$ is modified with a perturbation $\epsilon(\xi)$ such that the perturbed function also satisfies the constraint (27.83). This means that the perturbation of the constraint is given by

$$\int \left(\frac{\partial C}{\partial y} \epsilon + \frac{\partial C}{\partial \dot{y}} \dot{\epsilon} \right) d\xi = 0. \tag{27.84}$$

Using an integration by parts, as in Section 27.4, then gives

$$\int \left(\frac{\partial C}{\partial y} - \frac{d}{d\xi} \left(\frac{\partial C}{\partial \dot{y}} \right) \right) \epsilon(\xi) \, d\xi = 0. \tag{27.85}$$

This equation is equivalent to expression (27.75) for a finite-dimensional problem. With a similar argument, the minimization of the functional (27.82) leads to the condition

$$\int \left(\frac{\partial F}{\partial y} - \frac{d}{d\xi} \left(\frac{\partial F}{\partial \dot{y}} \right) \right) \epsilon(\xi) \, d\xi = 0. \tag{27.86}$$

The last two expressions state that the functions $\left(\partial C/\partial y - \frac{d}{d\xi} (\partial C/\partial \dot{y}) \right)$ and $\left(\partial F/\partial y - \frac{d}{d\xi} (\partial F/\partial \dot{y}) \right)$ are both "perpendicular" to the perturbation $\epsilon(\xi)$, just like ∇C and ∇V were perpendicular to the perturbation $\delta \mathbf{r}$ of the point on the sphere. Since this must hold for any perturbation $\epsilon(\xi)$, this means that these functions are "parallel" to each other:

$$\left(\frac{\partial F}{\partial y} - \frac{d}{d\xi} \left(\frac{\partial F}{\partial \dot{y}} \right) \right) = -\lambda \left(\frac{\partial C}{\partial y} - \frac{d}{d\xi} \left(\frac{\partial C}{\partial \dot{y}} \right) \right). \tag{27.87}$$

The is the analogue of equation (27.77). This argument is not rigorous, the quotation marks highlight a vague use of the concepts perpendicular and parallel. A more thorough (and more complex) derivation is given by Smith (1974). However, accepting expression (27.87), the constrained optimization problem is given by the equation

$$\frac{\partial (F + \lambda C)}{\partial y} - \frac{d}{d\xi} \left(\frac{\partial (F + \lambda C)}{\partial \dot{y}} \right) = 0. \tag{27.88}$$

This expression is nothing but the Euler–Lagrange equation for the function $F + \lambda C$. The constrained optimization problem can thus be solved by optimizing $F + \lambda C$. This introduces the additional parameter λ. This parameter can be found by solving the constraint (27.83).

In this chapter the constraint used is always satisfied exactly. For example, in this section the point always lies on the ball. The role of a constraint is different in Section 22.3; we add a constraint to the data misfit to define a solution to an ill-posed inverse problem. In expression (22.22) the constraint is minimized along with the data misfit, but the constraint is in that case not satisfied exactly.

27.8 Designing a can with an optimal shape

In this section we return to the problem of Sections 27.1 and 27.2 where we treated the design of the shape of a can that maximizes the content for a given surface area. We treat this problem here using the Lagrange multiplier. Let us consider the design problem in a slightly different way by assuming that the surface area S of the bottom and the top of the can is fixed. We noted in Section 27.2 that the volume of the can is largest when the circumference C is minimized for a fixed value of the surface S of the top and bottom.

The circumference is related to the radius $r(\varphi)$ by expression (27.14). The surface area of the top and bottom is given by

$$S = \frac{1}{2} \int_0^{2\pi} r^2 d\varphi. \tag{27.89}$$

According to the theory of the previous section, this constrained optimization problem can be solved by optimizing $C + \lambda S$, where the Lagrange multiplier λ follows from the constraint (27.89).

Problem a   Use expressions (27.14) and (27.89) to show that the constrained optimization problem is solved by solving the Euler–Lagrange equation for

$$F(r, \dot{r}) = \sqrt{\dot{r}^2 + r^2} + \frac{\lambda}{2}r^2. \qquad (27.90)$$

Problem b   Show that

$$\frac{\partial F}{\partial r} = \frac{r}{\sqrt{\dot{r}^2 + r^2}} + \lambda r, \qquad (27.91)$$

and

$$\frac{\partial F}{\partial \dot{r}} = \frac{\dot{r}}{\sqrt{\dot{r}^2 + r^2}}. \qquad (27.92)$$

Problem c   Take the total derivative of the previous expression with respect to φ and show that the Euler–Lagrange equation for this optimization problem is given by

$$\frac{\ddot{r}r^2 - \dot{r}^2 r}{\left(\dot{r}^2 + r^2\right)^{3/2}} = \frac{r}{\sqrt{\dot{r}^2 + r^2}} + \lambda r. \qquad (27.93)$$

Problem d   This is a nonlinear differential equation. However, we know from Section 27.2 that a constant radius is likely to be a solution. Show that the solution $r(\varphi) = constant$ reduces equation (27.93) to

$$1 + \lambda r = 0. \qquad (27.94)$$

Problem e   This condition implies that $r = -\lambda^{-1}$. Determine the Lagrange multiplier λ by inserting this expression into the constraint (27.89) to show that

$$\lambda = \sqrt{\pi/S}, \qquad (27.95)$$

and that this finally gives the solution

$$r = \sqrt{S/\pi}. \qquad (27.96)$$

This example shows that the theory of the Lagrange multiplier states that a cylindrical can has the largest volume for a given surface area.

27.9 Chain line and the Gateway Arch of St. Louis

In this section we apply the theory of Lagrange multipliers to the problem of finding the shape of a wire of length L that is suspended between the points $(0, 0)$ and

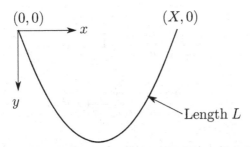

(0,0)   x   (X,0)

y

Length L

Figure 27.7   The geometry of a wire suspended between two points.

$(X, 0)$, as shown in Figure 27.7. Following equation (10.8), the total length of the wire is given by

$$L = \int_0^X \sqrt{1 + \dot{y}^2}\, dx. \tag{27.97}$$

This expression formulates the constraint that the wire has a given length L. In this section we use x as the independent variable and $y(x)$ as the dependent variable. The variable ρ denotes the mass of the wire per unit length. An increment dx corresponds to an increment in length given by $\sqrt{1 + \dot{y}^2}\, dx$, the corresponding mass is $\rho\sqrt{1 + \dot{y}^2}\, dx$, and the associated potential energy is given by $g\rho y\sqrt{1 + \dot{y}^2}\, dx$, with g the acceleration of gravity. The potential energy of the wire thus is given by

$$V[y] = g\rho \int_0^X y\sqrt{1 + \dot{y}^2}\, dx. \tag{27.98}$$

If the wire has no stiffness, it will assume a shape that minimizes its potential energy. The shape of the wire can thus be found by minimizing the potential energy (27.98) under the constraint (27.97) that its length is fixed. The curve $y(x)$ that minimizes $\int_0^X y\sqrt{1 + \dot{y}^2}\, dx$ also minimizes $g\rho \int_0^X y\sqrt{1 + \dot{y}^2}\, dx$. For this reason the constants $g\rho$ are left out in the following part of this section.

According to the discussion of Section 27.7, the solution of this constrained optimization problem can be found by optimizing $V + \lambda L$, and by using the constraint (27.97) to find the Lagrange multiplier λ.

Problem a   Use the theory of the previous section to show that the shape of the wire is found by optimizing

$$F[y] = \int_0^X \mathcal{F}(y, \dot{y})\, dx, \tag{27.99}$$

with

$$\mathcal{F}(y, \dot{y}) = y\sqrt{1 + \dot{y}^2} + \lambda\sqrt{1 + \dot{y}^2}. \tag{27.100}$$

Problem b   Show that

$$\frac{\partial F}{\partial y} = \sqrt{1 + \dot{y}^2},$$

(27.101)

and

$$\frac{\partial F}{\partial \dot{y}} = \frac{\dot{y}}{\sqrt{1 + \dot{y}^2}}(y + \lambda).$$

(27.102)

Problem c   Take the total derivative of the last expression to derive that

$$\frac{d}{dx}\left(\frac{\partial F}{\partial \dot{y}}\right) = \frac{1}{\left(1 + \dot{y}^2\right)^{3/2}}\left\{\ddot{y}(y + \lambda) + \dot{y}^2\left(1 + \dot{y}^2\right)\right\}.$$

(27.103)

Remember that in this derivative all variables that depend on x are differentiated.

Problem d   Show that the Euler–Lagrange equation for this problem can be written as

$$1 + \dot{y}^2 = \ddot{y}(y + \lambda).$$

(27.104)

This is a nonlinear differential equation that displays the complexity that often results in the solution of variational problems. There is no simple way to guess the solution, or to construct it in a systematic way. For this reason we simply state the following trial solution

$$y(x) = A \cosh\left(B(x + h)\right) + C,$$

(27.105)

where A, B, C, and h are constants.

Problem e   Insert this solution into the differential equation (27.104) and show that the result can be written as

$$(1 - A^2 B^2) = AB^2(\lambda + C)\cosh\left(B(x + h)\right).$$

(27.106)

The left-hand side of this expression is independent of x, while the right-hand side varies with x. This is possible only when the coefficients of the left-hand side and the right-hand side vanish. This implies that

$$1 - A^2 B^2 = 0 \quad \text{and} \quad AB^2(\lambda + C) = 0.$$

(27.107)

These equations are satisfied when $A = 1/B$ and $C = -\lambda$, so that the solution is given by

$$y(x) = \frac{1}{B}\cosh\left(B(x + h)\right) - \lambda.$$

(27.108)

This expression contains three constants. There are, however, three pieces of information that we have not used yet: the boundary conditions that state the wire goes through the points $(0, 0)$ and $(X, 0)$, and the constraint (27.97) that states the wire has length L.

Problem f Use the boundary conditions to show that $h = -X/2$ and $\lambda = B^{-1}$
$\cosh(BX/2)$, so that the solution is given by

$$y(x) = B^{-1} \left\{ \cosh\left(B\left(x - \frac{X}{2} \right) \right) - \cosh\left(\frac{BX}{2} \right) \right\}. \qquad (27.109)$$

Problem g Insert this solution into the constraint (27.97) and show that this leads to the following equation for the constant B:

$$\frac{2}{B} \sinh\left(\frac{BX}{2} \right) = L. \qquad (27.110)$$

This is a transcendental equation that cannot be solved in closed form, but it can be solved numerically. This completes the solution of finding the shape of a wire suspended between two points.

It follows from equation (27.109) that the wire is shaped like the cosine-hyperbolic. This curve is called the *catenary*; this word is derived from the Latin word *catena* that means chain. The catenary can be seen in the chain that is suspended in the background of Figure 27.8; for this reason the curve is also called the *chain line*. Knowing the shape of a suspended wire is more than a mathematical curiosity. A suspended wire has in general such a small stiffness that this property can be ignored. (In the analysis of this section the wire does not have any stiffness.) This means that the wire does not support any internal bending moments. The shape of the wire is dictated by the balance of the gravitational force and the net tensional force that arises from the curvature of the wire as shown in the left panel of Figure 27.9.

Let us suppose we turn the wire upside down and that we build an arch. In that case the gravitational force acts into the arch and the tensional forces of the left panel of Figure 27.9 are changed into compressional forces that balances the gravitational force, as shown in the right panel of Figure 27.9. This implies that an arch built in the shape of a catenary does not require any internal bending moments to maintain its shape. This is illustrated with the arch shown in Figure 27.8 that was built by students of the New Trier High School in Winnetka, Illinois. This arch consists of pieces of foam that are cut in a form to jointly give the shape of a catenary. The pieces of foam are not glued together. Since a catenary arch does

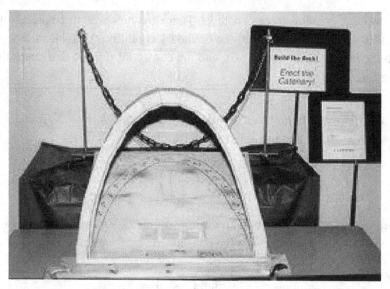

Figure 27.8   The chain in the background and the arch in the foreground have the shape of a catenary. The arch consists of blocks of foam that are not glued together. The figure is made by high school students and is reproduced with permission of the New Trier Connections Project.

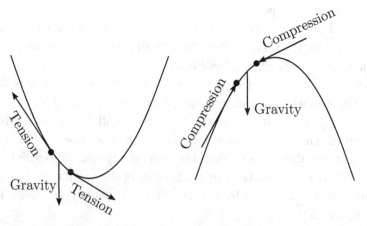

Figure 27.9   The balance of the tensional forces and gravity for a suspended wire (left panel), and the balance between tensional forces and the gravitational force for an arch (right panel).

not need any bending moment to support itself, there is no shear force that makes the blocks slide away from each other; all internal force acts along the curve of the arch itself. In this sense the arch is truly self-supporting. The Gateway Arch in St. Louis, with a height of 192 m, is built in the shape of a catenary. This allows for the slender design of this monument.

28

Epilogue, on power and knowledge

We all continue to feel a frustration because of our inability to foresee the soul's ultimate fate. Although we do not speak about it, we all know that the objectives of our science are, from a general point of view, much more modest than the objectives of, say, the Greek sciences were; that our science is more successful in giving us power than in giving us knowledge of truly human interest. (Wigner, 1972)

In this book we have explored many methods of mathematics as used in the physical sciences. Mathematics plays a central role in the physical sciences because it is the only language we have for expressing quantitative relations in the world around us. In fact, mathematics not only allows us to express phenomena in a quantitative way, it also has a remarkable predictive power in the sense that it allows us to deduce the consequences of natural laws in terms of measurable quantities. In fact, we do not quite understand *why* mathematics gives such an accurate description of the world around us (Wigner, 1960).

It is truly stunning how accurate some of the predictions in (mathematical) physics have been. The orbits of the planetary bodies can now be computed with extreme accuracy. Stephenson and Morrison (1995) compared the path of a solar eclipse at 181 BC with historic descriptions made in a city in eastern China that was located in the path of the solar eclipse. According to the computations, the path of the solar eclipse passed 50 degrees west of the site of this historic observation. This eclipse took place about 2000 years ago; this means that the Earth has rotated through about 2.8×10^8 degrees since the eclipse. The *relative* error in the path of the eclipse over the Earth is thus only 1.7×10^{-7}. In fact, this discrepancy of 50 degrees can be explained by the observed deceleration of the Earth due to the braking effect of the Earth's tides.

Another example of the accuracy of mathematical physics involves light. The light emitted by hydrogen atoms has discrete spectral lines, which are due to the fact that electrons behave as standing waves. Every electron is coupled to the field of electromagnetic radiation (light). There is a small chance that an electron will emit and reabsorb a virtual photon (Sakurai, 1978). (Photons are the light-quanta.) This leads to the so-called *Lamb shift* of the spectral lines of light radiated by excited hydrogen atoms. For hydrogen atoms the frequency shift for the transition between the 2s and $2p_{1/2}$ state is 1060 MHz, which is in very good agreement with observations. This frequency shift corresponds to a difference in energy of these two states of 4.3×10^{-6} eV. Compared to the difference in energy between the ground state and first excited state of a hydrogen atom (10.2 eV), the Lamb shift amounts to a relative change in energy of 4×10^{-7}. It is interesting to note that this prediction of the Lamb shift is based on second-order perturbation theory (Sakurai, 1978). This means that an approximate theory provides a stunningly accurate prediction of the Lamb shift. There are even more precise predictions of quantum field theory and the "standard model" for elementary particles: the magnetic moment of the electron is now explained with an accuracy of 1 part in 10^{-13} (Gabrielse, 2013).

A third example of the extreme accuracy of mathematics in the physical sciences is the perihelion precession of Mercury (Ohanian and Ruffini, 1976). According to the laws of Newton, a planet will orbit in a fixed ellipsoidal orbit around the Sun. The general theory of relativity predicts that this ellipse slowly changes its position; the point of the ellipse closest to the Sun (the perihelion) slowly precesses around the Sun. According to the theory of general relativity this precession is given by 42.98 arcsec/century, whereas the observed precession rate is 43.1 ± 0.1 arcsec/century. Note that this precession rate is extremely small, but that it is well predicted from theory.

Mathematics not only provides us with valuable and stunningly accurate insights in the world around us, it is also an indispensable tool in making technical innovations. The design and implementation of rockets, aircraft, chemical plants, water treatment systems, modern electronics, information technology, and many other innovations would have been impossible without mathematics. Mathematics and the physical sciences have created many new opportunities for mankind. For this reason one can state that mathematics and the physical sciences have greatly increased our power to modify the world in which we live; see also the quote of Nobel laureate Wigner at the beginning of this chapter.

The problem with releasing power is that it can be used for any purpose. Science is objective in the sense that a certain theory is either consistent with observations, or it is not. However, scientific knowledge does not come with the moral standard that tells us *how* the power that we release in our scientific efforts should be used. It is essential that each of us develops such a standard, so that the fruits of our knowledge can be used for the benefit of mankind and the world we inhabit.

References

Abramowitz, M., and Stegun, I.A. 1965. *Handbook of mathematical functions with formulas, graph, and mathematical tables. Applied Mathematics Series*, **55**, 1046.

Aharonov, Y., and Bohm, D. 1959. Significance of electromagnetic potentials in the quantum theory. *Physical Review*, **115**(3), 485.

Aki, K., and Richards, P.G. 2002. *Quantitative seismology*. 4th edn. Sausalito, CA: Univ. Science Books.

Aldridge, D.F. 1994. Linearization of the eikonal equation. *Geophysics*, **59**(10), 1631–1632.

Alvarez, L.W., Alvarez, W., Asaro, F., and Michel, H.V. 1980. Extraterrestrial cause for the cretaceous-tertiary extinction. *Science*, **208**(4448), 1095–1108.

Arfken, G.B., and Weber, H.J. 2005. *Mathematical methods for physicists*. 4th edn. Amsterdam: Harcourt.

Ashby, N. 2002. Relativity and the global positioning system. *Physics Today*, **55**(5), 41–47.

Backus, M.N. 1959. Water reverberations—their nature and elimination. *Geophysics*, **24**(2), 233–261.

Bajec, I.L., and Heppner, F.H. 2009. Organized flight in birds. *Animal Behaviour*, **78**(4), 777–789.

Bakulin, A., and Calvert, R. 2006. The virtual source method: theory and case study. *Geophysics*, **71**, SI139–SI150.

Barish, B.C., and Weiss, R. 1999. LIGO and the detection of gravitational waves. *Physics Today*, **52**(10), 44–50.

Barton, G. 1989. *Elements of Green's functions and propagation: potentials, diffusion, and waves*. Oxford, UK: Oxford University Press.

Beissner, K. 1998. The acoustic radiation force in lossless fluids in Eulerian and Lagrangian coordinates. *The Journal of the Acoustical Society of America*, **103**, 2321–2332.

Bellman, R., Kalaba, R., and Wing, G.M. 1960. Invariant imbedding and mathematical physics. I. Particle processes. *Journal of Mathematical Physics*, **1**, 280–308.

Bender, C.M., and Orszag, S.A. 1999. *Advanced mathematical methods for scientists and engineers I: asymptotic methods and perturbation theory*. Vol. 1. New York, NY: Springer Verlag.

Bennett, J.A. 1973. Variations of the ray path and phase path: a Hamiltonian formulation. *Radio Science*, **8**(8–9), 737–744.

Berry, M.V. 2013. Five momenta. *European Journal of Physics*, **34**, 1337–1348.

Berry, M.V., and Klein, S. 1997. Transparent mirrors: rays, waves and localization. *European Journal of Physics*, **18**(3), 222–228.

547

Berry, M.V., and Upstill, C. 1980. Catastrophe optics: morphologies of caustics and their diffraction patterns. *Prog. Opt*, **18**, 257–346.

Blakeley, R.J. 1995. *Potential theory in gravity and magnetics*. Cambridge, UK: Cambridge Univ. Press.

Bleistein, N., and Handelsman, R.A. 1975. *Asymptotic expansions of integrals*. Courier Dover Publications.

Boas, M.L. 2006. *Mathematical methods in the physical sciences*. Hoboken, NJ: John Wiley & Sons., Inc.

Brack, M., and Bhaduri, R.K. 1997. *Semiclassical physics*. Vol. 14. Reading, MA: Addison-Wesley Reading.

Brillouin, L., and Sommerfeld, A. 1960. *Wave propagation and group velocity*. Vol. 960. New York: Academic Press.

Buckingham, E. 1914. On physically similar systems; illustrations of the use of dimensional equations. *Physical Review*, **4**(4), 345–376.

Butkov, E. 1968. *Mathematical physics*. Addison-Wesley series in advanced physics. Reading, MA: Addison-Wesley Pub. Co.

Chopelas, A. 1996. Thermal expansivity of lower mantle phases MgO and $MgSiO_3$ perovskite at high pressure derived from vibrational spectroscopy. *Physics of the Earth and Planetary Interiors*, **98**(1), 3–15.

Choy, G.L., and Richards, P.G. 1975. Pulse distortion and Hilbert transformation in multiply reflected and refracted body waves. *Bulletin of the Seismological Society of America*, **65**(1), 55–70.

Cipra, B. 2000. *Misteaks... and how to find them before the teacher does: a calculus supplement*. 3rd edn. Birkhäuser.

Claerbout, J. 1976. *Fundamentals of seismic data processing*. New York, NY: McGraw-Hill.

Claerbout, J.F. 1985. *Imaging the earth's interior: Blackwell*. Oxford, England.

Coveney, P.V., and Highfield, R. 1991. *The arrow of time: a voyage through science to solve time's greatest mystery*. New York: Fawcett Columbine, 1991. 1st American edn., **1**.

Craig, G.M. 1997. *Stop abusing Bernoulli! How airplanes really fly*. Anderson, IN: Regenerative Press.

Curtis, A., Gerstoft, P., Sato, H., Snieder, R., and Wapenaar, K. 2006. Seismic interferometry – turning noise into signal. *The Leading Edge*, **25**, 1082–1092.

Dahlen, F.A. 1979. The spectra of unresolved split normal mode multiplets. *Geophysical Journal International*, **58**(1), 1–33.

Dahlen, F.A., and Henson, I.H. 1985. Asymptotic normal modes of a laterally heterogeneous Earth. *Journal of Geophysical Research: Solid Earth (1978–2012)*, **90**(B14), 12653–12681.

Dahlen, F.A., and Tromp, J. 1998. *Theoretical global seismology*. Princeton, NJ: Princeton University Press.

Davies-Jones, R. 1984. Streamwise vorticity: the origin of updraft rotation in supercell storms. *J. Atmosph. Sci.*, **41**, 2991–3006.

de Broglie, L. 1952. *La théorie des particules de spin 1/2: électrons de Dirac*. Paris, France: Gauthier-Villars.

DeSanto, J.A. 1992. *Scalar wave theory*. Berlin, Germany: Springer.

Dziewonski, A.M., and Woodhouse, J. H. 1983. Studies of the seismic source using normal-mode theory. *Earthquakes: observations, theory and intepretation*, edited by H. Kanamori and E. Boschi, North Holland, Amsterdam, 45–137.

Earnshaw, S. 1842. On the nature of the molecular forces which regulate the constitution of the luminiferous ether. *Trans. Camb. Phil. Soc*, **7**, 97–112.

Edmonds, A.R. 1996. *Angular momentum in quantum mechanics.* Vol. 4. Princeton, NJ: Princeton University Press.

Farra, V., and Madariaga, R. 1987. Seismic waveform modeling in heterogeneous media by ray perturbation theory. *Journal of Geophysical Research: Solid Earth (1978–2012)*, **92**(B3), 2697–2712.

Feynman, R.P. 1967. *The character of physical law.* Vol. 66. Cambridge, MA: MIT Press.

Feynman, R.P., Hibbs, A.R., and Styer, D.F. 1965. *Quantum mechanics and path integrals.* Vol. 2. New York: McGraw-Hill.

Fischbach, E., and Talmadge, C. 1992. Six years of the fifth force. *Nature*, **356**, 207–215.

Fletcher, C. 1996. *The complete walker III.* New York: Alfred A.Knopf.

Fokkema, J.T., and van den Berg, P.M. 1993. *Seismic applications of acoustic reciprocity.* Amsterdam, the Netherlands: Elsevier.

Förste, Christoph, Schmidt, Roland, Stubenvoll, Richard, Flechtner, Frank, Meyer, Ulrich, König, Rolf, Neumayer, Hans, Biancale, Richard, Lemoine, Jean-Michel, Bruinsma, Sean, et al. 2008. The GeoForschungsZentrum Potsdam/Groupe de Recherche de Geodesie Spatiale satellite-only and combined gravity field models: EIGEN-GL04S1 and EIGEN-GL04C. *Journal of Geodesy*, **82**(6), 331–346.

Fowler, C.M.R. 2005. *The solid Earth, an introduction to global geophyscis.* Cambridge, UK: Cambridge Univ. Press.

French, S., Lekic, V., and Romanowicz, B. 2013. Waveform tomography reveals channeled flow at the base of the oceanic lithosphere. *Science*, **342**, 227–230.

Gabrielse, G. 2013. The standard model's greatest triumph. *Physics Today*, **66**(12), 64–65.

Garcia, Raphael F., Bruinsma, Sean, Lognonn, Philippe, Doornbos, Eelco, and Cachoux, Florian. 2013. GOCE: The first seismometer in orbit around the Earth. *Geophysical Research Letters*, **40**(5), 1015–1020.

Goldstein, H. 1980. *Classical mechanics.* Reading, MA: Addison-Wesley Pub. Co.

Gradshteyn, I.S., and Ryzhik, I.M. 1965. *Table of integrals, series, and products.* Prepared by Ju. V. Geronimus and M. Ju. Ceitlin. Translated from the Russian by Scripta Technica, Inc. Translation edited by Alan Jeffrey. San Diego, CA: Academies Press.

Gubbins, D., and Snieder, R. 1991. Dispersion of P waves in subducted lithosphere: evidence for an eclogite layer. *Journal of Geophysical Research: Solid Earth (1978–2012)*, **96**(B4), 6321–6333.

Guglielmi, A.V., and Pokhotelov, O.A. 1996. *Geoelectromagnetic waves.* Vol. 1. Bristol, UK: Institute of Physics Publishing.

Gutenberg, B., and Richter, C.F. 1956. Magnitude and energy of earthquakes. *Annali di Geofisica*, **9**, 1–15.

Halbwachs, F. 1960. *Théorie relativiste des fluides à spin.* Vol. 10. Paris. Gauthier-Villars, Paris.

Hansen, P.C. 1992. Analysis of discrete ill-posed problems by means of the L-curve. *SIAM Review*, **34**(4), 561–580.

Hildebrand, A.R., Pilkington, M., Connors, M., Ortiz-Aleman, C., and Chavez, R.E. 1995. Size and structure of the Chicxulub crater revealed by horizontal gravity gradients and cenotes. *Nature*, **376**(6539), 415–417.

Holton, J.R., and Hakim, G.J. 2012. *An introduction to dynamic meteorology.* Waltham, MA: Academic Press.

Hutchings, Michael, Morgan, Frank, Ritor, Manuel, and Ros, Antonio. 2002. Proof of the Double Bubble Conjecture. *Annals of Mathematics*, **155**(2), pp. 459–489.

Ishimaru, A. 1978. *Wave propagation and scattering in random media.* Vol. 2. New York: Academic Press.

Iyer, H. M. 1993. *Seismic tomography: theory and practice*. London, UK: Chapman and Hall.

Jackson, J. D. 1998. *Classical Electrodynamics*. New York: John Wiley.

Jaynes, E.T. 2003. *Probability theory, the logic of science*. Cambridge, UK: Cambridge Univ. Press.

Jeffreys, H. 1925. On Certain Approximate Solutions of Lineae Differential Equations of the Second Order. *Proceedings of the London Mathematical Society*, **2**(1), 428–436.

Kanamori, H., Mori, J., Anderson, D.L., and Heaton, T.H. 1991. Seismic excitation by the space shuttle Columbia. *Nature*, **349**, 781–782.

Kanamori, H., Mori, J., Sturtevant, B., Anderson, D.L., and Heaton, T. 1992. Seismic excitation by space shuttles. *Shock Waves*, **2**, 89–96.

Kanamori, H.J., and Harkrider, D.G. 1994. Excitation of Atmospheric Oscillations by Volcanic Eruptions. *Journal of Geophysical Research*, **99**(B11), 21947–21961.

Kearey, P., Brooks, M., and Hill, I. 2002. *An introduction to geophysical exploration*. Oxford, UK: Blackwell.

Kermode, A.C., Barnard, R.H., and Philpott, D.R. 1996. *Mechanics of flight*. Harlow, UK: Longman.

Kline, S.J. 1986. *Similitude and approximation theory*. New York: Springer.

Kravtsov, Yu. A. 1988. IV Rays and Caustics as Physical Objects. *Progress in optics*, **26**, 227–348.

Kulkarny, V.A., and White, B.S. 1982. Focusing of waves in turbulent inhomogeneous media. *Physics of Fluids*, **25**, 1770–1784.

Lambeck, K. 1988. *Geophysical geodesy*. Oxford: Oxford University Press.

Lanczos, C. 1970. *The variational principles of mechanics*. 4. New York, NY: Courier Dover Publications.

Larose, E., Margerin, L., Derode, A., van Tiggelen, B., Campillo, M., Shapiro, N., Paul, A., Stehly, L., and Tanter, M. 2006. Correlation of random wavefields: an interdisciplinary review. *Geophysics*, **71**, SI11–SI21.

Lauterborn, W., and Kurz, T. 2003. *Coherent Optics: fundamentals and applications*. Berlin, Germany: Springer Verlag.

Lavenda, B.H. 1991. *Statistical physics: A probabilistic approach*. Wiley-Interscience.

Levenspiel, O. 2006. Atmospheric pressure at the time of dinosaurs. *Chemical Industry and Chemical Engineering Quarterly*, **12**(2), 116–122.

Lin, C.C., and Segel, L.A. 1974. *Mathematics Applied to Deterministic Problems*. New York, NY: SIAM.

Lin, F.C., Ritzwoller, M.H., and Snieder, R. 2009. Eikonal tomography: surface wave tomography by phase front tracking across a regional broad-band seismic array. *Geophys. J. Int.*, **177**, 1091–1110.

Lister, G.S., and Williams, P.F. 1983. The partitioning of deformation in flowing rock masses. *Tectonophysics*, **92**(1), 1–33.

Lomnitz, C. 1994. *Fundamentals of earthquake prediction*. New York, NY: Wiley.

Love, S.G., and Brownlee, D.E. 1993. A direct measurement of the terrestrial mass accretion rate of cosmic dust. *Science*, **262**(5133), 550–553.

Madelung, E. 1927. Quantentheorie in hydrodynamischer Form. *Zeitschrift für Physik A Hadrons and Nuclei*, **40**(3), 322–326.

Marchaj, C.A. 2000. *Aero-hydrodynamics of sailing*. 3rd edn. Easton, MD: Tiller Pub.

Marsden, J.E., and Tromba, A. 2003. *Vector calculus*. New York, NY: WH Freeman.

Merzbacher, E. 1961. *Quantum mechanics*. New York: John Wiley and Sons.

Meuel, T., Xiong, Y.L., Fisher, P., Bruneau, C.H., Bessafi, M., and Kellay, H. 2013. Intensity of vortices: from soap bubbles to hurricanes. *Scientific Reports*, **3**, 3455.

Middleton, G.V., and Wilcock, P.R. 1994. *Mechanics in the Earth and environmental sciences.* Cambridge, UK: Cambridge University Press.

Mikesell, D., and van Wijk, K. 2011. Seismic refraction interferometry with a semblance analysis on the crosscorrelation gathering. *Geophysics*, **76**, SA77–SA82.

Moler, C., and Van Loan, C. 1978. Nineteen dubious ways to compute the exponential of a matrix. *SIAM Review*, **20**(4), 801–836.

Morley, T. 1985. A simple proof that the world is three dimensional. *SIAM Review*, **27**(1), 69–71.

Mosegaard, K. 2011. Quest for consistency, symmetry, and simplicity – the legacy of Albert Tarantola. *Geophysics*, **76**, W51–W61.

Mosegaard, K., and Tarantola, A. 1995. Monte Carlo sampling of slutions to inverse problems. *Journal Geophysical Research*, **100**, 12431–12447.

Muirhead, H. 1973. *The special theory of relativity.* New York: Macmillan.

Nayfeh, A.H. 2011. *Introduction to perturbation techniques.* New York, NY: Wiley.

Nolet, G. 1987. *Seismic tomography: with applications in global seismology and exploration geophysics.* Vol. 5. Dordrecht, the Netherlands: Kluwer.

Ohanian, H.C., and Ruffini, R. 1976. *Gravitation and spacetime.* New York: Norton.

Olson, P. 1989. *The encyclopedia of solid earth geophysics.* Van Nostrand Reinholt. Chap. Mantle convection and plumes.

Özisik, M.N. 1973. *Radiative transfer and interaction with conduction and convection.* New York: John Wiley.

Paasschens, J.C.J. 1997. Solution of the time-dependent Boltzmann equation. *Phys. Rev. E*, **56**, 1135–1141.

Parker, R.L. 1994. *Geophysical inverse theory.* Princeton, NJ: Princeton University Press.

Parker, R.L., and Zumberge, M.A. 1989. An analysis of geophysical experiments to test Newton's law of gravity. *Nature*, **342**, 29–32.

Parsons, B., and Sclater, J.G. 1977. An analysis of the variation of ocean floor bathymetry and heat flow with age. *Journal of Geophysical Research*, **82**(5), 803–827.

Payne, R., and Webb, D. 1971. Orientation by means of long range acoustic signaling in baleen whales. *Annals of the New York Academy of Sciences*, **188**(1), 110–141.

Pedlosky, J. 1982. Geophysical fluid dynamics. Berlin, Germany: Springer-Verlag, **1**.

Peixoto, J.P., and Oort, A.H. 1992. Physics of climate. *Am. Inst. of Phys.*, New York.

Pissanetzky, S. 1984. *Sparse matrix technology.* London: Academic Press.

Popper, K.R. 1956. The arrow of time. *Nature*, **177**, 538.

Press, W.H., Flannery, B.P., Teukolsky, S.A., and Vetterling, W.T. 1992. *Numerical Recipes in FORTRAN 77: Volume 1, Volume 1 of Fortran numerical recipes: the art of scientific computing.* Vol. 1. Cambridge, UK: Cambridge University Press.

Price, H. 1996. *Time's arrow & Archimedes' point: New directions for the physics of time.* Oxford, UK: Oxford University Press.

Rayleigh, L. 1917. On the reflection of light from a regularly stratified medium. *Proceedings of the Royal Society of London. Series A*, **93**(655), 565–577.

Rhie, J., and Romanowicz, B. 2004. Excitation of Earth's continuous free oscillations by atmospere-ocean-seafloor coupling. *Nature*, **431**, 552–556.

Rhie, J., and Romanowicz, B. 2006. A study of the realation between ocean storms and the Earth's hum. *Geochemistry Geophysics and Geosystems*, **7**, Q10004.

Riley, K., Hobson, M., and Bence, S. 2006. *Mathematical methods for physics and engineering.* Cambridge, UK: Cambridge University Press.

Robinson, D.K., and Bevington, P.R. 1992. *Sample programs for date reduction and error analysis for the physical sciences.* New York, NY: McGraw-Hill.

Robinson, E.A., and Treitel, S. 1980. *Geophysical signal analysis.*, Englewood Cliffs, NJ: Prentice-Hall.

Rossing, T.D., Moore, R.R., and Wheeler, P.A. 1990. *The science of sound.* Reading, MA: Addison-Wesley.

Sakurai, J.J. 1978. *Advanced quantum mechanics.* Reading, MA: Addison Wesley.

Sambridge, M., and Mosegaard, K. 2002. Monte Carlo methods in geophysical inverse problems. *Rev. Geophys.*, **40**(3), 3.1–3.29.

Scales, J.A., and Snieder, R. 1997. Humility and nonlinearity. *Geophysics*, **62**(5), 1355–1358.

Schneider, W.A. 1978. Integral formulation for migration in two and three dimensions. *Geophysics*, **43**(1), 49–76.

Schulte, P., Alegret, L., Arenillas, I., Arz, J.A., Barton, P.J., Bown, P.R., Bralower, T.J., Christeson, G.L., Claeys, P., Cockell, C.S., Collins, G.S., Deutsch, A., Goldin, T.J., Goto, K., Grajales-Nishimura, J.M., Grieve, R.A.F., Gulick, S.P.S., Johnson, K.R., Kiessling, W., Koeberl, C., Kring, D.A., MacLeod, K.G., Matsui, T., Melosh, J., Montanari, A., Morgan, J.V., Neal, C.R., Nichols, D.J., Norris, R.D., Pierazzo, E., Ravizza, G., Rebolledo-Vieyra, M., Reimold, W.U., Robin, E., Salge, T., Speijer, R.P., Sweet, A.R., Urrutia-Fucugauchi, J., Vajda, V., Whalen, M.T., and Willumsen, P.S. 2010. The Chicxulub asteroid impact and mass extinction at the Cretaceous-Paleogene boundary. *Science*, **327**(5970), 1214–1218.

Schuster, G. 2009. *Seismic interferometry.* Cambridge, UK: Cambridge Univ. Press.

Silverman, M.P. 1993. *And yet it moves: strange systems and subtle questions in physics.* Cambridge, UK: Cambridge University Press.

Smith, D.R. 1974. *Variational methods in optimization.* Englewood Cliffs, NJ: Prentice-Hall.

Snieder, R. 1986. 3-D linearized scattering of surface waves and a formalism for surface wave holography. *Geophysical Journal International*, **84**(3), 581–605.

Snieder, R. 1996. Surface wave inversions on a regional scale. *Seismic modelling of Earth structure*, Eds. E. Boschi, G. Ekstrom and A. Morelli, Bologna: Editrice Compositori, 149–181. Amsterdam, the Netherlands: North Holland.

Snieder, R. 2002. Time-reversal invariance and the relation between wave chaos and classical chaos. Pages 1–16 of: *Imaging of complex media with acoustic and seismic waves.* Berlin, Germany: Springer.

Snieder, R., and Aldridge, D.F. 1995. Perturbation theory for travel times. *The Journal of the Acoustical Society of America*, **98**, 1565–1569.

Snieder, R., and Larose, E. 2013. Extracting Earth's elastic wave response from noise measurements. *Ann. Rev. Earth Planet. Sci.*, **41**, 183–206.

Snieder, R, and Nolet, G. 1987. Linearized scattering of surface waves on a spherical Earth. *J. Geophys.*, **61**, 55–63.

Snieder, R., and Sambridge, M. 1992. Ray perturbation theory for travel times and raypaths in 3-D heterogeneous media. *Geophysical Journal International*, **109**, 294–322.

Snieder, R., and Sambridge, M. 1993. The ambiguity in ray perturbation theory. *Journal of Geophysical Research*, **98**, 22021–22034.

Snieder, R., Wapenaar, K., and Wegler, U. 2007. Unified Green's function retrieval by cross-correlation; connection with energy principles. *Phys. Rev. E*, **75**, 036103.

Snieder, R.K. 1985. The origin of the 100,000 year cycle in a simple ice age model. *Journal of Geophysical Research: Atmospheres*, **90**(D3), 5661–5664.

Spetzler, J., and Snieder, R. 2001. The effect of small scale heterogeneity on the arrival time of waves. *Geophysical Journal International*, **145**(3), 786–796.

Stacey, F.D. 1992. *Physics of the Earth.* Brookfield, New York.

Stein, S., and Wysession, M. 2003. *An introduction to seismology, earthquakes, and earth structure*. Malden MA: Blackwell.

Stephenson, F.R., and Morrison, L.V. 1995. Long-term fluctuations in the Earth's rotation: 700 BC to AD 1990. *Philosophical Transactions of the Royal Society of London. Series A: Physical and Engineering Sciences*, **351**(1695), 165–202.

Strang, G. 2003. *Introduction to linear algebra*. SIAM.

Strichartz, R.S. 1994. *A guide to distribution theory and Fourier transforms*. Singapore: World Scientific.

Tabor, M. 1989. *Chaos and integrability in nonlinear dynamics: an introduction*. New York: Wiley.

Tarantola, A. 1987. *Inverse problem theory*. Amsterdam: Elsevier.

Tarantola, A., and Valette, B. 1982. Generalized nonlinear inverse problems solved using the least squares criterion. *Reviews of Geophysics Space Physics*, **20**, 219–232.

Tennekes, H. 2009. *The simple science of flight: from insects to jumbo jets*. Cambridge, MA: MIT Press.

Thompson, P.A., and Beavers, G.S. 1972. Compressible-fluid dynamics. *Journal of Applied Mechanics*, **39**, 366.

Tritton, D.J. 1988. *Physical fluid dynamics*. Oxford: Clarendon Press.

Tromp, J., and Snieder, R. 1989. The reflection and transmission of plane P-and S-waves by a continuously stratified band: a new approach using invariant imbedding. *Geophysical Journal International*, **96**(3), 447–456.

Turcotte, D.L., and Schubert, G. 2002. *Geodynamics*. Cambridge University Press.

Vallina, A.U. 1999. *Principles of seismology*. Cambridge, UK: Cambridge University Press.

Van Dyke, M. 1964. *Perturbation methods in fluid mechanics*. Vol. 964. New York: Academic Press.

van Wijk, K., Scales, J.A., Navidi, W., and Tenorio, L. 2002. Data and model uncertainty estimation for linear inversion. *Geophysical Journal International*, **149**, 625–632.

van Wijk, K., Channel, T., Smith, M., and Viskupic, K. 2013. Teaching geophysics with a vertical-component seismometer. *The Physics Teacher*, **51**(December), 552–554.

Vogel, S. 1998. Exposing life's limits with dimensionless numbers. *Physics Today*, **51**(11), 22–27.

Wapenaar, K., Slob, E., and Snieder, R. 2006. Unified Green's function retrieval by cross-correlation. *Phys. Rev. Lett.*, **97**, 234301.

Watson, T.H. 1972. A real frequency, complex wave-number analysis of leaking modes. *Bulletin of the Seismological Society of America*, **62**(1), 369–384.

Webster, G.M. (ed). 1981. *Deconvolution*. Tulsa, OK: Society of Exploration Geophysicists.

Weisskopf, V.F. 1939. On the self-energy and the electric field of the electron. *Physical Review*, **56**, 72–85.

Wheeler, J.A., and Taylor, E.F. 1966. *Spacetime physics*. San Francisco, CA: WH Freeman.

Whitaker, S. 1968. *Introduction to fluid mechanics*. Englewood Cliffs, NJ: Prentice-Hall.

Whitham, G.B. 2011. *Linear and nonlinear waves*. Pure and applied mathematics. Wiley.

Wigner, E.P. 1960. The unreasonable effectiveness of mathematics in the natural sciences. *Communications on Pure and Applied Mathematics*, **13**(1), 1–14.

Wigner, E.P. 1972. The place of consciousness in modern physics. Pages 132–141 of: Muses, C. and Young, A.M. (eds.), *Consciousness and reality*. New York: Outerbridge and Lazard.

Wu, R., and Aki, K. 1985. Scattering characteristics of elastic waves by an elastic heterogeneity. *Geophysics*, **50**(4), 582–595.

Yilmaz, Ö. 2001. *Seismic data analysis: processing, inversion, and interpretation of seismic data*. 10. Tulsa, OK: Society of Exploration Geophysicists.

Yoder, C.F., Williams, J.G., Dickey, J.O., Schutz, B.E., Eanes, R.J., and Tapley, B.D. 1983. Secular variation of Earth's gravitational harmonic J2 coefficient from LAGEOS and nontidal acceleration of Earth rotation. *Nature*, **303**, 757–762.

Zee, A. 2005. *Quantum field theory in a nutshell*. Universities Press.

Ziolkowski, A. 1991. Why don't we measure seismic signatures? *Geophysics*, **56**, 190–201.

Zumberge, M.A., Ridgway, J.R., and Hildebrand, J.A. 1997. A towed marine gravity meter for near-bottom surveys. *Geophysics*, **62**, 1386–1393.

Index

Printed in the United States
by Baker & Taylor Publisher Services